Springer Biographies

The books published in the Springer Biographies tell of the life and work of scholars, innovators, and pioneers in all fields of learning and throughout the ages. Prominent scientists and philosophers will feature, but so too will lesser known personalities whose significant contributions deserve greater recognition and whose remarkable life stories will stir and motivate readers. Authored by historians and other academic writers, the volumes describe and analyse the main achievements of their subjects in manner accessible to nonspecialists, interweaving these with salient aspects of the protagonists' personal lives. Autobiographies and memoirs also fall into the scope of the series.

Juan V. Mayoral

Thomas Kuhn

From Physics to Philosophy of Science

Springer

Juan V. Mayoral
Unidad Predepartamental de Filosofía
Universidad de Zaragoza
Zaragoza, Spain

ISSN 2365-0613 ISSN 2365-0621 (electronic)
Springer Biographies
ISBN 978-3-031-86511-4 ISBN 978-3-031-86512-1 (eBook)
https://doi.org/10.1007/978-3-031-86512-1

Parts of this book are based on the initial Spanish book "Thomas S. Kuhn: La búsqueda de la estructura", published by Prensas de la Universidad de Zaragoza.

Cover Credit Line:
Front Cover photo: Thomas Kuhn. "Courtesy of Special Collections, Johns Hopkins University, Sheridan Libraries".
Back cover photo: Thomas Kuhn, taken by Henry Perschak.

This Springer imprint is published by the registered company Springer Nature Switzerland AG
The registered company address is: Gewerbestrasse 11, 6330 Cham, Switzerland

For Yolanda

Acknowledgements

In preparing this book, some persons—friends, colleagues, and family—have been of really great help. As usually, I must begin with the most important and influential person of them all: Yolanda García Bejerano, without whom nothing of this sort would have been possible. I rely on her intelligence and common sense every day, and fortunately I have done so for the last 30 years now. This book is, of course, dedicated to her. I am also grateful to my family for always being there.

During the last years since my previous book, some people have helped me more than they probably know, and in many ways. I am thinking particularly of Paul Hoyningen-Huene, who was the first to seriously encourage me to prepare this book, and with whom I have had so many interesting conversations on many occasions for many years now. Paul also gave me access to his personal correspondence with Kuhn and to some pictures, a contribution for which I will always be indebted to him. Howard Sankey, Yafeng Shan and Alex Levine also count among those who have contributed to the success of this task. I am also indebted to Hanne Andersen, Vasso Kindi, Jouni-Matti Kuukkanen, Lori Nash, David Pérez Chico, George Reisch, Susana Gómez, Brad Wray, and my student Jesús Ortiz for very productive conversations on the themes of this book on different occasions. Philip Anderson, Bernard Cohen, John Heilbron, Roger Kuhn, Pierre Noyes, and Richard Rorty kindly and generously answered my petitions for details of Kuhn's life and thoughts and their interaction with him. I am deeply grateful to them. I am and always be indebted to Carlos Solís and Manuel Sellés for the greatest of graduate training and their friendship during and beyond that period. To the University of Zaragoza and to its philosophy students I owe my greatest thanks for providing a stimulating but peaceful context that has allowed me to work on this book intensely for the last year and a half.

I am grateful for the help obtained during the years since 2000 from many archival repositories in the United States and Austria: Distinctive Collections, Massachusetts Institute of Technology; the Harvard University Archives; the Bancroft Library, University of California, Berkeley; the Seeley G. Mudd Manuscript Library, Princeton University; the Indiana University Archives; the Hoover Institution Archives, Stanford University, California; the Rockefeller Archive Center,

Sleepy Hollow, New York; and the Karl Popper Archives, University of Klagenfurt (Austria). I thank Alison Gilchrist, archivist from Taft School, Connecticut, for her help with key documents related to Kuhn's period there as a student.

I thank Ute Heuser, Associate Editor of Springer, for her patience and encouragement during the last year and a half. She has been really helpful and supportive during these months. I would also like to thank Megan Watkins for the best editorial work with my always imperfect English.

A few extracts of documents have been reproduced in this book with the kind permission of the owners of their copyright. A brief phrase on Kuhn by P. W. Anderson in his interview with Alexei Kojevnikov in 1999 (Chap. 1) is here reproduced with kind permission of the Niels Bohr Library and Archives, American Institute of Physics. An extract of a letter by W. V. Quine to William Dennes (Chap. 4) is here reproduced with kind permission of Douglas B. Quine, Ph.D. A line from a letter from Karl Popper to Kuhn (Chap. 6) is reproduced with kind permission of the Karl Popper Archives, University of Klagenfurt, Austria (copyright © 2008 University of Klagenfurt/Karl Popper Archives. All rights reserved). Figure 8.4 is here reproduced by kind permission of Springer. I am indebted to all of them for kindly granting me permission to reproduce such extracts and figure. Of course, I am very grateful to the Distinctive Collections staff, Massachusetts Institute of Technology, for their help in this regard, too. I wish to thank Henry Perschak and the Sheridan Libraries, John Hopkins University, for granting me permission to reproduce the photos that appear on the back cover and on the front cover of the book, respectively.

This book has been written with the support of the grant PID2022-140892NB-I00 funded by MICIU/AEI/10.13039/501100011033 and by "ERDF A way of making Europe."

Competing Interests The author has no competing interests to declare that are relevant to the content of this manuscript.

Introduction

After graduating from university, I decided to go on to graduate studies until I obtained a Ph.D. in Philosophy. It did not take me much time to decide that Thomas S. Kuhn's philosophy of science was going to be the subject matter of my study. I had been attracted by his theory of scientific revolutions based on a fairly different concept called "normal science," whose main explanatory piece were the so-often repeated "paradigms." Kuhn talked about the dogmatic persistence of a group of scientists in articulating and improving the scope of those paradigms, as a sort of tool that needs steady maintenance work. Such scientists seemed to have changed their focus from the attitude that other thinkers such as Karl Popper had taught us to see as natural for anyone interested in finding out more about our surrounding nature from a rational, scientific point of view. Instead of looking beyond the received viewpoint, Kuhn's scientists seemed eager to improve the tools to do their job. Although some of those scientists boldly took the revolutionary step, the most fascinating, almost poetic, stance for me were those who preferred to know more of the world through the lenses that they had received, to improve those lenses, to add depth to the picture that was already visible for them thanks to the improvement of the lenses themselves. In part, that attitude seemed reasonable, but in a different sense, it was fascinating enough to delve deeper into. Attracted as I was by the figure of a revolutionary such as Nicholas Copernicus (*The Copernican Revolution* was the first book by Kuhn that I read), the vision behind the normal-science stance seemed to me the truly interesting puzzle that I wished to solve. How can the best explanation for the research work of most of the scientists in a mature discipline like physics or astronomy be that exploring and improving the same tool is, at the same time, the best way for grasping nature as it really is? On the one hand, the possibility seemed reasonable; on the other, the details of the answer were, at that time, challenging, and I wanted to get to the bottom of that view.

As part of my training, back in the late 1990s, I was advised to travel to the United States and to pursue a part of my dissertation in the same place Kuhn had lived until a few years before. It was 2000, and I began the fall term as visitor at the Center for Philosophy and History of Science at Boston University. Before moving to Boston, I came to know that the Massachusetts Institute of Technology (MIT) held the repository of Kuhn's personal documents, and it seemed only natural to spend some time investigating that material. I had bumped into Robert Merton's engaging study on Kuhn's intellectual development (I will never be able to thank my Ph.D. advisor, Carlos Solís, enough for pointing that text out to me).[1] The way he turned Kuhn into a subject of study from a sociological and biographical study really was the approach I saw as pertinent for examining Kuhn's views. He, as the good friend he was of Kuhn himself, went beyond the usual, normally critical, examination of Kuhn's theses in *The Structure of Scientific Revolutions* and shed light on the man behind the thinking and the social structure underlying both—Kuhn and his philosophy. I had not been trained in sociology, but I had been trained in history and the history of science, so I pursued that line of work. I went to the sources, and tried to find out who Kuhn was at the same time that I tried to reconstruct his ideas from a biographical point of view. For the rest of my stay at Boston and Cambridge, I spent most of my time in the MIT archives, exploring every letter, every note, and every manuscript. The final result of that work, around a quarter of a century later, is this book. Twenty years before, however, in 2004, a dissertation on Kuhn and the origins of *Structure* was enough for me to obtain a Ph.D. in philosophy. This book is the much improved and extended evolution of that initial (and partial) view.

Besides Merton, five more published results available at that time—though very different—were for me of great interest at that precise moment. They were, first, Jensine Andresen's 1999 paper for *Isis*, "Crisis and Kuhn," which helped me to see how Kuhn's best-known concepts could have roots in his personal experiences from high school. The second one was a draft that I had seen cited by Andresen in that paper: Karl Hufbauer's paper on Kuhn's early development as a historian of science, later published in 2012. I contacted the author at that time, who was so kind as to send me a hard copy of it and of a handout with a chronology of Kuhn's development, which was both extremely useful and, above all, inspiring. The third was Steve Fuller's book, *Thomas Kuhn: A Philosophical History for Our Times*.[2] At the time I first saw it, I was already in Boston and used to spend every day at the archives. I bought a copy in Barnes & Noble, and read it eagerly. I guess that anyone like me (a young apprentice doing his or her first serious research) would have felt as I did at that time. It seemed to me that many connections, sometimes hypothetical, about Kuhn and his relationship with Harvard president James B. Conant and the Harvard milieu could be contradicted on the basis of documentary evidence—a not very positive feeling, indeed. At the same time, though, Fuller had historicized Kuhn in a fascinating way. The overall approach was intellectually appealing, yet another sign that the time was ripe (as Conant and Kuhn would have said) for that kind of investigation.[3] Yet, its main thesis—that Kuhn had inadvertently worked for a conservative political agenda, obstructing the critical role of the history and the philosophy of science—seemed, to me, incompatible with the individuals and the

---

[1] See Merton (1977, 71–113).

[2] See Andresen (1999); Fuller (2000); Hufbauer (1997; 2012).

[3] See Conant (1947a, 9); Lowell 5, 44; GR 93.

context that I was getting to know better during this period.[4] I understood at that time that the same approach would shed light on Kuhn's work if the latter's personality and biographical itinerary were reconstructed before obtaining the conclusions. In the long run, George Reisch's *The Politics of Paradigms*, a very different book—in main thesis, method and style—approached the relationship between Conant and Kuhn, against the political backdrop of that time once again, and shed more light on their relationship.[5]

The fourth and fifth works that influenced me at that time were two books that were very different in conception and approach. They were mostly focused on the details of Kuhn's work in the philosophy of science, and became a part of my toolbox before going to the sources. The first of them was Paul Hoyningen-Huene's *Reconstructing Scientific Revolutions*, a necessary condition then and now for any serious research on Kuhn. The second was Kuhn's Festschrift, *World Changes*—edited by Paul Horwich, Kuhn's former colleague at MIT—which was full of interesting critical approaches to Kuhn's views, together with his responses.[6] If the second of these was a source of information, the first helped me not only to find out more about the underlying view on which Kuhn's perspective in *Structure* was based, but also to see that (and how) a partially global and methodical, partially developmental perspective could shed light on views about normal science and revolutionary science and on *Structure*'s difficult passages. The book is, however, unrepeatable. It is the result of systematic work on Kuhn's writings and of a weekly interaction with the subject of study in order to set the record straight about Kuhn's theses. Similarly to *World Changes*, this book sets the scene for anyone interested in getting deeply into Kuhn's underlying perspective—that which explains normal science and, accordingly, scientific revolutions and incommensurability.[7]

If we take into account such works, together with the extensive bibliography which was at that time already available about Kuhn, what was left to do with the sources? Furthermore, after 25 years of Kuhn scholarship since the beginning of the century, with a good deal of fresh insight into the thinker and his philosophy obtained from recent research (now that more than 60 years since the publication of *Structure* have passed by), what is left to do regarding Kuhn? Why is a book like this one worth the reader's time?

The answer is that there are probably few texts that attempt to provide a *full* intellectual biography of Kuhn from the many available original sources. James Marcum's two books are, to my mind, the first attempt to cover Kuhn's historiographical and

---

[4] See our Chap. 2, esp. Sect. 2.8; see also Fuller (2000).

[5] See Reisch (2019). I do not agree with every point Reisch makes there, but the book is the best comparison of Kuhn's and Conant's perspectives written so far. See Mayoral (2020) for my thoughts on that book.

[6] See Hoyningen-Huene (1993); Horwich (ed.) (1993).

[7] Alexander Bird's work on Kuhn was published in 2000. It was soon considered a groundbreaking work, and I now take it as such, but at that time I was looking for something with a greater biographical bias, which Bird himself said his investigation did not have (see Bird 2000, ix). So, it surely influenced me much less.

philosophical production in its entirety.[8] This book is my attempt to build one such biography. A great part of my time in these years has been devoted to reconstructing the successive steps Kuhn took when, first, he wrote *Structure*, and then, when he could make it clearer what the underlying philosophical point of view that supported the theses he had put forward in that book was. At first, my focus was only *Structure*. I first wrote a dissertation on the steps that led to *Structure*, and years later published a book in Spanish based on that dissertation.[9] But that was a half-biography at best. The rest of Kuhn's lifetime tells the full story and shows that, underneath the layer of ideas that were presented in his 1962 book, Kuhn worked hard on the foundations of his ideas for many years.

Kuhn started to try to expound a conviction as clearly as possible: that some solutions in science are caused by changes in viewpoint that affect the foundations of a science. Scientific revolutions were problem-solving devices. To prove that point, Kuhn played with the idea that the scientific method, as expounded by empiricist philosophers, and as assumed by his professors and colleagues in physics, should be revised. The Galilean method that the historian of science Alexandre Koyré used to describe was, for Kuhn, a case in point of the sort of scientific practice that the old methodological views could not explain. There were other examples, of course, which Kuhn sought and studied for years.[10] Examples like these were best explained by showing how the transformation of the foundations of a science leads (it is even required) to finding solutions to difficult problems, as is apparent in the case of Galileo. Similar, analogue transformations in other aspects of life, both inside and outside the sciences—say, in the psychologically analyzable aspects of ordinary life, in the sociological structure of communities, and in the history of science—helped Kuhn to ascertain that this vision revealed a discontinuist development of science that had thus far been hidden behind the veil of a formal representation of science as visible in textbooks. Kuhn elaborated on this view and these themes for more than a decade. During that period, the key role that scientific revolutions played for scientific development was clear for him, and he had the conviction that a specific kind of historiography of science was the better antidote for the disguising nature of current scientific method. On the road to *Structure*, and close to its publication, a new piece emerged: paradigms, the crux that helped him to explain how those revolutions required other periods of stability in which the behavior of scientists was, yet again, different from the account given by the philosophers of science—the already very famous Popper included. *Structure* emerged from that research.

That was the first focus of my dissertation and the resulting book. The other half of the story had to be recounted, too, and uniting both is the goal of this book. In this English version of my work, I have taken advantage of my former work. The first five chapters of this book are based on the eight chapters of the former. This

---

[8] See Marcum (2005; 2015). Reisch (2019) and Wray (2021) are not full biographies, but they are exemplary in that sense.

[9] Mayoral (2017c).

[10] See our Chaps. 2–4 for more details on Koyré's influence on Kuhn and for Kuhn's historiographical inquiries. See, of course, Koyré (1940; 1978).

Introduction and the four additional chapters—Chaps. 6–8 plus the Epilogue—have been written for this occasion.

One of the results of this long-lasting inquiry into Kuhn's writings, published and unpublished, has been the impression that he struggled with a handful of interrelated ideas discovered early in his life for the rest of his career. Out of that handful of ideas, he built up a whole structure of thinking that, in his last decade and a half, he had already established as a developmental alternative in philosophy of science against the static (as he used to call it in his later years), empiricist tradition.[11] Although he was fairly well-informed about contemporary philosophy, such trends never fitted well with Kuhn's views. In time, he would realize that perspectives such as those of Ludwig Wittgenstein, John L. Austin, and W. V. O. Quine were closer to his own positions. Obtaining that structure for his early intuitions took time, and I would say that he never managed to develop them in full. Kuhn, as shall be evident from the psychological features that occasionally emerge in the following chapters, used to fight hard with every idea, to be highly self-critical, and to fill that fight with emotions. His close friend Stanley Cavell offers testimony to this.[12] *Structure* provided that set, or handful, of interrelated ideas, sometimes intuitions, with new pieces. Yet, it was a benchmark result that, though satisfying for Kuhn for some years, served as a stepping stone for the subsequent development of the same set of ideas. A detailed exploration of the years that followed its publication show that he never stopped thinking about the ideas that filled its pages, and about the underlying structure of intuitions and rough conceptions that, in his mind and written words, fought to come to the fore. After the second, expanded, edition of *Structure*, he managed to show how his vision of the discontinuous progress of science could be best explained, and old and new ideas combined to make up an increasingly sound structure.

The second part of his intellectual biography helps us to see this late, post-*Structure* work, and also the connections with the pre-*Structure* part. A side effect of that examination is the impression that *Structure*, in spite of having been the work that made him a renowned author, and probably one of the best-known philosophers of science outside the discipline, conceals any other idea behind that work, that at best can be glimpsed by virtue of passing mentions. To put this into other words, for anyone approaching Kuhn's works after reading *Structure*, it is not uncommon to search for *precedents* of paradigms in works before 1962, and for elaborations of the idea of paradigm afterwards. The possibility of not paying attention to paradigms at all seems discarded at the outset. I don't aim to fault anyone with this marginal note. Rather, I am sure I have fallen prey to this tendency, and that I still do. *Structure* is a fairly convincing view, and friends and foes would be interested in finding out how paradigms—or other concepts of the same book, like normal science or incommensurability—appeared in an initial stage. It is sometimes more difficult to think that they are only a temporary piece of a long-standing investigation in which there were other, unrelated key concepts in the past playing their part in a slightly different vision, and that, after having transformed Kuhn's viewpoint irreversibly,

[11] See Chap. 8 and the Epilogue to this book for details.

[12] See Cavell (2010, 354–355).

paradigms (or other concepts) bequeathed their central place to other pieces. A full biography helps us to relativize the significance of such a key concept as that of a paradigm—and other central notions in the book—to only *some* of the periods that Kuhn's philosophy of science traversed through his lifetime.

In this book I have limited myself to intellectual sources, but have paid some attention to the local contexts in which he studied or worked, and from which he obtained training, principles, information and insights, and in which he suffered some frustrations as well. His education in his family and in progressive schools, his stay at the Center for Advanced Study in the Behavioral Sciences (Palo Alto), his years at Princeton University and at the Institute of Advanced Study (Princeton), or his later years, before his illness, count as a first and positive group of contexts. In contrast, the context of the University of California, Berkeley's Department of Philosophy in the late 1950s and early 1960s, especially as regards the senior philosophers in it, count as a good example of a context that produced mixed feelings in him at some points in his life. The orbits in which Kuhn's ideas move during his life have in this book a gravitational center in the ideas he gradually had and developed, in their interrelationships, in the influences (and also in his criticism) of other thinkers, and in that set of local contexts. Such a complex gravitational center changed during his lifetime, producing corresponding modifications in those orbits, and that is the connection under scrutiny and reconstruction here.

With two exceptions—progressive education and the Second World War—I have not attempted to locate Kuhn's development in a broader context; for example, in a sociological or political context with a national or an international dimension. In other words, I have not examined political or ideological factors that historicize Kuhn and his ideas, thereby explaining them on the basis of such broader contextual factors. The fact that it is possible to construct a compelling philosophical and locally contextual account of the development of his ideas without recourse to such broader factors is testimony to the autonomy of his philosophical development with regard to such kinds of influences (again, with the already mentioned exceptions), and I have attempted that kind of construction. This is not to deny the merit of projects such as Fuller's and Reisch's. On the contrary, I think they help us to find out what consequences Kuhn's work had in the surrounding intellectual context. In this sense, Fuller's approach was groundbreaking and worth further exploration. In my own case, though, I do not feel it is necessary, or even advisable, to rely on such factors for the descriptive and explanatory tasks of this intellectual biography in particular. Kuhn's ideas were developed on the basis of his own intellectual interests, particularly those related to the history, the sociology and the philosophy of science, with the aid of auxiliary fields such as psychology and linguistics. The role of such broader factors in that development was minimal in a fine-grained context and, in my humble opinion, whenever they are present, they look negligible, with the exceptions of progressive education and the war context in his college days. That's why I have not turned to such factors in search of an explanation. Of course, the idea that Kuhn innocently developed a philosophy of science that followed the agenda of a group of influential conservative thinkers of whose consequences he was not aware does not count among my working hypotheses.[13] This possibility seemed unlikely to me, once I saw how Kuhn struggled with every idea he used to put on the table.[14]

In this book I have reconstructed Kuhn's intellectual development in stages. So, Chap. 1 examines his earliest thinking in moments when, first, he still had no plans of becoming a philosopher and, later—with the abandonment of his vocational inclination to physics—in those other moments when he did start to dream of becoming one. Chapter 2 is mostly devoted to examining a context: the one created by the Harvard University's Society of Fellows, in which he turned to the history of science and to philosophy of science, and a man, his mentor, James Bryant Conant, former chemist and then President of Harvard. In those years, between his graduate studies and his early years as an instructor at Harvard, he had a significant experience that transformed his vision of science in significant ways. I shall try to show that he had to apply to the scientific field perspectives the had already been useful for him in literature, politics, and morals. Chapter 3 explores Kuhn's first synthesis of his ideas through a series of eight lectures delivered at the Lowell Institute, Boston. More than describing a stage, this chapter renders the contents of a text that acts as a hinge between his early days as a student and would-be historian of science and his professional dedication to that field. In contrast, Chap. 4 exposes a new stage, this time as an aspiring historian. The echoes of the Lowell Lectures and their philosophical views are displayed in *The Copernican Revolution.* In these works, his dedication to the General Education plan, as designed by Conant, is still present, but Kuhn shows interests that, as in the Lowell Lectures, are all his own. The exploration of historical cases in seminars at Harvard or his new relationships with sociologists of science, or philosophers interested in sociology, are visible at that point.

Those four chapters are intended to show who Kuhn was, that is, to gain a biographical profile of the individual, his ideas, interests, and relationship to the surrounding context. Chapter 5 is, in terms of the narrative, the first main station at which this train was intended to arrive. In that chapter, I describe the final steps that led Kuhn to a first full draft of *Structure*. The chapter is not designed to give a summary of the theses of *Structure* for the umpteenth time, nor to serve as a basis for further philosophical discussion of such theses. As in other parts of this book in which the section of Kuhn's work under examination is widely known, my underlying intention is to describe the creation of his master arguments. I have left the systematic exposition and the critical discussion of these for another kind of book. My only goal in this one is to show how he arrived at his main points (in *Structure*, in this particular case), and

---

[13] See Fuller (2000).

[14] A different explanatory hypothesis for Kuhn's itinerary is Merton's (1977) approach, that is, to pay attention to the social structure of the American academy and to show how Kuhn followed the fairly exclusive steps that used to lead a few scholars to the peak of such an academic structure. I have not followed any such enriching theoretical structure, though maybe because this is a biography without theoretical aspirations in sociology. Yet, I am sympathetic to such an approach, and, as noted, it probably influenced my original research. After all, the local contexts that I describe in this biography correspond to such contextual limits. I believe, as Kuhn used to say about science, that immediate professional contexts are powerful explanatory devices for the behavior of scientists. Such contexts are the limits of my attention in this particular book.

interpretive and critical literature shall only be mentioned whenever it is significant for the biographical narrative itself. Even in the Epilogue, when I revisit those steps once again, my intention is not to evaluate his theses in *Structure* and to discuss the contemporary interpretation of them, but to discuss the role the book played in Kuhn's entire intellectual development. The search for narrative coherence and the respect for a reasonable length (for a book that is already quite long) are the two main reasons that lie behind these decisions. Chapter 5 is in any case a description of a stage. It overlaps at some points with the previous description of his professional dedication to the history of science as shown in Chap. 4. Some more overlapping can be observed in other cases (for instance, between Chaps. 7 and 8), but that is inevitable when the stages to be reflected on must also be thematically complete. In Chap. 5, the elements that truly led Kuhn to *Structure* are included.

At that point, the contents of my earlier book are exhausted and the new stages explored seem, at first sight, diffuse though interesting. In Hoyningen-Huene's book, the works after 1969 are usually taken to be different from the main views of *Structure* on some specific points. Also, in Howard Sankey's exploration of the phases that the thesis of incommensurability traverses, the years after 1970 and before the early eighties are significant in themselves.[15] In this book, that year, 1970, is part of a wider fringe that communicates results explored in the previous years, right after having seen *Structure* published. As the development between 1962 (the original publication of *Structure*) and 1970–1974 (a period from the second, extended edition of *Structure*, to the year in which "Second Thoughts on Paradigms," another significant contribution, was published) has many details, I have explored it in two separate but consecutive chapters. So, in Chap. 6, I describe the stage of his thinking from the publication of *Structure*, a period in which Kuhn tried to defend his viewpoint from criticism, and in which he also re-examined his own theses critically and tried to go some steps forward from the point he had left them. The final part of that chapter leads us to the famous London colloquium at Bedford College in 1965, to his problems with Lakatos, and to his fruitful discussions with Margaret Masterman, to whose key contribution to Kuhn's views we shall return in the following chapter. Chapter 7 not only returns to that Kuhn–Masterman exchange, but also to the continuing work on the foundations of *Structure* and Kuhn's attempt to improve its basis. This process ends when he writes the three main pieces of his 1970–1974 new set of revisions of *Structure* in a row. From that point to the end of the chapter, we examine a different stage that followed this one: the research into the history and the sources of quantum theory, the contribution to the archive of sources of quantum theory from 1961 to 1964 (which is, of course, another example of overlap in my narrative), and the preparation of *Black-Body Theory and the Quantum Discontinuity, 1894–1912*, which was probably his best historiographical work.

Chapter 8 and the Epilogue are devoted to the second and third stations, respectively, in this journey. There is not a main station in the form of, say, a book or a set of chapters in Chap. 8. In its pages, I examine a series of writings, from lectures to lecture-series and some papers, and they describe, of course, a stage, but also

[15] See Hoyningen-Huene (1993); Sankey (1993).

a stage in development itself. As in the process that led us from Chap. 3, on the Lowell Lectures, to Chap. 5, on *Structure*, in this Chap. 8 there is a clear vision of gradual development and addition of layers to Kuhn's new vision. The works that Kuhn produced from 1976 to 1987, of a straightforwardly philosophical nature, are there, and the main theme is the development of his lexical account of scientific development and of incommensurability. The Epilogue is, in many senses, the final station of this itinerary. It is, I am afraid, imbued with some sadness because of the arrival of an abrupt end both to Kuhn's life and to his work. Yet, it also shows how he devoted his later years to a manuscript that, though unfinished, is a current of fresh air for all those interested in understanding and developing his view of science. The Epilogue not only shows this last, briefer stage, but also aims to set it in the context of Kuhn's entire development. The journey thus far must help to show not only his personal itinerary throughout his inquiry into scientific development, but also the point at which he left it, this latter signaled by his very last work. That biographical perspective shall prove its worth if it is able to show where he really left us thanks to *The Plurality of Worlds* and associated texts.

Contents

Abbreviations

Archival Repositories

HUA	Harvard University Archives, Cambridge, Massachusetts.
JBCP X	James Bryant Conant Papers, HUA, Cambridge, Massachusetts. X = box.
KPP X.Y	Sir Karl Popper Papers. The Hoover Institution Archives. Hoover Institution on War, Revolution and Peace. Stanford University, Stanford, California. X = Box; Y = Folder.
PHA	Correspondence between Kuhn and Paul Hoyningen-Huene. Hoyningen-Huene's personal archives.
PUTK	Thomas S. Kuhn, Faculty File. Princeton University Archives, Seeley G. Mudd Manuscript Library, Princeton, New Jersey.
TSKP X.Y	Thomas S. Kuhn Papers, 1922–1996. MC 240. Distinctive Collections, MIT Libraries, Cambridge, Massachusetts. X = Box; Y = Folder.
UCBTK	Thomas S. Kuhn Dossier. The Bancroft Library. University of California, Berkeley.

Books

Black-Body Theory	Kuhn (1987): *Black-Body Theory and the Quantum Discontinuity, 1894–1912*, 2nd edition with a new afterword ("Revisiting Planck"). Chicago: The University of Chicago Press (q. v.). 1st ed., New York: Oxford University Press, 1978.
Copernican Revolution	Kuhn (1999): *The Copernican Revolution: Planetary Astronomy in the Development of Western Thought*, 20th ed. Cambridge, Mass.: Harvard University Press (q. v.). 1st ed., 1957. Revised in the 7th ed.

Essential Tension	Kuhn (1977): *The Essential Tension: Selected Studies in Scientific Tradition and Change*, 1st ed. Chicago: The University of Chicago Press.
GR X	Kuhn (2021): *The Quest for Physical Theory: Problems in the Methodology of Scientific Research. Thomas S. Kuhn's Lowell Lectures of 1951.* G. A. Reisch (ed.), Amazon paperback. X = page.
Postscript—1969	Kuhn (2012): "Postscript—1969," in *Structure*, 2nd ed. (1970). Q. v.: 50th Anniversary Edition (2012).
Structure	Kuhn (2012): *The Structure of Scientific Revolutions*, 50th Anniversary Edition. Introductory Essay by Ian Hacking. Chicago: The University of Chicago Press (q. v.). 1st ed., 1962; 2nd. ed., 1970; 3rd ed., 1996.
The Last Writings	Kuhn (2022): *The Last Writings of Thomas S. Kuhn: Incommensurability in Science.* Edited by B. Mladenovic. Chicago: The University of Chicago Press.
The Road	Kuhn (2000): *The Road Since Structure: Philosophical Essays, 1970–1993, with an Autobiographical Interview*, J. Conant and J. Haugeland (eds.). Foreword by Jehane R. Kuhn. Chicago: The University of Chicago Press.

Communication with Roger Kuhn

CRK Juan V. Mayoral, communication with Roger Kuhn, 12–13 October 2001.

Documents

FR65	"Family Resemblance, 1965." TSKP 5.67.
Handy Notebook 1949	T[homas] S[.] K[uhn], Handy Notebook, 31 March–6 July 1949, TSKP, 1.7.
Lowell X	[Kuhn], *The Quest for Physical Theory—Problems in the Methodology of Scientific Research*, The Lowell Lectures, 3–30 March 1951, TSKP 3.11. X = lecture (1–8).
ML	[Kuhn], Lecture cards with "M. L." or "My Lib." at the top of the card. August 1943–August 1944. TSKP 9.
Rockefeller X	[Kuhn], "Seminar on Patterns in the Development of Scientific Knowledge, 1966–67," Rockefeller University, New York, TSKP 2.9. X = lecture (1–10).
SSR1A	Kuhn, "The Structure of Scientific Revolutions. Chapter I—Discoveries as Revolutionary," ca. 1959, TSKP 4.2.

SSR1B	Kuhn, "The Structure of Scientific Revolutions. Chapter I—What Are Scientific Revolutions?", ca. 1959, TSKP 4.3.
SSRD	Kuhn, *The Structure of Scientific Revolutions*, Draft, ca. September 1960–April 1961, TSKP 4.5.
T2E	"Towards a Second Edition, 1962–1965." TSKP 4.17.
TSK–SF	Thomas S. Kuhn, Student Folder. Faculty of Arts and Sciences, Harvard University. Reference: UAIII 15.88.10 1890–1968, Box 2724. HUA.

Interviews

Interview 1989	Skuli Sigurdsson, "The Nature of Scientific Knowledge: An Interview with Thomas Kuhn," *Harvard Science Review*, (Winter 1990): 18–25 (q. v.). Reprinted in *Shifting Paradigms: Thomas Kuhn and the History of Science*, A. Blum, K. Gavroglu, C. Joas, and J. Renn (eds.). Berlin: Max Planck Institute for the History of Science, 2016 (open access edition).
Interview 1995	Aristides Baltas, Kostas Gavroglu, and Vassiliki Kindi, "A Physicist Who Became a Historian for Philosophical Purposes: A Discussion between Thomas S. Kuhn and Aristides Baltas, Kostas Gavroglu, and Vassiliky Kindi," *Neusis: Journal for the History and Philosophy of Science*, 6 (Spring–Summer 1997): 145–200. Reprinted in *The Road*, 255–323 (q. v.).

Chapter 1
Family, Education, and Early Philosophy

1.1 Kuhn's Personality and Family

Thomas Samuel Kuhn was born on July 18th, 1922. The Kuhns were living in Cincinnati, Ohio at the time, but moved to New York just six months later. There, they lived on the Upper East Side, an area which fit their traditional and economically stable lifestyle quite well. Years later, they moved away from the city to the not-too-distant Croton-on-Hudson, New York, where the young Thomas Kuhn spent part of his childhood and adolescence, from the age of eleven to fourteen, before attending a preparatory boarding school for university.[1]

Kuhn grew up as a thoughtful boy, although also passionate, with a determined vocation to find the correct answer and eliminate any obstacles that stood in his way.[2] Paul Cruikshank, the Principal of the Taft School, in which Kuhn studied before going to college, as we shall see below, described him as an intelligent boy, more interested in the sciences and in mathematics than in sports, except perhaps basketball—being around six feet tall in his teens must have been of help in that regard.[3] Cruikshank added some aggressiveness to this picture (most probably in intellectual contexts) that Kuhn's classmates did not seem to like. However, he also noted that Kuhn had a good character.[4] This mixture of intelligence and aggressiveness is also present in the letter that his father, Samuel Louis Kuhn, wrote to the Dean of Students of Harvard College, Delmar Leighton. The young Kuhn had interests in mathematics and the sciences, but was also fond of writing essays; he had actually won a prize at school at that time. However, his reactions could be emotionally laden, and he could be vehement in intellectual debates. He seems to have not felt completely at

[1] See Interview 1995, 255–258; CRK; Andresen (1999, S44).

[2] Andresen (1999, S50). "The War and My Crisis," an essay by Kuhn from October 2nd, 1941 (TSKP 1.3) that we will deal with in this chapter, is a good example of this.

[3] See Samuel L. Kuhn (Kuhn's father) to Delmar Leighton (Dean of Freshmen, Harvard College), 15 August 1940, TSK–SF, 1.

[4] See Paul Cruikshank's "Principal's Report on Applicant," 22 February 1940, TSK–SF, 2.

J. V. Mayoral, *Thomas Kuhn*, Springer Biographies,
https://doi.org/10.1007/978-3-031-86512-1_1

home in social contexts; the forcefulness of his replies, his father thought, might well have been a defense mechanism against the former feeling. He was an orderly and reflective youngster with interests in pure learning, and in getting to the bottom of things rather than in pure data.[5] These traits are repeated in his adult behavior, as some of his students recounted.[6]

These characteristics are a faithful reflection of the behavior of his family members. This was a very close-knit group, made up of Samuel Kuhn, his father, Minette (Stroock) Kuhn, his mother, Thomas Kuhn himself (and the eldest son), and Roger, his younger brother. According to the latter, Roger, it seems to have been a group with somewhat traditional traits, although they had liberal and democratic political ideas, which, in Kuhn's case, leaned more to the left, mainly in his youth; debate and criticism were often present in daily life.[7] Cultural activity within the family, the appetite for the arts, literature, theater, and cinema, seems to have been habitual since Kuhn's childhood, and it continued in his youth and was part of his activities at university. Among his acquaintances were writers such as Lionel Trilling and Max Eastman.[8]

Minette seems to have promoted this appetite for the arts and literature in the family. We shall see how Kuhn maintained those customs with his wife and his own family.[9] Although Minette was not part of the bohemian cultural atmosphere of the city, more typical of Greenwich Village, she found herself at home in the world of arts and literature. The First World War had prevented her from graduating from Vassar College, which she left in 1919, but that did not stop her from developing her taste for the fine arts.[10] She cultivated her skills in the theater, at least during her brief college studies; her daughter-in-law, Kuhn's first wife, Kay, says that Minette had obtained the role of Joan of Arc in an audition where the poet Edna St. Vincent Millay was also present. Although she was not a writer, Kuhn's mother continued with her literary activity in the professional field and collaborated closely with writers in her environment, such as Karen Horney, a psychiatrist, and with Max Ascoli, a professor of political philosophy at the New School for Social Research and a determined opponent of fascism in his native Italy. She often used this knowledge to advise her son on writing matters.[11]

---

[5] See Samuel Kuhn to Leighton, 15 August 1940, TSK–SF, 1–2.

[6] See Buchwald (2010, 279–281); Kitcher, (1998, 3).

[7] For these features, see CRK; Interview 1995, 266.

[8] CRK. About the relationship between the Kuhns and Max Eastman, see the letter from Minette Kuhn to Eastman, 14 October 1956 (Max Eastman Manuscripts, 1892–1968, Indiana University Archives, Indiana University, Bloomington, Indiana, Correspondence), a condolence note for the death of his wife, Eliena Krylenko, where an old friendship between the two couples is shown. Also, see Andresen, (1999, S48); Forrester (2007, 785–786). About Eastman, see Diggins (1998, 191–192).

[9] See our Chap. 6.

[10] See Kuhn's "Application for Admission," Harvard College, 20 January 1940, TSK–SF, 1. See also [Kuhn], "Puzzles vs. Problems in Scientific Development," 4 November 1974, TSKP 5.9, 1, where he mentions his mother's studies at Vassar.

[11] For these details, see Andresen (1999, S48); CRK; Interview 1995, 260, 265.

Kuhn's seems to have had a closer relationship with one parent over the other. Through his comments, a greater relationship of respect and admiration for his father, Samuel, than for his mother can be discerned. She influenced him, according to his own impression, by her intellectual inclinations, which is why he chose to become a theoretical, not an experimental physicist. However, it seems that the adjective of "intellectual" that Kuhn applied to his mother ("sometimes in a somewhat flighty way," he added) did not include the same degree of admiration that he had for his father, who seems to have left a deeper imprint in him.[12] Kuhn argued with both, but the debates with his father seem to have marked him to a greater extent. According to Jensine Andresen, "Kuhn was very devoted to Sam."[13]

Samuel L. Kuhn, for his part, seems to have been a serious and upright man—a sharp, intelligent and hardworking individual who knew how to carve out a future thanks to his mind and his efforts. Like Minette, Samuel came from a financially well-off family from Cincinnati, Ohio (where Kuhn was born), and had also experienced some cultural activity. Samuel studied at Harvard and the Massachusetts Institute of Technology (MIT), where he obtained an AB (from Harvard University, 1916) and a master's degree (from MIT) in hydraulic engineering thanks to a program that allowed him to obtain both in five years.[14] During the war, Samuel Kuhn served in the army as an engineer and, after the war, he started to work in Cincinnati to help the family, as his father had already passed away. Kuhn's father's professional career seems to have progressively moved away from the subject he studied towards a dedication more typical of a financial advisor and a businessman. He collaborated, Kuhn said, with tobacco companies and with the Roosevelt administration, and he always seems to have been a pragmatic man. Not in vain, he advised Kuhn to dedicate himself to physics, not mathematics, since at that time the former of both disciplines had better professional prospects.[15] Similarly, when Kuhn was about to begin his freshman year, he told Leighton that his son should not completely disregard the possibility of studying Law at Harvard in case he didn't have the talent for the kind of career in physics he was aiming for. This caution does not appear to be indicative of a lack in confidence in his son's intellectual gifts—which he praised above his physical and social abilities in the same letter. It was rather a sign of fatherly concern about his son's near future, with a likely pragmatic alternative as a good solution in the worst case. In a similar fashion, Samuel asked Leighton to find for Kuhn a

[12] See Interview 1995, 259–260, for these data and adjectives. His brother, Roger, considered their father in the same way (CRK).

[13] Andresen (1999, S49; this is based on the testimony of Jehane Kuhn, our author's second wife); see also Andresen (1999, S48); CRK. Samuel Kuhn's letter to his son, Thomas Kuhn (July 14, 1957, TSKP 25.49, 1–2) is also significant. In that letter, he congratulates his son for *The Copernican Revolution* on the occasion of his upcoming birthday. This letter—formal, a bit critical, and paternal—is a brief praise for his book, his work, and his personal and intellectual maturity. Kuhn kept it with other correspondence and praise from his professional colleagues about that book. The letter is a good illustration of that mixture of respect and admiration that seems to have been usual between father and son.

[14] See Interview 1995, 259, and Kuhn, "Application for Admission," 20 January 1940, TSK–SF, 1.

[15] For these details about Samuel Kuhn, see Interview 1995, 259–261, and Interview 1989, 18–19.

freshman advisor with specific qualities: a smart man with a close, friendly attitude towards his son, somebody who could boost his interests in liberal education, not only in the sciences. Samuel's letter to Leighton is long, full of wishes and instructions for his son's future at college. It was an unequivocal sign of a concerned father.[16]

At home, Samuel and Minette created a traditional atmosphere, which included rectitude. There were service staff at home, and dinners were scrupulously respected. Samuel worked six days a week and the evening dinners were the proper moment to exchange impressions and debate on a variety of themes, from culture to politics. Kuhn and his father used to spend some time debating, his brother Roger recounts; more so than with the other two members of the family, Minette and Roger. As his students remember, Kuhn could pursue a question tirelessly, persistently, and this habit seems to go back a long way in time and already be present in those encounters between father and son. In a family that did not seem to have enjoyed vain amusements, these encounters shaped a conscientious, vehement, personality that had a love for truth.[17] As his student Philip Kitcher remembered, Kuhn "[m]ust have known from his youth that it was easy for him to be clever. Being right was far harder—but only that was ultimately satisfying."[18]

1.2 Liberalism, Reason, and Education

As noted, Kuhn's family had leanings towards liberalism. Their economically solid background could have easily placed them in a conservative political position but, in the Progressive era in the United States (1890–1930), Kuhn's family fit well into this latter current. In their household, movements in favor of civil rights and democratic and liberal principles were supported and the basis of this defense was a tradition of rational criticism cultivated in the family environment, as we have seen.[19] This context was a source of personal traits that the reflective Kuhn did not stop investigating from a very young age.

Kuhn addresses this last task in his early autobiographical essay "The War and My Crisis" (October 2nd, 1941), a Harvard student's term paper for his course in "English Composition."[20] In it, Kuhn dealt with three main traits: liberal ideals; the defense of

[16] See Samuel Kuhn to Leighton, 15 August 1940, TSK–SF, 1–2.

[17] These facts about the family's daily life come from Roger Kuhn (CRK), to whom this study is indebted.

[18] Kitcher (1998, 3).

[19] CRK; Interview 1995, 266; Kuhn, "The War and My Crisis," 2 October 1941, TSKP 1.3, 1.

[20] Andresen's paper (1999) is a rich source of information and interpretation about this interesting work of Kuhn in his youth. In the following analysis, I have highlighted some other features of "The War and My Crisis" in addition to those indicated by Andresen, but my debt to her work is clear. On the concepts of liberalism and reason, mentioned below as central in Kuhn's essay, see Andresen (1999, S50–S51). About Andresen, see a little further down, in this same section, and in our Epilogue. On this essay, see also Mayoral (2009, 176). Reisch (2019, 36–39) is another rich source of insight about this essay.

an education concerned with individual development and with the autonomy of that individual's formative stages; and full confidence in the use of reason. With respect to the first of these traits, Kuhn made it clear in his term paper that, for him, liberalism had always been a term with positive meanings, both ethically and politically. His representation of the liberal individual involved respect for tradition and for criticism in equal parts, and also distrust towards mere propaganda and dogma. An unqualified defense of reason was the fundamental basis of this liberal attitude, he also affirmed.[21]

Linked to this first trait is the second: an independent intellectual development, without impositions, and by phases, each of which represents an important and autonomous stage in the formation of the individual. This trait fits with the ideals of progressive education.[22] As we will see at the end of this chapter, something like this also fits with the anti–Whig perspective that Kuhn would defend both in his maturity and also during his brief studies in literature at Harvard. This vision is the origin of some of his more characteristic views about the contingent, historical, nature of scientific development. Concerning the roots of that vision, Kuhn emphasizes his parents' respect for such kinds of individual phases, and gives, as an example, his transition from juvenile literature (he mentions two authors of the genre that were well-known at the time, J. J. Farnol and S. Meader) to novels of greater complexity (like Marcel Proust and Ivan Turgenev). This was a process in which his parents seem to have opted to remain on the sidelines.[23]

However, the best example of this independent intellectual development marked by stages of certain autonomy (especially in terms of belief and norms) is the main narrative that runs through the entire essay. "The War and My Crisis" is an essay about the initial belief, in a Progressivist atmosphere, in pacifism and its active defense, followed by the transition, in a radically different context—the boarding schools, preparatory for the university, which we will talk about below—to an interventionist stance. These two beliefs were in conflict, of course, and Kuhn focuses on the latter, thus explaining the source of his main personal crisis.[24] The essay, moreover, conveys a narrative that prepares the reader for a final double conclusion: first, that beliefs, as sources of judgment, are contingent, relative to a human group; second, that reason is the only reliable source of evaluation, but it only acts in a context of beliefs.[25] Kuhn is thus the main character in an illustrative example of a conflict between personal beliefs. At the same time, insofar as these beliefs are a reflection of those shared by the other actors in the surrounding context, the Kuhnian conflict is an echo of a broader conflict (of a social nature in this case) very similar to that between scientific communities separated by a revolution, as he himself expounds in *Structure*. The inquiry into the sources of a conflict between beliefs, between stances—in this case concerning politics—and the experience of that conflict, was a main concern for him in these early years.

[21] Kuhn, "The War and My Crisis," 2 October 1941, TSKP 1.3, 1–2.

[22] See Sect. 1.3.

[23] Kuhn, "The War and My Crisis," 2 October 1941, TSKP 1.3, 1–2.

[24] Kuhn, "The War and My Crisis," 2 October 1941, TSKP 1.3, 3–6 (5–6 expound the conflict).

[25] Kuhn, "The War and My Crisis," 2 October 1941, TSKP 1.3, 8.

So, I agree with Jensine Andresen when she says that, in an essay like "The War and My Crisis"—and there are other examples like this—we find concepts and theses that are going to be repeated (and, of course, improved) in subsequent phases of his thinking.[26] This essay, in particular, shows some theses that would play an important part in Kuhn's core themes. For him, concepts, beliefs and norms only acquire full sense within the context in which they are used. These elements also evolve in limited settings that, when approached from a future point in time, are hard to understand or even unfathomable. We shall see below that, for him, this also happens with our scrutiny of past aesthetic tastes.[27] The autonomy of past belief and normativity within their own contexts, and its incompatibility with our current context are part of Kuhn's views in these early years, and would be—as Andresen rightly points out—preserved in the future, as is evident from his mature writings.[28] There are other important connections with positions that Kuhn would only discover shortly after, such as the productive relationship that can be established between this perspective on autonomy—which he applies here to his own educational phases—and Jean Piaget's more general and psychologically informed account of the autonomy of the educational stages in the individual, particularly in the child.[29]

Finally, I have to talk about a third feature that for Kuhn was essential in that liberal family environment in which he grew up and was educated; namely, his full belief in reason as the root of judgment and as a basis for the solution of any conflict, disagreement, or problem.[30] Here we must include the problem he deals with in this essay: the conflict, within the same individual, who constantly appeals to reason, between two beliefs that are similarly valuable for that individual (the initial pacifism, on the one hand, and the warlike action in the face of the advance of National Socialism in Europe, on the other). Kuhn's conclusion is that rational evaluation is the basis of the decision; yet, no judgment takes place without a background of beliefs. Every judgment arises in (and belongs to) a restricted context in which there are tacitly and unreflectively adopted base beliefs.[31]

"The War and My Crisis" is thus a representative text of a group of Kuhnian themes that are maintained and developed in greater detail in his mature work. On the other hand, it is also a manifestation of a conception of human beings, of their beliefs and of the norms that they adopt, which are indissolubly linked to the social and cultural environment in which they are educated and live. This conception has consequences for the idea of *understanding human behavior*—whether one's own or someone else's—which is another constant in Kuhn's thought. In other words, how we come

---

[26] See Andresen (1999, S43).

[27] See the current chapter, a bit below.

[28] Incommensurability is, Sarah Kuhn (our author's daughter) says, already present here—in a highly preliminary way, of course (see Andresen 1999, S47). In fact, Kuhn's suggestion of incompatibility between the different assessment criteria used in different phases of his life in "The War and My Crisis," 2 October 1941, TSKP 1.3, 4, is a good case in point.

[29] On Piaget's influence on Kuhn, see our Chap. 2.

[30] Kuhn, "The War and My Crisis," 2 October 1941, TSKP 1.3, 2–3, 5–8. About this permanent (and blind) belief in reason throughout his life, see Andresen (1999, S51).

[31] Kuhn, "The War and My Crisis," 2 October 1941, TSKP 1.3, 8.

to understand rational behavior and solve our *interpretive* problems depends, entirely, on a prior exploration of the context in which it unfolds. This conception is visible in Kuhn's educational context. "The War and My Crisis" is, therefore, a key piece for us, since—as Andresen already warns—it connects Kuhn's mature thought with the context in which he was educated. The following sections, before coming back to this, will provide us with a vision of this context sufficient to appreciate what other beliefs and ideas these early Kuhnian conceptions of understanding and the interpretive method faithfully reflect.

1.3 Aspects of Progressive Education

Kuhn says that his parents' concern for his and his brother's education was not overwhelming. They did not even seem to be fully aware of the details of the school learning of the two young Kuhn brothers.[32] However, this lack of concern for the particular outcome of education did not constitute a total disregard for the formation of their children's personalities. Quite the opposite. Kuhn recalled as early as 1941 that his parents had always taken care of the evolution of his ideas and personality, stimulating their progress and tempering their excesses.[33] In his case, this selective concern for the education of their children fit into a stance that, at that time, was also defended by progressive pedagogical currents, which had their moments of greatest splendor during these early decades of the century in the United States, mainly until the thirties.

Let's examine this current in more detail. As an introductory definition, we can follow the indications of Lawrence Cremin in his classic text on the movement. For him, "progressive education began as Progressivism in education: a many-sided effort to use the schools to improve the lives of individuals," which included improving the connection of the school with the family and the social environment (which implied, in turn, a greater awareness of the diversity of contexts involved), improving teaching techniques with the help of scientific research in "psychology and the social sciences," and allowing a more widespread access to culture.[34] Its influence was positive in various areas. For example, in the period 1890–1930, the general increase in schooling in the country owed much to the influence of the progressive movement, as historian John L. Rury asserts: "If one of the objects of Progressive educational reform was to improve school enrolment rates, educators in certain parts of the country realized a good deal of success."[35] Although it had critics, detractors, and of course failures, the movement left its mark on the development of education in the country. The Kuhns were faithful defenders of this reform, as evidenced by the practice of educating their children in reformist institutions.

[32] Interview 1995, 260.

[33] Kuhn, "The War and My Crisis," 2 October 1941, TSKP 1.3, 2.

[34] Cremin (1961, viii–ix).

[35] Rury (1985, 66).

The nature of this movement becomes clearer if we follow Cremin's indication and relate it to its political and ideological origin. Political Progressivism aimed to consolidate the idea of representative democracy, typical of the American democratic and liberal ideal.[36] This was an idea of liberal democracy in which the government of the majority should not result in a moral tyranny towards minorities. The social scenario of the time demanded this kind of plan. At the beginning of the twentieth century, the United States was formed by groups of different origins, with different religious creeds and ethnic roots, and with diverse moral and ethical principles sometimes opposed to each other, which often coexisted in urban environments associated with increasingly dense industrialization. Under these circumstances, the institutional hegemony of a single value scheme seemed counterproductive and a potential source of conflict. Not everyone thought this way, of course, but certainly those who did were the Progressives.[37]

For the Progressives, this complex social scenario required setting aside differences in creed and customs and supporting a common political and social goal, which, incorporated into the personal objectives of citizens, would produce an improvement in social welfare and a strengthening of the state. For the Progressives, this was only possible through a type of education contrary to the traditional and republican point of view, that is, an education accessible to all economic strata of society, without social, racial, religious or cultural discrimination, and, at the same time, adapted to the various moral and social environments, in which intellectual development did not imply the implementation of principles alien to those of the individual's vital context.[38] In summary, political Progressivism called for that kind of school connected to the social and family environment, sensitive to the individual's origin, and a genuine common gateway to knowledge, which, following Cremin, we mentioned above.

This type of education contrasted with the traditional one. This latter was widespread in the United States and had as a fundamental pattern of learning the memorization and recitation of lessons. The procedure usually came with the teacher's authoritarian treatment of students, who not infrequently found themselves with physical punishment as a result of their misbehavior (according to the former, of course).[39] For the reformists, this kind of teaching was inadequate for shaping individual personality. Instead, teachers and schools should follow J. H. Pestalozzi's, F. Froebel's or J. F. Herbart's reformist guidelines, which advocated for an education that bridged the gap between the teacher and the student, a less authoritarian education that did not involve a disturbing leap between the innocent world of the child and his encounter with school instruction. The early reformists didn't have the same social reform in mind that the Progressives later did, but undoubtedly the

[36] Croly (1909, esp. 193–206) is a good source for this ideal.

[37] See Ross (1991, 147–152). Ross also explains the role of Herbert Croly in the development of the progressive perspective. See also McGerr (1998, 548–550).

[38] See Croly (1909, 399–409); Rockefeller (1991, 290).

[39] See Reese (2001).

latter learned from the former.[40] At the beginning of the twentieth century, pedagogical reformism was a transformative current with notable strength, albeit perhaps heterogeneous and with not very well-defined doctrines.

At the turn of the century, the reformist current found the systematization that it had lacked for many decades. The architect of this work was the philosopher and psychologist John Dewey.[41] Although Dewey is one of the main philosophers of the early twentieth century, both in America and abroad, and later well-known to Kuhn, we are interested now in his pedagogical influence in the current we are studying, which was key—and important for Kuhn himself, who, indirectly, benefited from Dewey's thinking in this regard. Dewey had studied at the University of Vermont (1875–1879), in which, as in other universities like Yale or Princeton, religion had a strong foothold. Dewey did not like that orientation and preferred subjects with a greater social content.[42] He thus enrolled in the recently founded Johns Hopkins University, where that option was available.[43] Having already obtained a PhD in Philosophy, he worked at the University of Chicago with like-minded philosophers such as G. H. Mead, A. W. Moore and J. R. Angell.[44] As a result of his labor at the Laboratory School at Chicago, he wrote *The School and Society* (1899).[45] From his time in Chicago, and then, from 1905, at Columbia University, Dewey did not only become one of the main American intellectuals, but also a key theorist of the Progressivist movement in education. Dewey's canonical school model is that of a thumbnail real context, in which the teacher educates the child using the latter's experience as a vehicle. The teacher helps the children's behavior and skills to fit the scenario in which they are deployed and helps them to understand to what extent that behavior and those skills have consequences for the context they live in. The future of the society to which they belong largely depends on them. This is an assumption that must permeate the whole educational process.[46]

The Progressives adopted Dewey's educational model because it seemed to resolve the difficulties they had faced and solved step by step. However, the adoption of Deweyan views was not restricted only to them. Soon, all those institutions that aimed to replace traditional education with a more suitable way of cultivating intelligence adopted similar procedures—even if they were not Dewey's guidelines themselves.[47] Despite this influence, as W. J. Reese claims, this progressive reform was restricted to experimental institutions often sponsored by universities or private foundations, and was not followed by everyone. The economically and educationally less favored sectors did not follow it.[48]

---

[40] About these reformers, see Reese (2001, 8, 10–24).

[41] About Dewey and Progressive reform, see Reese (2001, 23); Cremin (1961, 115–126).

[42] See Dewey (1930, 153–154).

[43] See Reuben (1996, 83–84).

[44] See Ryan (1997, 77); Westbrook (1991, 72).

[45] See Cremin (1961, 135–142); Ryan (1997, 118 ff.). See Dewey (1899).

[46] See Dewey (1916, 28–30).

[47] See Cremin (1961, 116 ff.).

[48] See Reese (2001, 21–22, and 23 about Dewey).

The Kuhns fit into a relatively well-off social class. The economic possibilities they enjoyed and their political affiliation made it more likely they would adopt this educational reformism for their children, and the education that they indeed provided to the young Thomas Kuhn and to his younger brother, Roger, testifies to this. Given Kuhn's description in "The War and My Crisis" and the principles of progressive education that we have just examined in our historical excursion, we can see clearly that, in the education that Samuel and Minette sought for their children, strict order and discipline were replaced with the development of reasoning and character at the top of the hierarchy of educational principles. Rigor in educational matters was not lacking in Kuhn's family, of course, but the young Kuhn was pushed towards intellectual development and independence of thinking, rather than being led to a normalization of behavior and thought through the imposition of rigid, traditional values. He was induced to (re)think for himself the foundation of actions and ways of thinking from a very young age. His father wrote that his son did not take what the books said for granted.[49] "[F]reedom of criticism and action," Kuhn said that he had learned in his family, together with "respect for the opinion of others, admiration of the American democratic tradition tempered by a scepticism toward the text-book's phrase and the demagogue's promise."[50] This kind of skepticism towards textbooks (in general), and towards established doctrines, especially in philosophy, is a permanent feature in Kuhn's thinking. It seems to be firmly rooted in this education, both in his family and in the kind of school education they provided him. In that context, liberalism itself and the appeal to reason as principles of action and communication, were not exempt from scrutiny.

In this regard, it is worth recognizing that, although in the future Dewey's philosophical and psychological perspectives did not appeal to Kuhn, he was raised and educated in a pedagogical context that Dewey's work helped to promote and systematize. In other words, the *practical* consequences of Dewey's work, his influence on progressive educational currents, shaped some of the earlier contexts in which Kuhn was educated. "The War and My Crisis" is, as an essay, a representative result of someone who was, in his first years, educated according to Dewey's guidelines—ironically, guidelines that would lead him to doubt Dewey's views in the future.

[49] See Samuel Kuhn to Leighton, 15 August 1940, TSK–SF, 1.

[50] Kuhn, "The War and My Crisis," 2 October 1941, TSKP 1.3, 1. For him, these are "generalities which a child of a well-to-do family takes for granted," and they are related to the meaning he ascribes to the word "liberal" (1). Such attitudes towards tradition and criticism were not for him, therefore, exclusive of his family, which also suggests what the surrounding intellectual and educative milieu must have been like for him.

1.4 The Progressive Schooling of the Young Kuhn

The Kuhns moved to New York when young Tom was just six months old. He spent his childhood in New York City and later in Croton-on-Hudson. In New York City, Kuhn was sent to the Lincoln School. Of it, Lawrence Cremin says: "The typical Lincoln student came from a well-to-do family, he was college-bound and his IQ ran considerably above 100."[51] Kuhn fit into that category very well. Harvard University had been the alma mater of some of his relatives such as his grandfather, Solomon Marcuse Stroock, an uncle, Alan Stroock, and of his father, so, by tradition and personal taste, that was his goal.[52] The Lincoln School, therefore, fit well with the profile of the family and the student in question. The results of Lincoln students in colleges supported its priority choice as a prestigious training center in the New York area. Almost every student who had been tracked after leaving the Lincoln had comfortably got into college and had continued his or her studies with visible success.[53] The results endorsed the Lincoln School. Kuhn started at this school in 1927 and stayed there until fifth grade (1933). He started sixth grade at another school, the Hessian Hills School in Croton-on-Hudson, a less rigorous school but equally concerned with its students' intellectual and personal development from a progressive perspective.[54]

The Lincoln School was a primary and secondary school of well-earned fame, a symbol, says Cremin, of pedagogical progressivism and an experimental approach.[55] Such a peculiar combination is due to Abraham Flexner's design. As Cremin also says, "Emerson once remarked that every revolution is first a thought in one man's mind. Insofar as the Lincoln School wrought a revolution, Abraham Flexner was the man."[56]

Flexner was a reformer of American education—both primary and higher—whose ideas, plans, and criticisms became highly relevant.[57] One of his creations was, for example, the design for the Institute for Advanced Studies (IAS) in Princeton, from

---

[51] Cremin (1961, 287). Actually, in 1939–1940, Kuhn's IQ, according to the Taft School, was 136 (percentile 93). See Cruikshank's "Principal's Report Applicant," 22 February 1940, TSK–SF, 4.

[52] See Kuhn, "Application for Admission," 20 January 1940, TSK–SF, 3; Biddle to Gummere, 19 June 1940, TSK–SF, 2. George Biddle, from Croton-on-Hudson, wrote a warm handwritten letter of recommendation for Kuhn to the Chairman of the Committee of Admission of Harvard College, Richard M. Gummere. He was one of the three references Kuhn had provided in his application (3). On his grandfather, the lawyer Sol. M. Stroock, who died on September 11th, 1941—at the beginning of Kuhn's second year at Harvard—see Finkelstein (1947). Alan Stroock, also a lawyer, was the founder at Harvard of the literary journal *Hound & Horn*; see his obituary at *The New York Times*, 30 March 1985, 28, and Biddle's letter to Gummere, 2.

[53] See Cremin (1961, 287–288).

[54] For the data on Kuhn that are reflected in this paragraph, see Interview 1995, pp. 255–256, 259, 262; CRK; Andresen (1999, S46).

[55] Cremin (1961, 280, 287–288).

[56] Cremin (1961, 280).

[57] See Bonner (2002), for more details about his life and the impact of his work.

which Kuhn also benefited.[58] The Lincoln School was a previous project. Flexner had been a self-taught and entrepreneurial young man in his native Kentucky (he was born in Louisville in 1866). He studied at Johns Hopkins University, graduating after only two years (his financial resources did not allow for more). Once back in Louisville, he taught at a school for a while, and later opened his own school. The methods that Flexner applied were sometimes controversial. For example, he did not hesitate to fail an entire class, something that caused problems with the parents. At the same time, Flexner was highly efficient. His students learned their subjects and had no problems getting into college. He made no distinctions. His students might be problematic or highly efficient—it did not matter. Flexner was strict but flexible in that regard, and his teaching was tailored individually. Students got what they needed to develop their skills and their intelligence. It was the development of this latter that mostly interested Flexner, as he set out in his central text, *Universities*, from his Rhodes Lectures at the University of Oxford in 1928. For him, in a complex world that becomes more and more complicated as the individual progresses from school to university, intelligence is the appropriate tool for, in his words, the "comprehension of problems."[59]

For Flexner, Dewey's work had highlighted the close link that must be established between the school, the educational context, on the one hand, and the social environment, the group to which the individual belongs, on the other.[60] But Flexner was also concerned about the quality of teaching—that its result was a simple sum of knowledge from different disciplines and that any of them had the same weight as any other. From that kind of teaching, graduates would not obtain an education that, as noted, supported the development of their intelligence.[61] Flexner established his plans for the Lincoln with the intention of not falling into that trap, something that other progressive education plans did not attempt.[62] Flexner communicated his ideas in 1916 in a well-known article, "A Modern School," and from the plan outlined in this text the Lincoln School's structure of rigor and liberal education was built (it was founded in September, 1917).[63] At the Lincoln School, Flexner's ideas were applied somewhat experimentally—that is, from the point of view of someone who explores an unknown territory, makes plans, observes the results, and makes the proper corrections. As Cremin said, the Lincoln was highly experimental. One of its best-known teachers, H. O. Rugg, a former engineer who became an educator, said that the motto of Lincoln had been "Try anything once and see if it works."[64]

---

[58] See Regis (1987, 8–13, and 214 on Kuhn's time at IAS).

[59] Flexner (1930, 29–30). For the data in this paragraph, see Flexner (1960, 41–42); Bonner (2002, 28, Chap. 3); Regis (1987, 9–11).

[60] Flexner (1930, 49). See Bonner (2002, 41–42, 232–233) for his agreements and disagreements with Dewey. He reviewed *Universities*; see Dewey (1932, 331–332). Dewey's influence is also mentioned by Cremin (1961, 280, fn. 7).

[61] Flexner (1930, 47–54).

[62] See Flexner (1916, 466–468); Bonner (2002, 132–137).

[63] Cremin (1961, 280–282).

[64] Cited by Cremin (1961, 282). For that experimental character, on the basis of Rugg's views, see Winters (1967, 493–514, esp. 494–495). A good study in this regard is Perrillo (2016).

What would Kuhn have found as a student at the Lincoln? Cremin tells us that, at the Lincoln School, various disciplines were brought together to address the same issue. Kuhn was at the Lincoln until fifth grade (1927–1933), so Nell Curtis may have been one of his teachers. Curtis' teaching is an example of the Lincoln's pedagogical plan. Curtis carried out a historical, scientific, artistic, literary introduction to the mode of life around the Hudson River. With this, Curtis tried to make sense of numerous data and involved students in that search, stimulating their interest thanks, among other things, to the practical proximity of the context. Studying and reconstructing nearby environments was a useful tool in the education of the youngest students. Kuhn was one of them at the time.[65] Historian J. Perrillo describes a course prepared around 1936 for students that were older than Kuhn. Although it does not include the group to which Kuhn belonged (nor apply to the time when he studied there), it highlights the kind of teaching developed at Lincoln. In this course, "Living in a Machine Age," students experienced in situ ways of life typical of different environments, from the countryside to the city, and observed and used the different resources in each context. This was a learning method intended to understand current life and the surrounding world from within, and in practice.[66]

This progressive training continued and intensified after the Lincoln. In 1933, the Kuhns moved their residence to the town of Croton-on-Hudson. From the beginning of the twentieth century, some New Yorkers began to adopt this town as a picturesque location that allowed a certain relief from New York. The political spectrum of its new inhabitants was varied; however, a group of left-wing intellectuals and artists from Manhattan who began to live in an area of Croton-on-Hudson, Mount Airy, which came to be called "Red Hill," provided a hallmark. Max Eastman was one of those who lived in the area, for example.[67] The Kuhns followed this same path, possibly—Roger Kuhn conjectures—to offer their children a quieter environment than the populous city.[68] Their new school was Hessian Hills School, with a progressive orientation, like the Lincoln, although more leftist, which included its director and founder Elizabeth Moss's political attitude—like many other inhabitants of Red Hill. At Hessian Hills School, pacifist beliefs and attitudes such as those that can be observed in "The War and My Crisis" were promoted. As we already know, they influenced Kuhn's personal and intellectual development.[69]

Hessian Hills went further than the Lincoln as regards the reduction of student testing. The groups were small, so the students' relationship with the teachers was close, and the latter could get to know their students well. The development of one's

---

65 Cremin (1961, 283–285).

66 Perrillo (2016, 90–91).

67 See Kridel (2013, 279–295, esp. 291).

68 CRK.

69 For the data for this description, see Interview 1995, 256–257; CRK; Andresen (1999, S46, S48); Kridel (2013, 291). Lee Dick filmed a documentary (*School: A Film about Progressive Education*) about Hessian Hills in 1939, that is, not long after the years that Kuhn spent there. It constitutes an interesting document. About the school and its pedagogical approach, see Campbell (1989, 78–86); Kridel (2013). About all this, see Reisch (2019, Chap. 1), who explores the school in a political context in depth.

own points of view about problems or parts of particular disciplines prevailed over memorization and the acquisition of a strong disciplinary base.[70] Data were at the service of an active and comprehensive learning. In class at Hessian Hills, Kuhn read and debated C. A. Beard's *An Economic Interpretation of the Constitution of the United States*, for instance, and wrote essays on various topics. Among his term papers from 1936 there are writings on science ("Physics," or "Explanation of the Telegraph Relay," for example), on politics ("The U.S. as a World Power"), and also on literature, such as "Character Portrayal in *The Case of Sargeant Grischa*," dedicated to Arnold Zweig's anti-war novel.[71]

This last work represents well the emphasis on a comprehensive interpretation of the subject of study, typical of this kind of education. In this essay, Kuhn studies the way Zweig constructs some of his non-central characters, their relationship with the plot, and the detail he achieves in doing so, sometimes at the expense of a fluent reading of the novel itself. Kuhn highlights in a certain passage of the text that this technique allows us, by feeling empathy, to understand the characters in question and the situation they are going through. The variations in behavior with respect to expected conduct, he adds, help us understand the war context that causes and surrounds them—even better than if that context were made explicit.[72] He closes his essay by writing:

> This technic should be carefully noted, but one must be careful in its use not to allow it to get out of hand. In any case, it will build up a group of vivid characters around whom a great plot may be based. But if one is not careful to subordinate the minor characters to the major ones and to distinguish between historical background and plot, the novel must suffer. A vivid document may remain but hardly a great novel.[73]

At fourteen years old, Kuhn shows a remarkable sensitivity towards the nuances of novel construction and delves into them to explain specific aspects of a literary work (and one of great political significance for pacifist positions like this one). As James Marcum says, referring to this same essay: "This revealed his early ability to place himself within a text and explore the development of its characters—an ability that would serve him well when he shifted from science to its history."[74] I quite agree with Marcum. We will see this same ability that Marcum talks about again in action in his studies on literature at Harvard, where something else becomes evident—his anti-Whiggish stance. The most significant aspect of this essay for us at this point, however, is that it is a school essay. It shows the type of pedagogical activity developed at Hessian Hills, which promoted creative thinking, critical thinking, and objective judgment, and is good proof that Kuhn fit well into all of this.

With the same progressive goal of developing thought, personality, and understanding the surrounding world well, other seemingly playful activities were carried out at Hessian Hills, such as writing and performing plays, which were intended to be

---

70 CRK; Interview 1995, 256.

71 See Interview 1995, 256. See TSKP 1.2 ("High School, 1936") for the mentioned works.

72 See Kuhn, "Character Portrayal in *The Case of Sargeant Grischa*," ca. 1936, TSKP 1.2, 4–5.

73 Kuhn, "Character Portrayal in *The Case of Sargeant Grischa*," ca. 1936, TSKP 1.2, 6.

74 See Marcum (2015a, 5).

a sort of laboratory in which understanding the world and practicing social behavior were possible.[75] Kuhn took part in this. Among his writings from this time is a play in which he seems to have actively participated, "United We Stand."[76] In general, Kuhn remembered his time at Hessian Hills fondly.[77]

The last important nuance in this case, worth highlighting, is that it was at Hessian Hills that the pacifist beliefs that were key in the personal crisis that Kuhn describes in "The War and My Crisis" were established. Let's examine this key event in Kuhn's intellectual development again, through the contexts that triggered the problem in question, in this case the boarding schools where he received his pre-university education, which are also the subject of study in that early autobiographical essay.

1.5 Pacifism in Context

Kuhn makes it clear in "The War and My Crisis" how painful the loss of his pacifist stance was. He cherished the memories of those years, when his ideas were much clearer. And just because he felt them inappropriate at a certain moment (at the time of his defense of the United States' participation in World War II), he didn't regret the breakdown of his pacifist convictions any less. But he was forced to think and decide according to all the data of the context—to try to be coherent. When he could avoid incoherence in advance, he did.[78] Thus, for example, although at thirteen years old, and in that pacifist context, he wrote against war, he also refused to swear that he would never participate in it—an oath required to enter the American Student Union, about whose anti-war strike in April 1936 he wrote. Kuhn was aware that perhaps at some point he would be forced to break such an oath.[79] And he was not mistaken. In a short time, new details showed that making that oath would have, really, been a mistake.

Kuhn's behavior indicates a tendency towards independent reflection on problems and decisive action only at the opportune moment. Examples of this character are evident in two of his works. One of them is a brief essay written at Harvard on the *Autobiography* of John Stuart Mill. In that essay, Kuhn accused twentieth-century liberalism of alarming too easily and supported Mill's stance, who advocated for the containment of expression while waiting for the right moment for a completely

[75] Kridel (2013, 288).

[76] Kuhn et al., "United We Stand," TSKP 1.2. This work was compiled by Willison (ed.) (1940), who is also cited by Kridel (2013, 288). I thank George Reisch for providing me with this data.

[77] See Kuhn to Kathy Moos Campbell, November 9, 1982, TSKP 5.1. Moos Campbell was the granddaughter of the director and founder of Hessian Hills, Elizabeth Moos. See also Interview 1995, 256–259.

[78] In TSKP 1.2 we find notes on the pacifist strike of the American Student Union on April 22, 1936. In "The War and My Crisis" (2 October 1941, TSKP 1.3, 2), he says that he participated as a speaker in the pacifist strikes. See also Interview 1995, 257–258, 266, where he mentions his two periods of, first, greater, and then, lesser activism. See also Andresen (1999, S43–S45).

[79] Interview 1995, 257–258. See also "The War and My Crisis," 2 October 1941, TSKP 1.3, 2.

effective manifestation.[80] The second of them is an essay from October 1941, around two months before the Japanese attack on Pearl Harbor, in which Kuhn defended a strategic stance regarding the war which was fairly congruent with this perspective of decisive and effective action only at the appropriate time. The context was the debate about intervention in World War II, which Kuhn was already defending at that time. The war context never left him cold. He did not support a declaration of war without immediate military intervention, even if it appeased the international stage. War, he said, is not an end in itself, but a means to defend freedom, in this case against Hitler. War without military action was harmful to social cohesion. "A declaration of war in itself is nothing," he writes.[81] Two pages below, he says: "Used at the right moment, when it could be followed with truly militant action, a declaration of war could have tremendous effect, but to use it now as a step in a 'war of nerves' [...] would be to breed a discouragement and disillusionment which no amount of future American action could remedy."[82] Declaration and action, thus, must go hand in hand. Going against his pacifism in 1936, he seems to believe that it was time to intervene, but declaration and intervention must be very close to each other.[83]

The institutional and cultural support for the pursuit of intellectual independence disappeared from Kuhn's educational context when, at fifteen years old, he started boarding school. His parents moved again and settled in Peekskill. Kuhn was enrolled in Solebury, Pennsylvania (1937–1938), and then in the Taft School, Watertown, Connecticut (1938–1940).[84] Overnight, Kuhn found himself living and studying away from home. Taft was governed by very different criteria from the progressive schools that he had attended until then, so the pedagogical approach to which he was subjected had also radically changed from what he knew.[85] The conflict that "The War and My Crisis" narrates began to take shape with this change.

In boarding schools like Taft, especially in the most traditional and prestigious ones, among which this one was counted,[86] the average student came from the financial and political oligarchy of the country. The pedagogical foundation of this kind of institution was that disciplined behavior, adherence to Christian moral principles, asceticism, love of work, and a considerably austere lifestyle constituted the basis on which the adult individual would define their lifestyle. For these students' parents, this lifestyle would allow their children to overcome the evils of belonging to a favored social and economic class: moral laxity, the overconfidence in one's own abilities, etc. If those children were expected to serve their families and society at

[80] Kuhn, "Autobiography of John Stuart Mill," Summer of 1942, TSKP 1.3, 2–3.

[81] Kuhn, "International Morale and a United States Declaration of War," 19 October 1941, TSKP 1.3, 2.

[82] Kuhn, "International Morale and a United States Declaration of War," 19 October 1941, TSKP 1.3, 4.

[83] Kuhn, "International Morale and a United States Declaration of War," 19 October 1941, TSKP 1.3.

[84] Interview 1995, 258; Andresen (1999, S46). See also "Class of 1944," on Kuhn's previous itinerary and his entrance examinations, in TSK–SF.

[85] Interview 1995, 258–259.

[86] Saveth (1988, 367, 373–374, 383). See Synnott (1979, 24).

large in the way their parents and grandparents had done, this return to self-discipline and sacrifice, and this commitment to hard work were the right path. The children of many influential citizens around the country—politicians, industrialists, merchants, bankers—were sent to the most prominent group of this kind of institution with the hope that the way of life within them would make them value sacrifice, work, and other moral principles.[87]

The founder of Taft, Horace D. Taft (younger brother of William H. Taft, President of the United States, 1909–1913), had established this kind of education through a strict study regime from the beginning. The school's first catalog indicated that, when the school was opened in 1890, this was an institution where effort was rewarded and laziness and lack of application to study were punished (even with expulsion). Kuhn more than met these values. Additionally, as we have noted, the religious burden of teaching in these schools, mainly Christian, was considerable. At the end of the nineteenth century, when the school was founded, religious services were held daily, and were more intense on holidays, and religious education was part of school activities.[88] This aspect, however, does not seem to have been a problem for Kuhn, taking into account his Jewish background. He seems not to have had great interest in religion at large, something that worried his father a bit at the time of his application for Harvard College.[89]

Kuhn's transition to this strict kind of education can't have been easy. At Taft, he had less free time and lived away from his family environment, so he had very limited contact with that old context that applauded displays of intellectual independence and political activity. Instead, he spent his days adapting to the stricter, rigid, and uniform criteria of the boarding school. More than the difficulties of overcoming the educational minimums, the lifestyle to which he was subjected and the intellectual diet that constituted the strict and continuous disciplinary learning were hard. All of this contributed to a change in his behavior and way of thinking. His activism decreased, although he did not give up his past convictions. In addition, his detailed knowledge of certain academic disciplines improved. Kuhn does not speak well of the scientific teaching at Taft at that time, but he does of learning language and literature there. He studied Robert Browning at Taft, for example, an author that would be an object of his interest later, at Harvard (more on this below). The teaching in that kind of school was, he remembered, "more formal."[90]

Despite the consequent cutback in motivations to the expression of his own, original points of view, Kuhn adapted to the way of life at Taft. He graduated second in his class (123 students), and was awarded the Rensselaer Alumni Association Medal for his good academic results in science and mathematics and the Time Current Events

---

[87] See Saveth (1988, 371–374, 376–380).

[88] See *Mr. Taft's School for Boys, Pelham Manor, New York*, 1890, 2–3. See Saveth (1988, 373), for some more detail about these ideas.

[89] See Samuel Kuhn to Leighton, 15 August 1940, TSK–SF. At the time of his entrance at Harvard, he did not show any preference as to the religion of his roommate. He declared his Jewish religion in his application—a compulsory question, in any case. See the questionnaire in TSK–SF.

[90] Interview 1995, 258 (also 258–259, 266); see Kuhn, "The War and My Crisis," 2 October 1941, TSKP 1.3, 4–5.

Test Prize, which required being up to date with the problems of that time. The 1940 yearbook describes him as a young sportsman (sometimes given to fighting in that context), a lover of beauty, and a sincere, formal, meticulous young man, very active in debates. He was part of the Debating Club in 1939 and of the Debating Team in 1940. He was also a member of the theater club, *Masque and Dagger*, thus continuing with an activity started in Hessian Hills to which, we must remember, his mother had occasionally dedicated herself.[91] He seems to have gained a strong background in English, German, Modern History and Chemistry, as well as in Calculus, subjects that appear in his "Final Record" at Taft, the first four of which he proposed as his subjects for entrance examinations at Harvard.[92] He was part of the German Club and of the Chemistry Club in 1940.[93] However, during his time at Taft, his way of thinking and behaving had changed. He still loved intellectual independence, but the way he understood the world around him was different. "The War and My Crisis" describes his new concerns and perspectives; especially, his new attitude towards war and the basis for this change in stance.[94]

That probably was the root of the conflict he narrates in "The War and My Crisis." Kuhn had successfully navigated this new context, just as he had done when traversing the previous one, and in each of them, he had developed diametrically opposed, completely conflicting convictions about war. He establishes a parallel between his internal conflict and a Kantian antinomy.[95] Two ideas coexisted in his mind, and had emerged in a similar process of rational reflection, making use of his well-trained freedom of thought. Naturally, for a conscientious young man, a faithful pursuer of coherence and rationality, some questions were obvious: Does reason produce conflicting ideas in the same individual? Is this the inconsistent fruit of the use of freedom of thought and rationality?[96]

Kuhn narrates this conflict with some bitterness (and, perhaps, excess of drama) until reaching the already examined conclusion. Despite the doubts that the conflict casts on reason and liberalism, there is no place in his creed for the consideration of other principles of action. He had been educated in those years as a liberal democrat and a faithful rationalist and was not willing to renounce these convictions. But he is an appropriate embodiment of the results of the Deweyan and Flexnerian projects of progressive education. Engaged in his self-formation and aware of his responsibilities, our thinker knows his environment, knows how it has changed, and intelligently adapts to the situation. Thus, while presenting his conflicting perspectives, he also observes that each of them has its own context. One is that of his family and the progressive schools he has attended; the other, the new stage of his preparatory

[91] See *Taft Yearbook 1940*, 97; "Final Record," 20 June 1940 (signed by Principal Cruikshank), TSKP–SF; Andresen (1999, S53); Hufbauer (2012, 423).

[92] See his "Final Record," 20 June 1940, TSKP–SF; Kuhn, "Application for Admission," 20 January 1940, TSK–SF, 2. As that application says, English was compulsory in any case.

[93] See *Taft Yearbook 1940*, 97.

[94] Kuhn, "The War and my Crisis," 2 October 1941, TSKP 1.3, 4.

[95] Kuhn, "The War and my Crisis," 2 October 1941, TSKP 1.3, 5.

[96] See Kuhn, "The War and my Crisis," 2 October 1941, TSKP 1.3, 5–6.

education at Taft.[97] All of it leads to a conclusion that reminds us of the more mature Kuhn: "I still see reason as the only source of judgment," he says, "but I realize that all decision is composed of more than judgment and that my decision can be valid for others only in so far as I resemble those others in this irrational subjective element."[98] This perspective of contingency, typical of his later philosophical work, is already present in this text and in other student works that we will see next. The reference to irrational and subjective ingredients involved in human decisions reappears in his rough notes on methodology from 1949, as we shall see in Chap. 2. They are part of his early views on that matter in these formative years. Yet, as he approaches a more mature philosophical view, he strives to extricate his perspective from such characteristics.

In any case, the journey so far supports Andresen's thesis presented above, and helps us to understand it a little better. Let us remember that, according to Andresen, in "The War and My Crisis" there are ideas that resemble those that characterize *Structure*. It should be added that these ideas have their origin in the progressive education provided by his parents and in the liberal and rationalist environment of his childhood and youth. Part of what Kuhn defends in *Structure* ultimately mirrors part of what he inherits from his cultural and formative background. Moreover, all these ideas, let's not forget, appear clearly once he finds himself at Harvard. The college demanded from him the best of both worlds: intellectual independence, extra-academic activity—so central for the student's socialization—and also discipline and good academic results. Though they were a source of inner conflict and philosophical thinking, he skillfully combined these facets and became one of the most well-known young people at Harvard University. As Philip Anderson (later Nobel Prize in Physics) recalled, when they both graduated from Harvard in 1943, Kuhn "was the biggest man on campus."[99] Let's examine our thinker in that context.

1.6 Kuhn at Harvard

Entering and being educated at Harvard was not just a matter of following a regulated education and obtaining a degree. The extra-academic activity was also considered part of the educational heritage obtained from this university. Harvard's literary societies, its social or athletic clubs, fraternities, and university newspapers, such as *The Crimson* or *The Lampoon*, reveal the dense extra-academic activity that took place on campus. The student residences (at the time, relatively recent) were a form of grouping offered by the university itself and were also forms of student socialization

[97] See Kuhn, "The War and my Crisis," 2 October 1941, TSKP 1.3, 5.

[98] Kuhn, "The War and my Crisis," 2 October 1941, TSKP 1.3, 8 (also 2–3).

[99] Anderson (1999); cit. Mayoral (2011, 230–231). Reproduced with permission of the Niels Bohr Library and Archives, American Institute of Physics, College Park, MD, USA.

that enriched university life.[100] Staying on the sidelines of social life led to exclusion. James Bryant Conant—the brilliant chemist turned President of the institution, and, in a few years, Kuhn's mentor—looked askance at this stifling atmosphere of Harvard's old tradition and sought something else.[101] By the time Kuhn entered the college, socialization remained as desirable as it had always been. It still had many advantages, not the least of which were the purely intellectual ones.[102]

After the time he spent in boarding schools, less stimulating in creative terms, Kuhn felt intellectually encouraged again at Harvard. The debate forums in which he participated involved the use of his notable cultural background and of his dialectical abilities, and this provoked an intellectual reactivation that was very similar to the one he had displayed in progressive schools. In addition, for the first time, he felt integrated into the student community in a way that he had not been in the past.[103] Along with other physics students, Kuhn moved into the Lowell House, room E-43, in the summer of 1941, once the one-year period in a common residence (Thayer) had ended.[104]

Kuhn participated again in literary and political debate groups, as he had already done at Taft. One of them was the Debating Council, which held open discussions that were sometimes broadcast, and were not always calm, with other groups, from the Boston Socialist Party to Vassar College's students.[105] Another one was the Liberal Union, a group that defended liberal and democratic ideals, which was born in 1940, the year Kuhn started at the university.[106] He belonged to this group in his third academic year (1942–1943). Also, as an activity that he liked to highlight among others, he belonged to The Signet Society from his second year, and became its president in his third year. The Signet was a prestigious literary society. Both students and faculty participated in it. They held regular meetings where they discussed literature and other intellectual fields, and cultivated a culture of letters.[107]

[100] We must distinguish the student residences that C. W. Eliot built, which expanded a previously scarce offer in the second half of the nineteenth century, from the plan of Houses, more similar to Oxford's colleges, which was an initiative of A. L. Lowell and began to be carried out in 1929; see Bethell (1998, 22–26, 102–104); about the second of these plans, see Synnott (1979, 9, 33–34).

[101] See Hershberg (1993, 79–80). Synnott (1979, 22–24) describes the almost inaccessible character—due to its classism, traditionalism and even racism—of a few clubs like the Porcellian. Not every group, club or association intended to be so exclusive, of course.

[102] See *The Crimson Confidential Guide to Harvard Courses 1940* (Cambridge, Mass.: The Harvard Crimson, 1940), 6. *The Crimson* interviewed students about this at the end of the 1939–1940 course. 85% of the students admitted having participated in extracurricular activities; only 3.1% regretted it. 90 out of 141 students who did not do so considered it a mistake.

[103] See Kuhn, "The War and My Crisis," 2 October 1941, TSKP 1.3, 5; Interview 1995, 260–261.

[104] H. Pierre Noyes, communication with the author, 10 November 2002. Also see *The Nineteenth Forty-Three Harvard Album*, HUA, box 1674, 246. See these locations in Kuhn's letters and in the questionnaire for admission in TSK–SF.

[105] See *The Crimson Confidential Guide to Harvard Courses 1940* (Cambridge, Mass.: The Harvard Crimson, 1940), 7; also *The Nineteenth Forty-Two Harvard Senior Album*, HUA, 250.

[106] *The Nineteenth Forty-Three Harvard Album*, HUA, box 1674, 195.

[107] Interview 1995, 268. About the Signet, see also Hershberg (1993, 31), as Conant also belonged to it.

Kuhn's extra-curricular political and literary activity was completed with his election to the team of *The Crimson* in his second year, and he was chief editor in his third. *The Crimson* was known for its rigorous admission selectivity, insofar as only about 10% of applicants were chosen. His selection, therefore, is just one more indication of his quality as an essayist, writer, and critic, and of his intellectually restless temperament.[108]

However, Kuhn went to Harvard determined to become a scientist; specifically, a theoretical physicist. He recalled this in 1953: "A dominant interest in mathematics and scientific theory crystallized when I was in junior high school; I entered college determined to become a theoretical physicist."[109] Indeed, Cruikshank had certified his good background in the disciplines previously noted, from Chemistry to Calculus, together with high grades in Geometry and Advanced Algebra, and also Physics.[110] He chose physics to study at Harvard instead of mathematics, his original option, because Samuel convinced him that the career of a physicist had better prospects in the emerging field of physics in the United States. And, indeed, at that time there were good job prospects with a background in physics at some of the best universities in the country. Moreover, for theoretical physics itself there were more opportunities at that point than a few years earlier, when this branch of American physics was not considered among the most developed in the world (unlike the American experimentalists' research).[111] Kuhn could think of university physics departments, or, as his father pointed out, the Bell Laboratories, General Electric, the Bureau of Standards, and the Naval Research Laboratory as possible workplaces after his studies.[112]

Kuhn's beginnings in physics cannot be considered too cautious, nevertheless. Freshmen (first-year students) were encouraged not to dive decisively and definitively into the chosen field. Four subjects were enough to pass the first semester, although many took a fifth, which could be dropped if the student did not feel completely capable of keeping up the pace. However, it was emphasized that it was advisable to diversify with those four or five subjects, try other fields and choose subjects from disciplines far from their original interests to avoid late vocation changes. There was, therefore, a whole semester in which students could make mistakes and reaffirm their

[108] For all this extra-academic political and literary activity, see Interview 1989, 19; Interview 1995, 264–265, 268. The activity in *The Crimson* during the time that Kuhn went through there is collected in *Harvard Crimson Editorial Comments Book*, HUA, box 12, book: "Feb. 1941–Dec. 1942." About *The Crimson*, see *The Crimson Confidential Guide to Harvard Courses 1940* (Cambridge, Mass.: The Harvard Crimson, 1940), 7, and Hershberg (1993, 29–30).

[109] Kuhn, "Intellectual Autobiography," 22 October 1953, UCBTK, 1; edited as Appendix in Hufbauer (2012, 455).

[110] See Cruikshank's grades record for Kuhn, 22 February 1940, TSK–SF.

[111] P. A. M. Dirac had said in 1927: "there are no physicists in America"; see Schweber (1986, 55–56); see also Servos (1986) for more details about the improvement of theoretical physics (and, above all, about the necessary mathematical training for it) in the United States in the decades prior to Kuhn's arrival at Harvard.

[112] For the data from this conversation with his father, see Interview 1989, 18–19; Interview 1995, 261. About the job opportunities for physicists at that time, see Cassidy (2011, Chap. 3).

vocation. This could give a student the opportunity to realize that despite his or her dreams of becoming a doctor, it was actually French philology that he or she was really good at.[113]

Kuhn's career is a testimony to the wisdom of this approach and the folly of rushing, as it is true that late vocation realizations, like his own interest in philosophy, can appear during college studies at which time it is difficult to change direction. He mainly focused on physics and related disciplines. He chose the compulsory English course ("English 1") along with another in the history of philosophy ("Philosophy A"), taught by Raphael Demos.[114] Except for a course in Government during the summer, that was the extent of his diversification in his freshman year.[115] As a sophomore, he enrolled in two more courses in English: "English A-1," "English Composition: General Intermediate Course," and "English 7," on "American Literature," this latter led by P. G. E. Miller, and by F. O. Matthiessen—the author of *American Renaissance: Art and Expression in the Age of Emerson and Whitman.*[116] Kuhn recalls that he "admired [his American Literature professors] greatly."[117] During the summer of 1942, he enrolled in another course outside natural science, this time in History ("History S42b").[118] That was the extent of his diversification of his courses beyond physics during his three years as an undergraduate. The rest of the time, he devoted all his efforts to physics and related subjects, mostly mathematics and a course in chemistry. Yet, in these studies, Kuhn found his first opportunities to apply a perspective that, in his mature philosophy, would be of great importance: the idea that the past must be understood on its own terms. We have already seen that he had applied this principle to his own past, and that he had found out that understanding that past did not avoid the existence (and the persistence) of conflicting views—the past and the present views. We will return to this a little later in this chapter.

It is remarkable that Kuhn did not enroll in any other course in philosophy in those years, given his later interests, beyond that one in the history of philosophy. In 1940–1941, for example, Rudolf Carnap led two courses at Harvard: "Introduction to Analytical Philosophy," and "Principles of Empiricism"; at the same time, C. I. Lewis, in whose modal approach to logic, as we shall see at the end of this chapter, Kuhn seems to have been interested for a short while, also taught an "Introduction to

---

[113] *The Crimson Confidential Guide to Harvard Courses 1940* (Cambridge, Mass.: The Harvard Crimson, 1940), 14–15.

[114] See Kuhn's transcript at HUA, box 15, "UAIII 15.75.12"; see also the *Provisional Announcement of the Courses of Instruction Offered by the Faculty of Arts and Sciences During 1940–41, Official Register of Harvard University*, vol. XXXVII, No. 7, 4 March 1940, 176. (Available at <https://iiif.lib.harvard.edu/manifests/view/drs:467234954$1i>.).

[115] See Kuhn's transcript at HUA, box 15, "UAIII 15.75.12.".

[116] See Kuhn's transcript at HUA, box 15, "UAIII 15.75.12"; and the *Final Announcement of the Courses of Instruction Offered by the Faculty of Arts and Sciences During 1941–42, Official Register of Harvard University*, vol. XXXVIII, No. 54, 18 September 1941, 73. (Available at <https://iiif.lib.harvard.edu/manifests/view/drs:467235372$1i>.) See also Matthiessen (1941).

[117] See Interview 1995, 262–264, and 268 for Kuhn's expression. He also mentions K. Murdoch, but he does not seem to have led the course.

[118] See Kuhn's transcript at HUA, box 15, "UAIII 15.75.12.".

Symbolic Logic" and a "Theory of Knowledge," among others; that year, Quine also taught a course in "Logic" for undergraduates and two courses in "Mathematical Logic" for graduate students mainly, which he repeated the following year. The Department of Philosophy had, in short, a good number of courses that would have been interesting for Kuhn later on, some of them with philosophers that—such as Quine, for instance—later influenced him directly. It would have been a noteworthy situation to see Kuhn as a student of Carnap, indeed. Yet, he was not interested in any of them at that time. His main goal was physics, and only physics, and he devoted most of his time to that subject.[119]

Determined to become a professional physicist, he had to enroll in "Mathematics A," "Analytic Geometry. Introduction to Calculus," which had to be taken together with "Physics F," "Mechanics, Heat, and Sound," for those students that chose a major in physics. "Physics F" was taught by two professors, Jabez C. Street and Wendell Furry.[120] The latter had worked with J. R. Oppenheimer at Berkeley; while there, he had also met Niels Bohr when the latter was a visitor in 1933—a brief meeting that seems to have been productive for his own research.[121] His theoretical level was very high, and his classes were not easy at all. Kuhn liked what Furry taught, although it seems that he initially struggled to keep up with his lessons.[122] Kuhn followed a second course in mathematics during the summer of 1941, "Mathematics S2."[123] The sophomore year, he took his last course in mathematics with George Birkhoff and David Widder, a double course, "Mathematics 5a", and "Mathematics 5b," both on "Differential and Integral Calculus."[124] He struggled a bit with this course. "[W]e were doing multiple integrals and partial differentiation, and I couldn't quite see what was going on," he recalled many years later.[125] Kuhn understood both difficulties in terms that relate well to his mature work. According to his account, the difficulties with Furry and Birkhoff arose from not understanding the specific mechanics of problem-solving in each case; once he understood those, the difficulties disappeared. Sometimes, everything depended on a clue that allowed him to see the problem's set-up from a completely new point of view. Something like this happened with

[119] See the *Provisional Announcement of the Courses of Instruction Offered by the Faculty of Arts and Sciences During 1940–41, Official Register of Harvard University*, vol. XXXVII, No. 7, 4 March 1940, 177–182. On Quine's two courses, see also the *Final Announcement of the Courses of Instruction Offered by the Faculty of Arts and Sciences During 1941–42, Official Register of Harvard University*, vol. XXXVIII, No. 54, 18 September 1941, 167.

[120] See the *Provisional Announcement of the Courses of Instruction Offered by the Faculty of Arts and Sciences During 1940–41, Official Register of Harvard University*, vol. XXXVII, No. 7, 4 March 1940, 79, 86.

[121] See Schweber (1986, 88–89).

[122] Interview 1995, 261–262, 266. About the physics course, cf. *The Crimson Confidential Guide to Harvard Courses 1940*, 14–15, 30–35. About Furry, see Schweber (1986, 88–89).

[123] See Kuhn's transcript at HUA, box 15, "UAIII 15.75.12.".

[124] See *Final Announcement of the Courses of Instruction Offered by the Faculty of Arts and Sciences During 1941–42, Official Register of Harvard University*, vol. XXXVIII, No. 54, 18 September 1941, 140. See also Kuhn's transcript at HUA, box 15, "UAIII 15.75.12.".

[125] Interview 1995, 267.

Birkhoff. For Kuhn, these problems are, according to his mature work, "puzzles," so in his early years he would have experienced what he would later interpret as the process of integration into work under a paradigm. It is difficult to say to what extent this experience as a student was behind his own vision of science and not the other way around (although, as we will see later, his doctoral work does seem to have been fundamental for this reconstruction of scientific learning).[126]

His training in physics soon leaned towards those areas that could be more favorable in the immediate future, which at that time was marked by the war in Europe and the imminent participation in it of the United States. Before the attack on Pearl Harbor in December 1941, there was a clear awareness that the war was affecting the economic and political future of the United States, so participation was only a matter of time. Although criticized, sometimes harshly, by student newspapers, Conant publicly defended this stance.[127] In the early months of 1941, some members of Harvard's faculty decided to contribute to the war effort wherever their knowledge could be valuable, among them P. W. Bridgman, P. D. Bartlett, J. P. Denhartog, C. K. Drinker, A. R. Moritz and the aforementioned Street, who accepted appointments from the National Defense Research Committee (NDRC), one of whose directors was Conant himself.[128] The NDRC found Harvard to be a good location to set up the laboratories where weapons technology would be developed. Thus, the Radio Research Laboratory (RRL), dedicated to the research of radar countermeasures, and where Kuhn would end up working, was located at Harvard.[129]

Of course, part of the teaching and research work of the communication experts began to shift to issues related to the operation of radar systems. For example, some members of the Cruft Laboratory, specialized in communication technology, asked Conant to join the war effort that was beginning to develop in this academic setting.[130] Thus, the "Cruft Group," as it was known, began to offer courses for the army personnel who were going to be assigned to radar systems.[131] A large part of the Harvard community wanted to participate and, after Pearl Harbor, even the student body diminished its pacifist tone.[132] Thus, a former pacifist like Kuhn, who was now much less so, began to take steps for personal participation in the war effort. After Pearl Harbor, Kuhn remembered, "anybody in physics at Harvard was urged to

---

[126] For these two episodes, see Interview 1995, 261–262, 267. See *Structure*, pp. 45–47; *Essential Tension*, 305–307, for his perspective on normal science education.

[127] See Hershberg (1993, 131–132). For a sample of Conant's pro-interventionist harangue to Harvard students after Pearl Harbor, see "President Conant's Message to the Class of 1943 and 1944," in *The Nineteenth Forty-Three Harvard Album* HUA, box 1674, 8.

[128] The list (from March 15th, 1941) containing these names is found in JBCP 12, folder "OSRD–NDRC. NDRC 1940–1941." These names had appeared on a list from October 31st, 1940. We only list those from Harvard.

[129] See Karl T. Compton to Conant, 30 January 1942, JBCP 12, folder "OSRD–NDRC 1942–1944," 1; Guerlac (1987, Chap. 10, esp. 300).

[130] See Conant to Vannevar Bush, June 2, 1941, JBCP 12, folder "OSRD-NDRC. NDRC 1940–1941.".

[131] About this group, see Cohen (1999, 21–31).

[132] Hershberg (1993, 153).

concentrate in electronics, so as to prepare to help the war effort."[133] He concentrated the credits that were usually distributed over four years into only three and followed the courses of the members of the "Cruft Group," which meant abandoning training as a physicist dedicated to fundamental research for another route more akin to applied physics.[134]

In his sophomore year, Kuhn first followed two less-specific courses: "Physics G," part of the two-year plan that involved "Physics F" in freshman years, and led by J. J. Livingood, on "Electricity, Light, and Atomic Physics"; and "Physics 6," on "Heat and Elementary Thermodynamics." In the second case, both the course and its professor were significant for Kuhn in the future. Edwin C. Kemble became a close friend, and thermodynamics, its history in particular, a subject of Kuhn's interest, as we shall see in Chap. 4. He also took "Physics 4," and a course in chemistry, "Chemistry B," on "Inorganic Chemistry, chiefly of the Metallic Elements," and, during the summer of 1942, "Physics S21b," "Physics S24b," and "Physics S25a," the last three of them already specialized in themes from the Cruft.[135] In his senior year, 1942–1943, he followed 8 courses in physics, most of them with members of the Cruft Group. Six of them: "Physics 21a," "Physics 23a," "Physics 23b," "Physics 25a," "Physics 28a, "and "Physics 61." The first of them, for example, was on "Electric Oscillations and Circuit Analysis," led by E. L. Chaffee; the fourth was a "Laboratory Course in Electric Oscillations, Electron Tubes, and Acoustics," and the last one on "Electric Circuits and Electronics at Ultra-High Frequencies." The two other courses were different. "Physics 30" was an "Introduction to Mathematical Physics" by Herbert Jehle, and finally "Physics 31" was a course in "Dynamics" led by Philipp Frank.[136] This last case is remarkable for us, insofar as Frank played a key role, years later, when he encouraged Kuhn to pursue research in sociology of science around the time he received the offer to write *Structure* also in that context, as we shall see in Chap. 4.

Thus, in his third year, with two exceptions, Kuhn learned about electromagnetism, vacuum tubes, circuits, antennas, and everything related to research in electronics that he could develop on the Harvard campus. Except for a B in "English A-1," he made straight As in all of his courses.[137] In June 1943 he graduated Summa Cum Laude in physics. He was prepared to serve at the RRL, and he did during the war.

---

[133] Interview 1989, 19. See also Zangwill (2021, 30–32) for P. W. Anderson's experience (Anderson was, as we know, Kuhn's classmate).

[134] See Zangwill (2021, 30–32) for further details about that option.

[135] See *Final Announcement of the Courses of Instruction Offered by the Faculty of Arts and Sciences During 1941–42, Official Register of Harvard University*, vol. XXXVIII, No. 54, 18 September 1941, 38, 171–175.

[136] See the *Final Announcement of the Courses of Instruction Offered by the Faculty of Arts and Sciences During 1942–43, Official Register of Harvard University*, vol. XXXIX, No. 53, 23 September 1942, 167–172. (Available at <https://iiif.lib.harvard.edu/manifests/view/drs:467239072$1i>.) See also Kuhn's transcript at HUA, box 15, "UAIII 15.75.12.".

[137] See Kuhn's transcript at HUA, box 15, "UAIII 15.75.12.".

Yet, in the process, as he himself acknowledged, he drifted away from the regular training in theoretical physics.[138]

1.7 The Philosophy of Robert Browning

We have seen that Kuhn's intellectual interests went beyond the natural sciences—and physics in particular. He was interested in politics as a subject of study, as an aspect of his life, and as a debate topic. He also liked to write in student newspapers and to converse on arts, literature, cinema, and theater in The Signet Society, and he wrote some essays for his English courses, as we will see in this section. Of course, he was also attracted to philosophy, though it seems that this wasn't significant at this point in his life.[139] This last point is something we will examine in depth in subsequent sections.

Now, however, it is appropriate to delve into another series of ideas that will also echo in his future work and that can be traced back to these dates in his university education. These are ideas of a contextualist nature that we will find later in his life, in his own version of the hermeneutic method. Such ideas would not appear for the first time in Kuhn's mind as relating only to studies of the history of science. He would apply them first in essays on the history and criticism of literature—as well as in "The War and My Crisis." The essay we saw a few pages back, "Character Portrayal in *The Case of Sargeant Grischa*," from 1936, is—as James Marcum already pointed out—an early example of this intention to understand a work from the point of view of its creator, which, in the case of works from a more or less remote past, implies getting deep into sets of beliefs and aesthetic criteria foreign to the one defended by oneself and in one's own time. Kuhn was clear, as we saw in "The War and My Crisis," that his beliefs, his values, and even the meaning assigned to certain central terms in his vocabulary, tied him to a context which was chronologically, culturally, and geographically limited. The study of classic works by John Donne, William Shakespeare, and Robert Browning, among others, as he completed during his undergraduate studies, led him to adopt a comprehensive stance, a hermeneutic perspective, which would be the center of his historiographic method later on and, to a large extent, the root of his mature philosophy of science.

For this reason, we must dedicate at least a few pages to find out more about one of the origins of the Kuhnian opposition to the Whig historiography, that is, the kind of historiography that, to use Herbert Butterfield's words, led to the "ratification if not the glorification of the present"—a perspective always contrary to Kuhn's views.[140] Butterfield is one of the historians often mentioned by Kuhn, and mainly for two

---

[138] For the data used in this paragraph, see Interview 1989, 19; Interview 1995, 267–268, 272; Kuhn, "Intellectual Autobiography," 22 October 1953, UCBTK, 1–2, edited in Hufbauer (2012, 455); Kuhn, "Vita," 1 October 1955, UCBTK, 1.

[139] Interview 1989, 19; Interview 1995, 264, 268.

[140] Butterfield (1931, v).

reasons: first, for having reintroduced the history of modern science to history itself (as a discipline); second, and which was especially important for Kuhn, for having contributed to the study of the history of science from a contextual viewpoint, closer to post-positivist historiography.[141] What we can now verify is that Kuhn already had this anti-Whig perspective years before coming into contact with the historiography of science that Butterfield's works (and that of others such as Alexandre Koyré) showed.[142] Let's look at some examples of Kuhn's approach to the exploration and evaluation of the past in the case of literature, and observe the interesting discussion that this young Kuhn maintains with George Santayana's approach to aesthetic evaluation and its canons.[143]

In the first example, an essay titled "Nineteenth Century English Drama: Browning and Tennyson," Kuhn tried to contextualize a form of performing art that had traditionally been considered of low quality. The turn of the century, he said, had transformed the taste of the audience, and the audience itself, and even the material resources were not the same as in the previous century. Kuhn explains that decline on the basis of causes like these in the first eight pages of his essay—an excessively lengthy paper (26 pages, endnotes and bibliography included), according to his professor's commentaries—and then argued that such works were nevertheless of a high literary quality. Concerning "[t]he writers of the Victorian age," he says that, "if their plays were not successful on the nineteenth century stage, they are still the most interesting dramas of the period."[144] And he added: "Why did they fail? [...] It is the drama which failed, rather than that which succeeded, that presents the more interesting field for investigation."[145] The historical puzzle, so to speak, already seems to be the most attractive point of interest for Kuhn. For more than ten pages, he examined such works, especially those of Robert Browning and Alfred Tennyson. Kuhn showed that there were coincidences in the causes of the failure of the plays they wrote.[146] At the end of the detailed contextualized reconstruction that he gave of such authors and their plays, he concluded that, "in all fairness it must again be said that there is good in much of this work, good which would appear with far more clarity had the authors been less great."[147] Of *Strafford*, by Browning, and *Becket* by Tennyson, the main foci of his analysis, he said that they "deserve a better fate than they have received."[148] In this essay, Kuhn shows his interest in the

[141] See Kuhn, *Structure*, 85; *Essential Tension*, esp. xiii, 108–109, 131–132. In *Copernican Revolution* (283), Kuhn says that, among broad-spectrum works, *The Origins of Modern Science, 1300–1800* by Butterfield (1949) influenced his own text on Copernicus more than others. His card of the book (in TSKP 7) includes some praise. On Butterfield's influence on Kuhn, see Larvor (2003).

[142] On that historiography of science, see our Chap. 4.

[143] On these essays, see also Mayoral (2009, 176–177).

[144] [Kuhn], "Nineteenth Century English Drama: Browning and Tennyson," n. d., TSKP 1.3, 9.

[145] See Footnote 144.

[146] See [Kuhn], "Nineteenth Century English Drama: Browning and Tennyson," n. d., TSKP 1.3, 12, 19–20. Tennyson is analyzed in less than three pages (20–22), whereas seven and a half are devoted to Browning (12–19), who was more interesting for him, as shall be clearer below.

[147] [Kuhn], "Nineteenth Century English Drama: Browning and Tennyson," n. d., TSKP 1.3, 22–23.

[148] [Kuhn], "Nineteenth Century English Drama: Browning and Tennyson," n. d., TSKP 1.3, 23.

meaningfulness of the historical puzzle: how an anomaly like the failure of two great writers in similar aspects of their works, and for similar reasons, shows us the way to an underlying form of change. Browning and Tennyson, in Kuhn's description, are still two very different people and writers. The significant point for us is that Kuhn shows how such partially divergent individuals help the interpreter to make a common point with historical, not only biographical, significance. This feature is also visible later in many of his works on the history of science, from his book on Copernicus in 1957 to that on Planck in 1978.

In another essay, my second example, a paper for "English I," the compulsory course in English that he took as a freshman, he approached one of Shakespeare's more complex plays, *Coriolanus*. In Kuhn's essay, "The Plot Structure of *Coriolanus*," he refers to two main readings of that play, one of them political and the other one psychological. The conclusion of both is that the work has fewer virtues than defects. This very short essay (4 pages) gives little occasion for a deep examination. By contrast, this one was graded higher than the previous one (an A), and also praised more. Kuhn did not comment in depth on the traditionally argued defects of the play, and he did not relate them to the key readings of the play, either. Instead, Kuhn showed how such readings did not shed light on the main features of this play. Neither Coriolanus's relationship with the surrounding group nor his inner psychological conflicts seem to give a unified interpretation in which the defects seem to minimize. Kuhn prefers a reading based on the relationship of Coriolanus with his mother, Volumnia. He briefly argues for it in the last two pages of his essay.[149] Leaving aside the soundness of his arguments, the important point is Kuhn's wish to suppress alternative interpretations when they produce conflicting views and do not solve the main puzzle: how could none other than Shakespeare have written such a play? Is there a reading that minimizes the apparent defects? This is yet another feature of an attitude towards the past that Kuhn will later preserve in his analysis of past science. A similar question arose in his mind when he wondered how Aristotle could have developed such a defective version of an explanation of physical phenomena.[150] His next step was to realize that part of the answer was in himself and in his own categories. Kuhn was, at that time, playing the same role that the interpreters of Shakespeare had played with *Coriolanus* according to his own vision some years before.

If we think through these two examples, there seems to be a common feature in both. Kuhn does not choose texts that are easy to analyze and praise but rather problematic cases, that is, historical puzzles of low modern evaluation of some great writers' works that require some investigation and intent to discover the underlying cause of such evaluation and a new point of view from which such works are not so defective. One last example joins the two former ones and for the same reason: Kuhn looks for historical puzzles and their solution. Kuhn wrote another very short essay, 5 pages, on John Donne's well-known sonnet "Death, be Not Proud." Once again, Kuhn defends this sonnet from those interpretations (he does not mention any

---

[149] See Kuhn, "The Plot Structure of *Coriolanus*," n. d., TSKP 1.3.

[150] See our Chap. 2 for more details on this question.

in particular) that attack either its main theme or its style. The opening paragraph of his essay is significant for us:

> The modern reader is almost certain to ignore this poem as trite in subject matter, mediocre in versification, and, since it is not typical of Donne, lacking in historical significance. I cannot agree. This poem does represent a portion of the Metaphysical Poetry. In it Donne has taken an Elizabethan verse form, the sonnet, and adapted it to a new type of theme and a new type of versification. To evaluate this poem we must regard it both from the standards of modern poetry, and from the standards of the time it was written.[151]

Kuhn then proceeds to analyze the sonnet on the basis of its metric, in which, he argues, Donne bases his way to highlight his message, the trivialization of death. Once again, what I am mostly interested in is the last statement of the last extract: our evaluations of a piece of the past must be based on current and past standards. One could say—I would, in any case—that Kuhn fought to make sense of a statement like this (not just this one, of course) for the rest of his lifetime and with the force of his own philosophy. It summarizes the idea that underlies the essays that we have examined so far, which mix up a subject matter to be evaluated and the fact that it is detached from our modern world by time and by a strong cultural, aesthetic divide. Standards, in effect, are historically located, and the result of our analysis of works from the past—literary in this case, scientific, he would realize in a few years from now—must be aware of the divide that this historical locality marks, and of the effects it has for the applicability of standards. Otherwise, our diagnosis could be that the author in question prepared a failed work, almost inexplicably unusual in his or her distinguished career. If "The War and My Crisis" was already a sign of Kuhn's awareness of the contextual nature of belief, and of the not always objective basis of judgment, these literary essays show that he takes into account the contingency of standards. By this time, Kuhn seems to have had intuitions that a long career then helped him to make explicit, examine, and reconstruct in a systematic way.

These views are already clear in an essay from around 1940–1941, perhaps even 1942, heavily corrected by him and not graded. This essay is most interesting for us because he criticizes the Spanish philosopher George Santayana and the way he applied aesthetic standards of evaluation. The target of Kuhn's criticism is a paper by Santayana, "The Poetry of Barbarism," in which he criticizes Walt Whitman's and Browning's works on the basis of high and permanent aesthetic standards.[152] At the beginning of his essay, Kuhn writes:

> Santayana believes that the greatest poet of all times was Homer, and that other poets are to be measured by the degree to which the[y] approach his scope and ideality. He believes that the task of the poet is to find a universal character in man, his actions, and his emotions and to synthesize his characters from these abstractions so as to show the universal and necessary heritage of mankind in every action and every word. Thus far we might go with him and still find a narrow place for Browning, but Santayana does more. Upon this framework he superimposes a specific philosophy drawn from the classics, from Aristotle and perhaps even the scholastic philosophers to which he says every poet must ascribe. And here we can not

---

[151] [Kuhn], "An Analysis of the Metaphysical Lyric 'Death', by John Donne," n. d., TSKP 1.3, 1.

[152] Santayana (1900).

> agree. The standard is too dogmatic. Santayana has set up too limited a criterion upon which to judge of poetry; Browning clearly fails the test, therefore he is a barbarian poet.[153]

The features that Kuhn highlights of Santayana's points in his essay are enough to see why Kuhn's attitude about the latter are mostly critical. However, we must add to that that Browning, one of his favorite poets and writers—we have already seen an essay in which he tries to set his dramas in context and to understand him properly—suggests points that he shares. In the essay, he talks about the, as he calls it, "the fundamental tenet of Browning's credo: draw men as they are, do not try to idealize. Let each man find for himself the basic principles within the drawing, for no man by himself can set up principles which can be sound for all men."[154] Then he shows that the poet's goal is very different from the one he reads in Santayana. For Kuhn, the poet must depict human beings in all their variety, so as to help others to find their own identity in the variety. There is nothing eternal, so to speak. He then quotes Browning saying in "Rabbi Ben Ezra": "Now, who shall arbitrate? / Ten men love what I hate, / Shun what I follow, slight what I receive; / Ten, who in ears and eyes / Match me: we all surmise, / They this thing, and I that: whom shall my soul believe?"[155] These lines illustrate the kind of situation that Kuhn considers worth the effort of a poet as Browning. Perhaps it does not meet Santayana's standards, but Kuhn is clear: "poetry has many mansions."[156]

In previous essays we saw how Kuhn opposed the assumption, shared, in his opinion, by many critics and interpretive trends in literature, to consider eternal aesthetic values based on universal perspectives on human beings and on the structure of cultures and civilizations. Standards based on an assumption like that tend to conceal the genuine development of literature through history, so some features of literature itself, and of historical change, remain hidden, all of it owing to this. Browning represents the author that is aware of human variety, and Kuhn the evaluator that is conscious of the historical contingency of evaluation and standards. The two of them form a front—for Kuhn, a common standpoint—against the defender of a permanence of values and standards, in this case represented by Santayana. Whether Kuhn is right or not in this latter criticism is not the point I want to make here. I am more concerned with the vision that this young Kuhn, as yet an undergraduate student, has concerning the contingency of standards and the historicity of human culture and even of human character. If there is an original basis for some of its more central philosophical ideas, we can find it in this still restricted (at least as compared with his training in physics) humanistic formation. When, in 1958–1959, he turns to a comparative study of the development of science with that of fine arts, his viewpoint about the latter finds support from an old background such as this one just examined.

---

[153] [Kuhn], "In Defense of Barbarism," n. d., TSKP 1.3, 1.

[154] [Kuhn], "In Defense of Barbarism," n. d., TSKP 1.3, 3.

[155] Browning (1864, 816); [Kuhn], "In Defense of Barbarism," n. d., TSKP 1.3, 4.

[156] [Kuhn], "In Defense of Barbarism," n. d., TSKP 1.3, 4.

1.8 P. W. Bridgman: Philosophy and Physics at Harvard

Let's continue with our review of Kuhn's studies in subjects unrelated to physics. A mandatory stop is his first contact with philosophy before his graduation in 1943. That contact was very discreet in these early years. We saw that the Department of Philosophy at Harvard, thanks to distinguished members like C. I. Lewis and W. V. Quine, and to temporary members like Rudolf Carnap, was a good place for the cultivation of logic, philosophy of language, and contemporary epistemology. But Kuhn did not initially learn philosophy from contact with recent work, but through key figures of the European philosophical tradition such as David Hume and Immanuel Kant, whose work he studied in "Philosophy A" with Raphael Demos. In that course Kuhn began by examining some of the ideas of these two authors (among others from Plato onwards), such as the Kantian conception of the a priori and the problem of induction in Hume, with the point of view of a physics undergraduate. The following year, he prepared for "English A-1" an interesting essay that involved both points of view, and that we will examine below in this chapter. Kuhn sometimes said that Kant's ideas were the first to attract his attention to philosophy.[157] This does not imply that his reception was uncritical. On the contrary, we will be able to appreciate that he adopted a contrary position towards the a priori (only in its Kantian sense), and that the stance he took has close conceptual links with a well-informed critique of Kant.

His initial contact with philosophy was hard, owing in part to the instructor he had in Demos's course, Arnold Isenberg. His comments on Isenberg in his 1995 interview show that they were not on good terms. Isenberg left Harvard in 1941, and Kuhn continued without him for the rest of the course, and even "learned something more about how to study."[158] On Isenberg, in 1968, in the draft for the Isenberg Lecture at Michigan State University (at which the latter had spent the last three years of his life), Kuhn wrote:

> When I first entered Harvard College in the fall of 1940, I elected a course in the history of philosophy. The class was not regularly open to freshm[e]n, and I could claim no previous experience with philosophy nor even a conception of what it was about. As a result, I had a hard struggle with the senior Teaching Assistant before I was reluctantly granted admission. Arnold Isenberg was the man with whom I did battle, and I was lucky enough to have him also as the leader of my discussion section during the fall. The course through which I knew him proved to be the central intellectual experience of my first college year, and it redirected [m]y interests in ways that ultimately changed the shape of my career. Under

---

[157] About his initial contact with Kant (and other classic philosophers), see Interview 1995, 264. Later on, in his 1975 comments on *Theorie und Erfahrung* by W. Stegmüller (see our Chap. 8 for further details), Kuhn admitted that Kant had been important to him in times as early as those we are going to discuss here, and that he still was at that time. See Kuhn, "Excerpt from a letter of 20 January 1975 by T. S. Kuhn to W. Stegmuller Concerning *Theorie und Erfahrung*," TSKP 20.36, 14.

[158] Interview 1995, 263.

> those circumstances there can be no platform to which I feel more urgently called than that of the Isenberg Lectures.[159]

Kuhn later deleted that paragraph from the printed version of this lecture in *The Essential Tension*. However, the paragraph just quoted give us a fresh version of his beginnings in philosophy studies, and the difficult time that he had with his first instructor, even at the time of his admittance as student of the course. His acquaintance with philosophy was limited at that time. He had had some conversations with Adolph Oko, a Spinoza scholar and the husband of Samuel Kuhn's sister, according to Hufbauer. But that had been all. Later, he discovered in college that Spinoza was not his cup of tea—not like Kant, anyway.[160]

Kuhn had, in addition, other philosophical sources at hand at that time. One of the key contemporary authors on some problems of the methodology of science, especially on the meaning of scientific concepts, was Percy Williams Bridgman, whose writings in that field were not only close to those of the logico-empiricist philosophy of science, but also to the physicists themselves, especially to those who, one way or another, thought about their discipline in a more philosophical way, and sometimes wrote about it as well; as we shall see, Kemble, one of the physicists closest to Kuhn, was one of them. Bridgman's work is part of the context of philosophy of science at Harvard and so a key piece for somebody like Kuhn, who soon started to write about the philosophical problems in physics, as we shall see below.

Bridgman influenced several generations of American physicists. This experimentalist lived through the rise of the two great revolutionary theories of the time—the theory of relativity and quantum mechanics—and he strove to introduce himself to them and master them. It all started in 1914, when he had to replace the deceased Benjamin O. Peirce (father of the Pragmatist philosopher and logician Charles S. Peirce), and was forced to teach electrodynamics courses. Bridgman introduced himself to the work of Albert Einstein and began to teach some of the nation's first theoreticians. Kemble's Ph.D. thesis was supervised by him, as was that of John Slater; and Van Vleck, although himself supervised in turn by Kemble, also learned from Bridgman. As historian Silvan Schweber says, "Bridgman deeply touched the lives of the leading theoreticians of the generation that matured between 1915 and 1925. Edwin Kemble, John Van Vleck, John Slater and Robert Oppenheimer, all Harvard bred, and Gregory Breit, who was an NRC fellow at Harvard during 1923–24, all came into contact with Bridgman."[161] In summary, Bridgman was a master of experimental physics who had the courage to face the new theoretical challenges of his time and managed to master them to influence the following generations of his era. His students, especially those mentioned, had fond memories of his teaching.

---

159 Kuhn, "The Relations between History and Philosophy of Science," typewritten draft, Isenberg Lecture, Michigan State University, 1 March 1968, TSKP 23.10. It was later published for the first time, modified, in *Essential Tension* as essay 1 (Kuhn 1976c).

160 See Interview 1995, 263–264; Hufbauer (2012, 424).

161 Schweber (1986, 62).

Bridgman would have liked, he confessed, to have had more free time to better understand the theoretical terrain.[162]

From Kemble to Van Vleck, and including Slater and Oppenheimer among others, Bridgman's influence was felt beyond the teaching of physics itself. From 1914, with the beginning of his theoretical teaching, Bridgman had been developing his own point of view on the scientific method in physics, which he ended up presenting in an orderly manner through the so-called "operational analysis," whose most complete presentation can be found in *The Logic of Modern Physics*, his classic text.[163]

Operational analysis asserts that the concepts used in physics derive their meaning exclusively from the set of independent operations that determine the numerical values that these concepts refer to.[164] Thus, for example, the measurement of the length of an object or space by a standard measure, which is also based in turn on standard units, is the only thing that gives meaning to the concept of length. On the other hand, from the moment we assign different groups of operations to different measures of length, we are also talking about different meanings for that same term, so we are dealing with disparate concepts. There is, for example, a great difference between the measures of length that are made using measuring rods and other tactile and optical instruments at the same time, and the purely optical measurement that is made in astronomy or cartography. Despite the use of similar terms and psychologically close notions, as is the case with the idea of length, physical theory must discern between the concepts that are used in different observation and measurement contexts. Despite the terminological similarity, there can be serious confusion created by the changing meaning of the same term when it is used in such different contexts. The concept in question may differ radically when it passes from one context to another, so we assume consistency in a concept although really this is not the case.

Operational analysis was intended to be a useful tool for physicists to clarify the meaning of physical concepts, especially given Bridgman's aim to eliminate any theoretical aspect not directly related to observations, to phenomena. In a science like physics, so marked by experimentation, scientists must adhere to the use of concepts whose correspondence with the results of experimental measurement is clear. If they do so, the correction of theories will not affect the fundamental concepts (that is, the basic quantities to which they are reduced). The exploration of phenomena involves the use of concepts and models that are not in direct correspondence with this kind of experimental work (that is, that do not strictly arise from it). For Bridgman, this is not only legitimate, but also common. In such cases, subsequent experimental work will have to corroborate whether the use of concepts is truly legitimate or not.[165]

[162] In this regard, see the letter from Bridgman to L. Silberstein, 10 May 1922, in Schweber (1986, 71). Bridgman confesses there this frustration. For the data of this brief profile of Bridgman, see Kemble (1962); Slater (1963); Kemble, Birch and Holton (1970); Van Vleck (1971, 4–5); Moyer (1991, 249–250).

[163] Bridgman (1927).

[164] See Bridgman (1927, 5–7).

[165] See Bridgman (1927, 7; 1938, 123–124).

Outside of this more exploratory plane, Bridgman's motto, as Maila Walter states, is to avoid the kind of fundamental conceptual change generated by revolutions like that of the theory of relativity, which Bridgman very much had in mind when writing *The Logic of Modern Physics*.[166]

This kind of change, which we should not be afraid to call "revolutionary," is also the one that concerned Kuhn in his early years at Harvard. Despite the conceptual changes in physics, like those Einstein had recently provoked, Kuhn asked: Can the concepts of physics be preserved at any stage of its development? And how do these same concepts relate to experience? Moreover, how must they relate to each other so that such radical changes do not occur? These questions, as we shall see later in this chapter, represent the concerns that Kuhn would develop as a physics student at Harvard before and after the war. In this respect, Kuhn already shared a philosophical goal with Bridgman. He would share only a few more things, as we will see, because his diagnosis was, ultimately, anti-Bridgmanian. For Kuhn, a conceptual analysis as succinct as Bridgman's reveals little of the real mental plane that underlies the theoretical plane of physics. His complete answer to Bridgman wasn't yet available, of course, but it would soon be. Its first manifestation was in the Lowell Lectures in 1951. However, it is significant to see that, in part (although a quite essential part), Kuhn's work is a response to concerns he shared with an empiricist and physicist like Bridgman.

A few more common traits become visible if we examine a passage from Bridgman's main work. Let's remember that, for him, the physical theory depends entirely on measurement and its results, and that this limits the scope of application of concepts. There is, for example, no problem in using our notions of time, space, or cause within the scope of phenomena typical of classical mechanics, but there are problems when transferring them to the domain of, say, quantum mechanics. In *Logic*, Bridgman refers to those theories that, like special relativity or the aforementioned quantum mechanics, used traditional concepts in empirical areas in which these did not fit. The realm of high speeds or of the very small is prohibitive for the accustomed use of the most basic classical concepts—space and time, mainly. In these situations, Bridgman says, conceptual research is equivalent to "solving a cross-word puzzle." This expression appears in a passage in *Logic* that illustrates this important idea—one that reminds us so much of the use of the term "puzzle" by Kuhn himself:

> Not only are concepts hazy around the edges and so incapable of fitting nature exactly, but there is always the chance that there are concepts other than those which we have adopted which would fit our present phenomena. Finding concepts to fit nature is much like solving a cross-word puzzle. In the puzzle there may be some parts of the pattern which we fill completely and easily, but sometimes we find parts in which we can fill in everything except one or two obstinate definitions, so that we are sure we are on the right track, and rack our brains for the missing words, when with a flash of inspiration we see that the obstinate words can be fitted in by a complete change in those which we have already accepted. It may be that we are soon to witness a similar change in our concept of the nature of light. An important difference between the cross-word puzzle and nature is that we can never tell when we have

---

166 Walter (1990, 107–109).

filled in all the squares in any of the parts of nature's puzzle; there is always the possibility of new phenomena which our present scheme does not touch.[167]

There are three ideas in this passage that remind us of similar ones in Kuhn, as I have indicated before: first, the mention of crossword puzzles and puzzles in general as an analogy for the organization of an explanatory (or conceptual) framework for a certain theory; second, the related idea that the perspective that allows us to find the solution to the puzzle sometimes involves a rethinking of the already established concepts; and third, that all this arises from a "flash of inspiration," in Bridgman's words. We will see later that Kuhn uses similar analogies from the beginning to explain the nature of scientific problems and their solution; he even talks about something similar to these "flashes of inspiration," explainable in psychological terms. It should be noted, however, that Kuhn only uses the analogy of the solution of scientific problems with the solution of puzzles shortly before publishing *Structure*.[168] Yet, the idea that scientific concepts order the world in a restrictive, interrelated and rigid way, and that certain solutions need to alter such a limited conceptual structure through a change in the accepted point of view appears as early as the Lowell Lectures.[169] Even the idea that our concepts are "hazy around the edges," as he says at the beginning of the previous extract, is again visible in Kuhn in different moments in his life. However, the point of view from which he uses such ideas and vocabulary is, however, different from that of Bridgman. Appreciating such differences and their origins is part of what I will try to do in the next chapters.

Let's return one last time to the previous paragraph from Bridgman's *Logic*. The empirical support of well-established theories is therefore assured, but that is not the main problem that Bridgman is concerned with. He is worried about the preconceptions that push us to predict, without an empirical basis, how new experimental data act in the modification and substitution of concepts. For Bridgman, there are no established methods in this regard. Theories depend on data within an irreducible measure of uncertainty. For this reason, the appropriate methodological and theoretical research tools are, for him, as strict and restrictive a resource as operational analysis—or dimensional analysis, too.[170] Also for this reason Bridgman discarded a primary role for what he called "special views of nature," that is, those perspectives that projected some form of ontological framework for natural phenomena, although they could be useful in research. Ideas about simplicity and determinism were, in particular, two examples of those "special" perspectives.[171]

Bridgman's viewpoint fit very well with the ideals of scientific research that began to develop among theoretical physicists from the mid-twenties and took root in American physics. It is not surprising that Bridgman became a spokesperson and systematizer of an attitude that students and colleagues shared, because it was consistent with the changes generated by the arrival of quantum theory and its results. A sample

[167] Bridgman (1927, 202–203).

[168] See our Chap. 5.

[169] See our Chap. 3.

[170] See Bridgman (1922) for more details on dimensional analysis.

[171] Bridgman (1927, Chap. IV).

of this collective agreement is very illustrative. Let's examine such a sample before looking more closely at Kuhn's opinions about Bridgman and operationalism, and before examining our author's first set of reflections on philosophy of science.

1.9 A Philosophy for Physicists

In *The Logic of Modern Physics* Bridgman had maintained a wait-and-see attitude towards the need to modify the concepts of space and time in the field of quantum mechanics.[172] In a footnote at the beginning of the section where he dealt with this, he said that, after writing those pages at the beginning of 1926, his position had changed thanks to quantum mechanics. However, nothing mentioned had to be essentially corrected "because many of the developments actually taken in the new mechanics follow the lines that it is here urged they ought to take, and in so far afford interesting confirmation of the point of view of this essay."[173] For Bridgman, indeed, the change of scale implied an alteration of the concepts of space and time in operational terms, especially if the number of operationally independent concepts is greater or less than in the ordinary scale of classical mechanics, so that these concepts, in their ordinary sense, would cease to apply in the quantum field. All this then depended on further experimental investigations. What Bridgman did assume was that the possibility of requiring new concepts in the new field of the quantum view of the atom had to be left open. For Bridgman, the task of conceptual creation is complicated and often postponed, not always for the worse (quite the contrary).[174] (We know that Kuhn theorizes about the same resistance to change.) But, for Bridgman, if there were occasions that required taking that step, this was possibly one of them. The creation of concepts is produced (as he said in the paragraph reproduced above) by the need to make the necessary theoretical pieces appear so that the gaps in the crossword puzzle posed by the phenomena are filled. This usually comes with a change in perspective and often implies that we will have to alter in some way those concepts already established. What makes no sense is to insist on keeping the words already written in the crossword, even if we cannot find a way to face the challenge of the quantitative anomaly—that is to say, of the sign coming from phenomena that something does not fit in what we already have. As pointed out in *Logic*, it may happen that ordinary mechanics is not so fundamental and insisting on formulating quantum conditions in classical terms would then make no sense.[175]

In 1938, Edwin Kemble gave a lecture at the Franklin Institute in Philadelphia on the appropriateness of the operational method in research based on quantum

[172] See Bridgman (1927, 186–195).

[173] Bridgman (1927, 186, fn. 1).

[174] Bridgman (1927, 195).

[175] Bridgman (1927, 191–195).

mechanics.[176] In it, Kemble shows that scientific knowledge consists of the interaction between theory and experience in a verificationist way. Although we all consider the basic components of our environment to be completely real, they are as theoretical as those that populate the conceptual world of physics. Simply, the latter are more elaborated, thanks, among other things, to their continuous comparison with experience.[177] The advancement of physical theory, adds Kemble, sometimes consists of altering this theoretical landscape, its basic ingredients. In the process of reinterpreting changes in a way akin to common-sense realism, we may encounter "apparent paradoxes," says Kemble.[178] Kemble, however, is explicit when talking about these:

> My first thesis today is that most of the paradoxes are not really problems for the physicist although they may be of considerable interest to the professional or amateur philosopher. This is because they do not really involve any conflict between theory and experiment. They have to do with elements in the theory which do not look right from the standpoint of common sense and mark the divergence between new constructs and old. But the concern of physics is the correlation of experiences. Its domain is the domain of the experimentally verifiable. Therefore the job of the physicist is to describe the experimental facts in his domain as accurately and simply as possible, using any effective procedure without regard to such *a priori* restrictions on his tools as common sense may seek to impose. His one semi-philosophical concern must be to clarify fully the procedure whereby he passes from his sense-perceptions to the world of his constructs and back again.
>
> In so doing he will of necessity adopt the method of operational reasoning recently advocated by Bridgman and closely related to the positivism of Mach and the pragmatism of the American philosopher and scientist, Charles S. Peirce.[179]

As we know, in his mature work, Kuhn makes clear that some of these "paradoxes" do have an important role in scientific research (and in the way it develops), and that they will lead to ignoring them or to trying to solve them, creating two different scenarios: normal science and revolutionary science. But this is mainly because, for Kuhn, these "paradoxes" are generated not simply between the stage of intuition and the more limited one of physical theory, but, in a more general way, between scientific theories that, at a certain moment, enter into a process of (failed) comparison; this not only affects the relationship between scientists and non-scientists, but within the former group, when science is practiced according to different paradigms. But all this will come much later, of course, and even now, in these early years, Kuhn does not have a very different view of the method from that exhibited by Kemble, as we will see a few pages further ahead.

Kemble emphasizes this perspective rooted in Bridgman's operational analysis with a significant phrase: "I assert that the province of the physicist is not the study

---

176 Kemble (1938). This lecture was part of a symposium in which J. C. Slater and E. U. Condon also participated. About this symposium, see Schweber (1986, 65–67), where some of the extracts from Kemble's text and, later, from Condon's (that we will see below) are also reproduced.

177 Kemble (1938, 264–266).

178 Kemble (1938, 266).

179 See Footnote 178.

of an external world, but the study of a portion of the inner world of experience."[180] The realist commitment completely disappears from his perspective and in its place the interaction between theory and experimental result takes priority. For him, "there is no reason why the constructs introduced need correspond to objective realities."[181] This emancipation from common sense frees physical theory from its commitments to intuition and from the brake that intuition sometimes imposes on it. The result is a point of view which is heir to Ernst Mach's positivism, whom he quoted at the beginning of his lecture,[182] in which questions about the external world cease to make sense. About the external world, Kemble makes his position clear: "I can no more deny the existence of such a world than you can. [... At the same time] I admit I do not know exactly what I mean by saying that an objective external world exists. [...] On the other hand it seems to me that the formulas of present day physics do not describe such an external world."[183] Kemble, therefore, limits the content of physical theory to its relations with experience, without realistic implications. It is these relationships that the physicist must understand, not these implications. Given these limitations, operational analysis is the most suitable method for generating good results.

Also relevant is the opinion of theoretical physicist E. U. Condon, who invoked Dewey at the same meeting as Kemble to affirm with the philosopher's words that "[t]he function of intelligence is therefore not that of copying the objects of the environment, but rather of taking account of the way in which more effective and more profitable relations with these objects may be established in the future."[184] Regarding the current situation in contemporary physics, Condon concluded:

> Of course I do not mean to insist that the present form of quantum mechanics *is* ultimate nor that possible future developments which go beyond present ideas would be unsatisfactory or superfluous from the point of view taken here. [...] But I also believe that these developments will come not by seeking for classical mechanical models of the universe but by more extensive exploration of the possible forms of mathematical model which are capable of coordinating all our past and future experience of the physical world.[185]

It is worth highlighting the link established between the stance represented by the Bridgmanian ideas used by Kemble and Condon and Kuhn's own stance. We will see a little later that, though Kuhn does not disagree with the relationship between experience and theory that Kemble expresses, his concern for realism (which Kemble, on the other hand, could share aside from this defense of positivism *in the practice* of physics) is a subject of reflection at the beginning of his university studies.

Before getting to his early reflections on realism, however, we should know what contact Kuhn had with Bridgman's work. More specifically, did Kuhn read Bridgman at the beginning of his career? In a letter to E. T. Robinson from 1965, quoted below,

---

[180] Kemble (1938, 274).

[181] See Footnote 180.

[182] Kemble (1938, 264–265).

[183] Kemble (1938, 275).

[184] Dewey (1931, 30); cit. Condon (1938, 256), and Schweber (1986, 65).

[185] Condon (1938, 261).

Kuhn admits that, from Bridgman, he had only read *The Logic of Modern Physics* during his service in the Second World War, something that we can also verify through his reading card of that text, dated September 3, 1943. Kuhn was then working at the RRL, as we will see again below in this chapter.[186] There is also a reference to Bridgman's book, especially to the section on scientific explanation of his second chapter, among Kuhn's cards from that time.[187]

Regarding the impact of the book on his own thinking, Kuhn harbored doubts that the book had deeply influenced him. In 1965 he told Robinson that "positivistic ideas, including Bridgman's specifically operational approach, were very much in the air at Harvard during my undergraduate days, and I can not have found very much in the book that was terribly new to me in spirit."[188] Kuhn considered, therefore, that for him and his contemporaries, Bridgman's philosophical work was a sort of assumed starting point.[189] "I am sure that, like most of my contemporaries, I found it a liberation at the start though my recollections of that period are now very dim."[190] Despite this, Bridgman's influence on his philosophy of science did not seem too plausible to him.[191] Thus, recalling what must have been his opinions after reading Bridgman's book at that time, Kuhn said:

> My impression is [...] that I liked Bridgman's general approach a good deal but was not quite sure it could be put so far as he wanted to take it. Even then, I believe I was a bit skeptical about operationalism. In any case, I do remember quite clearly that I found a number of his specific analyses (particularly, I believe, the analysis of distance and space) rather naive.[192]

In the reading note mentioned above, Kuhn endorses these reservations about *The Logic of Modern Physics* with certain comments that, although a little cryptic, do not hide his lukewarm reception of the text. In 1943, he criticized the book, saying that it "[f]ails finally both Physically and Philosophically by not going far enough in either direction: Is not always consistent with operational phil[osophy]. See use of concept."[193] In short, Kuhn seems to suggest that Bridgman did not have a significant influence on his thinking, unlike other authors who appeared later. However, we will see later that some of his criticisms in his early texts (for example, in the Lowell Lectures) target Bridgman's operational analysis.[194] Thus, while Bridgman did not have positive influences on Kuhn's thinking (although some typical ideas and concerns in him are already present in Bridgman, as we saw above), he seems to have acted as a reference for a vision of science against which Kuhn tried to present an alternative image. For this reason, Bridgman, like Kemble and especially the (in

---

[186] See ML.

[187] See ML. Kuhn notes pp. 37 and following of *The Logic of Modern Physics*.

[188] Kuhn to E. T. Robinson, 8 March 1965, TSKP 4.14, 1.

[189] P. W. Anderson (communication with the author, 12 November 2002) confirmed this point to me.

[190] Kuhn to E. T. Robinson, 8 March 1965, TSKP 4.14, 2.

[191] See Footnote 190.

[192] Kuhn to E. T. Robinson, 8 March 1965, TSKP 4.14, 1.

[193] [Kuhn], "Bridgman, P.W. The Logic of Modern Physics," 3 September 1943, ML.

[194] See Chap. 3.

several senses) Bridgmanian Conant (whom we will see in detail in Chap. 2), helps us understand Kuhn a little better.

In the following pages we will see that Kuhn wonders about the relationship of physical theory with the external world, about the way in which we can affirm that the conceptual correlation of data tells us something about that question before which Kemble declares himself agnostic: What is the external world like? Does physical theory tell us anything about it? Or as Kuhn asks: Does physics have a metaphysical value? Throughout his career, Kuhn's choice will be to ignore that question—the way Kemble indicates—while he ventures into the realm of experience and its internal dynamics. He would never be in a position to *deny* that external world, nor the idea that its interaction with the mind (socially organized in his case) *in* experience generates scientific theory.[195] But, like Kemble, his speculation about the external world would be reduced to the bare minimum to avoid problematic positions like idealism.

The difference with Kemble, with Bridgman, and with empiricism in general, is his search for a general theory of conceptual change. Where the defenders of operational analysis saw concepts as a mere aid to relate facts, Kuhn observed a complex dynamic organization whose components he had to discriminate if he wanted to better understand science. As he progressively perceived (and even more so since 1947), in order to study that organization, his own vision of non-scientific belief, of its historical contingency, and of its social roots—something that he had observed in the field of politics, ideology, and art—shed light without casting shadows. It is in this search, in this plan, that we find the point at which Kuhn's work departs from the point of view of the physicists among whom he was educated. We will return to this idea later in this chapter.

1.10 The Metaphysics of Theoretical Physics

At Harvard, Kuhn wrote a (for us) interesting essay: "The Metaphysical Possibilities of Physics" for "English A.1," already in his second year. The essay reflects his point of view about physics from the point of view of a sophomore majoring in that discipline, but also the results of his previous acquaintance with philosophy through the course in the history of philosophy in his freshman year. In this essay, Kuhn affirmed something significant: "Too seldom is it realized that physics is also a philosophical science searching for absolute truths concerning the nature and structure of the universe."[196] Kuhn also complained that physicists had neglected the philosophical implications of their work, turning it into an intellectual and applied exercise where practical results and the difficulty of language had become inherent features of such an endeavor. Recall Kemble's agnostic attitude a few pages back. For Kuhn, physics

[195] See our Chap. 6.

[196] Kuhn, "The Metaphysical Possibilities of Physics," ca. 1941, TSKP 1.3, 1. On this essay, see also Mayoral (2009, 177–179); Hufbauer (2012, 425); Isaac (2012b, 99); Reisch (2019, 156).

had one more important feature: being a means of obtaining knowledge of reality. If these metaphysical capabilities were beyond its reach, Kuhn seems to want to know that too. This is the ultimate perspective of this essay. Kuhn had not yet contemplated that normative contingency could encompass the realm of scientific thought, as he himself would argue in a few years, so there is no reflection of such an idea here. In this essay, Kuhn tried to show how the belief that the language of physics transcends its pragmatic efficacy (and the realm of experience, as Kemble claimed) could be justified, and tried to find out if it could be affirmed that its propositions and theories offer knowledge of that external nature.[197] It is already possible to affirm that for someone with these ideas, the professional exercise as a physicist as previously summarized must have been far from satisfying. What is said in the essay is a good summary of what Kuhn *did not find* in the practice of physics but glimpsed in philosophy. We will also see that his questions and forecasts distanced him from the methodological reflections of some physicists like Bridgman or Kemble (whom he nevertheless does not mention in the essay).

"The Metaphysical Possibilities of Physics" assumes an inductivist point of view with more similarities to David Hume's empiricism and Bridgman's operationalism than any other philosophical current. In this essay, Kuhn shows us that data is obtained from measurement procedures and that concepts have a purely fictitious character that allows us to link groups of data together. Concepts are, thus, an initially arbitrary starting point. There is always, however, a qualitative link between the concepts and the data they serve. Concepts, in any case, are not a component of scientific theory based on the structure of the human mind itself. Concepts are not necessary, in a sense that we could consider close to Kant (whose work Kuhn knew through Demos's course). From the concepts of space and time to those of matter and energy, the concepts of physics are postulated to be able to introduce order into the data. The source of concepts is sensory experience, Kuhn shows, and not the permanent structure of the mind. Thus, "time is derived from endurance and change, cause and effect from repeated interaction, and mass and energy from tactual impressions." These, he adds, "are not the concepts of Kant's *Critique of Pure Reason*. They are not imposed upon sense data by the nature of the mind itself."[198] In other words, the concepts of this early Kuhn are neither a priori nor necessary. They are a sort of arbitrary convention qualitatively derived from the most basic requirements of the data and the need to create order in the tangle of measurement data.[199]

Kuhn's goal in this essay, however, is to consider what we would now call *scientific realism*. He tries to find out if it is possible to ensure some conceptual framework, *a single* conceptual framework, to the exclusion of its alternatives, so that the representation thus offered of the world has full guarantee and constitutes *the* knowledge of

---

[197] See Kuhn, "The Metaphysical Possibilities of Physics," ca. 1941, TSKP 1.3, 1.

[198] Kuhn, "The Metaphysical Possibilities of Physics," ca. 1941, TSKP 1.3, 3.

[199] See Kuhn, "The Metaphysical Possibilities of Physics," ca. 1941, TSKP 1.3, 1–4, for Kuhn's stance summarized in this paragraph.

the external world.[200] However, he does not seem ready to infer too easily, too immediately, that a realistic understanding of theories best explains the success of many sciences such as physics. This skeptical attitude against a later well-known argument on behalf of scientific realism is based on the idea that concepts may be deemed fictional resources, and that there could be a plurality, maybe even an infinitude, of alternative explanations of the same data.[201] This attitude matches Bridgman's operational approach, which, faced with incorrect unifications, prefers, as a normative message, that scientific concepts do not exceed the local contexts in which they are synonymous with well-defined operations. But Kuhn wonders about what remains beyond the operationalist caution and the fictionalist anti-realism. Is it possible to overcome the skeptical attitude typical of this empiricism (clearly influenced by Hume), of the positivism so far defended? Kuhn considers that this question—is it possible to prove the truth of theories?[202]—affects physics research, and that the answer may be within the physicist's reach:

> Do the fictions of physics approach nearer and nearer to the truth, and will the ultimate fiction thus correspond to the truth? [...] This has been the dilemma of the physicist: he searches for a truth which, though found, cannot be proved. We can only be glad that he has not been so discouraged by it as to give up the attempt, for further investigation shows possibilities of a less discouraging nature.[203]

Kuhn wonders, in the rest of his essay, if this skeptical attitude would be countered through an investigation of the relationship between data and concepts. The crux of the question is that information may be finite even though the amount of data may grow infinitely.[204] As he says, "[d]ata is infinite, but concepts are derived not from an individual datum, but from the common qualities of all data. The infinity of data is the infinite of all points in a finite line: the concept is the line itself, the common attribute of all the points."[205] So, he shows, if the amount of information obtained from data is finite, perhaps we could find out how many concepts are required to put them in order. A possibility for explanatory pluralism remains, according to him. Concepts "are [...] derivative," he recalls; each concept has features that depend on the data, on the qualities that these latter exhibit.[206] However, there could be alternative explanations, based on different concepts, of the same data. In this case, only with a program intended to determine if conceptual reduction—if elimination of duplicities—is practicable, could we go from the initial pluralism to the possibility of a unique explanation of all data. At that point, he says, we could rely on the possibility of a unique "image of reality," and in that sense in the possibilities of scientific realism, we could say.[207]

---

[200] See Kuhn, "The Metaphysical Possibilities of Physics," ca. 1941, TSKP 1.3, 1, 6–8, 11.

[201] See Psillos (1999, Chap. 4) for a detailed introduction to scientific realism.

[202] Kuhn, "The Metaphysical Possibilities of Physics," ca. 1941, TSKP 1.3, 6.

[203] See Footnote 202.

[204] See Kuhn, "The Metaphysical Possibilities of Physics," ca. 1941, TSKP 1.3, 9.

[205] Kuhn, "The Metaphysical Possibilities of Physics," ca. 1941, TSKP 1.3, 10–11.

[206] Kuhn, "The Metaphysical Possibilities of Physics," ca. 1941, TSKP 1.3, 10.

[207] See Kuhn, "The Metaphysical Possibilities of Physics," ca. 1941, TSKP 1.3, 11.

Note that Kuhn is not suggesting a philosophical investigation into conceptual reduction once the number of possible theoretical explanations of data are already on the table, but rather a more abstract exploration of the necessity of concepts given a limited amount of information obtained from an infinity of data. He is discussing something like the underdetermination of theories by evidence, and trying to find out if there is a way out towards scientific realism, that is, not only towards a sort of explanatory monism—which an empiricist thinker, to some extent agnostic, would accept. Kuhn is, instead, speculating about the "metaphysical" potentiality of physics. This is highly significant, because, first, he seems to be considering whether an explanatory pluralism is the only reliable position or not, and second, because he is wondering if, as a physicist, he must settle for the empiricist, operational, attitude that we have seen in Kemble and Bridgman; an attitude that is also well-represented by logical empiricism. After the war, we will also see how his vision of a scientific theory is much more generous with the conceptual dimension of physics than the latter current, and how he has in mind a perspective on physics as a discipline that is able to say something about notions of a more metaphysical than physical nature. We examine that more below. At this point, we can conclude that Kuhn has *metaphysical concerns*, and this is something that the physics he knows does not help him to respond to—and, given what we have seen in Bridgman and in Kemble, not even to confront. Let us now see what happened between him and physics once he abandoned Harvard College in June 1943.

1.11 From Theoretical Physics to the Philosophy of Physics

From the summer of 1943 to that of 1945, Kuhn worked at Harvard's RRL and was sent around England and then France during the entire second year. The RRL pursued research on radar countermeasures. His professional career as a physicist began with these investigations applied to war weaponry and his familiarity with the community of physicists also increased in this context; thus, for example, his future thesis director, J. H. Van Vleck, was his supervisor at the RRL.[208] After studying how to optimize the use of countermeasures at Harvard, he served intelligence in England; then he went to Rheims (France), where he supervised the use of countermeasures in the 9th Bombardment Division; he also went to Holland and Germany, where he inspected the radar installations that the German army was abandoning as it retreated, and

[208] On the RRL, see Guerlac (1987, 287, 300); Buderi (1996, 207); Brown (1999, 293). On Kuhn's activity at the RRL, see Interview 1995, 268–269. Kuhn's work during this war time is documented in his various technical reports. See Kuhn, "Preliminary Considerations Concerning Radar Echoes and RCM Parameters in Naval Operations," n. d., ca. 1943–1944, TSKP 1.8; "Notes on Jamming Effectiveness Measurements," n. d., ca. 1943–1944, TSKP 1.8; Kuhn and P. J. Sutro, "Theory of Ship Echoes as Applied to Naval RCM Operations," n. d., ca. 1943–1944, TSKP 1.8 For more details about this period and work, see Mayoral (2011, Sect. 4).

interviewed German scientists about their own research.[209] An interesting notebook entry shows us a glimpse of his stay in England and his subsequent flight to France. About the former, on August 20, 1944, he wrote the following:

> 20 August [1944]:
>
> It took considerable effort, but I managed to get away from base by 5:15 PM and drove into London with Don Taylor, Dan D. and a British Artillery Captain. Uneventful except that the latter turned out to be an Oxford grad[uate] in Ancient History and Phil[osophy]. He did not break down much, but a few interesting remarks were drawn forth. Among them comparison of Cambridge and Oxford, etc. [...] Now for a bit of *Under the Bridge* and to bed.[210]

Further notes attached to the diary and other entries show how he landed in France five days later, after some days in England. Writing from his hotel room in Paris, he describes his arrival on the Cherbourg Peninsula, from where he drove with others to Rennes via Saint Lô, which was a "shambles," he says, and from there to Le Mans. "All along the road," he wrote, "people had waved to us. The V sign is a password." On 26 August, as he also described in the interview in 1995, they got to Paris in time to witness Charles de Gaulle's Victory Parade. "This day is probably the greatest I have seen," he wrote from that room. He confirmed that feeling 51 years later in Athens.[211]

After this year in Europe, in the fall of 1945 Kuhn obtained permission from the RRL to return to study physics at Harvard and thus obtain his doctorate. His plan was to combine work and research. Except for his journey through Europe during the war, he had remained at Harvard since he started university. During the war period, he had been in touch with the members of Harvard's Department of Physics. He had strong ties with Harvard physics, so he stayed there, chose Van Vleck, who he was close to, as his thesis advisor, and opted for solid-state physics (Van Vleck's main field) as his own subject of research.

During the following predoctoral stage, Kuhn delved into theoretical physics as much as his already diminished appetite for the subject allowed.[212] A scholarship from the National Research Council obtained in February 1946 helped him with economic support, and in June of that same year he got his AM.[213] During his thesis preparation he squeezed Harvard's educational offer a little more by following a physics course with Julian Schwinger, "Physics 47," which dealt with "Special Topics in Theoretical Physics," and which that year dealt in particular with "Advanced

---

[209] For the details of his European itinerary, see Interview 1995, 269–272; the letter from J. N. Dyer to H. S. Vandenberg, 17 January 1945; L. S. Taylor, "Assignment of Thomas S. Kuhn to Operational Research Section," 19 January 1945; Kuhn and Lusted, "Activities of Radar Investigator Attached to Navy Trip #105", 3 June 1945, esp. 1–2. The correspondence between Kuhn and Dyer and Kuhn's reports (in TSKP 1.9) have many data of interest in order to better understand this itinerary. See Mayoral (2011, 235–236).

[210] [Kuhn], [Notebook, 25 August–9 December 1944], TSKP 1.10, 1.

[211] [Kuhn], [Notebook, 25 August–9 December 1944], TSKP 1.10, esp. 5, 6, 9, for the quoted phrases. Kuhn described this also, and in more detail, in his Interview 1995, 270.

[212] See Interview 1995, 269, 272–274.

[213] Letter from Ross Harrison to Kuhn, 12 February 1946, TSKP 14.17; Kuhn, "Vita," 1 October 1955, UCBTK, 1.

Nuclear Physics."[214] Kuhn prepared two essays for Schwinger: "Application of Integral Iteration Methods to Nuclear Force Problems" for fall term, and "Group Theoretical Investigation of the Saturation of Exchange Forces" for spring term.[215] Not much later, however, Kuhn received an interesting proposal to join Conant's General Education group, which redirected Kuhn's attention towards such tasks and forever changed his professional trajectory. We will talk about this proposal a little later.[216] From then on, a Ph.D. in Physics became a necessary step (although probably useful given his potential future dedication to the history and philosophy of science and general education) that he had to get as soon as possible.

At this point, it is worth mentioning that none of the options chosen by Kuhn in this period—solid-state physics, Harvard and its Department of Physics, and Van Vleck himself—would have been his first choice if he had opted for physics as his main field. Kuhn was already trying to distance himself from that future as a physicist, which was no longer appealing for him. Thus, opting for Harvard, for Van Vleck, and for solid-state physics was the result of pragmatic consideration. However, as sometimes happens, in the long run it was a wise choice, as it kept him tied to Harvard, which in turn allowed him to start a fruitful professional and intellectual relationship with Conant and become a historian and philosopher.[217] Besides that, Kuhn's research as a physicist also had other side effects. Years later, he mentioned that some central aspects of his idea of normal science not only had their roots in his scrutiny of the history of science, but also in his own research as a physicist.[218] This took place both at the RRL and in his predoctoral work with Van Vleck. He also said later that it was from Van Vleck he obtained the idea that a scientist only faces the problems that he or she will be able to solve and that, thanks to this, science progresses at the good pace that it does.[219] This is a central idea in Kuhn's conception of normal science.

In any case, Kuhn was tired of physics. For two years (1943–1945), he had carried out technical work that, needless to say, had little to do with working on the frontier of physics (countermeasures did not go that far). When he remembered his stay in Bushy Park, London, he said: "that was hard. I was having trouble, sort of fitting in and getting interested in what I was being asked to do. But it was all right and there was some fun involved."[220] Shortly after he said more forcefully that

---

[214] See the *Final Announcement of the Courses of Instruction Offered by the Faculty of Arts and Sciences During 1946–47, Official Register of Harvard University*, vol. XLIII, No. 21, 3 September 1946, 132. (Available at <https://iiif.lib.harvard.edu/manifests/view/drs:467240026$1i>.) Although this course appears in this catalogue as a fall-term course alone, Kuhn's essays, as shown below, are prepared for *two* courses, one per term.

[215] Kuhn, "Application of Integral Iteration Methods to Nuclear Force Problems," 19 January 1947; "Group Theoretical Investigation of the Saturation of Exchange Forces," spring term 1947; both in TSKP 1.5.

[216] See Chap. 2.

[217] On these pragmatic considerations, see Interview 1995, 272–274.

[218] *Essential Tension*, 226; Kuhn (1983b, 28).

[219] Kuhn, "Puzzles vs. Problems in Scientific Development," 4 November 1974, TSKP 5.9, 11.

[220] Interview 1995, 269.

> as all of this was going on[221] I increasingly realized that I was not all that interested in radar work. This gave me a somewhat bad taste of what it was going to be like to be a physicist. It was of course totally misleading. A number of my classmates, in somewhat similar positions, instead of going on to radar and radar countermeasures, wound up at Los Alamos. And I don't think it's out of the question that if I had gone to Los Alamos I might still be in physics. I doubt it, I mean I think there were too many other factors involved, but certainly an increasing distaste—which is already too strong a word—but an increasing number of doubts as to whether this was for me began to pile up.[222]

The predoctoral research with Van Vleck does not seem to have improved his impression of what awaited him in physics. Let us briefly look at the project Kuhn pursued with Van Vleck. Eugene Wigner and his student Frederick Seitz had obtained an important method in solid-state physics. The two had applied quantum mechanics to calculate the cohesion energies of sodium. Their method was based on the compartmentalization in cells of the network formed by the atoms that make up the metal. Lillian Hoddeson and Vicki Daitch summarize the innovative work of Wigner and Seitz as follows: "Since the birth of the quantum in 1900, the physicists who studied solids had limited their work to ideal materials. [...] But until 1933—until the work that Wigner and Seitz were just then doing—the theory was incapable of dealing with *real* solids."[223] A problem with this method was, nevertheless, the difficulties of calculation, especially of a mathematical order. Avoiding them and improving the results were interesting challenges for a physicist. Perhaps they did not constitute revolutionary work (in the Kuhnian sense), but they did require talent, ingenuity, and creativity. Van Vleck entrusted his doctoral student, the young Tom Kuhn, with this research work as a thesis topic.[224]

Kuhn carried out this work during the predoctoral period and developed a couple of methods for this purpose. One of them, later published in *Physical Review* along with Van Vleck, was called "Function Matching Method"; the other was an application of the WKB method based on the work of the Japanese I. Imai. Kuhn's research also involved more purely mathematical work, and he published some relevant results in that field.[225] His work with Van Vleck in this field improves the manageability of the Wigner–Seitz cell method and allows a closer proximity of the theory to experimental results. In other words, Kuhn's work enhances the virtues of a theory that was already in use. It is difficult to avoid tagging this work in terms of his later philosophical thinking as pure *normal* science.[226] As I previously noted, for Kuhn

---

[221] Kuhn is referring to his field inspection of German radar.

[222] Interview 1995, 271–272; Interview 1989, 19.

[223] Hoddeson and V. Daitch (2002, 58–59; see also 62).

[224] For what is said in this paragraph, see Sopka (1988, 258, 281); Hoddeson and Daitch (2002, 53–56, 58–62); Mayoral (2011, 240–242). Among the sources, see Wigner and Seitz (1933; 1934). See also Lillian Hoddeson and Paul Henriksen's interview with Seitz, 26 January 1981, Niels Bohr Library and Archives, American Institute of Physics, College Park, Maryland. (Available at <www.aip.org/history-programs/niels-bohr-library/oral-histories/4877-1>.).

[225] See Mayoral (2011, 237–246) for more information about these methods.

[226] Mara Beller (1999) suggests the opposite. For her, Kuhn's notion of normal science seems inspired by his limited first-hand experience with physics as a doctoral student (1999, 306). See Mayoral (2011) on Beller's viewpoint. Hamlin (2016, esp. 303) alludes to something similar.

himself, important aspects of the concept of normal science were inspired in part by his experience as a scientist.[227]

Be that as it may, Kuhn's predoctoral work enabled him to obtain his PhD and publish three papers, two of them in physics (one of them in collaboration with Van Vleck) and another one in mathematics. He also made at least a couple of presentations at conferences.[228] Not without some pride, Kuhn pointed out in his "Intellectual Autobiography" of 1953 that "Since I left the field [i.e., physics], the methods which I originated have been further studied and improved by Professor Harvey Brooks of Harvard. It now appears that my brief career in physics may have provided the basis for significant new techniques in the computation of solid-state parameters."[229] However, once that work and that period were over, Kuhn officially left physics. I say "officially" because unofficially his mind was already elsewhere; specifically, in the philosophy of physics. The journey from physics to philosophy of science had already begun some time before.

Philosophy was a field in which he soon felt comfortable. We have already seen a sample of his writing on philosophical topics at the beginning of his undergraduate studies. In 1943, at the age of twenty-one and with a degree in physics, Kuhn confessed in a letter to his aunt Emma K. Fisher, that physics was no longer as attractive to him as philosophical reflection. He wrote:

> I have a feeling that I hinted at much of this in my last letter, and so will not again drag you through the tortuous machinations. The dreaming goes on with somewhat more direction, but with considerably less energy. I do not know that the sacrifice of motive power is worth the gain in clarity, but since the former was inevitable, I am rather happy to find the latter a concomitant. (Can that be a noun; if not, it is now.) I lean, in short, increasingly toward the academic, and [...] philosophy of and through science rather than to pure physics. But I am to[o] far from knowing where it will end and to waste your time here.[230]

From room G–13 of Lowell House, our budding philosopher also told Fisher about his plan to work with Bertrand Russell and Alfred North Whitehead's *Principia Mathematica* as soon as he could. He confessed to be reading Sigmund Freud's *Introductory Lectures on Psychoanalysis*. He was not clear about the basis of Freud's more theoretical claims and therefore psychoanalysis did not fully convince him, although he did not distrust its therapeutic value, which he considered more intuitive

---

[227] On Kuhn's work in physics, see Galison (2016, 42–50). This is Beller's point; see footnote 226.

[228] See Kuhn, *The Cohesive Energy of Monovalent Metals as a Function of their Atomic Quantum Defects*, PhD Dissertation, Department of Physics, Cambridge, Mass.: Harvard University, 1949 (available at the HUA). Chapters II–IV deal with the first of the methods, while chapters V–VI discuss the second. Also see Kuhn and Van Vleck (1950); Kuhn (1950; 1951a). "Atomic Quantum Defects Applied to Solid State" (9 May 1951) and "Computing Metallic Cohesive Energy without the Use of Potential Functions" (15 June 1951) were two public presentations of his work, both in TSKP 12.33.

[229] Kuhn, "Intellectual Autobiography," 22 October 1953, UCBTK, 1; edited by Hufbauer (2012, 455).

[230] Kuhn to Fisher, June 27, 1943, TSKP 12.33, 2–3. Hufbauer (2012, 426–427) and Isaac (2012b, 98) also mention this letter.

than theoretical, anyway.[231] "The Whitehead and Russell," he told her about the other volume, "has not yet arrived which I take to mean that it's probably 'on order' from England. In this case it may take a couple of months, but there's no harm done as I'm sceptical about the possibilities of getting a good start at it before fall."[232] He began to read classics of philosophy with the desires and objectives of someone who is already planning to make it more than a fleeting intellectual pursuit.

Between July 1943 and April 1944 (in May 1944 he left for England, towards his post at the Telecommunications Research Establishment in Great Malvern, Worcestershire), Kuhn dedicated his free hours to reading some philosophy of physics and some of the popular books on science at the time. He was simply following the inclinations he had informed his aunt of months before. We can even take a tour of some of his readings at that time.

In August 1943, Kuhn was reading *The Nature of the External World* by Arthur Stanley Eddington, a book that, in his opinion, did not find the right balance between popularizing science and properly addressing problems. In August and September 1943, Kuhn finally read *The Logic of Modern Physics* by Bridgman, which we have already discussed. In October 1943 he read Nietzsche's *The Antichrist*, a reading that attracted him—although he did not spare some criticisms about its incoherence—and also the Bampton Lectures in Oxford in 1884, delivered by Bishop Frederick Temple and titled *The Relations between Religion and Science*, a text known for its attempt to reconcile Darwin's work with the Christian religion, whose treatment of certain specific philosophical problems pleased Kuhn.[233] In December 1943, Kuhn was immersed in *The Varieties of Religious Experience* by William James, a book with which he said he had little in common, but which he nonetheless praised, as well as its author. After this "trilogy" of readings on the relationships between science, philosophy, and religion, Kuhn read *Daedalus, or Science and the Future*, by J. B. S. Haldane, in January 1944—which is also a well-known lecture read to "The Heretics" at Cambridge in February 1923. Finally, Kuhn read and commented very positively on *Experiment and Theory in Physics* by Max Born, a recent lecture given by the quantum physicist at King's College in Newcastle-upon-Tyne on May 21, 1943.[234]

Kuhn's mind continued to dwell on the same themes during his stay at Harvard. Unfortunately, we do not know much about his period in England, and whether he had time to continue his readings. Roger Kuhn pointed out to me that his brother had read more philosophy while in England.[235] After the war, his preference for philosophy had not diminished. In 1945, while discovering that his limited academic knowledge of philosophy was an obstacle in his intention to focus only on that field and to abandon physics, he enrolled in a couple of graduate courses in philosophy

---

[231] Kuhn to Fisher, June 27, 1943, TSKP 12.33, 3. Many years later, Kuhn still remembered a similar impression from Freud's work; see Interview 1995, 280.

[232] Kuhn to Fisher, June 27, 1943, TSKP 12.33, 4.

[233] See Temple (1884). About Temple, see Reuben (1996, 32–33); Cosslett (ed.) (1984).

[234] See Born (1943). For Kuhn's dated note cards on the authors and books mentioned, see ML. See also Mayoral (2009, 178, 183).

[235] CRK.

and logic with Donald C. Williams and Henry M. Sheffer, respectively. They were the courses "Philosophy 9," devoted to "Metaphysics," and "Philosophy 8," titled "First Course in Relational Logic."[236] I am particularly interested in the work he prepared for "Philosophy 9," as he dedicated an essay to causality in science. In any case, I will also review the paper he wrote at that time about logic, a field he would never revisit in his entire lifetime, if only because it now helps us establish connections between Kuhn and other philosophers at Harvard, like Lewis, and the kind of philosophical point of view that could have been found in that milieu (as also happened with Bridgman).

1.12 Causality and Metaphysics

"An Analysis of Causal Connexity" is the title of the work that Kuhn wrote for Williams's course in the fall of 1945.[237] Kuhn's intention is to examine a concept, causality, whose function in modern physics, he asserts, has been minimized to almost completely exclude it.[238] Bertrand Russell and his critique of the notion of cause is a case in point. His diagnosis on causality is well known: "The law of causality, I believe, like much that passes muster among philosophers, is a relic of a bygone age, surviving, like the monarchy, only because it is erroneously supposed to do no harm."[239] Kuhn does not believe, however, that the notion of causality is not part of scientific thought, although it is possible to dispense with metaphysical ingredients that play no role in that concept. He says that it is possible to reconstruct it with only the strictly necessary elements. This is what he tries to argue in his essay. For him, finding out more about the uses of causality in physics enriches our philosophical perspective on this concept, without this implying a reduction of the latter to what the former can contribute.[240]

Kuhn discusses this at the beginning of the essay, where he also briefly comments on his sources. Besides mentioning Russell, he cites Hume's *Enquiry Concerning Human Understanding* as a source for his study.[241] It is worth remembering that the essay was written for Williams, whose study of induction, published a few years later, includes interesting discussions of Hume's perspective and of versions of it offered by more recent philosophers; it also points to significant connections with *Mind and the World-Order* by C. I. Lewis, an author we will return to later.[242] Kuhn also mentions

[236] See the *Announcement of the Courses of Instruction Offered by the Faculty of Arts and Sciences During 1945–46, Official Register of Harvard University*, vol. XLII, No. 8, 31 March 1945, 100–101. (Available at <https://iiif.lib.harvard.edu/manifests/view/drs:467240026$1i>.).

[237] On this essay, see also Mayoral (2009, 181–183).

[238] Kuhn, "An Analysis of Causal Connexity," fall 1945, TSKP 1.3, 1.

[239] Russell (1912–1913, 180).

[240] See Kuhn, "An Analysis of Causal Connexity," fall 1945, TSKP 1.3, 1–2.

[241] Hume (1902, esp. secs. III–VII) is the edition that Kuhn handled along with Hendel (ed.) (1927).

[242] See Williams (1947, 184–195); Lewis (1929).

Williams's "eight lectures on causality in his Harvard course on metaphysics" as a source of his reflections in this essay.[243]

The first point that Kuhn debates in "An Analysis of Causal Connexity" has to do with the relationship between causality and determinism. Regarding this latter concept, we must distinguish, as he does, between the determination of future events and the ability to determine them from our knowledge of present circumstances. Against positivism, he asserts, he trusts that any proposition about the future will have a defined truth value. It does not seem, he makes clear, that science can be practiced without this kind of assumption. On the other hand, if we link this determinism to our predictive capabilities, we arrive at the perspective of Pierre-Simon de Laplace and question who or what conscious entity would be capable of knowing all the values of the variables necessary to obtain a prediction of the future that could be considered strictly true.[244] Our ability to determine future events is undermined by our limited epistemic possibilities—our ability to predict future events is only approximate—so talking about it does not make complete sense. Consequently, talking about determination does not make sense either, because how do we know that the future is determined if we cannot predict it? The epistemic barrier thus found prevents us from inferring that statements about the future have a defined truth value from anything other than an a priori foundation. This, of course, goes against a positivistic perspective. Kuhn, however, believes that this ability to determine is an integral part of science, although it is also appropriate to consider it within its limits, as real science does. As we can see, his anti-positivism grows in this essay, both in his starting assumption about scientific metaphysics and in this additional thesis. He makes his anti-positivistic attitude explicit here.[245]

For him, the notion of cause includes the ability to predictively determine the future, although he considers that this is not the only thing involved. A form of uniformity, whose basis we could catalog as intentional, underlies the development of the prediction mechanisms used by science. We use scientific methods to predict future states, but we do so within certain expectations about uniformity that are based on the experience we already have of the relationship between the present and the future, an experience encoded, in the case of physics, in laws such as, for example, those of Newton's dynamics. This perspective on uniformity is an inseparable part of the notion of causality, as it is used in science. For this reason, Kuhn will criticize the, otherwise enlightening, Russellian process of reduction of causal thinking to its mere predictive scheme by adding an intentional dimension that he considers inseparable from the scientific causality. He excludes Russell's eliminativism about causality from his own program.[246] Let's look at Kuhn's argument in a little more detail.

---

[243] See Kuhn, "An Analysis of Causal Connexity," fall 1945, TSKP 1.3, 2.

[244] See Laplace (1814, 2–3). Kuhn does not quote Laplace, but he does mention what we know as his "demon"; see Kuhn, "An Analysis of Causal Connexity," fall 1945, TSKP 1.3, 4–5.

[245] See Kuhn, "An Analysis of Causal Connexity," fall 1945, TSKP 1.3, 3–5, for the ideas summarized in this paragraph.

[246] Kuhn, "An Analysis of Causal Connexity," fall 1945, TSKP 1.3, 12–13, 16–18.

Let's start by looking at what Russell says in his article. His main criticism has to do with the vagueness of the usual expressions of causality, with its philosophical definitions, some of which he reviews at the beginning of his article. Russell is left with two main elements of causal sequences: their invariability, or necessity of the relationship between events in a causal sequence, and the temporally-mediated relationship between these. So, he considers that the definition we could reconstruct is based exclusively on the relationship of invariable succession between two events, with a time interval in between. Russell considers that this relationship is established in the sciences, say, in the mathematical statements of physical laws like Newton's, alluding to very specific characteristics of such events; usually, he shows, their spatio-temporal location is central. We should not add too many features to the sequence, or else we will over-particularize the phenomena involved, which would move us away from the sought generality. The intention of these expressions is to serve as a tool to predict the future behavior of objects or natural processes within a certain degree of variation. There is, finally, one more important nuance that we should highlight. For Russell, science does not simply seek to ascertain causal sequences—they must serve to help us understand natural phenomena. In the prediction of phenomena, he finds the necessary ingredients for a scientific explanation of natural behavior. "That bodies fall is a vague qualitative statement; science wishes to know how fast they fall," says Russell.[247] Given this point of view, Russell's stance consists in replacing old ideas about causality with a notion closer to the practice of science, in which a more specific mathematical language replaces the old vague relations between causes and effects.

In his essay, Kuhn leads us to another point. His argument can be summarized as follows. The "how" to which Russell alludes can be a goal, but it is also a result, and this result does not seem to say anything about the theoretical and intentional context from which it arises. This nuance is important in Kuhn, as it represents a change that we will later see become explicit in his work. To talk about science philosophically is not to limit this discourse to expressions with predictive functions, but to expand it to include everything that underlies them. The well-known Kuhnian distinction between "textbook science" and "creative science" that we will see later,[248] is already latent in this criticism of Russell and is also part of Kuhn's stance against an operational point of view. A reconstructed language eludes all the purely theoretical and intentional background of the language used and experienced, and the language of science is no exception. This thesis is already implicit in this criticism. Kuhn here refers to the basic ontological assumptions (for example, referring to the nature of space and time) that lead to specific visions of uniformity; or to the mathematical tools that, like calculus, are employed, given these specific visions; or to the format of the laws that rely on both things; or to the way in which all this conditions the type of predictions that *we expect* to obtain from scientific theory.[249] Kuhn's aim in this essay is to broaden our

---

[247] Russell (1912–1913, 188; and see also 181–194 for these ideas). For a brief history of this Russellian stance, see Frisch (2014, 1–7, 18–21).

[248] See *Structure*, 136 ff., for the idea of "textbook science." See also our Chap. 3.

[249] See Kuhn, "An Analysis of Causal Connexity," fall 1945, TSKP 1.3, 8–13.

view of a scientific concept in order to observe in it other relevant nuances without which we do not fully understand it. To do this, he adds more contextual elements to his analysis (like the ones we have just mentioned), something which fits with the point of view that we have already seen him deploy in other areas, in essays not far removed from this one *in time*—although they are as regards the *topic* discussed (for example, those on literature).

Let's return to that level of detail. Indeed, we can limit our understanding of physical phenomena to a description based on identity criteria that only use characteristics of, say, location in space and time. But this is not a direct path to the "how"; at least, it is not a direct route to understanding the "how" that is handled in science. Kuhn shows in his essay that behaviors described by means of Newton's laws are *limited*; that the range of properties used in each theory, for each group of laws, involves restrictions. An intention underlies the specific representation.[250] He shows that, unless we take these boundary conditions into account, the change in identity criteria promoted by Russell does not guarantee that the result of our reconstruction of causality illuminates real science, and so the authentic notion of causality used in it. We have seen before that Russell said that the goal of science is to find out, for example, the particularities (that is, an analysis) of free fall and not just to confirm that the phenomenon occurs. Kuhn would agree, but he would also add that, in achieving that goal, our description of that phenomenon must adhere to a certain pattern of uniformity. However, there is nothing about this in the perspective of the British philosopher.[251]

Kuhn agrees with Russell that our mathematical reconstruction of a natural behavior depends on defining the events involved according to properties that come into play in the mathematical formulation of the laws that govern such behavior. If we study a body's motion in the vicinity of the Earth, for example, the behavior in question responds to Newton's laws and, in this regard, we are interested in features of the events such as the position of the moving body, its speed at a moment, and its acceleration. Kuhn shows that, in mathematical terms, the last two are defined as derivatives (first and second, respectively) of the position with respect to time. There are more *possible* derivatives, he shows, and considering them would modify the natural behavior that we wish to reflect. But, since it is this behavior (a body's motion according to Newton's laws) that we wish to represent mathematically, our reconstruction of the objects involved does not reach them (or other features that they represent).[252] Russell is surely aware of this kind of restriction. But what Kuhn tries to show here is that this process of reconstruction of the object or event involved in a natural phenomenon, which is based on a behavior governed by laws, is an integral part of the idea of causality dealt with in science. In summary, not everything in the notion of cause dealt with in science is restricted to expressions that reflect the result of that selection process without making it explicit.

---

[250] See Kuhn, "An Analysis of Causal Connexity," fall 1945, TSKP 1.3, 11.

[251] See Kuhn, "An Analysis of Causal Connexity," fall 1945, TSKP 1.3, 9–13, 17 ff.

[252] See Kuhn, "An Analysis of Causal Connexity," fall 1945, TSKP 1.3, 11–13, 17.

Kuhn talks about the way in which this reconstruction of the physical phenomenon is produced using the aforementioned example of the movement of a body according to Newton's theory: "If the motion of the body at an instant is an event, and all the derivatives of the motion [i.e., characteristics as velocity, acceleration, etc.] are contained in the event by intension, the laws of motion actually limit the types of events which can occur by placing conditions upon the intention of physically admissible events."[253] Thus, Kuhn shows that we limit our possibilities of representation of natural phenomena by reference to the behavior that our theory and the laws that integrate it describe. To do this, we set limits for potential, *possible* properties and we look among them *only* for those that allow us to define physically representative events, that is, those that fit well with the laws and with the uniform behavior described in the terms of such laws.[254] He makes it clear that these laws adopt a specific theoretical, mathematical format, which in turn depends on other, more basic, ontological assumptions—for example, those related to the nature of space, time, or the phenomena involved.[255]

In this 1945 essay, Kuhn offers a stance of disagreement with the positivistic views of science that Russell's perspective epitomizes.[256] It should be emphasized that it is not yet a stance affected by any experience with the history of science, since Kuhn only looks at physics and from the point of view of a physicist. It is, in any case, a stance in which the goal of *understanding* nature (always seeking an image of uniformity that allows the prediction and explanation of phenomena through groups of laws that govern objects with known characteristics) is a central ingredient of scientific thought. Accordingly, Kuhn argues that, assuming that the *causal connection* is part of science, such a connection is characterized by two properties, neither of which can be ignored when describing it. One of them is "predictability"; the other, everything that surrounds the establishment of a pattern of uniformity and that is characterized by distinguishing between legitimate events and "excluded" events.[257] Kuhn explains it as follows:

> [T]here are two portions to the notion of cause. One is the notion of predictability expressed in the quotation from Russell.[258] It has been observed that under certain circumstances the validity of this notion may be analytically demonstrable. The other notion is simply that of

[253] Kuhn, "An Analysis of Causal Connexity," fall 1945, TSKP 1.3, 11.

[254] Kuhn gives a second example with the behavior of gases according to Boyle's law; see Kuhn, "An Analysis of Causal Connexity," fall 1945, TSKP 1.3, 14–17, 20–21.

[255] See Kuhn, "An Analysis of Causal Connexity," fall 1945, TSKP 1.3, 8–13, 18–19.

[256] It seems appropriate to call such a stance positivistic if, as Mathias Frisch has pointed out, the Russellian conception "repeated many of [Ernst] Mach's criticisms" (Frisch 2014, 5).

[257] Kuhn, "An Analysis of Causal Connexity," fall 1945, TSKP 1.3, 8 ff., 13.

[258] Kuhn must be referring to the quotation from Russell's paper "On the Notion of Cause" that he makes a few pages before (p. 6): "Given any event e_1, there is an event e_2 and a time-interval t such that, whenever e_1 occurs, e_2 follows after an interval t" (Russell 2012–2013, 183).

> the removal from the ontologically meaningful space-time field of certain types of event-particles which are conceivable and describable and therefore subsistent but which, if this analysis is correct, can never be existent.[259]

Kuhn does not perform an analysis of the ontological and epistemological nature of the "excluded particle-events,"[260] but it seems clear that such a class of event is a mere possibility that remains outside the world whose uniformity is described by a certain number of laws and under a certain intentional approach. The idea of establishing that pattern of uniformity in which certain events are excluded has similarities with the perspective of another of the authors mentioned in this work, Alfred North Whitehead. In *The Concept of Nature*, a book to which Kuhn refers in a note, we find a similar idea and, above all, a similar vocabulary.[261] It would not be very relevant, were it not for the fact that there are further coincidences a little later.[262]

Kuhn's point of view about the laws is, therefore, modal. They are designed from a conception of the possible and the real—a conception that, for that reason, is partially based on a priori ingredients.[263] With all these components, he is offering us a perspective of scientific thought that is based on an idealized reconstruction of phenomena and that includes a function not only for prediction, but also for the understanding of nature based on its intentional reconstruction. This is, in fact, the Kuhn that we will see develop in the following chapters and that will enrich this perspective with other intellectual sources, from the history of science to the new currents of psychology, among others. However, the dependence on a priori conditions for physical theory should not lead us to conclude that he approaches Kant in this essay. Rather, he adheres to Hume's skeptical diagnosis about causal connection and induction, although he adds that there is a certain search for uniformity in science, or a certain intention of representation, as we have been saying, that conditions causal connections, their format, and their scope.[264] Kuhn does not mention anything in this essay about the origins of this search for uniformity, about its epistemological nature and function. He merely notes that this search, that this intention, conditions our tools of scientific prediction and, with it, shapes the idea of cause in science. We will have to wait a while longer for Kuhn to delve deeper into such origins.

However, there is a perspective that remains in his work in later years and, it could be said, throughout his philosophy of science: the scientific theory is studied taking into account the perspective of the phenomena adopted by the scientist (by

[259] Kuhn, "An Analysis of Causal Connexity," fall 1945, TSKP 1.3, 12. I have corrected a misspelling in "analysis.".

[260] Kuhn, "An Analysis of Causal Connexity," fall 1945, TSKP 1.3, 13.

[261] Whitehead (1920, esp. 186). In "An Analysis of Causal Connexity" (fall 1945, TSKP 1.3, 8, fn.), Kuhn refers to this book (Chapters II–III) as a useful reference for his definition of event.

[262] See our next chapter, Sect. 2.2.

[263] Kuhn, "An Analysis of Causal Connexity," fall 1945, TSKP 1.3, 5–6; Kuhn mentions the idea of the a priori on p. 11.

[264] For his claim of Hume, see Kuhn, "An Analysis of Causal Connexity," fall 1945, TSKP 1.3, 18–21.

the observer, we could say), and it is analyzed by envisioning how the objects and phenomena involved are reconstructed at the same time as the symbolic expressions that represent the laws are elaborated. This ontological reconstruction is, in other words, difficult to separate from the construction of a theoretical language. They are mutually interrelated. If we add to this that everything depends on a certain epistemic stance in which some a priori ingredients come into play, we find some of the basic components of Kuhn's later perspective here—that interrelationship and the a priori. They are not yet developed, of course, but they appear to be the seeds of important aspects of his later theory.

1.13 Lewis, Sheffer, and Logic

In the previous discussion of Kuhn's work, we have observed a point of view related to the application of a modal perspective and of a vocabulary for semantics that refers to the intension of terms in a language. The notion of intension is an aspect of the logical analysis of the terms that he likely became familiar with at that time from various sources.[265] But all this also directs our attention to a logician and pragmatist philosopher actively working at Harvard at the time when Kuhn took these metaphysics and logic courses at that university: Clarence Irving Lewis. To tell the truth, there is no indication that Kuhn knew Lewis personally at any time; we cannot confirm that it did not happen, but there isn't any evidence of it. However, Kuhn did read Lewis's logical work through his course with Henry M. Sheffer, "Philosophy 8," just as he did *Mind and the World-Order*, Lewis's best-known book (which he liked, though we do not know when he read it). Moreover, he found similarities between this work and his own epistemological perspective.[266]

The significance of Lewis's work for a better understanding of Kuhn's philosophy (and even of its biography) has already been emphasized in previous works by other authors (and by myself, too).[267] However, in many of these cases, the parallels established are between Kuhn's mature philosophy as represented by *The Structure of Scientific Revolutions* (and other later texts) and certain passages from *Mind and the World-Order*, which takes us away from our thinker in this period of his youth. However, the philosophical point of view that Kuhn adopts in the essay previously discussed seems to apply expressions and perspectives that are also characteristic

[265] For example, Cohen and Nagel (1934, 30–33), discuss this notion and it is a text that Kuhn knew; see Kuhn, "A Comparison of the Logic of Propositions with that of Ascriptives," fall 1945, TSKP 1.3, 2, fn. 2.

[266] Kuhn, "A Comparison of the Logic of Propositions with that of Ascriptives," fall 1945, TSKP 1.3, 8 (fn. 9), 15 (fn. 15), 17 (fn. 16), cites Lewis (1918) and Lewis and C. H. Langford (1932). His appreciation of *Mind and the World-Order* (Lewis 1929) can be seen in TSKP 7, Section "Gen. Philos.".

[267] See Fuller (2000, 266–280; 2006, 359–360); Rosenthal (2007, 53); Mayoral (2009; 2017); Abbott (2016, 171, 181).

of Lewis's work. There are, thus, apparent points of contact at this early moment in Kuhn's life.

Let's focus on one of those points—one that is particularly remarkable. We have seen that, for Kuhn, our representation of the world is manifested in patterns of uniform, causal behavior, which depend on certain groups of laws and on the subject's selection of significant objects or events, whose defining characteristics are in intensional relation. In other words, the term that refers to an object or event has a meaning that depends on our perspective of natural behavior in which it plays a role. Lewis expresses something very similar by saying that "the intension of a term represents our intention in the use of it."[268] These words belong to *An Analysis of Knowledge and Valuation*, Lewis's Paul Carus Lectures delivered at the University of California, Berkeley, in December 1945, that is, around the dates when Kuhn was preparing the essay we discussed earlier. Additionally, his interpretation of the proposition in Chapter III ("The Modes of Meaning"), from which I have extracted the previous words, "is," says Lewis, "the same in essentials—though the terminology and some details are different—as that which my colleague Professor H. M. Sheffer has for some years presented to his classes."[269] And Kuhn, as I have already pointed out, was a student of Sheffer precisely at that time. We will return to this connection later.

In Lewis's view, the meaning of a term refers to its intension, and this represents a mental criterion for the application of the term in question. In *Analysis*, Lewis makes it clear that he "wish[es ...] to identify" the "intension" or "meaning" of a term *A* ("the concept of *A*") "with the *criterion in mind* by which it is determined whether the term in question applies or fails to apply in any particular instance."[270] Kuhn also assumes a relationship between semantics and cognitive stance that is very similar to Lewis's in those lines. For Kuhn, the selection of the properties that constitute the meaning of a term for an object or event that is describable (and possibly explainable) in science depends on the intention to represent a specific type of behavior by means of the scientific theory (for example, the behavior that is governed by the laws of that theory) on the part of the observer, and with this observer's "criterion in mind," to repeat Lewis's words. The epistemological perspective of both thinkers is as close as Kuhn himself observed. Talking about the influence of Lewis's perspective on Kuhn, whether direct or indirect, is therefore plausible. Let's look at some other contextual elements that add a bit more plausibility to this interpretation.

As noted, there is an additional *indirect* biographical connection between Lewis and Kuhn, a likely influence whose bridge is Sheffer. Lewis's interpretation of propositions resembles and depends on Sheffer's research in logic, in particular on what is known as ascriptival logic. Kuhn wrote an essay on this logic for Sheffer while he took Sheffer's course at Harvard in parallel to Williams's course in metaphysics.

[268] Lewis (1946, 43).

[269] Lewis (1946, xii).

[270] See Footnote 268.

The most relevant aspect of Kuhn's essay on ascriptival logic for us is that it reveals his familiarization with Lewis's logic and epistemology.[271]

Sheffer himself deserves more attention than I can dedicate to him at this moment. He had the aura of a genius and a complex personality. His work on logic was highlighted by Russell in the second edition of *Principia Mathematica*.[272] Sheffer's list of published articles is nonetheless short. He devoted his time to teaching, but little to the communication of his ideas.[273] His colleague and friend Horace Kallen encouraged him to write an introductory presentation to his thinking, but Sheffer never followed the advice.[274] In the end, his philosophical work remained in the dark. For Quine, Sheffer was opaque. "He talked," Quine wrote, "of the elements of *Principia* and of a nebulous theory of formal systems about which he was proprietary and conspiratorial."[275]

Kuhn attended his classes and Sheffer's logical ideas seem to have been attractive and useful to him. In "A Comparison of the Logic of Propositions with that of Ascriptives," the essay he wrote for Sheffer, Kuhn studied the advantages of a reformulation of the logic of the *Principia Mathematica* in terms of Sheffer's pre-assertive entity: the ascriptive.[276] The ascriptive had been useful to Lewis. His intensional logic is based on a logic of terms that he later applies to propositions through an analogy whose foundation was Sheffer's notion of the ascriptive. Lewis explains it as follows:

> As [Sheffer] has for many years explained to his classes, there is for every statement, '*S*,' that which '*S*' asserts, and which the interrogation "Is it the case that *S*?" inquires about, and the postulation "Let us assume that *S*" puts forward as hypothesis. What all three of these entertainments entertain, in their various ways, is the *proposition* in question. It can be identified, for any statement '*S*,' with the expression 'that *S*,' which is what '*S*' affirms. For convenience of reading, it can also be identified with the expression which results from replacing the verb in '*S*' by its participial form; thus the proposition asserted by the statement "John is tall" is 'that John is tall' or 'John being tall.'[277]

Lewis continues, indicating in the same place that "so considered, a proposition turns out to be a kind of term," which, he adds, "signifies some 'state of affairs'" that may or may not be the case. "The intension of this propositional term," Lewis adds, "includes all propositions which are deducible from it and would coincide with the

---

[271] Kuhn knew other logic texts, such as Cooley (1942), Eaton (1931), Langer (1937), Russell (1924; 1929; 1938) and Russell and Whitehead (1925). See Kuhn, "A Comparison of the Logic of Propositions with that of Ascriptives," fall 1945, TSKP 1.3, 1 (fn. 1), 2 (fn. 3, 4), 3 (fn. 5), 4 (fn. 7), 10 (fn. 12) and 18; Kuhn, "An Analysis of Causal Connexity," fall 1945, TSKP 1.3, 22.

[272] Russell and Whitehead (1925, xv). See Van Evra (1966, 194–195, and fn. 8, 9).

[273] See the testimony of one of Sheffer's students in Kallen's preface to Henle, Kallen and Langer (eds.) (1951, v–vi).

[274] See Kallen to Sheffer, 10 March 1962, Horace M. Kallen Papers, Jacob Rader Marcus Center of the American Jewish Archives, Hebrew Union College, Cincinnati Campus, Jewish Institute of Religion, box 25, folder 27.

[275] Quine (1985, 82). See also Morton White's testimony (White 1999, 100–101).

[276] On Sheffer's ascriptive, see Wood (1953, 401).

[277] Lewis (1951, 27).

conjunction ('and'-relation) of these."[278] Sheffer's perspective is one of the bases of Lewis's intensional logic, so its palpable utility in his work must have been evident. For Kuhn, it certainly was, as we are going to see below.[279]

In his essay, Kuhn examined the logic of ascriptives and defended its usefulness. If we follow Lewis, an ascriptive would be the proposition separated from its corresponding assertion or inquiry, etc. It resembles the proposition in that it is expressed by a (grammatically correct) combination of words or symbols, but their respective combinations are different in the sense that Lewis indicates. Kuhn also expounds the difference as well as the notational changes. These notational changes are supposed to make clearer the distinction between the ascriptive, the assertion, and the logical or empirical criteria used in the latter.[280] Regarding his defense, Kuhn claimed that Sheffer's system would improve the system of the *Principia Mathematica* on a philosophical level, and briefly addressed the details of the corresponding reformulation.[281] Another significant advantage is that, for Kuhn, the distinctions introduced in Sheffer's system contributed to a modal approach to logic such as that inherent in Lewis's work. When Kuhn discussed this in his essay, he showed some familiarity with Lewis's approach, which supports the connection between them that I previously suggested.[282]

Kuhn's essay for Sheffer, just like his previous paper for Williams, contributes to our acquaintance with the philosophical concepts and views he was developing (and learning) at that time, and with the readings in logic and philosophy that were part of Kuhn's brief philosophical training in the mid-1940s. It also helps us to grasp the point of view from which he would approach a view on logic and its contribution to the study of science. In this regard, we should not be surprised that this point of view was somewhat heterodox, as his brief formal education in the field was, too. Above all, it reveals contacts with the Harvard philosophical context aside from the long shadow that Quine's figure casts over it. These contacts explain details of his later work, especially those that do not quite fit with the Quine's perspective, despite the influence the latter had on Kuhn. Let's now leave logic and philosophy at Harvard and see, in Chap. 2, what paths the turbulent course of Kuhn's career took in these crucial years: who influenced him and what context helped him to shift from physics to philosophy of science.

---

[278] See Footnote 277.

[279] See Murphey (2005, 257–258). Lewis offers a very similar version in *An Analysis of Knowledge and Valuation* (Lewis 1946, 48–49), which includes the mention of Sheffer.

[280] For more details, see Kuhn, "A Comparison of the Logic of Propositions with that of Ascriptives," fall 1945, TSKP 1.3, 3–7. Kuhn's description of the ascriptive is different from Lewis's. For Kuhn, the proposition "includes a judgment as to the correspondence of the hypothetical state of affairs represented by the symbols to the actual situation," and the ascriptive lacks that judgment. For Kuhn's description of the ascriptive, see pp. 3–5 of his essay; the quotation is from p. 3.

[281] See Kuhn, "A Comparison of the Logic of Propositions with that of Ascriptives," fall 1945, TSKP 1.3, 10–12, 15–16.

[282] As noted above, Lewis' *Survey* (1918) and *Symbolic Logic* (his work with Langford: Lewis and Langford 1932), were known to Kuhn.

Chapter 2
James Bryant Conant and Harvard University

2.1 A Timely Invitation

At the beginning of 1946, Kuhn seems to have begun to seriously reconsider his professional career. He had studied logic and metaphysics with Sheffer and Williams, and understood that he had too many gaps in his philosophical knowledge to dedicate himself full-time to that field, so he continued studying physics. But despite having decided not to go on into philosophy, he does not seem to have had a well-defined plan. His own memories of 1946 show bitter disillusionment, uncertainty, and lack of direction.[1]

The proposal to collaborate on Conant's experimental education course changed his life. We do not know for certain how the invitation arose, who first thought of it and why (although Kuhn was a prominent student at Harvard), but there are certain documents and evidence whose chronological order would show a possible itinerary. The reconstruction I can offer is as follows. On February 3, 1947, the then president of the Committee on General Education at Harvard, Benjamin F. Wright, summoned Kuhn with a brief note in which he told him that he wanted to talk to him in a couple of days to propose something for the following year that he might like.[2] Possibly, at that meeting, Wright offered him the opportunity to join Conant's group. The first mention of Conant's experimental course in the meetings of the aforementioned committee took place only fifteen days after Kuhn's meeting with Wright, and Wright said there that Conant would start his course the following fall with Fletcher G. Watson as

[1] See Interview 1995, 273–274.

[2] Wright to Kuhn, 3 February 1947, TSKP 5.84. Hufbauer (2012, 429) already mentions this invitation. The note seems to have the meaning here ascribed with a high probability. It would be extremely rare, in fact, that Kuhn would have preserved such a brief, cryptic note in the documents he himself deposited in the MIT archives if it had been devoid of any biographical meaning. However, there is always the possibility that it was an irrelevant note slipped into a folder by mistake. Therefore, this reconstruction is only probable. I thank Karl Hufbauer and George Reisch for very enlightening conversations about this interpretation. The reconstruction offered here is indebted to Hufbauer's version (2012, 428–430). See also Reisch (2019, 45).

J. V. Mayoral, *Thomas Kuhn*, Springer Biographies,
https://doi.org/10.1007/978-3-031-86512-1_2

his assistant.[3] They went on to discuss some formalities of the course. Kuhn really does not appear at any time in these minutes. But this makes sense, given that he assumed an occasional assistantship and his main obligation was his research in the Department of Physics, where, from February 1946, he was a predoctoral fellow of the National Research Council. In fact, he had to request a temporary exemption to take that position alongside Conant and Watson for four months. He later returned to his predoctoral fellowship, which lasted until he obtained his doctorate in March 1949, when he already was a junior fellow of the Society of Fellows.[4] Of course, Kuhn collaborated with Conant and Watson during this period of official relationship and later, preparing programs, making small contributions, and attending the meetings held by the professors and fellows of the course. In fact, he prepared, with Watson, the first draft of the program that was discussed at the meeting held with Conant, Roller, Nash and Watson on May 2, 1948.[5]

Even if it was Wright who invited Kuhn, we cannot know for sure who suggested Kuhn to Wright or Conant as a possible assistant. However, the reconstruction could continue as follows. Two years earlier, Dean Paul S. Buck had already asked Kuhn to summarize and comment on the "Red Book," common designation for the Harvard report on General Education—the well-known *General Education in a Free Society*. One of its authors, along with Buck, was Wright himself. By then, Kuhn's name must have been heard often in that context. According to Kuhn, both proposals were due to his previous extra-academic work, that is, his journalistic activity in *The Crimson* and his participation in The Signet Society, etc. Moreover, in the Department of Physics, they probably knew about Kuhn's inclinations towards other disciplines outside of physics; after all, he had followed Williams's and Sheffer's courses with their permission. Kemble, for example, participated in Harvard's General Education program and taught a course in it. So, there are several options regarding the source of Kuhn's recommendation as Conant's assistant alongside Watson, although none holds more weight than any other.[6]

---

[3] See *Minutes of the Committee on General Education*, meeting of 18 February 1947, HUA, Ref.: UA III 10199.2, folder: "1946–1951 NOV."

[4] For Kuhn's NRC scholarship, see the telegram from R. Harrison to Kuhn, 7 February 1946, and from B. D. van Evera to Kuhn (renewal), 14 June 1948. For the four-month exemption (from 1 October 1947 to 1 February 1948), see the letter from W. A. Noyes to Kuhn, 12 June 1947 and the letter from the National Research Council secretary to Kuhn, 11 September 1947; all of them in TSKP 14.17.

[5] See [Watson and Kuhn], "Tentative Outline of Natural Science 4," n.d., and "Outline of Natural Sciences 4 as agreed to tentatively at a conference held in Cambridge on May 2, 1948, attended by Messrs. Conant, Roller, Watson, Nash, and Kuhn," both in JBCP 23. The first outline must have served as the basis for the second, in which Kuhn appears in the course plans without assigned material; he would only lead a couple of course exercises along with Watson; see "List of Documents for Natural Science 4," JBCP 23, exercises 1, 2. Kuhn attended these first meetings with some initial ideas that we will examine in the next section.

[6] For this data, see Interview 1995, 273, 275; *General Education in a Free Society: Report of the Harvard Committee* (Committee on the Objectives of a General Education in a Free Society 1945); [Kuhn] (1945) (which is a reprint of what appeared in the September 22, 1945 issue of the *Harvard Alumni Bulletin*). I thank Karl Hufbauer (private communication, 6 February 2003) for his

In any case, Kuhn accepted the job, and this is what is truly significant, as in this job he found a suitable occupation for his professional preferences and also for his education. Kuhn's knowledge about physics was useful in this educational work. Additionally, the job allowed him to move away from his work in physics, and make use of his philosophical reflections on science and on its metaphysical and logical structure. It was an acceptable solution for everyone.

2.2 Ideas for a General Education in Science

In May 1947, Kuhn prepared an outline for a series of meetings that were going to take place with the course leaders, including Conant and Watson among others. The outline, titled "Objectives of a General Education Course in the Physical Science," was only intended to be a collection of useful ideas for the meetings.[7] We have seen that the first draft for the future course emerged from these meetings. In these notes, however, we also observe Kuhn's initial ideas. He warned that, most probably, an elementary course could not deal with all that he included in his outline. Here we find a synoptic image of his conception of science at that time, which already includes the influence of reading *On Understanding Science*, authored by Conant himself.[8] As regards this book, a reading card shows Kuhn was drawn to the first case study on Boyle, which, unlike the rest of the cases, was (in his opinion) "full enough" and seemed useful. The "[g]eneralizations about science—listed in the back—are less useful," he then added.[9] Despite this opinion, the influence of some of the contents of this book on Kuhn's ideas is evident, as shall be apparent later. In the outline, Kuhn also hoped that a General Education for non-scientists would help them to understand the different aspects of his image of science.[10]

Kuhn's outline includes three parts. Part I describes the "process which is physical science in terms of its stages," and shows how a science like physics consolidates, has a well-established conceptual framework, and has confirmed predictions. Part II has a better-defined title, "The Roles of Theory and of Experiment," and deals,

---

indication of the connection between Kuhn and Kemble. The latter's name appears in discussions about general education courses in physics from 26 February 1946. See *Minutes of the Committee on General Education*, meetings of 26 February 1946 ff., HUA, Ref.: UA III 10199.2, folder: "1946–1951 NOV." In addition, he collaborated on a volume expressly dedicated to general scientific education; see Kemble (1952). As already mentioned in Chap. 1, Kuhn and Kemble maintained a good relationship for years, as evident in the correspondence between them; see a brief example related to Kuhn's *Copernican Revolution*, in our Chap. 4, Sect. 4.6.

[7] Kuhn, "Objectives of a General Education Course in the Physical Sciences," May 1947, TSKP 1.4, 1 and note stapled to the outline. About this outline, see Hufbauer (2012, 430); Isaac (2012a, 214–215; 2012b, 100).

[8] See Interview 1995, 275; Hufbauer (2012, 430).

[9] See TSKP 7.

[10] Kuhn, "Objectives of a General Education Course in the Physical Sciences," May 1947, TSKP 1.4, 1.

in the first place, with the separate functions of theory and experimentation, and then goes on to examine their mutual relationship. Part III, finally, is also clearly titled "Symbols, Logic, and Mathematics as the 'Language' of Science," and is concerned with studying logical and mathematical languages as scientific vehicles of communication, of development and of reflection on the theory, and their combination with observation in inductive inference.[11]

In this outline, Kuhn distinguishes between two different classes of objects treated by scientific thought: "perceptual objects" and "scientific objects."[12] Scientific objects are those that the theory deals with—the referents for its terms, we can say, and mostly for their theoretical terms. At this point, we can find a likely link between Kuhn's views and Alfred N. Whitehead's philosophy of science. The latter also speaks about "scientific objects" in his own work, and there are similarities between his and Kuhn's usage of that concept, as will be shown below.[13]

According to Kuhn, the process by which scientific objects are obtained is an ascension from the most basic objects of perception in their "pure classification"—that is, according to the features that allow us to group them immediately and to start to produce empirical generalizations about them—to other objects relevant to the theory, its scientific objects. The latter allow us to be more precise in terms of their generalizable features thanks to the selection of "essential similarities in divergent phenomena," which also includes "the rejection of superficial similarities." For Kuhn, this is a process of "Successive Abstraction," whose ultimate results are the aforementioned scientific objects.[14] The similarities with Whitehead's views are based on this process. Whitehead says in *The Concept of Nature* that:

> [S]cientific objects [...] embody those aspects of the character of the situations of the physical objects which are most permanent and are expressible without reference to a multiple relation including a percipient event. Their relations to each other are also characterised by a certain simplicity and uniformity. Finally the characters of the observed physical objects and sense-objects can be expressed in terms of these scientific objects. In fact the whole point of the search for scientific objects is the endeavour to obtain this simple expression of the characters of events.[15]

There is in Kuhn's views at this early time a close similarity in vocabulary and conception with Whitehead's perspective.[16] However, as in the case of the connections with C. I. Lewis' work, this similarity should not be taken too far, to the point of a central and definitive influence in his thinking, as Kuhn seems to be using Whitehead's general point of view as a way of expressing his own ideas. Some terms of the

[11] Kuhn, "Objectives of a General Education Course in the Physical Sciences," May 1947, TSKP 1.4, 1–2.

[12] Kuhn, "Objectives of a General Education Course in the Physical Sciences," May 1947, TSKP 1.4, Pt. I, Secs. 1, 3, respectively.

[13] See Whitehead (1920, 158).

[14] Kuhn, "Objectives of a General Education Course in the Physical Sciences," May 1947, TSKP 1.4, Pt. I, Secs. 1–4.

[15] Whitehead (1920, 158).

[16] Fuller (2000, 302 [fn. 89], 423) already considers the connection between them. Before Fuller, George R. Lucas, Jr. (1989, 133) also did it so. See also Mayoral (2008a) for other details.

reproduced paragraph would need further clarification within Whitehead's thought, but something like this would only show a very approximate knowledge of his work on Kuhn's part. We can neither ask nor expect more from a physics student, who philosophically reflects on this in his spare time.

However, Kuhn knew a bit more about Whitehead than this modest connection indicates. He confessed to the biologist C. H. Waddington that the (for the latter, somewhat unforgivable) absences of any mention of Whitehead in *Structure* were due to his readings of his work, which dated back to the moments before the development of his current perspective on science. That confession fits with his reference to Whitehead in the 1945 essay. Besides that, Kuhn had never been attracted to key notions of the English mathematician, logician and philosopher, such as "prehension."[17] Kuhn knew other works by Whitehead, such as *Science and the Modern World*, which he sometimes quoted, as well as *Axioms of Projective Geometry* and *Symbolism: Its Meaning and Effect*, which seem to have pleased him. It does not seem that *Process and Reality*, if he ever read it, had the same fate.[18]

The constitution of scientific objects through a process of abstraction from the objects given to perception fits well with the design of the theory that we saw in Kuhn's essay on causal connection and, in this way, with the vision that he also seems to be developing, which is in contrast to the perspectives of empiricism and operational analysis.[19] On two points in this scheme, Kuhn opposes considering that the function of the experiment is the one assigned by the operational analysis.[20] For him, experimentation contributes to establishing the meanings of terms in a definite way thanks to their interaction with scientific objects. For him, the experiment is "the architect of scientific definition," though he adds that it is "[n]ot operational philosophy, but the historically demonstrated connection of experiment and 'scientific objects.'"[21] Hence, the experiment helps to refine the information coming from observation and so to a selection of features that leads to "the discovery of regularity."[22]

This nature of the meaning of terms that is not purely operational fits well into Kuhn's view of induction, something he refers to at the end of Part III. There, he asserts that his view of inductive inference is not understood as an inference from

[17] Kuhn to Waddington, 24 October 1963, 1; Waddington to Kuhn, 7 March 1963, 2; both in TSKP 4.16.

[18] See Whitehead (1913; 1925; 1927; 1929). Kuhn has cards dedicated to the first and third in TSKP 8–9 and mentions *Science and the Modern World* in, for example, *Copernican Revolution* (123, 280). Whitehead's famous sentence, "A science which hesitates to forget its founders is lost," from *The Organisation of Thought* (Whitehead 1917, 115), is quoted by Kuhn in *Structure*, 138, but there are no guarantees that Kuhn would have obtained it from that text, as he does not cite the source and there is a slight variation (he writes a "that" instead of the original "which").

[19] That's why I do not totally agree with Reisch (2019, 47), when he says that "[t]his is a classic picture of science.".

[20] See Kuhn, "Objectives of a General Education Course in the Physical Sciences," May 1947, TSKP 1.4, Pt. II, Sec. 1, b; Pt. III, Sec. 3.

[21] Kuhn, "Objectives of a General Education Course in the Physical Sciences," May 1947, TSKP 1.4, Pt. II, Sec. 1, b.

[22] Kuhn, "Objectives of a General Education Course in the Physical Sciences," May 1947, TSKP 1.4, Pt. II, Sec. 1, a.

particular observations to pure empirical generalization, but as a process of abstraction and of hypothesis-generation in which imagination and creativity are present.[23] Kuhn assigns another role to the experiment as a means of "[c]onfirmation of a new and primarily deductive or speculative theory," and mentions general relativity theory and J. W. Gibbs' thermodynamics as examples of hypotheses thus generated and, consequently, later subjected to confirmation processes.[24]

Kuhn, therefore, contributes to the emancipation of theory with respect to its own empirical base. He does not dispense with the role of observation and experimentation in science (in sciences such as biology and geology, he says, the former fulfills the role of the latter).[25] However, the domain of scientific objects is constructed in the theory, and laws are, in that theory, something different from pure simple generalizations from observations.[26] The theory also uses mathematical vocabulary that is a tool, as we said, for communication and research. Logic, he adds, plays a restricted role, mainly dedicated to avoiding contradiction. It is remarkable to spot "[t]he distinction between logical incompatibility and 'unreasonableness,'" in his words, adding briefly: "The cure of the latter by time."[27] Two years later, he would consider the relationship between the logical and the psychological aspects of science, and these schematic annotations would start to make sense. He seems to distinguish between a purely logical contradiction and the impossibility of considering some statement or hypothesis reasonable, where the latter seems to be more related to psychology than to logic, and probably altered by the passage of time, as he also suggests. This latter reconstruction is pretty speculative, but it is coherent with annotations from 1949.

In this early view, Kuhn also shows that the theory collaborates with the experiment in obtaining theories that, as interrelated sets of concepts, have their own dynamics: experimental results, he shows, do not always immediately affect these sets, and sometimes these act as a brake on the advancement of science.[28] This last perspective is opposed to Popper's (whom Kuhn did not yet know, at least personally), in which counterexamples to a scientific hypothesis are unavoidable in good science—they contribute to its refutation—and is more akin to Conant's views, who in *On Understanding Science* already speaks of the obstacle that well-established theories (or conceptual schemes) present to progress based on their substitution.[29]

---

[23] Kuhn, "Objectives of a General Education Course in the Physical Sciences," May 1947, TSKP 1.4, Pt. III, Sec. 3.

[24] Kuhn, "Objectives of a General Education Course in the Physical Sciences," May 1947, TSKP 1.4, Pt. II, Sec. 1, c, ii.

[25] Kuhn, "Objectives of a General Education Course in the Physical Sciences," May 1947, TSKP 1.4, Pt. II, Sec. 1, d.

[26] Kuhn, "Objectives of a General Education Course in the Physical Sciences," May 1947, TSKP 1.4, Pt. II, Sec. 2, a, b.

[27] Kuhn, "Objectives of a General Education Course in the Physical Sciences," May 1947, TSKP 1.4, Pt. III, Secs. 1–2.

[28] Kuhn, "Objectives of a General Education Course in the Physical Sciences," May 1947, TSKP 1.4, Pt. I, Sec. 4; Pt. II, Sec. 3, b.

[29] See Conant (1947a, 88–96, 103). We shall return to the concept of conceptual scheme below.

Next, we will discuss Conant's views in greater depth. However, it should be noted that Kuhn's scheme shows us a young physicist fond of philosophy whose image of science is halfway between his own non-purely empiricist and operationalist convictions (although some of that vocabulary still remains) and a historical perspective, recently obtained from Conant, which shows him that his concern for the way in which a theory allows us to understand, and, in a way, modify our perception of phenomena, is not mistaken. In the following pages, we will observe how Kuhn consolidates this vision, which is contrary to empiricism and to Bridgman's operational point of view. It is striking, however, that the first to offer him a way out of this latter vision of physics is someone as close to operational analysis as Conant was. It is precisely this figure of American science and politics, so influential in Kuhn, that we will examine in Sects. 2.3–2.7.

2.3 Conant's Views

We should start with a brief biographical description of Conant. Given the importance of this figure for scientific research (especially chemistry), for science policy, and for science education in the United States, Conant's work has been the subject of many studies, some of them books. Here, I will only attempt to give some highlights of his life, especially those that are of interest for our own study of Kuhn. In what follows, I will combine such highlights with some of his key ideas in order to provide the reader with a portrait of the man and of his ideas.[30]

Conant was born in Boston on March 26, 1893, and grew up in Dorchester, one of its neighborhoods. He did not belong to Boston's upper and more traditional class. He had nothing to do with the atmosphere of Back Bay or Beacon Hill. However, he was a talented young man who held a tenured position as a Professor of Organic Chemistry at Harvard University from 1927 to 1933. He obtained his Ph.D. in 1916 and, in 1919, he was made Assistant Professor at Harvard. He was President of that institution from 1933 to 1953.[31]

Conant's talent for chemistry manifested itself in his preparatory school, the Roxbury Latin School, where he studied from the age of eleven to seventeen (1904–1910). There, Conant met Newton Henry Black, a science instructor who had obtained his Master's degree a few years earlier at Harvard, and who guided his steps in the field of physics and chemistry. Conant showed not only interest but a lot of talent in these subjects. He also spent many hours in the lab with Black, who looked after his students. At Black's suggestion, Conant was soon prepared to pass the entrance exams for Harvard College. Black introduced him to Theodore William Richards,

[30] See Hershberg (1993) and Jennet Conant (2017); Conant (1970) is Conant's autobiography. Among the shorter essays on Conant, see Hamlin (2016); Wray (2016; 2021, Chap. 2). Reisch (2019) studies Conant and Kuhn in parallel.

[31] See Bartlett (1983, 91, 93–94, 97, 102). See also Hershberg (1993, 16–17, 27–28), for the data and features handled in this and the next paragraph.

the renowned chemist, and took care of the young Conant's early steps in his college education.[32] When Conant entered Harvard in the fall of 1910, he had been attending Richards' classes for some time—some of them were intended for advanced students in chemistry. The famous chemist had started to be aware of the young aspiring chemist's talent.[33] In turn, Conant found in Richards a new mentor of great ability, and a temperament that, in some respects, was exemplary for him. In particular, Conant liked Richards' pursuit of autonomy as a researcher, something we shall see repeated for Conant himself. Eventually, he made another, more unexpected, connection to the distinguished chemistry professor: in 1921, Conant married Richards's daughter, Grace Thayer Richards, and his mentor also became his father-in-law.

In the latter, he had not only a mentor and a father-in-law, but also a canonical example of a certain class of scientist about which he would theorize years later. For years, and especially in his key book, *On Understanding Science*, Conant defended the difference between the "strategist" and the "tactician" as models of scientist. They were two extremes in a continuum of positions and scientists were usually more drawn to one pole than to other. A strategist, he said, usually approached his or her science from the point of view of its long-term transformation and improvement, and was not scared of changing its very foundations. The tactician, by contrast, was used to dealing with specific problems with the tools that he or she had inherited from the discipline. This contrast represents a distinction between psychological stances that seem to be implicit in the choices that a given researcher makes during his or her career, with the theory–experiment interaction as the backdrop and a pool of concepts and generalizations that help to describe such choices. This point of view is more individually centered, and so it does not count as another way to voice the normal/revolutionary difference Kuhn would later talk about. Neither does Conant have in mind such a deep frontier between forms of practicing science for a whole community or subgroup, nor are scientific revolutions in Kuhn always the direct outcome of the effort of a strategist (or some of them). Conant's distinction primarily reveals a difference in attitude (together with some generalizations about the usual practice of science). In that respect, Richards was a clear example of a good strategist.[34]

Conant worked as a chemist for the army during World War I. He began in the fall of 1917 in Washington, where he collaborated with J. F. Norris, and with E. P. Kohler. This latter became a second mentor to him at Harvard. He faithfully exemplified the figure of a very good tactician, according to Conant.[35] Kohler and Norris supervised the production of poisonous gas, in which Conant participated. After working for two more years in the manufacture of chemical weapons, he was appointed Assistant Professor of Chemistry at Harvard in the fall of 1919, as already mentioned. From 1929, he occupied the Sheldon Emery Chair of Organic Chemistry, and two years

---

[32] See Conant (1970, 17–18). On Richards, see Wray's recent paper on his scientific collaborations (Wray 2024), and of course Servos (1990, 78–82).

[33] See Hershberg (1993, 17).

[34] See Conant (1947a, 16–17, Chapter IV, esp. 100; 1970, 27–40).

[35] Conant (1970, 31, 33); see also Hershberg (1993, 28–29, 33, 36).

later headed up the Department of Chemistry.[36] Indeed, Conant's academic career was quickly on the rise.

Those years as a chemist at Harvard were, in his own words, "the best years of my life."[37] The sheer number of his academic publications was astounding. From 1916 to 1933 he published 117 articles and five manuals, alone or in collaboration.[38] He achieved international recognition in physical chemistry and organic chemistry. His colleagues noticed his virtues as a researcher—a mixture of intuition and imagination that made him a successful problem-solver.[39] For Conant, experimental zeal, imagination, and individual creativity were the basic (and necessary) weapons of the scientist. For him, individual talent *alone* constituted the foundation of scientific progress. There is a clear difference with Kuhn's views in that respect—particularly with Kuhn's mature views in *Structure*—which is at the root of their future disagreements.[40]

This latter conviction had practical consequences. In his teaching work, Conant used to make it clear that the students' or researchers' academic talents were the only forces that should elevate them to the top of their professions. Paul D. Bartlett, a student of Conant and later a Harvard professor, remembered that Conant did not usually outline specific work plans for his disciples; rather he seemed to let the students themselves figure out which path to follow. Bartlett illustrated this as follows:

> A visit to a coworker in the laboratory would often open with "What's new?" If something interesting was reported, he rarely prescribed the next experiments, but was more likely to ask: "What are you going to do next?" The implied expectation that the student would have good ideas of his own was a constant stimulus toward its fulfillment.[41]

The stimulus that Bartlett speaks of seems to underlie the individualistic belief (also visible in Richards) that the researcher, through freedom and at the same time control of action, and his or her own pursuit of results, should hold the reins of scientific work.

In fact, for Conant, mere production without initiative should not be the basis of a successful career in chemistry and a tenured position at the university. His university policy later was marked by this conviction that a position at Harvard was only justified by *creative* scientific production, valuable, and originating from an active mind eager to acquire knowledge. Under Conant's presidency, only the exceptional genius usually remained at the university after a period as instructor. The intention was that those scholars who returned to Harvard as professors did so after a previous career in which they had shown their worth. He did not want to dispense with exceptional talent. In fact, he defended the scholarship system for

---

[36] Hershberg (1993, 45–55); Bethell (1998, 114).

[37] Conant (1970, 59); cit. Bartlett (1983, 107).

[38] One of them with N. H. Black: *Practical Chemistry* (1920); see also Conant and Black (1937). On Conant's publications, see Hershberg (1993, 54), and Bartlett (1983, 113–124).

[39] Hershberg (1993, 54, 790).

[40] See the last pages of our Chap. 5 in that regard.

[41] Bartlett (1983, 106).

gifted individuals who lacked the required economic resources to study at Harvard. He saw to it that access to the university, as a student or as a professor, was based solely on mental abilities and the display of intelligence and creativity. Candidates in both cases had to prove their worth.[42]

For the same reasons, Conant never looked favorably upon the idea that research in science should be completely planned, except in the *temporary* conditions of wartime projects (it is suitable to emphasize the word "temporary" here). Strict plans, like those necessary in such projects, hindered careful, reflective, sometimes tentative, and always necessarily autonomous work on open problems that the scientists themselves found attractive. His questions to collaborators that Bartlett mentioned are consistent with this point of view, which he shared with Richards and, as we will see, with Bridgman. It is worth remembering that Conant spoke from experience, as—on top of his work as a chemist in World War I—in World War II he had been in charge of the Division B of the National Defense Research Committee (NDRC), devoted to chemical weaponry, and later was the chairman of the entire committee. Along with Vannevar Bush, he also organized the scientific research that led to the Manhattan Project.[43] That is, when it came to discussing these special conditions for research, Conant spoke from experience.

In contrast with such projects, the university, for Conant, should be governed by its own interests. In his words, "[a] university [is] an independent, self-governing community of scholars concerned with professional education, the advancement of knowledge, and the general education of the leading citizens."[44] As he also makes clear, if the university serves as an advisory board to the larger society, the independence of opinion and of decision regarding the projects undertaken within the university is part of the very essence of its activity.[45]

One of the obligations of the university, and one of the most important for Conant, was the generation of "accumulative knowledge," which for Conant was a synonym for "science."[46] Thus, he says that "Accumulative knowledge is distinguished from the other two divisions of human learning [philosophy and poetry] by the fact that we can readily recognize a progressive increase in our capacity to answer questions which have attracted the attention of inquiring minds for generations."[47] The development of this type of knowledge was incompatible, in practice, with the planning that had happened during the war campaign. In the same talk from which we have obtained the previous excerpt, Conant established a subtle parallelism with military campaigns, as he would later do to illustrate the way the scientist works in *On Understanding Science*:

---

[42] See Bethell (1998, 114, 128–130, 153); Hershberg (1993, 107–110).

[43] See Bartlett (1983, 99–100); Bethell (1998, 132–136, cap. IV); Zachary (1997, 112–117, 129–131); Cassidy (2011, 74–77).

[44] Conant (1947b, 282).

[45] Conant (1947b, 284, 286). See Warner (1984, 223–225).

[46] Conant (1944, 399–400). See also Conant (1947b, 285).

[47] Conant (1944, 400; he alludes to philosophy and poetry on p. 399).

> Some of the really difficult problems in science are like old-time fortresses which will yield only to long siege. Only a few troops are needed for the continuing encirclement. What are needed are mobile reserves to exploit a breakthrough in the defenses when one occurs, which translated means: We need a national supply of first-rate scientists who are free to work in whatever field they choose.[48]

For him, the resolution of a problem corresponding to that cumulative knowledge did not depend on economic and human expenditure, or on detailed planning and scheduling, but on the time invested with total freedom by a handful of men of true talent:

> [I]n every section of the entire area where the word *science* may properly be applied, the limiting factor is a human one. We shall have rapid or slow advance in this direction or in that, depending on the number of first-class men who are engaged in the work in question. If I have learned anything from my experience in Washington as chairman of the National Defense Research Committee, it is that ten second-rate men are no substitute for one first-class man. It is no use pouring second-class men on a problem, even if you are under the greatest pressure for a solution; second-class men often do more harm than good. So, in the last analysis, the future of science in this country will be determined by our basic educational policy.[49]

The decisions Conant made in administrative matters after the Second World War and during the Cold War period, like his statements, made his opposition to the military imperative dominating scientific work very clear.[50] Military research grants were substantial, so they were tempting for many scientists, but there were also those who preferred independence in their work.

Bridgman—like Richards, too—was a defender and a typical case of a scientist who loved independence and individual work, not organized or planned, and whose research was a direct result of this methodological motto.[51] Time to meditate, to plan for oneself, to develop an action plan, to execute it, to observe problems and confront them, to make mistakes, to become one's own critic, to explore promising alternative paths—all of this, Bridgman mentioned, was behind the research work he had carried out. Organized, excessively planned work involved allocating time and personnel, specifically scientific talent, in sterile organizational and administrative tasks.[52] Bridgman's conclusion regarding the needs of science is, thus, very similar to Conant's, and can be well summarized in a phrase like the following: "[T]he existence of science itself, which I think all conceded to be a social good, is impossible without scientific freedom."[53] If Bridgman held Conant in high esteem, it was also partly due to his defense of this independence. Bridgman said:

---

[48] Conant (1944, 404–405).

[49] Conant (1944, 401–402).

[50] See Hershberg (1993, 554–577).

[51] On Richards' case, see Kevles (1995, 82). It is also relevant Richards (1898, 726–727), cit. by Servos (1990, 78), which is also a good source on this personality features of that chemist; and see also Conant (1977, 257, 263).

[52] On all this, see Bridgman (1947, 129–130).

[53] Bridgman (1947, 131).

> There are many threats to our universities; among the most formidable of these is domination by Government. This is especially true in President Conant's own area, the physical sciences. All of us, scientists and nonscientists, will continue for many years to be indebted to President Conant's insistence that Government support for projects in this [Harvard] University be kept to a minimum.[54]

However, Conant is aware that scientists, despite their independence, must rely on instances outside their field of expertise, whether to request funds or to be accountable; or even, in a more general sense, to gain support from the whole of society for his knowledge enterprise. In this interaction between scientists and non-scientists, Conant is already contemplating evident communication problems. In his Foreword to I. B. Cohen and F. G. Watson's collective book *General Education in Science*, Conant comments on one such partial-communication situation.[55] The situation in question is that of making decisions about which projects should find financial support from the government or from private companies. Decision-making is hard in these cases if we take into account that the information that the applicants provide is also hard to understand by itself. The administrator cannot always be knowledgeable, sometimes even superficially, about the disciplinary fields in which such projects are framed or the specific problems they deal with. However, Conant corrects, in such cases, administrators can surround themselves with certain types of advisor, the figures he calls "expert on judging experts"[56]: someone capable of understanding the experts' intricate way of reasoning and of communicating among themselves. This figure is "somebody who over the years has time and again faced such problems and learned how to elicit by careful examination the premises both explicit and hidden in the argument of each adviser and who knows how to obtain information about the past reliability of each expert's prognosis."[57] In a previous paper, Conant shows that the same communication problem exists in the other mentioned area: that of scientists with the general public.[58]

This communication problem, which shows a rupture in the relationship between scientists and non-scientists, poses a real challenge for Conant. It is of foremost importance to us, because Conant's attempts to solve it provided Kuhn with the context in which he could leave research in physics and could enter the field of general education in science for non-scientists—and, with it, the history and the philosophy of science. Let's see how Conant faced the communication-breakdown problem that we have just observed, because a direct result of that effort was his influential—for Kuhn in particular—*On Understanding Science.*

---

[54] Bridgman, in a 1953 faculty meeting, cit. Hershberg (1993, 575, 896).

[55] See Conant (1952).

[56] Conant (1952, xiii).

[57] Conant (1952, xiii).

[58] See Conant (1944, 406).

2.4 How to Understand the Scientist's Mind

Conant began to think about that issue—the communication problem—at the end of World War II (late 1944).[59] In the early months of 1947, when Kuhn agreed to be his assistant, Conant had already established a General Education program for Harvard students that would be very influential throughout the country, and that was an appropriate medium for his image of the scientist—the one he shared with Bridgman—to be adequately described. Moreover, there was a general concern among scientists about the consequences of large-scale scientific projects during the war period for the public image of science. For this reason, Conant's program enjoyed wide resonance outside Harvard.[60] *On Understanding Science* was the official presentation of his project in both pedagogical and theoretical terms.

In *On Understanding Science*, Conant introduced a few case histories based on some main characters from the history of science in the seventeenth and eighteenth centuries. Conant was clear that those recent centuries represented a maximum chronological limit that no instructor should exceed without lessening the prospective students' understanding of the historical cases.[61] His intention was not to make a general list of great scientists, classic works, and notable discoveries—a clear example of a kind of historiography that neither he nor Kuhn adopted. None of the elements in such a list would help students understand the mental and material process of the scientist in action, which was the ultimate goal of his educational plan. He was much more focused on describing, and helping students understand, the *processes* of discovery and problem-solving. Kuhn is a faithful follower of this aspect of Conant's plan. Conant was interested in describing the intellectual process and in the kind of steps that the scientist takes in order to achieve a solution to a problem.[62]

He showed that his approach offered an image of science opposed to the one usually narrated by previous philosophers such as Karl Pearson in *The Grammar of Science*—he also included Ernst Mach in this category. For Conant, Pearson's vision was that scientists could apply their ingenuity and be imaginative when developing laws and theories, but that this was preceded by a much more methodical work of pure data-ordering. For the former Harvard chemist, this thesis could give a misleading indication of what scientific research truly was.[63] For Conant, this thesis is inaccurate, and he points out: "Many of the 'Facts' with which modern physics and chemistry deal might well deserve the name 'artifacts.'"[64] In his book, Conant tried to develop a much more complex description of the scientist's intellectual activity, of the creative

---

[59] See Conant (1944, 406, 454).

[60] See Rudolph (2002, 47–50, 130–131).

[61] Conant (1947a, 17–18).

[62] The kind of historiography of science that Kuhn abhors is well represented in our Chap. 4. Nevertheless, his anti-Whig stance in general is visible in his criticism of Santayana in our previous chapter.

[63] Conant (1947a, 14, 114–115).

[64] Conant (1947a, 115).

impulse, and of the determined attitude of someone who tries to find a solution to a problem amidst a multiplicity of results and hypotheses opposed to his or her own. There is little classifying inertia involved in all this and, for him, this was something that needed to be talked about.[65]

Rather than the simplistic anatomy of the induction process that Pearson and Mach seemed to offer, Conant rendered an image of scientists as individuals who feel compelled to anticipate their steps. It is no surprise that he spoke of the tactics and the strategy of the scientific researcher, as previously mentioned.[66] Of course, in talking about tactics and strategy, we refer to different kinds of plans, of anticipations. By setting a strategy, the scientist establishes a long-term plan and goal, while when the researcher sets a tactic, the goal involved is much closer. The scientist, in this case, tries to take the right steps when carrying out certain compulsory research operations.

Among the historical cases of *On Understanding Science*, a good example of this difference in attitudes was the comparison of the French chemist Antoine-Laurent de Lavoisier's working methods with those of the British pneumatic chemist Joseph Priestley, in light of the solution to the combustion problem.[67] As is well known, in the first six decades of the eighteenth century, there was a well-established chemical theory that employed an "earth" called "phlogiston" as one of the few principles by virtue of which chemical substances could be analyzed. As an explanatory resource, the phlogiston bestowed metals their metallic properties and was also a combustion principle that explained the combustion process pretty well.[68]

The example of its main anomaly originated from laboratory experimentation with the combustion of metals. It is very often mentioned and usually taken as the starting point for the so-called Chemical Revolution: the eighteenth-century Revolution in Chemistry. The phlogiston theory explained that the result of the calcination of a metal was its loss of phlogiston, which in turn left the residue of that metal as a substance without the corresponding metallic qualities. The metal was recovered when burning its calx (that residue) in presence of carbon, which was a substance rich in phlogiston. Though this process was satisfactorily explanatory in qualitative terms, it was anomalous when paying specific attention to the quantitative relationship between the reagents and the products of the reaction. It had already been established as a fact when Lavoisier began his own attempt at explaining it. According to the phlogiston theory, the calx should weigh less insofar as it had lost its metallic principle—the phlogiston. The evidence from the laboratory showed, however, that it actually weighed more. Current chemical theory easily explains why thanks to the advancement that brought about this revolutionary episode. So, the phlogiston-based

[65] Conant (1947a, 15). See Pearson (1900, 37). About Pearson, see Porter (2004). Conant also mentions Mach (1882); see Conant (1947a, 114–115).

[66] Conant (1947a, 5–12).

[67] See Conant (1947a, 75–97; 1948c).

[68] On the details of the historical case of the Chemical Revolution that is expounded from this point to the end of this section, see Conant (1948c, 70–74, 87); Guerlac (1961, 113–125, 129–131, 136, 141); Brock (2000, 78–84); Solís and Sellés (2005, 713–738).

explanation of the combustion of metals suffered a serious setback in terms of that quantitative anomaly.

This will do as a quick summary of the main anomaly that Lavoisier confronted on the way to the development of his substitute for the, at that time, well-established chemical theory. In the hands of Conant, who gave a good and influential summary of that episode in *The Overthrow of the Phlogiston Theory*—one of his case histories for general education in science for non-scientists—Lavoisier represents the strategist in his personal classification of the scientific stance towards problems and their solutions. A strategist like Lavoisier skillfully worked with a "broad working hypothesis," that is, with the kind of hypothesis that was able to provide a phenomenon with a certain explanation based on a given theory; in this case, an alternative to that based on phlogiston.[69] For Lavoisier, to be more precise, the gain in weight was due to the adherence of a certain substance during combustion, not to a loss. That, Conant specified, was the essence of that broad working hypothesis. The laboratory experiences that helped to prove that point were decisive for starting the revolution that Lavoisier consciously and explicitly led.[70]

In that process, a new figure appeared that represents the other extreme in Conant's classification of kinds of scientific attitude—the tactician. Conant recalls that Lavoisier was able to prove his point on the basis of a substance whose calx could be reduced—that is, the metal could be restored to its original condition—without carbon, to wit, the *mercurius calcinatus per se* (at that time also called "red calx of mercury," HgO in its modern chemical notation). Yet, the broad hypothesis also involved that there was a loss in that process (i.e., as explained by Lavoisier) that involved a particularly pure sample of air; of course, nothing to do with an air rich in phlogiston. The procedure to check the resulting sample owes much to Joseph Priestley, a great experimenter who showed Lavoisier that the air emitted in the reduction was *particularly* pure. Priestley, Conant showed, was the prototype of a tactician. He had designed a method for proving the purity of a sample of air based on the solubility in water of a sample of "nitrous air" (i.e., nitric oxide, or NO nowadays).[71] The latter, the nitrous air, was not soluble in water, but it would be if combined with pure air. Experiences with the air that resulted from the reduction of the red calx of mercury showed, in particular thanks to Priestley's own experimental zeal, that such air was particularly pure, as one could test its purity again and again and the repetition did not show a decrease in the air's purity.[72]

Conant saw Priestley in the character of good tactician, in contrast with Lavoisier, owing to the fact that Priestley had provided Lavoisier with two results that confirmed "limited working hypotheses" that were deduced from the broader one.[73] For one, Priestley had talked Lavoisier about the red calx that led him to check his hypothesis about the irrelevance of phlogiston in combustion (Priestley told Lavoisier about

---

69 Conant (1948c, 72–73).

70 See Conant (1947a, 99–100; 1948c, 72–73, 87).

71 See Conant (1948c, 74).

72 See Conant (1948c, 74–78).

73 See Conant (1948c, 87, 104).

it in Paris in 1774); for another, he corrected Lavoisier's first opinion about the purity of the resulting sample of air, emphasizing it. Priestley was particularly acute when it came to performing a certain experiment to prove or disprove a specific (i.e. "limited") hypothesis. Lavoisier, by contrast, used to have a far-reaching view about his science, and knew how to build an argument on its behalf. Accordingly, while Priestley interpreted such results in the framework, or *conceptual scheme* (I shall return to this notion below), of the phlogiston theory and considered such a sample of pure air "dephlogistized air," Lavoisier developed a full, new explanation that his experiments helped to support, based on a new kind of principle, that of oxygen, for a new view on gases, which, in his new explanation, were no longer simple but composed substances.[74]

In Conant's views, the roles of the strategist and the tactician are clear. For him, a revolution in science, whenever it occurs, requires both roles, but a scientist like Lavoisier is best represented as the former. He or she is aware of facts, more particularly of those *anomalous* facts, in need of an explanation, and he or she seeks for a powerful alternative to account for that and other alternative facts. To do so, the strategist sometimes needs to overthrow a well-established theory. As the title of Conant's case history reads, in finding the solution to the riddle, Lavoisier provoked "the overthrown of the phlogiston theory" itself.[75]

2.5 The Controlled Experiment and Conceptual Schemes

We should now delve into *On Understanding Science* and some of its main concepts in order to find out which of these concepts left a clear mark on Kuhn. To do so, I should first start with a concept that is not reproduced in our author's work, but that is nevertheless central to understanding Conant's vision and its connection with Bridgman.

Conant's historical cases have a lower temporal limit that is fixed by what he considers to be the very beginnings of the science he intends to illustrate for the public. Here he fully agrees with Bridgman and probably with many of his contemporaries: modern science is perfectly represented by what both call "the controlled experiment." In Bridgman's words, which Conant partially reproduced in *On Understanding Science*:

> The first scientific epoch was initiated by what was in essence a new trick of intellectual technique—the controlled experiment. This is now accepted so much as a matter of course that it requires a forceful act of imagination to recover the point of view of the early days. In those days nature was hopelessly complex, for it had not yet been envisaged that laws of mechanics and chemistry even existed. It was, therefore, not at all obvious that it was possible artificially to create simple situations which would be significant because they repeated themselves in more complex situations. In fact, it was not obvious what constituted simplicity in a situation. [...] The final result of decades of trial and failure, cogitation and

[74] See Conant (1947a, 101–102; 1948c, 74–75, 88–89).

[75] See Conant (1947a, 90–91, 100; 1948c).

> instruction, was the development of a new tool for use by intelligence, which alone has made possible present science and technology.[76]

And if we ask ourselves what Bridgman is referring to with "the first scientific epoch," he gives us an illustrative answer shortly before: "Two epochs may be distinguished. The first runs from Galileo and Newton to the first part of the present century. [...] The second epoch begins with this century, with the revolutions in scientific thinking involved in relativity theory and quantum mechanics."[77]

For Conant, the appearance of the controlled experiment is a significant theme featured in the course he introduces in his book.[78] Thus, he limits his stories to what he calls "the early days in the evolution of the modern discipline."[79] Something like this includes, he shows, physics and chemistry in their respective revolutionary and post-revolutionary periods, which takes us back to the centuries of Galileo, Newton, and Lavoisier, and also to other disciplines such as biology and geology at key stages of their development.[80] He does not go further back than this. That way, the empirical and mathematical knowledge that is required to understand the historical episodes is alleviated, and it allows students to observe the effort of scientists to understand that "hopelessly complex" nature—as Bridgman calls it in one of the previous extracts—through the first forms of controlled experimentation.[81] Despite the simplification, says Conant, the study of these historical cases requires from the student the "act of imagination" that Bridgman talked about (although this is very pedagogical):

> The student in the proposed course would be required to perform to the best of his or her ability just such a forceful act of imagination. And it is my contention that for a vast majority of students only by performing such an act can they hope to understand what Bridgman describes as "the development of a new tool for use by intelligence, which alone has made possible present science and technology."[82]

For Bridgman and Conant, we are so accustomed to the isolation involved in experimental control that they assume that it will be difficult for the student to understand which developmental phases led to it, and how. However, for Conant, it helps us to understand what this "new tool for use by intelligence" (in Bridgman's words again) is and what the foundations of modern science are. Conant specifies what the controlled experiment is in *On Understanding Science* as follows:

> The essence of the controlled experiment is, of course, the control of the relevant variables such as temperature, pressure, light, and presence of other materials, particularly small amounts of air and water. This is always a question of degree, and ways of estimating the effect of lack of control are of first importance. Progress in a given area often depends

---

[76] Bridgman (1945, p. 451). Conant (1947a, 115, n. 6) reproduces the first two sentences.

[77] Bridgman (1945, 451).

[78] Conant (1947a, 115, n. 6; see also 18).

[79] Conant (1947a, 17).

[80] See Conant (1947a, 18).

[81] See Conant (1947a, 18).

[82] Conant (1947a, 115, n. 6). Conant's last quotation from Bridgman can be found in one of my previous extracts from this physicist.

> on recognizing what are in fact the relevant variables and devising methods of measuring and controlling them. Errors frequently spring from the overlooking of important variable factors.[83]

It is not possible to read these passages from Conant and Bridgman without noticing that the idea that the current mind must strive—or be imaginative, they say—to put itself in the same point of view as the pioneers of experimentalism reminds us of Kuhn's future messages about the kind of effort that is demanded of the historian of science to understand science in its own terms, to overcome one's own categories, as he will say, and (he would add) to delve into another vision of the world. We know that Kuhn read *On Understanding Science* in proofs,[84] so it is hard not to think about the influence that this viewpoint about the difference between points of view that surrounds experimental work according to Bridgman and Conant had on Kuhn's thought. However, we also know from the first chapter that this idea is not new for Kuhn, as his essays in literary history, and even his description of his own intellectual development in "The War and My Crisis," assume a rupture between different world-views, and between groups of assessment criteria.[85] Yet, Bridgman and Conant present the possibility that something like this also happens in scientific research. This would not be the only source of this important idea for Kuhn. His experience of understanding Aristotelian physics in its own terms and Alexandre Koyré's work are his oft-mentioned root of that point of view. But we must consider Bridgman's and Conant's views in that respect as another source—and a quite plausible one.

Along with the idea of the controlled experiment and this distinction of historical phases, there is one more notion that is valuable to us when tracking the influence of Conant on Kuhn. In fact, it could be the most valuable of all those that we will see here. I am referring to the notion of a *conceptual scheme*. When Conant talks about concepts, we can interpret what he says in a Bridgmanian, operational key. Conant refers to the concepts in use when interpreting the outcomes of early examples of controlled experimentation. Thus interpreted, it is possible to say that he is giving an operational interpretation for such concepts. Actually, Conant states that, in *On Understanding Science*, he wants to show

> the difficulties which attend each new push forward in the advance of science, and the importance of the new techniques: how they arise, are improved, and often revolutionize a field of inquiry. I should hope to illustrate the intricate interplay between experiment, or observation, and the development of new concepts and new generalizations; in short, how new concepts evolve from experiments, how one conceptual scheme for a time is adequate and then is modified or displaced by another. [...] I should hope that almost all examples chosen would show the hazards which nature puts in the way of those who would examine the facts impartially and classify them accurately. The "controlled experiment" and the planned or controlled observation would be in the forefront of every discussion.[86]

---

[83] Conant (1947a, 56).

[84] Interview 1995, 275.

[85] See our Chap. 1.

[86] Conant (1947a, 18).

One of the examples that best illustrate this development comes from the history of pneumatics from 1638 to 1660, from Galileo Galilei, Vincenzo Viviani and Evangelista Torricelli to Otto von Guericke and, above all, to Robert Boyle. Conant dedicates the second chapter of the book to this example because of the various aspects of his vision of science that it helps to convey: in particular, the interaction between conceptual change and experimentation and also the social structure of science.[87] Regarding the latter, Conant here launches a new critique at the "ardent advocates of a greater degree of national planning."[88] For him, the institutional structure of science and its resources allow sufficient access to the enormous volume of published research without the need for that kind of centralized intervention. Another objective of his General Education program is to contribute to a better understanding of this latter point, and Conant also makes it quite clear here.[89]

Boyle's work helps Conant to explain what those concepts are that, in his perspective, generate or are an essential part of a conceptual scheme.[90] For Conant:

> Boyle clearly distinguishes between his concept of air as an elastic fluid and the *explanation* of this elasticity in terms of pictures and models. This distinction is a useful one in studying the "Tactics and Strategy of Science." The explanations offered for the behavior of gases were at first little more than unbridled speculations, useful as pedagogic devices (as Boyle used the analogy with lamb's wool to show compressibility), but not a necessary part of a conceptual scheme.[91]

Shortly after, Conant praises Boyle who, in his opinion, was a visionary when he showed his inclination towards the empirically confirmable. In his opinion, the reflection on scientific method in the twentieth century is also worthy of praise; here he adds: "This might well be the appropriate point in the course to make reference to the operational point of view clearly stated by Professor Bridgman in his *Logic of Modern Physics*."[92] For Conant, the notion of a conceptual scheme assumes operationally well-defined concepts, which make the theories that employ them susceptible to experimental confirmation. The rest is a mental game, a pedagogical resource that is not part of the "big game," so to speak, to which only the controlled experiment and the operationally well-defined concepts are invited.[93]

---

[87] Conant (1947a, 30–31).

[88] Conant (1947a, 60).

[89] See Conant (1947a, 60).

[90] Conant (1947a, 47). As we will see later in this chapter, the notion of conceptual scheme already appears in L. J. Henderson's writings; for example, in his "Sociology 23" classes, which Kuhn also seems to have read (see the corresponding card in TSKP 8–9). See Henderson (1970, 87); Isaac (2012a, 69); and our Chap. 3.

[91] Conant (1947a, 47).

[92] Conant (1947a, 48).

[93] See Conant (1948a, x–xiv) for this view of the close relationship between concepts and experiments that confers clear operational virtues and confirmation for theories based on them.

2.6 A Conceptual Scheme for Kuhn

Karl Hufbauer has shown the similarities and differences between Conant's proposal in *On Understanding Science* and Kuhn's first general perspective on scientific theories and their development, as it appeared in his 1947 scheme that was discussed at the beginning of this chapter. As regards the differences, Hufbauer underlines, for example, that Kuhn did not develop his early view from the point of view of Conant's social objectives. We can add that Kuhn did not do so later either, as his interests remained tied to his philosophical objectives.[94] In addition, we must not forget that Kuhn qualified the role of the experiment in scientific research more than Conant did. It is difficult to think, of course, that Kuhn considered that the empirical basis of theories did not play a central role in their development, but we have already seen how little he was inclined towards the operational approach for the analysis of the meaning of theoretical terms. If there was any important role of a historical nature for the controlled experiment in his own view, his main interest was to frame it, and laboratory work at large, in a more complete vision of scientific research in which such a component of scientific research occupied a place that was not more privileged than the one assigned to theory. Kuhn, as we know, increasingly tended to observe how both components interacted, "dialogued," so to speak, in scientific research and how a clear separation was an illusion of empiricist methodology.[95]

Hufbauer has also pointed out the similarities between Kuhn's views in that early outline and Conant's *On Understanding Science*. The influence of a historical perspective is, Hufbauer says, visible in Part I of the scheme, when Kuhn talks about the importance of studying science in terms of its developmental "stages," as well as when he talks about "conceptual structures."[96] I have doubts that Kuhn's reference to such "stages" in that part of his outline truly refers to an historical approach at precisely that point, as Hufbauer suggests; however, I agree about the developmental perspective involved (though perhaps in more abstract terms), and that is probably also Conant's influence. But reference in that outline to the "continued interplay of theory and experiment," and to the effects of experimental results on such "conceptual structures" are definitely Conant's ideas and expressions, too.[97] These points in common are much more visible if we recall Kuhn's mature views and then read the following passage of *On Understanding Science*:

> Usually our conceptual schemes grow by an evolutionary process, by the gradual incorporation of a series of amendments, so to speak. In this case a completely new idea came along and rendered obsolete the older one. [...] We can put it down as one of the principles learned from the history of science that a theory is only overthrown by a better theory, never

[94] See Interview 1995, 275–276.

[95] See Hufbauer (2012, 430). See also Kuhn, "Objectives of a General Education Course in the Physical Sciences," May 1947, TSKP 1.4, 1 (Part I, Sect. 4).

[96] See Hufbauer (2012, 430).

[97] See Kuhn, "Objectives of a General Education Course in the Physical Sciences," May 1947, TSKP 1.4, 1–2, for such phrases within quotation marks. Concerning such "interplay," see Conant (1947a, 18).

> merely by contradictory facts. Attempts are first made to reconcile the contradictory facts to the existing conceptual scheme by some modification of the concept. Only the combination of a new concept with facts contradictory to the old ideas finally brings about a scientific revolution. And when once this has taken place, then in a few short years discovery follows upon discovery and the branch of science in question progresses by leaps and bounds.[98]

Many pages later, he repeats that thesis by saying that "a conceptual scheme is never discarded merely because of a few stubborn facts with which it cannot be reconciled; a concept is either modified or replaced by a better concept, never abandoned with nothing left to take its place."[99] A little later he says that "it takes a new conceptual scheme to cause the abandonment of an old one; when only a few facts appear to be irreconcilable with a well established conceptual scheme, the first attempt is *not* to discard the scheme but to find some way out of the difficulty and keep it."[100] Thus, the basic scheme of change in which the dynamics of change involve a contraposition between two different conceptual schemes, almost a clash between them—and not merely a refutation of hypothesis by virtue of a fact and without alternative—is visible in both, and that scheme is clearly influenced by his mentor. We can include two more theses from *On Understanding Science* that reappear in Kuhn's writings: (1) the concepts to which scientists are used to may prevent the latter from adopting another, perhaps more useful, one; (2) the "experimental discovery must fit the times," or, in other words, the availability of experimental results does not immediately lead to its "acceptance."[101] Passages like these confirm Hufbauer's point once again.

In short, Kuhn's perspective on the difficulties that facts and concepts find in becoming part of a hegemonic science can already be found in Conant's writings—and in this book in particular. Recall that this vision of the nature of closed structures of concepts in physics and its problems with adaptation to changes is also visible in Bridgman's *Logic of Modern Physics*, too. There is a significant background of expressions and perspectives on the conceptual frameworks in physics and chemistry Kuhn could draw on when looking for ways to describe the phenomena of theoretical change in the development of science. Yet, Conant's influence stands out from other likely sources of insight, as Hufbauer also points out.[102]

Despite that, the view that Kuhn would develop from this point of view goes well beyond this view of a development of science explained in terms of clashing conceptual schemes and of the kind of hegemony that they adopt over their likely competitors. The developmental view of science and the process of problem-solving that involves this interaction between great conceptual schemes affected Kuhn in his early synthesis—the Lowell Lectures of 1951. At the same time, though, he also delved deep into their functioning and into the cognitive conditions that made it explainable. This kind of exploration on Kuhn's part made a considerable difference

---

[98] Conant (1947a, 36–37).

[99] Conant (1947a, 84).

[100] Conant (1947a, 89).

[101] Conant (1947a, 74); see also Lowell 5, 44; GR 93; *Structure*, 62, 64–65.

[102] See Hufbauer (2012, 430).

with Conant's perspective. By the time the ex-president of Harvard read the result of his former disciple's inquiry in 1961, he didn't recognize the image of science that he himself had attempted to convey back in 1947. In Kuhn's hands, fifteen years later, the cognitive commitment of a conceptual scheme and the divide with their neighboring competitors, whenever there is any, is much deeper than Conant had ever dreamed of. The distance with Conant increases as Kuhn better understands what working and seeing the world from a conceptual scheme truly involves—both psychologically and, later on, sociologically as well. In the next three chapters we shall study that process. But we must begin with a significant early step.

2.7 Physics in Aristotle's Time

In principle, the view about the origins of science that Kuhn found in Conant's plan should not have made him exceed the historical limits that we have observed in Conant's views, namely, those based on the controlled experiment and its implications. Conant's plan had, as we have seen, a lower limit in the Scientific Revolution of the seventeenth century. However, the task that he entrusted to Kuhn led him to a new, broader historical perspective; one, as we have said, more rupturist and discontinuist. Kuhn himself would recount this in a letter to Conant years later, in June 1961, in which he responded to Conant's criticisms of his first version of *The Structure of Scientific Revolutions*:

> You opened *On Understanding Science* by discussing cumulativeness as the distinguishing feature of science. You then sent me off to look at pre-Newtonian dynamics. I returned from that assignment convinced that science was not cumulative in the most important sense. Newton was not trying to do Aristotle's job better; rather Aristotle had been trying to do a different job and one that Newton did not do well. Would you say that home industry was *merely* a less effective way of doing what the factory system later did?[103]

The critical and even reprobative tone of the letter as a whole is not as noticeable in this short extract. Perhaps it is a little clearer in the last line, but is perfectly evident in the whole missive, which was a reply to Conant's criticism of an early full draft of *Structure*, something that Kuhn did not appreciate very much. We will find out more about this in a later chapter.[104] For the moment, our main goal is to observe how Kuhn came to such conclusions in the period of his collaborative work with Conant. Indeed, Conant commissioned Kuhn to prepare certain examples from the history of physics for the General Education course that he was about to teach. Conant wished to apply his plan, that is, to show his main messages by virtue of examples from the history of chemistry, but he also wished to supplement these with cases from the history of physics and astronomy. Kuhn remembered that: "At our first meeting, Conant turned to me and said 'I can't imagine a General Education course in science that doesn't have something about mechanics in it. But I'm a chemist, I can't *imagine*

---

[103] Kuhn to Conant, 29 June 1961, TSKP 25.53, 5.

[104] See Chap. 5.

how to do that! You're a physicist, go find out!' So I went out to learn something about the history of mechanics [...]."[105]

Kuhn and Fletcher Watson were in charge of these supplements in the course that they both taught with Conant, "Natural Sciences 11a," in the fall term of the 1947–1948 academic year.[106] Kuhn, for example, taught lessons on hydrostatics and aerostatics on October 3rd, 1947. They played a comparative role quite well, as they came after Conant's first lessons on the material for the second chapter of *On Understanding Science*, which addressed Torricelli's and Boyle's works. Kuhn, incidentally, highlighted the dynamics of change in conceptual schemes in the notes he took from Conant's classes, drawing attention to the latter's use of that notion—conceptual scheme.[107] Later on, after almost two months of a course that went from Boyle to Joseph Black and Lavoisier, on November 21, Kuhn taught another lesson on atomism.[108] Meanwhile, a few classes, possibly taught by Watson, were dedicated to the history of astronomy and may have constituted Kuhn's first contact with the subject of his first book, which he published ten years later, *The Copernican Revolution.*[109] Another historical example by Conant, this time about nineteenth-century chemistry, came after Kuhn's lesson.[110]

In preparing these classes, Kuhn had a famous experience that he recounted over and over again throughout the years. To understand the episode in question, we will first take a small detour that allows us to understand a little better what he understood as a result of it. Kuhn remembered (for example, in 1989) having prepared some classes on Galileo during the preceding summer of 1947, for which he had to understand what ideas about nature had been hegemonic before Galileo's mechanics; in other words, he wanted to know first whom Galileo was responding to. We already know that Kuhn was not unfamiliar with this kind of question—he had found some examples of a similar kind while studying the history of literature, for example. The answer to his current question was simple and well known: Galileo's conversation partner was evidently Aristotle.[111]

---

[105] Interview 1989, 20.

[106] The course had the title "The Growth of the Experimental Sciences," and only Conant and Watson were listed as the teaching faculty involved. See the *Final Announcement of the Courses of Instruction Offered by the Faculty of Arts and Sciences During 1947–48, Official Register of Harvard University*, vol. XLIV, No. 25, 9 September 1947, 23. (Available at <https://iiif.lib.harvard.edu/manifests/view/drs:467240622$1i>.).

[107] See [Kuhn], "Natural Sciences 11(a)," 24 September–28 November 1947, TSKP 1.11, 1–2; for the case developed by Conant, see pp. 2–9, as well as 10 for Kuhn's class (of which there is only a brief mention).

[108] See [Kuhn], "Natural Sciences 11(a)," 24 September–28 November 1947, TSKP 1.11, 10–20 and 25–46; Kuhn's class is mentioned on p. 46.

[109] See [Kuhn], "Natural Sciences 11(a)," 24 September–28 November 1947, TSKP 1.11, 21–23.

[110] See [Kuhn], "Natural Sciences 11(a)," 24 September–28 November 1947, TSKP 1.11, 47–57.

[111] Interview 1989, 20.

At that time, however, Aristotle made no sense to him—Kuhn still had the perspective of a physicist pursuing his doctorate and as yet active in the field.[112] "One memorable (and very hot) summer day" in Cambridge, Massachusetts, he "suddenly" (Kuhn will often repeat this adverb in his descriptions, as can be observed below) understood what the Greek philosopher intended to express.[113] He, for example, remembered many years later that:

> I was sitting at my desk with the text of Aristotle's *Physics* in front of me and with a four-colored pencil in my hand. Looking up, I gazed abstractedly out the window of my room—the visual image is one I still retain. Suddenly the fragments in my head sorted themselves out in a new way, and fell into place together. My jaw dropped, for all at once Aristotle seemed a very good physicist indeed, but of a sort I'd never dreamed possible.[114]

Although this version was first published in 1987, a former, very similar one appears in his notes for the Foerster Lecture at the University of California, Berkeley, in 1976, so almost at the same time as the parallel account he gave in the Preface for *The Essential Tension.*[115] And in 1957, he had already mentioned a "shocking experience" of that kind at the beginning of his career in the history of physics, which perfectly matches his later description.[116] Undoubtedly, at that moment something sudden and quite significant happened to him, and he used to refer to it as a premise in his arguments on behalf of his thesis. For his daughter, Sarah, "[t]he importance of the Aristotle story to my dad's views can't be overstated, since we heard it many times in numerous iterations."[117] Clearly, it was a turning point in his early career. Moreover, it represents the kind of historical puzzle that promotes a historiographical inquiry that his philosophical views are meant to clarify, as he seems to suggest in *The Essential Tension*, where he also says that he had concluded that describing such an experience is a better option for displaying the nature of his "historical practice," and its relationship with his philosophy, than his finished historical papers.[118]

Kuhn found further evidence for this historical barrier to understanding in the behavior of some of the main mathematicians and natural philosophers from the Middle Ages to the seventeenth century regarding their attitude towards Aristotle. Kuhn told his history of science classes about this as early as 1953. Thinking

---

[112] *Essential Tension*, x–xiii.

[113] See *Essential Tension*, xi, for the quoted expressions.

[114] Kuhn (1987, 16).

[115] See Kuhn, "Does Knowledge 'Grow'?", Foerster Lecture, UC Berkeley, 29 November 1976, TSKP 5.13, 4; see also Kuhn (1987). That latter paper (Kuhn 1987) was based on the first Perspective Lecture, which he gave at the University of Notre Dame in 1980 (see *The Road*, 13), where it is possible to find a very similar description as well. For that lecture, see Kuhn, *The Natures of Conceptual Change*, University of Notre Dame, 17, 19, 21 November 1980, TSKP 5.36, Lecture 1, 6; on these lectures, see Mayoral (2023b) and Melogno (2023). On this episode, see Wray (2021, Chap. 1) and Reisch (2019, esp. Chap. 4). For the text of Kuhn's Foerster Lecture and a commentary on it, see Kuhn (2024), and Mayoral (2024b).

[116] See Kuhn, "A Historian Views the Philosophy of Science," 16 April 1957, TSKP 3.12, 2.

[117] Sarah Kuhn, conversation with Jensine Andresen, 2 April 1999, quoted by Andresen (1999, S56; see also S55).

[118] *Essential Tension*, x. On the preparation of that book of essays, see our Chap. 8.

about main historical figures such as the philosophers of medieval scholasticism and then other early modern thinkers—philosophers and mathematicians—like Nicholas Copernicus, Galileo Galilei, Tycho Brahe, René Descartes, Christiaan Huygens, Robert Boyle, or Isaac Newton, Kuhn found a clear difference between the last four (and some of their contemporaries) and the previous group. Although with a clearly critical objective, the former expressed themselves in a conceptual and categorical scheme partially shared with Aristotle, while the latter had completely turned their backs on him in that sense.[119] Sometimes, as Kuhn recalls in *The Essential Tension* and as he told his classes in 1953,[120] even someone like Descartes treated Aristotle and his work on motion with clear disdain.[121] Among the cards I have included in the group abbreviated here as ML, Kuhn left one, undated, titled "Historical Questions," whose second point it makes sense to include in this discussion, because he wondered why the categories handled by Galileo are so close to those of Aristotle and, at the same time, there was much less proximity between this and those of Newton and Huygens.[122]

It is simple and appropriate to conjecture that, to understand Aristotle, Kuhn applied the same approach he had used years before to adopt the point of view from which Donne's, Shakespeare's or Browning's most controversial literary texts made complete sense. We can also assume that, just as in his early years at Harvard he had disregarded Santayana's dismissive opinions on this matter—aesthetic judgment—he now began to alter his modern attitude towards science when reading Aristotle, and instead tried to find out what kind of physicist and philosopher was the Aristotle who was a solid interlocutor of Galileo. Kuhn's later views effectively respond to those initial views. Arguably, his theory—even the most mature, post-*Structure* view—responds to such evidence of change. After all, his lexical theory, which we will have the opportunity to discuss in our last chapter, expounds scientific change in terms of the categorical and linguistic organization of facts in the world handled by historical actors. Kuhn's way of interpreting Aristotle in his own terms in his class and private notes, as in those that we have just mentioned, is not very different from that later perspective. It is no surprise that reconstructing Aristotle's physics was a preferred example for him to introduce as evidence of revolutionary change—a sure pathway to incommensurability—from 1976 onwards. *The Plurality of Worlds* itself (Kuhn's last unfinished book) begins with an improved version of a reconstruction of Aristotle's physics that Kuhn had started to use in talks and published papers almost twenty years before.[123]

---

[119] [Kuhn], "History of Science 105—Lecture 25—12/5/53," 5 December 1953, TSKP 1.21, 4.

[120] See [Kuhn], "History of Science 105—Lecture 25—12/5/53," 5 December 1953, TSKP 1.21, 4.

[121] *Essential Tension*, xiii.

[122] Most of the cards in this group (ML) are dated 1943, as evident through Chap. 1. However, the thematic divergence and the dates of Kuhn's first contacts with these scientists and philosophers of the past suggests that some of these bibliographic cards may have slipped among a group written years before.

[123] Compare, for example, Kuhn's reconstruction in *Last Writings*, 135–141, with his "Does Knowledge 'Grow'?" (29 November 1976, TSKP 5.13), 2–10, or with *The Road*, 15–20 (which corresponds

Of course, Kuhn achieved, for himself, the rehabilitation of Aristotle as a full-fledged scientist—although with a very different perspective of nature, his own and disparate ontology, a well-articulated set of concepts that was difficult to relate to our own set, and epistemological and methodological precepts which were also different from ours. Aristotle's closed and hierarchical cosmos, his way of privileging a qualitative explanation of phenomena directly related to his overall metaphysical vision, or the importance he gave to teleology in explanation, along with a large number of factors in his philosophy—all of which deviated from Kuhn's twentieth-century-physics point of view—were what appeared before Kuhn's eyes when examining Aristotle closely and seeking answers to his apparent incoherence. Alexandre Koyré undoubtedly promoted both this change and the previous perspective on Descartes, for, as he says in his *Études galiléennes*,

> As one goes from Galileo to Descartes, from the *Dialogue Concerning the Two Chief World Systems* to *The World*, one has a strange feeling, a feeling that could be described, inadequately no doubt, by saying that we experience an abrupt change of intellectual atmosphere. [...] The era of battle, the era of struggle, has been left far behind. [... Descartes is not] interested in criticising Aristotelian physics, in analysing its foundations, its weaknesses and contradictions: it is enough to make occasional fun of its primary matter or of the imaginary space of the philosophers. For Descartes, traditional physics is not only dead, but buried as well. One does not bother with it any more. What needs to be done, what Descartes confidently sets out to do, is to replace it. It is to establish and develop a new, and true, physics, to offer a new picture of the world, which is to say, in particular, a new concept of matter and a new concept of motion.[124]

The Aristotle experience was, in a nutshell, central for Kuhn. We shall see him returning once and again to that moment, to that "sudden" feeling of strangeness with regard to the past, and, especially in his older years, we shall also see him trying to get the most of it, philosophically speaking. Koyré was inspiring; Conant, by contrast, does not count among the most inspirational sources of historiographical insight, especially concerning the method of the historian of science. In that respect, Kuhn, to some extent, designed his own method.[125] In 1995 he said:

> [The] Aristotle experience was terribly important. Conant, in case histories of his own and in his teaching, never I think saw to the extent that I did the need to say what people had believed *before*. He would always start in more or less with the beginning of the work. There would be something about it, but there was little preparation for getting to the person. I always felt you had to do more; and that meant you had to do a stage set, within another conceptual framework, in order to get at these things. And that was what this did for me. [...] I used to think—forgive me—that with the possible exception of Koyré, and maybe not with the exception of Koyré, I could read texts, get inside the heads of the people who wrote them, better than anybody else in the world. I loved doing that. I took real pride and satisfaction in doing it.[126]

---

to "What Are Scientific Revolutions?", i.e. Kuhn 1987). Identical introductions to that case can be found in other series of lectures during the 1980s; see Chap. 8 for further details.

[124] Koyré (1978, 251); see also Koyré (1940, 318).

[125] However, he often confessed Koyré's influence. See, for example, *Essential Tension*, xiii, n. 3; Interview 1989, 20; Interview 1995, 285; Kuhn (1983b, 27; 1984, 361).

[126] Interview 1995, 275–276.

In that passage, in a part that I have not reproduced in the previous extract, Kuhn recognizes that, just as other historians did not take him to be a historian *tout court*, he saw himself as one "of a rather special narrow sort."[127] Although I am not sure there is a canon to fit, I agree that the kind of history of science that he pursues tries to understand, with scrupulous attention to detail and timing, the process by which scientific beliefs and concepts are altered or replaced with others in order for the scientists involved to find a first or a better solution to a problem. Very often, in doing so, he himself reproduces that process at the history-of-science level, and so he attempts to change our ideas about, say, Aristotle, Copernicus, Carnot, or Planck, and their proper roles in the history of some episode of change in order to obtain a better interpretation of the historical facts. At both levels, a philosophical view is latent. It helps to explain the first kind of transition—the one at the scientific level—and it supports his specific interpretive shift in focus whenever it happens. For the rest of his lifetime, from those early days trying to understand early modern physics, he tried hard to turn that latent view into an overt and detailed system of philosophy of science that could be helpful to others, with a similar responsiveness to change and to the effects that understanding it properly produces in our view of science. And again, he promoted a shift in focus in that philosophical field, too.

2.8 Kuhn, Harvard Fellow

Conant's General Education group increased in the 1948–1949 course, which followed Kuhn's first collaboration in the course "The Growth of the Experimental Sciences"—better known to us as "Natural Science 11a." This course was then listed as "Natural Science 4," or "Nat Sci 4," and had the same title as its predecessor.[128] In January 1948, there were already plans to invite Duane Roller, a physicist from Wabash College (Crawfordsville, Indiana), who, from July of that year and for a course, would develop his activities with the group. This was also the case with Leonard K. Nash, a member of Harvard's Department of Chemistry, who accepted the offer in February 1948.[129]

---

[127] Interview 1995, 276.

[128] Conant and Watson, together with Duane Roller and Leonard Nash (about whom I shall say something below), are the only faculty included; Kuhn is not mentioned yet. See the *Final Announcement of the Courses of Instruction Offered by the Faculty of Arts and Sciences During 1948–49, Official Register of Harvard University*, vol. XLV, No. 24, 15 September 1948, 19. (Available at <https://iiif.lib.harvard.edu/manifests/view/drs:467241064$1i>.) Kuhn shall already appear in the course catalogue for the first time in 1950–51, still ascribed to the Society of Fellows; see the *Final Announcement of the Courses of Instruction Offered by the Faculty of Arts and Sciences During 1950–51, Official Register of Harvard University*, vol. XLVII, No. 23, September 1950, 22. (Available at <https://iiif.lib.harvard.edu/manifests/view/drs:467242132$1i>.)

[129] For this data, Conant to B. F. Wright, 30 January 1948; Nash to Conant, 20 February 1948; both in JBCP 23, folder: "Nat. Sciences 4, Correspondence 1948–49." For Roller's appointment, see the *Minutes of the Committee on General Education*, meeting of 3 February 1948, HUA, Ref.: UA III

The course plan divided the students into five sections of about twenty or more students, three of which would be led by Roller, another by Nash, and still another by Watson and Conant. The work plan was not very different from that seen in "Natural Science 11a," albeit at a slower pace, with a lighter workload, and without practical work.[130] Additionally, two scholarships were offered. They were intended to help with evaluation and research. Aspirants needed to hold a PhD in Physics and would be allowed to continue their work while helping with Conant's course. Two collaborators, young physicists, were appointed in May 1948; one of them, from the Palmer Physical Laboratory of Princeton, was Eugene P. Gross, later a distinguished physicist; the other was Charles L. Clark. In the instructions that Conant gave them, he talked about the nature of the course from the position of his views on general education. The course was not intended to simply inform about scientific data but would rather deal with historical or philosophical issues. Conant also mentioned the presence of Kuhn, who was already a Harvard Fellow.[131]

Kuhn was still part of the group, indeed, but after his assistance to Conant during the General Education course, he had asked him for nothing less than his support for admission to the exclusive Harvard Society of Fellows.[132] At the meeting on May 17, 1948, the Society approved Kuhn as a Junior Fellow from the following July 1. Yet, he was granted a three-month leave, later four, in order to finish his studies and get his PhD. During that time, he continued as a fellow of the National Research Council. After obtaining his PhD in theoretical physics, he became a fellow of this exclusive and meritocratic society for the next three years, until July 1951.[133]

Admission to the Society was no small achievement in the career of an American student like Kuhn.[134] Since its foundation in 1933, the Society had been the meeting point for a select number of promising young people from within and outside Harvard College. By the time he joined it, John Bardeen, Garrett Birkhoff (son of Kuhn's calculus professor), Henry Guerlac, W. V. Quine, Paul A. Samuelson, Arthur M. Schlesinger, Jr., B. F. Skinner and E. B. Wilson had already been Junior Fellows of the Society and some of them were now Senior Fellows.[135] Each academic year, new candidates were recommended (they never applied for themselves), and were chosen by the senior members. As a rule, they would not accept more than eight new members, and the whole group did not have more than twenty-four members.[136]

10199.2, folder: "1946–1951 NOV."; Conant to Wright, 16 February 1948, JBCP 23, folder: "Nat. Sciences 4, Correspondence 1948–49."

[130] See Conant to Wright, 30 January 1948; Nash to Conant, 20 February 1948, JBCP 23, folder: "Nat. Sciences 4, Correspondence 1948–49"; *Minutes of the Committee on General Education*, meeting of 29 January 1948, HUA, Ref.: UA III 10199.2, folder: "1946–1951 NOV."

[131] See Conant's letters to Gross and Clark of 24 May 1948, JBCP 23, folder: "Nat. Sciences 4, Correspondence 1948–49," where the scholarship notice can also be found.

[132] See Interview 1995, 276, 278.

[133] For Kuhn's appointment as a fellow, see the certificates of 9 June and 4 October 1948 in TSKP 14.10.

[134] About the Society, see Merton (1977, 81–85); [Anonymous] (1948); Heyl (1968); Isaac (2012a).

[135] [Anonymous] (1948, 596–597).

[136] [Anonymous] (1948, 594–595, 598); Interview 1995, 276, 279.

Kuhn, however, admits that he tried a somewhat more direct and, perhaps, less orthodox procedure: "I went to Conant—I had developed a relatively warm relationship with him, to the extent that people had warm relationships with him. [...] I asked him whether he was willing to sponsor me for the Society of Fellows. Officially that's a question you didn't ask; but I felt able to, and he did, and I got in."[137]

Both the sponsor and the candidate had to get through a meeting with the Senior Fellows, which was not always a positive memorable experience. Thus, Bardeen, the future double Nobel Prize winner in Physics, and fellow since 1935, had to face Whitehead, Lawrence J. Henderson, and the historian Samuel Eliot Morison. "I was placed before this very distinguished group of people who asked me questions," the physicist remembered, and added: "I think I was too scared to hardly say a word." Van Vleck also attended the interview. "I'm sure it was Van Vleck that got me in," Bardeen also said.[138] Henderson was a particularly intimidating figure. According to sociologist G. C. Homans, Junior Fellow at that time and Henderson's disciple, "At the hands of others, this experience was unnerving enough, but at Henderson's it might make the interviewee a candidate for the looney bin, not the society."[139] A well-known anecdote supports what Homans said. On one occasion, Henderson told a candidate: "This is not an examination. No one in this room is competent to examine you. The purpose is for us to get acquainted, and the best way to do that is to talk. So talk!"[140] No wonder, the candidate did (or could) not talk, Homans said, though he later became a fellow.[141]

As we have said, the Society was very selective. Homans himself was left out in the first selection process the first year the Society was in operation. He was not accepted as a poet (in his own opinion, quite correctly), but he was as a sociologist the following year.[142] Apparently, the interview was the most important part of the selection process. The academic merits or writing an essay especially for the occasion were not as important for becoming a fellow as they were for the Prize Fellowships at Trinity College, University of Cambridge, which was, however, a model for The Society of Fellows just as the All Souls College at the University of Oxford or the Fondation Thiers in Paris had been.[143] The Society of Fellows was founded so that its fruits would specifically come from direct contact between its individual members. It was the type of "microenvironment" that Merton spoke of, artificially created on the basis of its founders' vision of (and aspirations to) the possible results of such direct and continuous interaction—excellence and independence in research. As Merton also shows, it promoted independent study, not marked by the pursuit of a higher degree. The Society aimed to push every researcher beyond the limits of current knowledge, so contributing to the advancement of every disciplinary field beyond

[137] Interview 1995, 276.

[138] Hoddeson and Daitch (2002, 64).

[139] Homans (1986, 121).

[140] See Footnote 139.

[141] Homans (1986, 121); [Anonymous] (1948, 598).

[142] Homans (1986, 121–122); Heyl (1968, 321).

[143] See [Anonymous] (1948, 595–596, 598).

the state in which such researcher had found it, perhaps recombining it with other fields as well.[144]

So, the Society's ultimate goal was to create an environment for free intellectual inquiry. Nevertheless, Henderson seems to have emphasized the collaborative spirit; a free mind and ambitious initiative to change current knowledge did not involve a solitary work. The Society had the intention of creating a social environment that led to productive interaction, one that would be fertile for the emergence of new ideas. So, the ideals we have examined early in this chapter in Bridgman's and Conant's thinking, independence and freedom, found their place in the Society. Joining it implied that, in the interview, one was able to show what in Conant's vocabulary would be a strategist's viewpoint.[145] Along with the disinterested pursuit of knowledge, good education, willingness to collaborate, and integrity, the Society defended the relentless pursuit of a great vision. Thus, its "Declaration of Principles" said:

> You will seek not a near but a distant objective, and you will not be satisfied with what you may have done. All that you may achieve or discover you will regard as a fragment of a larger pattern which from his separate approach every true scholar is striving to descry.[146]

In this book, we can see Kuhn's unyielding willpower to find a unique, solid and detailed explanatory theory of scientific development—a goal to which he devoted his entire lifetime. It was intended to change the common-sense and the philosophical images of science, and he, indeed, saw his advances as pieces of a larger picture that he pursued until the end of his life. So, in this way, he pursued the Society's ideal throughout his life. Despite this, I honestly doubt that he first learned such ideals from the members of this group of scholars—it was a personal goal. In any case, the Society was a good place for him because they had similar goals.

Life in the Society was as its founders had wished. Junior Fellows would meet on Monday evenings with the Senior Fellows for dinner, to enjoy relaxed conversation and learn something from them. During the nine years that Henderson was able to attend these dinners (he died suddenly in 1942 from complications in a surgical procedure) his presence was the focus of the conversations, as he never went unnoticed. Regarding his political ideals, the phrase of G. C. Homans and O. T. Bailey is often repeated: "Henderson's beard was red but his politics were vigorously conservative."[147] Henderson was the son of a grocer established in Salem, although he was born in nearby Lynn, both in Massachusetts. Bernard Barber makes it clear that his time as a student and his subsequent professional dedication (until his death) at Harvard University (he graduated in 1898 and went through its medical school, from which he graduated in 1902) did not make him a stranger to foreign culture. His affinity towards France and its culture was fueled by his summers on the French-speaking island of Saint Pierre, near the Canadian island of Newfoundland, and by his

[144] See Merton (1977, 76, 81–83).

[145] See [Anonymous] (1948, 598) for a good description of those goals.

[146] [Anonymous] (1948, 597).

[147] Homans and Bailey (1959, 3), cited by Barber (1970, 7); the quotation also appears in [Anonymous] (1948, 596).

postgraduate studies at the University of Strasbourg, where he studied biochemistry. Henderson was a good physiologist and biochemist, well known for his book *Blood: A Study in General Physiology* (1928), but also for his fondness for philosophy and, above all, the social sciences.[148] He was a staunch defender of Vilfredo Pareto's theory in this regard, and developed his perspectives on the latter's theory in some of his writings. Most well known are his seminar on the author and his "Sociology 23" course at Harvard, which he taught from 1938 to 1942, and which, as I note below, even Kuhn himself seems to have read.[149] Part of this sociological vision is significant for our reconstruction, as it contributes a key ingredient to the Society's atmosphere.

Henderson combined his affinity towards a well-constructed discourse, devoid of arbitrariness and with justified assertions, with a sociological perspective of the origin and utility of language and with an interest in the conceptual foundation of scientific thought.[150] For example, regarding the first affinity, Quine told Rudolf Carnap in 1935 that he saw Henderson every week when the Society members dined together (Quine was then a Junior Fellow) and that he had spoken to Henderson about Carnap: "He readily became interested [in your work], because his own attitude is extremely antimetaphysical and skeptical."[151] Besides his affinity for conceptual clarity, he had a perspective on the origin of certain terms and concepts that allowed him to differentiate between their conceptual clarity and their utility. The terms we use and the concepts we assume sometimes have a minimal and dubious empirical basis, but their legitimacy is grounded in their function in behavior; they also express emotions and habits. This function (not so much the meaning) should be an object of study for the social scientist.[152]

This idea was not alien to Kuhn, as we saw in "The War and My Crisis,"[153] since his diagnosis about the basis of his own personal crises implied a dependence on a state of shared belief, marked by vague emotions and visions, sometimes not well founded, and, in summary, not fully rational. This proto-sociological vision in the Kuhn of 1941–1942 finds a more finished manifestation in works like Henderson's, which Kuhn read ("Sociology 23" can be found among the texts reflected in his reading notes) and, it seems, with some interest, both in relation to this aspect and to the one I mention next.[154]

---

[148] See Homans (1986, 89); Barber (1970, 2–3, 6).

[149] See Barber (1970, 5, 39 ff.); Homans (1986, 108–116); see footnote 90.

[150] For that contrast, see Barber (1970, 23–25); see also Henderson (1970, 86).

[151] Quine to Carnap, 31 March 1935, in Creath (ed.) (1990, 163). Quine, however, also added in parentheses that Carnap would see that Henderson's "state of mind [is] unsatisfactorily confused and naïve." (Creath [ed.] 1990, 163). He also says there that Henderson had read parts of Carnap's *Logical Syntax of Language* (Carnap 1934) and that he had told Harvard's philosophers that Quine should teach about Carnap's work.

[152] See Henderson (1970, 83–84, 92–99).

[153] See our Chap. 1.

[154] [Kuhn], "Henderson, L. J: Introductory Lectures to Sociology 23; privately distributed by the Harvard Business School," n. d., TSKP 8–9.

We cannot go into the details of Henderson's Paretian perspective here.[155] Suffice it to say that this view of the function of language responds to a more complex approach to the social sciences, for which Pareto's social theory offers what Henderson repeatedly calls a "conceptual scheme."[156] And he defends its usefulness in the social sciences. "Sociology 23," for example, is a course intended for the education of social scientists with that scheme as its basis.[157] The idea of a conceptual scheme is again a familiar notion for Kuhn, especially if he read "Sociology 23" after the proofs of *On Understanding Science*, something which is plausible but difficult to prove (his card lacks a date). It seems that ideas like the ones we just examined are recurrent in the modes of expression of some authors influenced by Henderson's sociology. Some of those authors shared a table with Kuhn at the Society's dinners. Two names immediately come to mind: George C. Homans, whose contact with Kuhn we will discuss shortly; and Crane Brinton, the historian who led the Society during the time our author was Junior Fellow and with whom he maintained a good relationship in the future.[158] As Barber shows, Brinton frequently used a Hendersonian vocabulary, and more specifically the notion of conceptual scheme.[159] Even if Kuhn did not learn such a concept from Henderson and the others mentioned above, its use must have been reaffirmed in such a milieu.

We cannot be sure of other influences except for Quine's, which according to Kuhn was a serious influence on him. He was, he remembered in 1995, concerned with the nature of meaning and with the idea that knowing what a term means does not depend on having grasped definitions or common rules for its use. Quine's work on the criticism of the analytic/synthetic distinction was developed at that time, he says, and that influenced him. Other people did not, and he barely recalled them years later. "I don't think there was anybody that I talked to in the Society," he said, "who was terribly important to my development."[160]

Steve Fuller interpreted Kuhn's *Structure* as the result of an indoctrination into the Society, which unconsciously led him to justify that scientific communities were legitimate in isolating themselves from the wider lay public, thereby turning them into a sort of self-governing elite.[161] Kuhn truly believes that a growing distance from the public is a sign of maturity (to be more precise, a side effect of paradigm-based specialization), but he does not seem to suggest that this public is coerced to stop trying to understand science. Conant's program is intended to produce the contrary effect, in fact. And indeed, as Kuhn's theory matured, the goals of Conant's

---

155 See Barber (1970) and Henderson (1970) for more information. Also see Cross and Albury (1987).

156 See Henderson (1970, 75–76, 86–88), among other passages. For the elements of Pareto's conceptual scheme according to Henderson, see Henderson (1970, 99).

157 Henderson (1970, 106).

158 The good relations between Kuhn and Brinton are observed in their private correspondence; see TSKP 14.15. See our Chap. 6, too.

159 Barber (1970, 38, 51–53). See Brinton (1952, esp. Chap. 1).

160 Interview 1995, 279.

161 See Fuller (2000, xii–xv, 33–35, 164–171, 180–182, and Chapter IV).

program became much more difficult to achieve. And they were indeed difficult, yes, but not impossible. Historians of science, as interpreters of past science, have hard work to do before being able to convey that past to modern readers; and modern science, on the other hand, possibly needs another kind of interpreter, and possibly of general education, too. Despite that, science only seems *difficult* to understand when mature, in Kuhn's hands. It does not seem to be an armored vehicle in *Structure*, and much less in future works. Kuhn, on the other hand, does not seem to have been *successfully* indoctrinated in the Society—if he was at all; his own memories contradict that suggestion and the documents do not support it, either. Additionally, the amount of conservativeness in his theses in *Structure* is in the eye of the beholder. Fuller's interpretation, therefore, is not wholly convincing.

2.9 Life in the Society

The Society met for Monday dinners in the rooms reserved for them in Eliot House, an oak-paneled lounge and a dining room with views of the garden and with Edwardian motifs, to Henderson's liking.[162] As Quine recalls, Henderson left a mark on those evenings, just as he used to do in the entrance interviews.[163] At these dinners, fellows talked among themselves in accord with the Society's goals—that is, to nourish their own cutting-edge research through first-hand interaction with the no less outstanding work of other members. Quine recalls that, in 1933, "[w]e Junior Fellows were alarmingly privileged that first year, for the distinguished Senior Fellows outnumbered us at table seven to six."[164] He then provides a bit more detail about those evenings:

> The dinners were for conversation. Speeches and papers were banned. There was late lingering over port and madeira, which moved along the table in a silver cart presented by President Lowell. The table was said to have been the very breakfast table of Oliver Wendell Holmes, and it was presided over in an ornate highbacked chair by our own autocrat, L. J. Henderson. He was bald and had an ample, reddish beard and a penetrating cold, pale eye. Vilfredo Pareto was his guiding light. He was quick to detect or to suspect rationalization and self-deception, and to bear down on the young offender. He was a force for good.[165]

Regarding what that conversation included, Homans and Bailey provide a very illustrative image applicable to the first fifteen years:

> The conversation is understandably lively, and goes on in knots up and down the table. Here there is a discussion of the foreign policy of Afghanistan, from which unlikely country a Junior Fellow whose field is Indic Philology has just returned. There the question turns to the operational definition of concepts, and the degree to which it can be applied in the social sciences. Here a defense of Hugh O'Neill, the great Earl of Tyrone, ends in an explanation of

[162] See [Anonymous] (1948, 598); Interview 1995, 279.

[163] Quine (1985, 109).

[164] See Footnote 163.

[165] See Footnote 163.

> Elizabethan expansion as the result of a prize squeeze on the gentlemen of England. There *Totem and Taboo* is tabooed, with anthropological reasons. Here some pellet-counters thrash out the relative merits of the rat and the hamster as laboratory animals. There the probable next moves of the Rubber Workers' Union are mapped. There are never any speeches, set papers, or set topics. Junior Fellows talk about their own work only as it comes in naturally. Many of the Fellows and guests remain at the table long after the dinner is over. The rest return to the parlor, where they pull up chairs, to continue a subject already begun or to join a group that is starting something new.[166]

Who attended such dinners? Founding members, Senior Fellows, probably Conant, and also Junior Fellows and some guests, as indicated in the previous paragraph. In Homans's time, Quine, B. F. Skinner, Conrad Arensberg, Birkhoff, Stanislaw Ulam, R. B. Woodward and Bardeen were among them. In Kuhn's time, J. R. Levenson and Hao Wang were among those who joined him in 1948.[167] While Quine spoke of Carnap's logic and philosophy, Skinner might talk about behaviorism and Arensberg (Homans recalls) explored Bridgman's operational analysis. Years later, when Kuhn joined the Society, Homans, by then an "occasional guest at the dinners" (his tenure was 1934–1939), recalls having come into contact with him in that same context.[168] Henderson had passed away by that time, but the traditions already established—solemn dinners and relaxed lunches of the junior members—were still there. Homans remembered in his autobiography his contact with Kuhn and his "new view of the way science grows."[169] In those post-war years, Homans struck up a friendship with Kuhn and at the former's request, Kuhn made a few recommendations of scientific texts:

> Simply for the good of my soul, I asked Tom [Kuhn] to suggest books for me to read that treated classical physics rigorously. He first suggested Ernst Mach, *The Science of Mechanics*, a history of the oldest branch of physics, and later Max Planck, *Treatise on Thermodynamics*. [...] Kuhn, through Mach and Planck, helped me appreciate what physicists had certainly recognized as good theories for their times, though no theory is forever. At his urging I read other classic works of science, notably D'Arcy Thompson, *On Growth and Form*.[170]

As Homans recounted, the reading of Mach suggested by Kuhn helped him understand the mathematical aspects of Pareto's and Henderson's theories.[171]

As Homans points out, the Society was an environment that invited the display of one's own argumentative talent, which privileged the discursive temperament of young people who was eager to defend their own arguments and, in the process show how well-informed they were about almost everything—especially about highly specialized fields—and that they were *better* informed than the rest, too.[172] This was a behavior for which Kuhn was also well trained.

---

[166] Homans and Bailey, as quoted in [Anonymous] (1948, 598).

[167] The list is available at <www.socfell.fas.harvard.edu>.

[168] Homans (1986, 126–127).

[169] Homans (1986, 127).

[170] Homans (1986, 321–322).

[171] Homans (1986, 109–110).

[172] Homans (1986, 126).

2.10 A Personal Program in Philosophy

From March to June 1949, Kuhn delved into a good number of fields simultaneously. More specifically, he began by reading several works of a strictly philosophical nature and even belonging to logic and analytical philosophy. I will divide them into three main groups to establish some order. Some of them involve readings of specific books by certain authors from whom he later read other works; in some of these cases, I shall not be able to specify the date. I will also comment in this section on those further readings (that might have come first, however) carried out at an undetermined date, when they contribute to increasing our information about the knowledge that Kuhn could obtain from such authors. I will close by briefly describing other influences outside philosophy that were foremost for Kuhn's intellectual development in these years: Ludwik Fleck's and Benjamin Lee Whorf's works.

A first group of readings in philosophy are classic works of its history. Here we must include the *Critique of Pure Reason*, the *Critique of Practical Reason* and the *Critique of Judgment* by Kant. We know that at the beginning of his career he had studied Kant in Demos' course.[173] However, we may also suppose that perhaps he did not go deeply into Kant, and that now, with enough time, he wished to do it properly. Also mentioned is *A System of Logic* by J. S. Mill, an author with whom he maintained an ambivalent relationship; the British philosopher represented a vision of the scientific method that Kuhn criticized, but it would influence him for a period of his most mature stage.[174]

A second group of readings include American authors and, more specifically, pragmatist philosophers. Before moving on to the classic pragmatists, Kuhn's reading of Susanne Langer and her *Philosophy in a New Key* stands out. Concerning those classic pragmatists, Kuhn read William James' *Essays in Pragmatism*, which complements his previous reading of the *Varieties of Religious Experience* during the war.[175] Perhaps more significant is his reading of books by John Dewey. In Chap. 1 we noted a likely indirect influence of Deweyan pedagogy on Kuhn's thinking,[176] as it served as a basis for the progressive educational reforms in some of the institutions which Kuhn attended, and where he obtained some characteristic aspects of his thinking. If he now examined works by Dewey, it is worth asking if the indirect influence became *direct*. Let's examine this possible source in a little more detail.[177]

---

173 See our Chap. 1.

174 *Handy Notebook 1949*, 31 March 1949, 1. On Mill's passing influence in his mature stage, see our Chap. 8.

175 *Handy Notebook 1949*, 31 March, 12 April, 1949, 1, 9.

176 See especially the closing paragraph of Sect. 1.3.

177 On Kuhn and Dewey and other pragmatist thinkers, see Mladenovic (2017, Part III). For a study on the connections and differences between Dewey's and Kuhn's works, especially devoted to the comparison between the former's *Logic* (Dewey 1938) and Kuhn's books and papers, see Mayoral (2022).

Years later, Kuhn told Giovanna Borradori in an interview that "I had read some Dewey on pedagogical issues, but I have never been all that enthusiastic about pragmatism as a philosophical position."[178] At the end of March 1949, as a Junior Fellow, Kuhn read *Reconstruction in Philosophy*, but this reading does not seem to have left any mark on him.[179] A decade later, he confessed to Harley C. Shands, Professor of Psychiatry at the State University of New York, his disillusionment with Dewey. Shands had told Kuhn in a letter that, after reading *Structure*, he missed a mention of Dewey's work with Arthur Bentley on transactional psychology (that is, *Knowing and the Known*), something that underlies the investigations of the Hanover Institute that Kuhn mentions in his book.[180] Kuhn's response makes clear his limited affinity for Dewey's work despite the contacts we are observing. For Kuhn, reading Dewey and Bentley allowed him to see that there were common interests. However, *Knowing and the Known* did not quite convince him due to the lack of clarity he saw in it, and to the doubts he shared with others about the real contribution of the book.[181]

Despite this, Kuhn read two more books by Dewey, although in these two new cases we know nothing about the time period in which he did so. These are *Human Nature and Conduct* and *The Quest for Certainty*, of which we only have a bibliographic card in each case. In the case of *Human Nature*, Kuhn marks a passage that is noteworthy, as it enriches our vision of the relationship between their respective points of view.[182] If we go to page 131 of that book, we see what Kuhn highlights—a characterization of the pragmatism inherent in the classification process. Early in this chapter, we discussed his 1947 scheme and the importance he gives to the process of classifying information and obtaining scientific objects from a process of theoretical elaboration. Kuhn gives importance to scientific classification in the Lowell Lectures, and in post-*Structure* work he will return to this issue.[183] Therefore, Dewey's statements on that page show a point of coincidence that has to do with an idea that is and will remain significant for Kuhn. Dewey says there: "To classify is, indeed, as useful as it is natural."[184] Establishing definitions, using labels, he adds, are "acts" that, "like other intelligent acts are performed for a purpose, and the accomplishment of purpose is their only justification."[185] However, Dewey says afterwards:

> When we assume that our clefts and bunches represent fixed separations and collections *in rerum natura*, we obstruct rather than aid our transactions with things. We are guilty

---

[178] Borradori (1994, 157–158).

[179] See *Handy Notebook 1949*, 31 March 1949, 1. There is also a bibliographic card for this text in TSKP 8–9.

[180] *Structure*, 112. See Shands to Kuhn, 30 January 1963, TSKP 4.15, 1. See also Dewey and Bentley (1949).

[181] Letter from Kuhn to Shands, 12 March 1963, TSKP 4.15, 2.

[182] For the relevant reading cards, see TSKP 8–9.

[183] See our Chaps. 3 and 7.

[184] Dewey (1922, 92).

[185] See Footnote 184.

> of a presumption which nature promptly punishes. We are rendered incompetent to deal effectively with the delicacies and novelties of nature and life.[186]

In the Lowell Lectures we will observe a similar attitude towards the impositions on categorization that are born from a language that has passed through the sieve of logical formalization or operational clarification. Kuhn's message is very similar to the one Dewey offers here. Sometimes, we assume that classification is a faithful reflection of the reality that we must progressively establish rather than an adaptive tool in an environment which produces novelties, which in turn force us to modify such classification. If we do so, the main function of that classification—to be such a tool to deal with the environment—is completely lost due to its inefficiency. In this idea, Kuhn and Dewey agree, and the kind of perspective that Kuhn develops both in the Lowell Lectures and, years later, after *Structure* testifies to this.[187] There are other interesting coincidences between their respective works. Kuhn could be seen as a thinker that supplements some of Dewey's points on scientific research, for example.[188] Despite those coincidences, however, Kuhn's explicit position is usually of disagreement with (and even of some disdain for) Dewey.[189]

A third and final group is composed of authors of the analytical tradition and the logico-positivist school of philosophy.[190] We can start with Bertrand Russell, to whom he again dedicates some time; he read *The Scientific Outlook* before March 1949. It is appropriate to include here the *Readings in Philosophical Analysis*, compiled by H. Feigl and W. Sellars, which includes papers from classic authors of positivism and logical empiricism as well as from C. I. Lewis and from W. V. Quine, among others.[191] From Quine, he precisely records the reading of his *Mathematical Logic*. As we know, during these years Quine wrote "Two Dogmas of Empiricism," a central text in the critique of logical empiricism that would influence Kuhn.[192] In his corresponding note card, he praises it, and characterizes it by its holistic, pragmatist features and criticism towards the analytic/synthetic distinction and towards verificationism. Likewise, Kuhn read "Truth by Convention" possibly at this time, since he ascribes the text to the aforementioned *Readings*. Be that as it may, the text broadens his familiarity with Quine's work, whom he also met at this time, and from whom he received help during his period in the Society.[193] Other articles by Quine that appear among his readings are "Ontology and Ideology"—Kuhn points out a connection

---

[186] See Footnote 184.

[187] Again, see our Chaps. 3, and 6–7 (especially the last one).

[188] See Mladenovic (2017, Part III) and Mayoral (2022) for further details and discussion in that respect.

[189] See our Chap. 5, especially, in Sect. 5.9, Kuhn's reply to a Conant letter of June 1961 after the latter's reading of *Structure*.

[190] On the connections (both biographical and philosophical) between Kuhn and logical empiricism, see Reisch (1991; 2005, 229–233), Earman (1993), Irzik and Grünberg (1995), Friedman (2003), Richardson (2007), and Irzik (2012). See also our Chaps. 4, 5 and 7.

[191] *Handy Notebook 1949*, 31 March 1949, 1.

[192] *Structure*, xli.

[193] Kuhn, "Advanced Studies and Research," n. d., TSKP 25.53, 2.

between Quine's paper and his own seventh Lowell lecture—as well as "Identity, Ostension, and Hypostasis," "On What There Is," and "The Problem of Interpreting Modal Logic." There are note cards and comments on each one of them.[194]

With respect to the authors of logical positivism and empiricism, two of them stand out from the group. The first is Hans Reichenbach and his *Experience and Prediction*, the book that led him to Ludwik Fleck's book during his period in the Society.[195] The second is Rudolf Carnap. At this time, Kuhn annotated in his diary that he had read Carnap's *Introduction to Semantics* before the end of March 1949.[196] He seems to have read three more texts by Carnap—*Der logische Aufbau der Welt*, *Scheinprobleme in der Philosophie* and *Foundations of Logic and Mathematics*—but we do not know anything about when he did so.[197] However, Kuhn said that, until the 1990s, he did not know about the Carnap piece that would have interested him most, to wit, "Empiricism, Semantics, and Ontology," which was published in 1950, that is, while our author was in the Society.[198] This ignorance is surprising, given that Quine cites and discusses this Carnap text in "Two Dogmas of Empiricism."[199] This article by Carnap, not minor in his extensive work, offers a vision of the linguistic frameworks and the instrumentalism associated with them that is very similar to Kuhn's later one, and which would have been of great value to him. Given that it seems unlikely that he was interested in the *Aufbau* but not in that paper by Carnap, it is plausible that he read all these positivist texts more or less at the same time and that his interest in them (and in Carnap's work at large) decreased, or even disappeared, at the same time, too. We shall see in Chap. 7 that Carnap's work was not as interesting for him (or maybe even as close to his own thoughts, or as fruitful) as, for example, Hempel's or Nagel's philosophy of science—who were, in addition, also very close to him in personal terms, particularly in the 1960s and later. So, maybe around 1950, his interest in Carnap's work was already equivalent to that of Koyré, who, after knowing that his student J. E. Murdoch was reading *Logical Syntax of Language*, told him that he "had much better work on [his] Greek."[200]

Kuhn was not as opposed to the authors of this current as Koyré,[201] but he does reveal in his comments and assessments that their formalist approach to the analysis of science did not seem fruitful in understanding it, and that is, of course, a constant

---

194 See the corresponding cards in TSKP 8–9. The mentioned essays by Quine appear as chapters, or parts of them, in *From a Logical Point of View*; see Quine (1980, 169–170) for the description of the origin of the essays.

195 Kuhn (1979c, viii).

196 *Handy Notebook 1949*, 12 April 1949, 9.

197 See the cards for these books in TSKP 8–9.

198 See Carnap (1950). In *The Road*, 227, and in a letter to G. Irzik, October 6, 1995 (reproduced in Irzik 2012, 39–40), Kuhn confesses his ignorance of this later Carnap. It seems that the articles by Earman (1993) and by Irzik and Grünberg (1995) were the ones that informed him of this connection. See Mayoral (2017a, 121–122). On his acquaintance with Carnap's work beyond the year in which *Structure* was published, see our Chap. 7.

199 See Quine (1951, 45–46).

200 Murdoch (1987, 74).

201 See Quine's recommendation letter, partially reproduced at the beginning of our Chap. 4.

in his thought during his entire career.[202] When writing private note cards about Carnap, he only showed some fascination for the *Aufbau*, though always with a cautious attitude. Carnap's *Scheinprobleme in der Philosophie* attracted him much less, because of its, again, formalistic approach. Kuhn seems to be pursuing, already at this point in his life, a more naturalized vision of scientific thought. Of the two remaining books by Carnap, he offers no assessments. Reichenbach's *Experience and Prediction*, however, with its critique of positivism and its expository clarity, seems to have been a more attractive book to him. As we shall see in Chap. 7, he also read *The Rise of Scientific Philosophy* at some point, though we cannot be sure of the date. He did not like it very much, though he wrote many notes on the book. Reichenbach represented the kind of logico-empiricist philosophy of science that privileged rational reconstruction (an expression that Kuhn nevertheless considered applicable to his own approach, once its meaning was appropriately adapted), and that excluded the historical, psychological, and sociological factors of interest that Kuhn considered essential. However, he at least took some time to put his critical thoughts on paper.[203]

Kuhn read other texts by authors associated with this current. In this respect, *Introduction to Logic* by Alfred Tarski and *Linguistic Aspects of Science* by Leonard Bloomfield left no mark on him, while for the work of J. H. Woodger, *The Technique of Theory Construction*, he showed only *some* interest.[204] The first clear sign of attention towards a logico-positivistic work is the one he exhibits towards Alfred J. Ayer's classic primer, *Language, Truth and Logic*.[205] Kuhn's interest (and criticism) on Ayer's presentation while in the Society of Fellows is worth examining a little more carefully. This is what we are going to do in the next section.

Finally, it is worth mentioning two critics of this current: Karl Popper and Ludwig Wittgenstein. Regarding the former, at this time, rather than reading his work, Kuhn attended the William James Lectures that Popper gave at Harvard, *The Study of Nature and of Society*, from February 16th to April 27th, 1950.[206] The interaction between them before their famous meeting in London in 1965 does not seem to have been very productive. Kuhn did not agree with the way Popper talked about the progress of theories. "Popper," said Kuhn, "was constantly talking about how the later theories *embrace* the earlier theories, and I thought that was not just going to

[202] See, in particular, our Chap. 7.

[203] Again, see our Chap. 7, especially Sect. 7.3, for further details on Kuhn's criticism of Reichenbach.

[204] See *Handy Notebook 1949*, 31 March 1949, 1. The comments on Woodger are found in TSKP 8–9.

[205] See *Handy Notebook 1949*, 29–31 March 1949, 1–8.

[206] See Popper, *The Study of Nature and of Society*, William James Lectures, 16 February- 27 April 1950, Sir Karl Raimund Popper Papers, 1928–1987, Hoover Institution Archives, Stanford University, Stanford (Cal.), box 39, folder 3. See *Essential Tension*, 267, fn. 2.

work out quite that way."[207] So, aside from suggesting that he read Émile Meyerson's work and advising him, their relationship was not close.[208]

Regarding Wittgenstein, Kuhn acknowledged in an interview with D. G. Cedarbaum that he had read a draft version of *The Blue and Brown Books*.[209] As we will see in Chap. 3, traces of this reading can be seen in his Lowell Lectures. The influence of Wittgenstein's *Philosophical Investigations* and, later, of *On Certainty* would become even more serious for Kuhn's philosophical research, both for *Structure* and for later works. In subsequent chapters, I will deal with such influences and their origin in detail, so I won't discuss them further here. Suffice it to say that Kuhn's readings of Wittgenstein are as significant for him as those of Quine's papers and *Word and Object*—which were a main influence in his thinking. But Wittgenstein's influence is clearer when observed in its proper contexts.[210]

As indicated at the beginning of this section, two other influences from outside philosophy were of primary significance for Kuhn. One of them was Ludwik Fleck's little-known book, *Entstehung und Entwicklung einer wissenschaftlichen Tatsache: Einführung in die Lehre vom Denkstill und Denkkollektiv*, published in 1935. If Conant had showed Kuhn a pattern for the dynamical interaction of the experimental and the theoretical, or conceptual, work, Fleck would add further ingredients to that pattern and to Kuhn's conceptual baggage, and the latter reused some of this Polish physician's terms and notions in his own presentation, first in the Lowell Lectures and then in his own drafts for the early, though never published, Chapter I of *Structure*.[211] Then, as is well known, he mentions it in his Preface to *Structure* as "an essay that anticipates many of my own ideas."[212] Closer to the moment in which he read it first, Kuhn wrote in a card for his own records that Fleck's monograph was "[t]he best and most important methodological work of this century, so far as I know." He added that Fleck's historical research was "interlarded with methodological generalizations which almost precisely parallel my own thoughts derived from similar research in physical science." Kuhn praised the work highly in the sociological dimension, but he also saw some "gross overemphasis on the vaguest generalities" in that field of science.[213] The book, he said years later when he wrote the foreword for its English

207 Interview 1995, 286.

208 See Interview 1995, 286–287; Borradori (1994, 159), where Kuhn states something very similar. For the difficulties of the relationship between Kuhn and Popper, see Heilbron (1998, 510). About Popper's memories of his encounter with Kuhn at that time, see Popper (1983, xxxii). See also the last sections of our Chap. 6.

209 See Cedarbaum (1983, 188); Wittgenstein (1958).

210 Chaps. 3, 7–8, shall provide such contexts.

211 See our Chaps. 3, 5 for further details. See, of course, Fleck (1935) for the original book and Fleck (1979) for its English translation. I shall return to this latter in Chap. 8.

212 *Structure*, xli.

213 [Kuhn], "Fleck, Ludwik: Entstehung und Entwicklung einer wissenschaftlichen Tatsache: Einführung in die Lehre vom Denkstill und Denkkollektiv, Benno Schwabe & Co., Basel, 1935," card, n. d., TSKP 7, Sec. "Philos. of Sci.".

version, and in his 1995 interview, helped him to check that there was at least a precedent for his views and that he was reassured by it.[214]

The second influence that needs to be mentioned in this group was Benjamin Lee Whorf, the comparative linguist, whose work on linguistic relativity is also key for Kuhn. It also played a leading role as a source of some of his ideas in the Lowell Lectures and, of course, later.[215] He read the papers collected in *Four Articles on Metalinguistics*, from 1950, including "Science and Linguistics," "Linguistics as an Exact Science," "Languages and Logic," and "The Relation of Habitual Thought and Behavior to Language."[216] The notion of "thought world" in such writings[217] and its relationship with language and behavior are very similar to the ideas about parallel notions in Kuhn's Lowell Lectures, as we shall see in the next chapter.

Let us now return to Kuhn's readings in those days and to the theses he developed from his reflection on them. Our first stop on this more detailed journey is, however, pretty typical for a philosopher of science at that time: it involves a return to the basics of the logico-positivistic view of science. His critical attitude is already present there, though. Let us see what Kuhn had to say about A. J. Ayer.

2.11 Kuhn and the Empiricist Criterion of Meaning

In *Language, Truth, and Logic*, Ayer provided a primer to the logico-positivistic anti-metaphysical attitude and tried to explain and clarify what one of its main theoretical tools for extricating language from meaningless expressions was: the *empiricist criterion of meaning*. The criterion had gone through some changes, and Ayer himself had contributed some of them. If a would-be philosopher like the still young Kuhn aimed to offer a critical treatment of current philosophy of science, logical positivism was an unequivocal starting point, and Ayer's book, whose second edition had been published a few years before, in 1946, was a popular primer.[218]

Kuhn's main perspective was critical, which should not surprise us, but perhaps more conservative than we might expect. According to him, we should take into account that, among the propositions to be saved by this criterion, we find those whose relationship with the facts (that is, whose satisfaction of a verification criterion) would

[214] See Kuhn (1979c, viii); Interview 1995, 283. See also Oberheim (2005, 367), who emphasizes that the notion of incommensurability, as applied to two historically separated concepts, is visible in Fleck, too (see Fleck 1979, 62).

[215] See, again, our Chap. 3.

[216] See Whorf (1950); and see TSKP 8, for the corresponding note card. I am indebted to George Reisch for showing me that source. They were later reprinted with more of his papers in Whorf (1956).

[217] See Whorf (1956, 147–150).

[218] See Ayer (1946), and Hempel (1965, Chap. 4) for a thorough study of such changes in the criterion. See also Friedman (1999, 3, fn. 3, 5, fn. 9) for a criticism of the version involved in that (i.e. Ayer's) and other views on that school of philosophy that, as Friedman remarks there (esp. 3, fn. 3, on the basis of Rorty's views), influenced Kuhn's reading.

be uncertain, but through which we obtain others that are unequivocally verifiable. These latter have a closer relationship with experience, while the first ones (I will phrase this according to the spirit of his notes) describe facts in a way that is at best "indirectly verifiable" (Ayer's expression).[219] They produce a benefit for him, though: that the uniformity of the world image that the whole set of propositions helps to build is not upset or broken. Psychological and logical consistency is the foundation for the subsistence of such propositions. Thus, although he seems to suggest that this is a mere modification of the empiricist criterion of meaning that Ayer provides, his suggestion seems more akin to a holistic viewpoint about language, perhaps halfway between the (not very distant in time, but still not present) stance of Quine and a contextualist conception of meaning such as the one that Hempel expounds.[220]

Kuhn adds in these brief notes some ideas in favor of his position. Statements of a metaphysical nature, even the religious ones, he says, are not clearly different from certain statements of scientific law. We can keep them as long as they respect the aforementioned new version of the criterion that he suggests—that is, the one based on psychological and logical coherence, which ultimately refers to logical relationships in our language and in our view of the world.[221] He thus says that every statement with specific predictive charge, such as, for example, the position of a celestial body for a certain date, if verified by the observations, is verified only within a certain margin of error. For him, what we mainly develop with verification is *reliability* of the proposition for a future use. The verification of a statement makes the propositions that have allowed us to obtain it reliable. Reliability is of course *subjective*; we lean towards such propositions even emotionally, he says.[222] In fact, he seems to be trying to say here that we turn them into principles of action for science.[223] Scientific laws or statements thus treasured have the same nature as beliefs in reason or freedom as he saw them in "The War and My Crisis." They are reliable for the future, not because we check their direct correspondence with reality, but because we will not hesitate to deploy our future behavior around them. Later on, in *Structure*, his revision of the idea of rule and of its relationship with the notion of scientific law echoes this perspective.

In these notes, Kuhn seems to get close to the thesis of the *underdetermination of theories by evidence*—or at least to a very initial conception of that thesis. He considers that we can obtain the same verifiable statements from two more general propositions (which are only indirectly confirmable) that lead to divergent views. He realizes here that the meaning of a proposition goes well beyond its strict relations

---

[219] See Ayer (1946, 13).

[220] *Handy Notebook 1949*, 31 March, 3–6.

[221] *Handy Notebook 1949*, 31 March, 5.

[222] *Handy Notebook 1949*, 31 March, 4–6.

[223] See *Handy Notebook 1949*, 31 March, 6. Kuhn says there that we do this with propositions from which we do not deduce verifiable or confirmable propositions, such as those of a *moral* nature. He is also aware there that pragmatism and positivism treat moral propositions differently: the first one does not exclude them from meaningful expressions, while the second one excludes them from that group. He wishes to find the difference between meaningless and meaningful expressions, but without leaving aside the group of moral propositions, which for him are not semantically empty.

with experience, which is again a hint of holistic thinking like the one mentioned before.[224]

Kuhn's notes on Ayer and the empiricist criterion of meaning show the direction that he seems willing to take as regards the logico-empiricist doctrines concerning meaning. Logical relationships may at first lead the way in an initial approach to methodological matters, but the considerations regarding holism, or about the uniformity of the global view of nature, on which other uses of language—for example, the predictive ones—are based, showed him another pathway that, rather than being ignored as the logico-empiricist doctrine used to defend, must be explored in detail. The new pathway involves a better acquaintance with psychology, with the scientific method in particular disciplines—especially the social sciences—and with the evidence from the history of science. All this prompted Kuhn's response in the form of an approach to other disciplines such as the ones just mentioned.

2.12 The Influence of Jean Piaget: The Genesis of the Two-Worlds Perspective

One of the main authors in which Kuhn was interested during this period was the psychologist Jean Piaget. He read books and papers from the psychologists of the Gestalt school, and a book by Heinz Werner. He noted his reflections on the latter's book, *Comparative Psychology of Mental Development* in his notebook, in which he also uses the term "Gestalt."[225] We shall see in our next chapter that the term "behavioral world" is present in the Lowell Lectures in 1951,[226] and that it is also visible in Kurt Koffka's *Principles of Gestalt Psychology*, a book that, together with a paper by Wolfgang Köhler and Richard Held, count among his registered readings in psychology.[227] It is difficult to say exactly when he read these works, just as it is with other readings, for example of the works of the Hanover Institute, the Dartmouth Eye Institute, or of Jerome Bruner and associates.[228] Piaget's influence on Kuhn is

[224] See Footnote 221.

[225] See *Handy Notebook 1949*, 15 June, 22–24 on Werner, and 22 June, 27, for the use of "Gestalt.".

[226] See Lowell 6, 21 ff.; Koffka (1935, 31).

[227] [Kuhn], "Koffka, K: Principles of Gestalt Psychology, Harcourt Brace, New York, 1935," n. d.; "Köhler, W. & Held, Richard: 'The Cortical Correlate of Pattern Vision', Science, Oct. 21, 1949," n. d., both in TSKP 7, Section "Gen. Psych." On Kuhn's readings in psychology, see also Kaiser (2016, 76–82).

[228] On these latter, see the following cards: [Kuhn], "Ames, Adelbert, Jr: 'Nature and Origin of Perception' unpublished Lab manual prepared to accompany the demonstrations of the Darmouth Eye Institute," n. d.; "Hastorf, Albert H: 'The Influence of Suggestion on the Relationship Between Stimulus Size and Perceived Distance.' Jour. of Psych., 1950, *29*, 195–217," n. d.; "Bruner, J. S. & Postman, L: 'On the Perception of Incongruity[: A Paradigm]' Journal of Personality, vol 18[,] 2, (Dec. 1949) 206ff," n. d.; "Bruner, Jerome S., Leo Postman and John Rodrigues, 'Expectation and the Perception of Color', The American Journal of Psychology, Vol. 64, No. 2, 216–227, Apr. 1951," n. d.; "Bruner, 'Personality Dynamics and the Process of Perceiving', Chap. 5 in Perception—An

also clear, but, in this case, the moment in which he read and reflected on this psychologist's work is, just like in Werner's case, very easy to observe and date. Through Piaget's work, Kuhn managed to develop (more clearly in the Lowell Lectures) the ontological foundations for his epistemological vision, which was necessary for his perspective on science and its method. He also managed to give these foundations a dynamic and even a genetic character, also based on Piaget.

Kuhn became interested in Piaget's idea of the evolution of cognitive stages through the improvement of their functional character. According to the Swiss psychologist, such cognitive states enjoyed autonomy when categorizing incoming information. In Piaget, he found an empirically based theory with which he could give theoretical shape to his impressions about autonomy and change in the psychological and conceptual world. How did Kuhn come across Piaget? Kuhn's *Handy Notebook* again gives us some clues.

In his 1995 interview, Kuhn said that he had started reading Piaget's texts after stumbling upon a mention of the English version of *Le Jugement et le raisonnement chez l'enfant* (i.e., *Judgment and Reasoning in the Child*) in "Science, Technology and Society in Seventeenth Century England" by Merton (reading date, May 13th, 1949).[229] His notebook accurately backs up this version. Between Merton's text and his reading of *Judgment and Reasoning in the Child* there are seven days. Between them, Kuhn records *The Seventeenth Century Background* by Basil Willey (reading date, May 15th) and everything seems to indicate that, at least on the dates recorded in the *Handy Notebook*, he did not finish reading this latter book.[230] Four days later, he notes the complete reading of Piaget's *Les Notions de mouvement et de vitesse chez l'enfant*, in this case the original French.[231] This book and *Le Causalité physique chez l'enfant* (again, in its English translation, *The Child's Conception of Causality*) are the texts by Piaget that Kuhn would cite more often in the future.[232]

With Piaget, Kuhn found empirical support to set up the proper division between two worlds that he had already seen as relatively independent, despite the constant genetic interaction between them: the *psychological world* and the *real world*. Kuhn thought that Piaget helped him generate his own point of view.[233] Given that on June 14th, 1949, Kuhn notes basic details of his perspective, we can set this date as one of the reference points in the early exploration of a perspective that he would present in more detail a few years later in the Lowell Lectures, and which serves as an ontological basis for *Structure*.[234] This point of view consists of two realms or worlds.

Approach to Personality, by Blake & Ramsey, Ronald Press Co. New York, 1951," n. d.; "Bruner, Jerome et al.: A Study of Thinking (New York, 1956)," n. d.; all of them in TSKP 7, sec. "Gen Psych.".

229 Interview 1995, 279. See Merton (1938, 580); Piaget (1928).

230 Willey (1934).

231 *Handy Notebook 1949*, 9.

232 See *Structure*, xl (n. 2); *Essential Tension*, 21–23. See Piaget (1927; 1930).

233 See *Handy Notebook 1949*, 14 June, 10, 13–16.

234 See *Handy Notebook 1949*, 14 June, 10–18. See also Isaac (2012a, 216–217) and Galison (2016, 51–54), who give passages from Kuhn's diary and appropriate descriptions of them.

On one hand, there is a *physical* world. In it, we include the causal substrate of our perception, which we transform and classify through our cognitive categories and our languages and conceptual schemes. On the other hand, there is a *psychological* world, the approximately well-ordered and selective environment in which that causal substrate takes shape, our perception and language unfold, and with respect to which we deploy our behavior. The idea that the second classifies the first is already present here, so that we behave according to the components of the second. Novelty occurs nonetheless, and we sometimes perceive that there is something in the physical environment that is not taken into account in the psychological world. However, he also adds that, to deal with novelty, it must be perceived and categorized and that this involves the psychological, cognitive environment once again. In other words, it is in this latter world that we deploy our conscious way of life, so even disturbances must be taken into account and dealt with *in it*.[235]

It is inevitable that we observe a certain parallelism of this rough viewpoint with Kant's much more elaborate perspective, although Kuhn bases himself here on Piaget (which, anyway, does not exclude the previous parallelism). No wonder, then, that Hoyningen-Huene's magisterial reconstruction of Kuhn's perspective in the 1980s according to a similar two-worlds ontology had a Kantian basis and the resulting picture was very close to Kuhn's perspective as previously expounded.[236] It is also possible to allude to the Gestalt psychology—Kuhn actually uses the term "Gestalt" here—as another source of insight for him as regards some details of the two-worlds perspective. And we should not forget other nearby influences, like C. I. Lewis and his pragmatist epistemology, with its relationship between the given and its interpretation, which Kuhn probably already knew, and that is also similar in some points. Kuhn's vocabulary is also close to that local source of philosophical insight.[237] All of them can shed some light on the sources of Kuhn's perspective in that regard, and they are mutually compatible, at least at that initial, rough level of philosophical approach.

So, in Kuhn's view, cognitive development can be understood as an expansion of the subject's competences to face the world, the surrounding environment in which behavior is deployed. It is so for Piaget's children, and Kuhn considers that it is also an appropriate approach for understanding science and the scientist's mind. Just as in the infant stage, the early stages of a science are marked by juxtaposition and syncretism, Kuhn says. In these stages, many things are seen and sometimes used in a somewhat naïve and contradictory way. Generalizations are based on the most

[235] See *Handy Notebook 1949*, 14 June, 13–15.

[236] See Hoyningen-Huene (1993). See also Solís (1997). Friedman (2003, 27–34) also offers a Kantian reading of Kuhn.

[237] See *Handy Notebook 1949*, 14 June, 13, 15; 15 June, 19, 21. See also Mayoral (2009; 2017b) and our Chap. 1, especially Sect. 1.13.

immediate and simple aspects of sensory experience.[238] The overcoming of egocentrism in the beginnings of a science originates in the need to establish something similar to the new relationships of more logical order of which Piaget spoke.

To better understand the analogy that Kuhn establishes with Piaget, it is useful to first outline Kuhn's view of science at that moment, and to show its similarities and differences with child learning. For Kuhn, science pursues goals of functional classification of the information that reaches our senses, as I previously mentioned. An auxiliary goal in this classification is functionally efficient simplicity, which Kuhn considers (in comments from June 15 on the theory outlined on June 14) an *economical* version of the world, of a scientific nature.[239] Science contributes to transform the individual's initial, intuitive, psychological environment. It uses experiment (here the trace of Conant and Bridgman can be seen) to achieve this improvement and selection, which is mainly conceptual (or *intentional*, Kuhn qualifies), and above all seeks to increase either the internal coherence of the psychological environment in conceptual or logical terms, or its external consistency with the information coming from the outside.[240] Here, Kuhn is already considering the utility of *thought experiments*, which usual methodological theories do not take into account, but that he considers in some of the key figures of science, such as Galileo Galilei and Albert Einstein.[241] For Kuhn, the transformation of the initial psychological environment of the individual into more simplified, economical, coherent and empirically consistent versions is what a scientific theory truly seeks.

There is, of course, a marked difference between the processes of learning and the development of science. In 1995, Kuhn said that:

> I read a good deal [of Piaget], beginning with his *Mouvement et vitesse*. And I kept thinking, my, these children develop ideas just the way scientists do, except—and this was something I felt Piaget did not himself sufficiently understand, and I'm not sure that I realized it early—they are being taught, they are being socialized, this is not spontaneous learning, but learning what it is that is already in place. And that was important.[242]

The process by which the individual begins, from childhood, to make progress in his or her own psychological environment until the latter becomes a more functional, socially effective, version of its primitive form is not exactly what the scientist must do to make the theory, or the psychological and theoretical contexts that are established with it, advance in a similar way. The resources available to the researcher in a scientific discipline are more effective; Kuhn refers here to the experimental and formal resources.[243] In addition, as indicated in the passage we have just seen,

---

[238] *Handy Notebook 1949*, 14 June, 10–11. On Piaget's concepts, especially those mentioned in this section (e.g., egocentrism, juxtaposition, syncretism), see Piaget (1928, Chaps I and V), and Chapman (1988, 35 ff.).

[239] *Handy Notebook 1949*, 14 June, 15–16; 15 June, 19.

[240] See *Handy Notebook 1949*, 14 June, 10–11, 13–16; 15 June, 19.

[241] *Handy Notebook 1949*, 14 June, 17. On his classic paper on thought experiments (Kuhn 1964), see our Chap. 8.

[242] Interview 1995, 279.

[243] See *Handy Notebook 1949*, 16 June, 23. Kuhn notes this idea in his commentary on Heinz Werner's text in psychology (Werner 1948).

sometimes such a researcher starts out from a source that is not a pre-established social, conceptual, and linguistic environment.

There are three more ideas worth highlighting that appear in his comments on the outline of the theory that he prepared on June 14 (the comments are from the following day, June 15). First of all, for Kuhn there is already an attention to revolutionary scientific change, which is not strange after reading Conant. Kuhn's account of change is not so much based on the perception of a problem that is internal to a certain theoretical perspective, but on the *disposition* to perceive the change. He refers to a catchphrase that appears frequently in Conant's writings, and that we have already emphasized: "a scientific discovery must fit the times."[244] With this phrase, Kuhn tries to show that the perception of a theoretical problem—or, more generally, of a problem in the psychological environment in which the scientific theory is framed—involves an *expectation* of such a problem. In Chap. 3, I discuss how Kuhn deals with the hypothesis about the role of youth in scientific discovery, and we see that the basis of his views on that connection is the scientist's relationship with the established theory. For him, the perception of the problem, its acceptance and that of its solutions depend on a less close relationship with the theory or psychological environment that such solutions lead to correct, or even to abandon.[245]

Secondly, Kuhn considers that the most typical tools of traditional scientific method, such as its formal resources and its appeal to experiment, are part of what drives change in science. However, he also indicates that the importance of psychological coherence, which refers to what seems reasonable or not to us, is often downplayed, and that the search for the latter plays a much more central role in theoretical advancement than has traditionally been granted.[246] Finally, he introduces a distinction between the natural and the social sciences that is not based on methodological issues, but on a certain difficulty on the part of the latter to perceive their own internal inconsistencies.[247] He would make further comments on the method in the social sciences later in his career at different moments, some of them very significant, for instance while preparing some key arguments for *Structure*, in which the contrast between both groups of sciences became important. Although he would never pay as much attention to the social sciences as to the natural sciences, he discussed the former group every so often and in different contexts.[248]

Towards the end of June 1949, however—more specifically on June 20, that is, a few days after these annotations on Piaget—Kuhn would take a few notes on the methodology of social sciences as a result of a few readings of Max Weber's work in that field. These readings enriched his vision of these disciplines and their method. In turn, they also increase our understanding of the ideas that he developed in these first months in the Society. In particular, those notes help us understand some aspects

[244] Conant (1947a, 83; see also 74).

[245] See *Handy Notebook 1949*, 16 June, 20–21.

[246] *Handy Notebook 1949*, 16 June, 22.

[247] *Handy Notebook 1949*, 16 June, 18–19.

[248] See our Chaps. 5 and 7 in particular.

of his own vision on the natural sciences. That is why I shall turn to them in the next section.

2.13 A Kindred Spirit: Kuhn on Weber

Kuhn considered his correction of the empiricist methodology as a "transition to subjectivity in the evaluation of physics."[249] In this project, he found new parallels in other methodological works, such as that of Weber. The German sociologist was well known at Harvard—especially in the environment of the Society. Talcott Parsons had promoted his work upon returning from the University of Heidelberg in the late 1920s after obtaining a doctorate and a good knowledge of Weber's work; thus, he translated *Die protestantische Ethik und der Geist des Kapitalismus* in 1930, and dedicated the third part of his *Structure of Social Action* to his work.[250] Merton was also one of the authors who, in "Science Technology, and Society in Seventeenth Century England," cited the relationship that Weber had established between puritanism and science.[251] It is difficult to say how Weber's work ended up attracting Kuhn's attention. Yet we know the book by Weber that Kuhn read around July 20, 1949, and that he commented on in his notebook: *Max Weber on the Methodology of the Social Sciences*, which had been published that same year.[252] The context of the Society of Fellows is, again, a factor to take into account here.

What parallels did Kuhn see between Weber's methodological work and his own program? Let's look at an excerpt from Weber's book that appears on the same page (p. 115), which Kuhn cites to support the parallelism. At the beginning of "Critical Studies in the Logic of Cultural Sciences," Weber says:

> [... H]itherto the attempt has usually been made to define the nature of the social sciences by distinguishing them from the "natural sciences." In this procedure there is always the tacit assumption that history is a discipline which devotes itself exclusively to the collection of materials, or if not that, is a purely descriptive discipline which in fortunate cases drags in "facts" which serve as the building materials for the intellectual work which "really" begins only after the historical work has been done. And what is more, even the professional historians, unfortunately, have contributed not a little to a strengthening of the prejudice that "historical work" is something qualitatively different from "scientific work" because "concepts" and "rules" are of no concern to history.[253]

As Kuhn writes in his notebook, the "transition to subjectivity in the evaluation of physics" that he puts forward is comparable to the one that Weber defends and that goes from that conception of history as a "collection of materials" to his own methodological point of view.[254] Let me first briefly examine those aspects of Weber's

[249] *Handy Notebook 1949*, 21 June, 26.

[250] See Treviño (2001); Weber (1930).

[251] See Merton (1938, 418 ff.); references to Weber are abundant.

[252] Shils and Finch (eds.) (1949).

[253] Shils and Finch (eds.) (1949, 114–115).

[254] *Handy Notebook 1949*, 21 June, 26–27.

perspective that will later help us understand the ideas that Kuhn finds in common between them.

For Weber, in history and in the social sciences in general, the empirical basis is built on the meaningful behavior of individuals. His stance, the well-known *methodological individualism*, resorts, as Kuhn says, to the subject's point of view to find the kind of concepts with which to construct his explanation of social phenomena—in this case, of events and behaviors. As Weber said in 1921

> Sociology (in the sense in which this highly ambiguous word is used here) is a science concerning itself with the interpretive understanding of social action and thereby with a causal explanation of its course and consequences. We shall speak of "action" insofar as the acting individual attaches a subjective meaning to his behavior—be it overt or covert, omission or acquiescence.[255]

To accomplish the interpretive task, we put ourselves in the place of the subject—we try to *understand* the basis of his or her actions. Since all behavior in a social sense is the interaction of an individual's action with the action of others, the setting that the social explanation tries to cover is that of the interaction of a social nature.[256]

Now, how do we obtain such an explanation? Weber makes use of the concept of *ideal type*, a core notion of his vision of the method in the social sciences. As noted in the extract above, Weber advocates for the search of understanding of *subjectively meaningful* individual action. Ideal types constitute internally consistent sets of beliefs, ideas, and concepts that mark a specific form of behavior in society and that, accordingly, contribute to the explanation of social behavior. They are central pieces in the development of starting hypotheses that help to achieve that end. It must be emphasized, as Weber tries to make clear, that those pieces are *idealized*. In his own characterization, such an ideal type would be a "utopia."[257] As in other idealizations, we learn by trying to verify the consequences of the hypotheses that had been posited. We assess the applicability of the concepts used in the elaboration of the model. Its purpose, therefore, is to explain particular events by virtue of ideally typified behaviors, something that Kuhn wouldn't find alien to the field of natural science from his own experience of it. As Weber mentioned in one of the articles read by Kuhn ("'Objectivity' in Social Science and Social Policy"), "The goal of ideal–typical concept-construction is always to make clearly explicit not the class or average character but rather the unique individual character of cultural phenomena."[258]

Finally, we should emphasize that the selection of the components of such explanations, the ideal types, also involves another kind of subjective view—that of the social researcher. It is from this view that we select the elements that constitute our abstraction. Such elements, indeed, are for Weber, value-related, as he explains in the same text that I have just quoted:

---

[255] Weber (1978, 4; 1988, 542).

[256] On Weber's methodological individualism, see Gordon (1991, 470–471).

[257] Shils and Finch (eds.) (1949, 90).

[258] Shils and Finch (eds.) (1949, 101; on the perspective offered here, see also 89–101).

> The concept of culture is a *value-concept*. Empirical reality becomes "culture" to us because and insofar as we relate it to value ideas. It includes those segments and only those segments of reality which have become significant to us because of this value-relevance. Only a small portion of existing concrete reality is colored by our value-conditioned interest and it alone is significant to us. It is significant because it reveals relationships which are important to us due to their connection with our values. Only because and to the extent that this is the case is it worthwhile for us to know it in its individual features.[259]

This, however, does not involve a diminished amount of objectivity. Weber says that: "In the *method* of investigation, the guiding 'point of view' is of great importance for the *construction* of the conceptual scheme which will be used in the investigation"; then, however, the result of the ideal configuration can only be evaluated like any other hypothesis, "[f]or scientific truth is precisely what is *valid* for all who *seek* the truth," Weber adds.[260]

The points of coincidence with the Kuhnian vision now and later are quite evident to anyone familiar with *Structure* and, above all, with Kuhn's later texts.[261] In the mature stages of his thought, Kuhn would make it clear that science requires an idea of truth that, without depending on the more traditional correspondence theory, must fulfill a function of discrimination like the one we usually concede in everyday life and in science. He says so despite the fact that he continues to defend a relativistic point of view for languages and conceptual systems. In this defense, he does not move away from Weber. In other aspects of his early thinking, which we will examine shortly, he also agrees with the German sociologist.

Kuhn mentioned, in his notes, passages from the three essays collected in the Weber volume. His reading of the volume was complete, and its impact was also notable. In these notes, he considered that Weber showed ideas about the methodology of science that were very similar to his own. He began his comments on Weber by praising one of his ideas concerning the function of scientific laws; referring to these, Weber said: "The establishment of such regularities is not the *end* but rather the *means* of knowledge."[262] This perspective certainly fits very well with Kuhn's views. He does not see science as the *result* of an investigation, but as the *vehicle* that allows us to better understand nature through an exhaustive exploration of the theoretical apparatus, of that tool of representation of (and also of practical interaction with) nature that it produces. He was aware from, at least, 1945 that every physical theory is based on a modal approach to nature that presupposes conditions of natural uniformity that are intentionally, subjectively introduced, on whose basis predictions are propounded, and from whose testing we obtain keys for further inquiry. In these notes, Kuhn underlines that even the kind of statements that an empiricist would discard plays a role as a cause of changes in the scientific view of the world.[263]

---

[259] Shils and Finch (eds.) (1949, 76).

[260] The last two excerpts from Weber come from Shils and Finch (eds.) (1949, 84).

[261] Mladenovic (2017, 43–53) interprets Kuhn's *Structure* in terms of its parallels with Weber's ideal types.

[262] Shils and Finch (eds.) (1949, 80). See *Handy Notebook 1949*, 21 June, 24–25.

[263] *Handy Notebook 1949*, 22 June, 27.

Indeed, this latter option is far from being what the empiricists take into account; even someone like Bridgman. For Bridgman, revolutionary change is a failure of scientists to find the correct combination of concepts; the product, above all, of a lack of operational caution when acting.[264] Kuhn does not see it as an eliminable evil, but as the product of the natural course of scientific research—understanding historical development from a Piaget-inspired point of view. To see that kind of change this way, we should understand that real research work with scientific theories is not an enigmatic set of activities later tamed by the justification procedure, but the only context in which all the relevant procedures for science—normativity included—take place, so we must examine and reconstruct *real* science. And real, historically observable science is an instrument, a research vehicle. Perhaps Kuhn would not have agreed with Dewey in general, but Dewey would have agreed with him, at least as regards this idea—scientific knowledge is not found in books, but in the action of researchers. And their theories and laws are a vehicle for this, not the ultimate goal, which is a real control of nature in practical and epistemic terms—its understanding. In Weber, Kuhn finds, in this way, a kindred spirit, at least as regards their vision of their respective sciences.

Kuhn shows in these notes his confidence in Weber's notion of ideal type, to the point of suggesting that the main field that Weber had chosen for this methodological concept, the social sciences, is not the only one appropriate for it: the field of the physical sciences, which Kuhn knows so well, is equally suitable. For him, the scientific description of the physical world employs a similar series of assumptions, definitions, and generalizations that are not exclusively obtained from experience, but also from what we take the nature we experience to be. So, the proximity of that scientific description to the Weberian perspective based on the notion of ideal type is greater than we might have anticipated. The discrepancies in the confirmation of hypotheses in each field should not be seen, Kuhn shows, as the product of a completely divergent theoretical architecture, but can be accounted for on the basis of the particularities of the domain of phenomena that is explored in each case.[265] Kuhn, in this way, embraces Weber's point of view in that respect without hesitation. He even takes Weber's point of view (which he partly assumes as identical to his own) beyond the latter's area—i.e., to the natural sciences. As he would recall years later, in 1989:

> Forty years ago, when I first began to develop heterodox ideas about the nature of natural science, especially physical science, I came upon a few pieces of the Continental literature on the methodology of social science. In particular, if memory serves, I read a couple of Max Weber's methodological essays, then recently translated by Talcott Parsons and Edward Shils, as well as some relevant chapters from Ernst Cassirer's *Essay on Man*. What I found in them thrilled and encouraged me. These eminent authors were describing the social sciences in ways that closely paralleled the sort of description I hoped to provide for the physical sciences. Perhaps I really was onto something worthwhile.
>
> My euphoria was, however, regularly damped by the closing paragraphs of these discussions, which reminded readers that their analyses applied only to the *Geisteswissenschaften*, the

[264] See Bridgman (1927, 6–7).

[265] *Handy Notebook 1949*, 21 June, 25–26.

> social sciences. "*Die Naturwissenschaften*," their authors loudly proclaimed, "*sind ganz anders*" ("The natural sciences are entirely different"). What then followed was a relatively standard, quasi-positivist, empiricist account of natural science, just the image that I hoped to set aside.[266]

From *An Essay on Man* by Ernst Cassirer, Kuhn noted on a private reference card precisely what he said in 1989 about both (Cassirer and Weber): that the distinction between natural and social sciences was, in his opinion, unjustified.[267] Kuhn seems eager to bring social sciences back to the space already occupied by the natural sciences, physics in particular, as a model of appropriate methodological behavior, although traditional visions of the scientific method—and we can think here, for example, of Hempel's and Popper's views—contributed to distancing them from it.[268] However, we should not infer that this position involves a defense of the method in the social sciences on his part. It is rather a critique of the empiricist vision of the methodology of the natural sciences. We shall examine the comparison that Kuhn made from 1958 onwards between both fields, and at that point the natural sciences, and specifically physics, would exemplify again a (now social) pattern of learning and scientific research that he considered mature and more suitable for scientific progress. We shall return to this point in Chap. 5.

2.14 Publication Plans

Kuhn's notebook ends on July 6, 1949. I have referred mainly to those of his readings that are relevant to methodological issues, but he also dealt with other things during this period. His readings on the history of science attest to this. For example, he noted having read Martha Ornstein's *The Role of Scientific Societies in the Seventeenth Century* on April 30 and Basil Willey's *The Seventeenth-Century Background* on May 15.[269] Of course, the examples that he displayed two years later in the Lowell Lectures about Aristotle, the Scholastics, Galileo Galilei, Robert Boyle, Isaac Newton or Antoine-Laurent de Lavoisier, and his plans to write a book that would rely on them, also show that he began to cultivate this part of his new profession at this time. Something similar happens with logic. Quine seems to have helped him, and this is easy to verify by the connections between their respective thoughts, but Kuhn did not gather information in this regard, apart from a few brief notes on reading cards.[270]

---

[266] Kuhn (1991b, 216–217).

[267] [Kuhn], "Cassirer, E: An Essay on Man, Yale UP, New Haven, 1948," n. d., TSKP 7, Sec. "Gen. Philos.".

[268] Consider Hempel (1942) or Popper (1994).

[269] Conant included them in his bibliographies; see Conant (1947a, 119) for Ornstein; Conant (1948b, 63) for both; Kuhn, "Natural Sciences 11(a)," 24 September 24–28 November, 1947, TSKP 1.11, 17 (class of 10 October 1947). See *Handy Notebook 1949*, 30 April and 15 May, 9. See Willey (1934) and Ornstein (1928).

[270] See [Kuhn], "Advanced Studies and Research," n. d., TSKP 25.53, 2.

During the rest of 1949, Kuhn continued to develop his point of view, and planned his following steps in order to correct defects and better establish his arguments. He also seriously thought about future publication possibilities. Before the end of 1949 we already have three consecutive plans to publish his theory in book form, as well as the titles of some future articles. On July 5, 1949, still in his notebook, Kuhn shows us an index of what was initially going to be called *The Process of Physical Science*:

Part I: Language and Logic: The Tools of Thought:

1. Language

2. Logic and Math[ematics]

3. Non Formal $^{\text{(Linear)}}$ Linguistic Modes

Part II: The Scientific Function

1. The Physical Real World and The Psych[ological] Real World

2. The Problem for Science

3. The Emergence of Explicit Tools

Part III Science at Work

Examples from the history of science

Appendix: Rel[ations] to Other Sciences Including Social.[271]

The next day, Kuhn changed his mind a little about the organization. He considered it easier to include sections I.2 and II.1 in a first part, along with a third section in which he would show how to apply what he had described before to science. A second part would include II.3, Part III and the Appendix. Most probably, the remaining, unmentioned sections would be included in the new ones. In any case, these are work schemes that in the end, were not developed as presented here.[272]

We still find one more scheme, a summary of contents, among his notes from 1949.[273] In that summary, there is a notable alteration as compared to the two lists that we have just seen. In this one, it is possible to see an organization that is very similar to that on which his Lowell Lectures of 1951 will be based. It seems to have been prepared before receiving such an invitation, something that, as we will see in the next chapter, would take place in March 1950. In previous lists of contents, his main focus was on the elements seen in the previous section: the physical and psychological worlds, the relationship between them, and the nature of science in relation to the new emphasis on the psychology of the subject. He wished to talk about his newly discovered subjective generation and correction of this world of mental images and concepts, about the tools used to pursuit such goals, and especially about the logic and experimentation (real and mental) used in scientific thinking in the search for psychological coherence. He wanted to add historical examples to the text, and also to show the common methodology that he envisaged for the social

---

[271] *Handy Notebook 1949*, 5 July, 31. I have modified the text by converting uppercase to lowercase letters and completing some minor abbreviations. Punctuation has also been unified. "(Linear)" is located over "Formal" in its entirety in the original. The scheme I partially reproduce a little further down has been modified in this way, too.

[272] *Handy Notebook 1949*, 6 July, 32.

[273] TSKP 1.6 ("Incomplete Memos and Ideas, 1949").

and the natural sciences, based on his theory. In the new scheme, however, history has come to the forefront. He introduces the reader to the problems for philosophy of science from extended cases in the history of science about "The Foundations of Dynamics," from Aristotle to Newton, and then on atomism and on subtle fluids, as he will do in the first Lowell lectures right after the introductory one—just as in this case. He also wished to add a chapter on a "Modern Case (If possible)." He plans to do this in six chapters, the introduction included.[274] Then he plans to reconstruct the image of science that follows from such historical selection with his new perspective:

> Chapter VII—Science as a Process (The role of speculative elements; their successive redefinitions & reprecissions; the stages of science, explanation)
>
> Chapter VIII—Technics of Reorientation (the hitch, experiment & theory, logical untangling, etc.)
>
> Chapter IX—The Psychology of Reorientation (examples of parallel processes from psychology, ethnology, and math[ematical] problems, etc.)
>
> Chapter X—The role of formalism: Logic and Math[ematics]
>
> Chapter XI—The Role of Formalism: Language
>
> Chapter XII—Textbook Science and Creative Science (The limitations of logical reconstruction of texts; the normative fallacy; Positivism, etc. Relation of history and philosophy)
>
> Chapter XIII—Canons of Constructive Research (Relation of science to other sorts of work; the reconstruction of science as a cumulative discipline)[275]

A look at the Lowell Lectures in the next chapter shows the resemblance between the plan for this book and the list of such lectures given in 1951. Two chapter titles, "Canons of Constructive Research" and "The Role of Formalism," reappear then as the two last lectures; the dynamic perspective that refers to "Reorientation" and to its underlying "Psychology" is a key piece of the lectures' argument; the dynamic vision of "Science as a Process" with "stages" is central in them, too; and the difference between "Textbook Science and Creative Science" appears there as the central aspect of the introduction. In other words, the Lowell Lectures, as we shall study in the next chapter seem to adapt this plan, which seems to be one of the earliest versions of a list of contents for the book he envisioned in these years.

Alongside this typewritten list of contents, there are handwritten lines in which he lists four articles that were supposed to be written soon. One of them planned to expose the common points of Piaget and Anneliese Maier—who was another of his historiographic influences.[276] Two of them referred to the idea of textbook science and logical reconstruction, belonging to the last chapters of the book. Finally, he also mentioned a study on chemistry and atomism in the seventeenth century, a paper that ended up being his well-known 1952 article "Robert Boyle and Structural Chemistry in Seventeenth Century."[277]

---

[274] [Kuhn], "The Book," n. d. (ca. 1949), TSKP 1.6.

[275] See Footnote 274.

[276] See *Structure*, xl; Maier (1949).

[277] Kuhn (1952).

Kuhn took some notes on natural language, formalization and meaning on loose sheets, without a title, for a couple of days, whose dates were not recorded.[278] In these notes, he defends the natural state of ordinary language against the plans for its formalization—a thesis that would endure in his thinking for his entire lifetime. For him, natural language must preserve what he calls "fringe of vague meaning" if it is to remain the effective tool for cognitive adaptation and social cohesion that it already is.[279] This fringe also occurs in the domain of the sciences, but the formalization plans threaten to dispense with it.[280] The previous criticism of Ayer takes on a renewed momentum in the pages that we are going to examine in the next section. These pages help us glimpse the Kuhn that we then see in the Lowell Lectures in the next chapter.

2.15 The Adaptive Vagueness of Languages

For Kuhn, our language is built on past experience. Regarding this experience, we find two already known modes of talking about the meaning of the terms of our language. He introduces the traditional distinction between the extension of a term of our vocabulary and its intension, though he refers in this case to the distinction denotation/connotation, which is equivalent to the previous one. Meaning is expressed by reference to cases of previous application of the term or by a definition obtained through the application conditions, which involves the terms that are in intensional relation with it. However, as he will later do in the Lowell Lectures, in *Structure* (especially in Section V) and later on after *Structure*, he aims to go beyond this equating of meaning with either of these two methods. For him, we use them in a combined way because *in practice* (which is the situation that really interests him), meaning summarizes our previous experience in the use of a term and our predictions about its future uses, all of which depends in turn on the function of language in past and future behavior.[281]

And it is precisely in practice where, from Kuhn's perspective, we find the kind of language that needs to be preserved. Hence his criticism of the formalization of languages, especially of the scientific ones—which we have already seen in the previous criticism of Ayer and in his defense of Weber's point of view. To observe what kind of language Kuhn defends (he takes natural language as a model), we should first look at in what kind of practical scenario he needs it. In these brief notes, which summarize his starting points for his approach to science (his future book), he also deals with this last point.

---

[278] [Kuhn], [Memorandum 1949], n. d. (ca. 1949), TSKP 1.6. The notes from the first day extend over five and a half pages. On the sixth page a change of day is indicated and the notes continue until p. 9.

[279] [Kuhn], [Memorandum 1949], n. d. (ca. 1949), TSKP 1.6, 2.

[280] [Kuhn], [Memorandum 1949], n. d. (ca. 1949), TSKP 1.6, 8–9.

[281] See [Kuhn], [Memorandum 1949], n. d. (ca. 1949), TSKP 1.6, 2, 7, 8.

Our old experience of objects is of course not exhaustive; for Kuhn, we must count on new experience, and on the appearance of events and objects so far unexpected but plausible. Remember, for example, that in the psychological environment in which, in his view, our daily life or scientific research take place, there are novelties in perception that constitute a form of indirect relation with the causal substrate of our conceptual schemes. Such novelties force us to reform our psychological perspective of the world. So, the changes that education motivates, described by Piaget, where novelty and the requirements of a socialized use of language lead the individual from the egocentric stages to the logical ones, are another good example that Kuhn mentions here.[282] Here, he only focuses on the consequences of his new view for the meaning of terms, that is, on the linguistic counterpart of this previous idea about perception, experience and its conceptualization, which was an epistemological vision. What he tries to provide is a more appropriate vision of meaning that responds efficiently to the constant flow of experience and of novelties occurring in it. If we take as a starting point the numerous discrepancies between the experiences of objects that different individuals have, the meaning of a term is not reduced to its denotation nor to its connotation. Given these two ways of assigning meaning to a term, its relationship with objects and with the conditions of applicability (which do not form an exhaustive and stable definition) will remain as it is until the emergence of novelties that put that relationship in doubt. All the terms in a language respond to a coherent categorization of the world of experience, which is at times challenged by novelties and is thereby forced to change.[283]

In later chapters (especially in Chap. 8) we will have the opportunity to examine his lexical theory, which aims to explain the dynamics of scientific change in *Structure* in an alternative way. It is worth mentioning that the first version, still underdeveloped, of that mature theory has its origins in these dates, 1949–1951, as I shall repeat often in later chapters. Its most finished version for these earlier three years is in the Lowell Lectures of 1951, which we will see in the next chapter. Now, in 1949, we have the first manifestation of such ideas. As we will observe in Chap. 3, the idea that science is a complex set that consists of psychological, linguistic, and historical dimensions (he did not talk about its social dimension until *Structure*) can already be seen here, where the second of them, the one related to language, is highlighted and reflected upon.

In these reflections, the plan for formalizing language fares quite badly. The limitation that our past experience imposes, and the commitments we nevertheless acquire with it, make the most orderly representation of language, in which its connotative and denotative aspects are clear, not as effective a tool as our linguistic practice requires—especially with the prospect of future novelties in experience.[284] From the *pragmatic* perspective that he offers here on language, the classifications of objects

[282] See [Kuhn], [Memorandum 1949], n. d. (ca. 1949), TSKP 1.6, 4.

[283] See [Kuhn], [Memorandum 1949], n. d. (ca. 1949), TSKP 1.6, 1–2, for the contents of these notes summarized in this paragraph.

[284] For this critique of the formalization of language, see [Kuhn], [Memorandum 1949], n. d. (ca. 1949), TSKP 1.6, 4–6.

are established by virtue of a combination of methods based on the knowledge of the intension created around terms and, at the same time, on the perceptual contact with instances of their extension. He does not say anything very different (though it is better structured and more detailed) in "Second Thoughts on Paradigms," nor in the preparatory work for that paper.[285] Therefore, it is easy to see that some of the modifications and explanations of his Wittgenstein-inspired understanding of language are quite a few years earlier than *Structure*. But Kuhn adds something more to all this. Not only do we combine the methods of ascribing meaning by denotative and connotative means, but in our language we usually host a margin of uncertain meaning, a *fringe of vagueness*, in which there is room for those referents whose applicable terms can create doubts among users.[286] Later, in the Lowell Lectures, we are able to appreciate to what extent this "fringe of vague meaning" is important for him.[287]

This view of language is well illustrated in his third page of notes from 1949. There, he shows how language is constructed by the combination of methods mentioned above, and how the variations between its uses (which are motivated by previous experience) from one individual to another are more the rule than the exception:

> [...] since no two objects in the world are absolutely identical (or let's suppose so), we have an infinity of possible different sets [of] connotations for each box (supposing its membership fixed in advance) such that the choice of any one connotation will be sufficient to characterize every member of the box and to exclude the members of every other box. This is to say that for a finite experienced world, fixing the denotation of words does not at all fix their connotations. In practice, at least in a given society, our desire to make the connotations as simple as possible (for pragmatic reasons) leads to a considerable measure of agreement as to the connotations in actual use. (Of course no individual specifies his limited set of connotations. On the contrary he emphasizes some, expects others to follow as a matter of course, and is totally unconscious of others.) This "relative agreement" still permits sufficient leeway in the choice of different individuals (who we suppose would agree about past applications of the word in question), so that their fringes of vague meaning would display decided differences.[288]

This paragraph is nicely illustrative of the vision of meaning that Kuhn showed from that point onwards. We can express his ideas about the relationships between the mind and the world that we have been examining in this chapter by saying that, for him, our language responds to a mental representation of the world, which is an ordering of the incoming information (the one that reaches our senses) in the form of a specific presentation of experience and its objects. That representation is concise and economical, although it does not mean that it responds to everything that might be real beyond itself. Conceptual and linguistic change often depends on

---

285 See our Chaps. 6 and 7.

286 In Chap. 6, we shall see how Kuhn shall later (in the mid-1960s) revise his opinions about the use of a term like "vagueness" as applied to the meaning of terms in natural language; see also, Kuhn (1974, 316–318, n. 21) on that new attitude. At this moment, in 1949–1951, his opinions are more positive, though.

287 See [Kuhn], [Memorandum 1949], n. d. (ca. 1949), TSKP 1.6, 2.

288 [Kuhn], [Memorandum 1949], n. d. (ca. 1949), TSKP 1.6, 3.

the willingness to overcome that mismatch. The language that corresponds to that representation has part of the meaning of its terms fixed around some regular aspects of common experience, but there is room for vagueness around the meaning of such terms as well. This is an important factor of variation, as he says in the previous paragraph. The experience that each speaker has of the objects and events to which terms are applied is diverse, so there are margins of vagueness in the meaning of terms. Those margins can vary from one individual to the next, and they affect their behavior—linguistic or otherwise. In summary, in Kuhn's picture of language, there are certain fixed aspects of meaning and certain others, those that belong to the margin of vagueness, that are more or less variable.[289]

In the Lowell Lectures, especially in the seventh and eighth lectures, Kuhn develops this point of view on language. We can also appreciate, as here, his critique of the formalist project. Here he already advances his critique as follows: "[...] any attempt to logicize language along the lines now contemplated is doomed to failure—in so far as it expects to emerge with a language which could function instead of natural language. For such a language would have to remove the fringes which would involve a precise specification of the meaning of its symbols."[290] Note that attempts to formalize language in the terms suggested from logical empiricism and Bridgmanian operational analysis directly threaten the fringe of vague meaning. Kuhn is already very aware of this shortcoming and launches his attack on such plans as follows:

> If we wish to logicize a given vague word which has heretofor[e] been adequate for a natural language we have an infinity of choices of connotation which will do the job. Among these we may distinguish two extreme procedures. One, the "narrow" choice, takes a connotation which restricts denotation to those cases in which the word has already been used and to cases identical with these: that is, it restricts the possible membership of the box to its actual membership in historic time. It thus leaves vast regions of the world of possibles outside of the language process. The other, the "broad" choice, would preserve the actual content of each box, but would extend the possible content so as to exhaust the world of possibles. Then there are many intermediate cases. (Of course, the two above are theoretical limits—not practically available).[291]

For Kuhn, the individual's relationship with the perceived object and with the ways he or she applies a term to it are varied and are marked by behavioral, pragmatic conditions. Equating the meaning of a term to the previous experience that subjects have of the objects that have been cataloged and named by it, and to the beliefs that have allowed and will allow them to apply that term, does not exhaust, in Kuhn's view, all that they should qualify in it. Aside from individual variations (which already involve a considerable alteration), they must envisage the future application of terms, which depends on a previous experience that varies from individual to individual. All this is the content of the fringe of vague meaning, and it is a heterogeneous and somewhat uncertain content—experiences with an object are sometimes

[289] See [Kuhn], [Memorandum 1949], n. d. (ca. 1949), TSKP 1.6, 3, 7–8.

[290] [Kuhn], [Memorandum 1949], n. d. (ca. 1949), TSKP 1.6, 4.

[291] [Kuhn], [Memorandum 1949], n. d. (ca. 1949), TSKP 1.6, 7.

nothing more than passing relationships and not entirely recorded in memory, as noted. However, for Kuhn, if the meaning is circumscribed by the experience in the use of a word and, depending on this, by the forecasts of use in the future, then *all* that content of the fringe of the vague meaning *expresses the concept*, imprecise in its margins, enclosed under the term, on which a certain community of speakers depends for its future uses. Meaning is not just mental content or a linguistic relationship for Kuhn, but a certain historical and cultural heritage with a practical nature, with a behavioral focus. If, from a Deweyan point of view, knowledge depends on the practical context in which we use it, for Kuhn meaning can be interpreted in similar terms, too—which is not so far from Dewey's views, either.[292] In the next chapter, we will see that Kuhn tries to offer a bridge between both worlds, that of knowledge and that of language, that of beliefs and that of meaning, in the same pragmatic terms.[293] For all these reasons, it should be no surprise that, for him, neither of the two pro-formalist solutions mentioned in the last quoted passage are at all useful for making improvements in language. What we must find is a balance, but that is not new in our natural use of language. Kuhn says:

> The intermediate cases will be better, as compromises, but they can only be set up on the basis of our expectations of the future, and can only be judged on the basis of the actual correctness of these expectations. It appears that what we actually do is to leave the boundaries of our boxes fluid (we need not and probably ought not let them overlap in doing so); The hard denotative core is more or less common and the central connotative elements are more or less so (at least for one society—differences in central connotation for the same denotation corresponding to differences in culture will normally be followed by differences in behavior toward the same objects, etc.). But the "fringe" lying between the narrow and broad choices is characterized (metaphorically) by a weighting function, differing a good deal from individual to individual, which represents our expectation about the likelihood of an actual occur[r]ence in that part of the fringe. In this manner we achieve an eternal (but ever shifting) compromise[294] between the limitations of the two choices without ever making a particular choice. Our language remains capable of adapting itself to experience without discontinuity in its changes; any logicization would deprive it of this ability unless the whole world were given in advance of the particular logicization, and even then one would require a language of great complexity or an advanced knowledge of those aspects of the given world with which one were going to be called upon to deal, for even the world given in advance would have to be simplified for purposes of a language with which we could operate.[295]

Kuhn does not intend to say here, nor later, that we should completely dispense with formal tools, as they are useful and sometimes central, particularly in those languages (or precise aspects of them) with which there is no other way to work (formal languages), or which can benefit from their use. The logical clarification of the vocabulary of a formal language such as geometry, or of a language with empirical content, such as Newton's mechanics, are illustrative examples for Kuhn of this

---

[292] On Kuhn and Dewey, see Sect. 2.10, in this chapter.

[293] It will not be the last moment in which he discusses that bridge; see our Chap. 8 for more on it.

[294] In the original it reads: "an eternal (but ever shifting compromise).".

[295] [Kuhn], [Memorandum 1949], n. d. (ca. 1949), TSKP 1.6, 8. On the idea that a formalized language is only possible in a given world in advance, also 4–5.

constructive, positive function of the formalist program, which he addresses in the seventh Lowell lecture. Beyond that point, a general formalization of languages, the basis for the application of empiricist criteria of meaning, is a program that, for him, is excessive because it promotes the stagnation of a language and the suppression of its beneficial properties of adaptation to novelty in experience.[296] With a formalist plan, language loses the characteristics that make it the kind of vehicle for social communication and a medium of adaptation that has been so beneficial to our species.[297] I will talk about all this in the next chapter, when discussing Kuhn's Lowell Lectures.

[296] See Reisch (2019, 192) on the limited number of philosophers that would be willing to pursue such a project.

[297] See [Kuhn], [Memorandum 1949], n. d. (ca. 1949), TSKP 1.6, 5, 9.

Chapter 3
Kuhn's First Philosophy of Science: The Lowell Lectures

3.1 A New Invitation

On March 8, 1950, Ralph Lowell invited Kuhn to give a series of lectures during the following academic year at the Lowell Institute. Kuhn, of course, accepted. The courses at the institute had the prestige of having had Charles S. Peirce, William James, Bertrand Russell, and Alfred North Whitehead, among their lecturers. The Harvard fellows often gave them. However, when Kuhn participated, he stated that they were no longer what they once had been.[1] Lowell told Kuhn that he would give a couple of lectures a week. It was usual to offer a cycle of about six or eight of them. The specific period to do so would be up to Kuhn, but it needed to be between October 1950 and March 1951 in any case. Kuhn, as usual, did not received any remuneration, but it was a great opportunity to organize his own thoughts about science and present them publicly—and he worked hard to do it properly, he recalled in 1995.[2] He delayed the lectures as much as he could, and managed them so as not to affect his General Education classes.[3] His plans for the lectures, as he indicated to Lowell, were as follows:

> If agreeable to you, I should like to devote the series to the discussion of several aspects of scientific behavior, or method, as they may be abstracted from the history of scientific ideas. The lectures would probably divide almost evenly between illustrative historical material, and philosophical or methodological elucidation. The latter portion would be particularly concerned with a preliminary isolation of certain non-logical, perhaps even psychological, characteristics of creative research in physical science. The series might appropriately be titled "The Creation of Scientific Objects," though unless you require such information in the immediate future, I should like to leave both title and length of the series open.[4]

---

[1] Interview 1995, 289. About the Institute, see Croce (1995, 95–96).

[2] See Lowell to Kuhn, 8 March 1950, TSKP 3.10.

[3] See Kuhn to Lowell, 19 March 1950, TSKP 10.1.

[4] Kuhn to Lowell, 19 March 1950, TSKP 3.10.

J. V. Mayoral, *Thomas Kuhn*, Springer Biographies,
https://doi.org/10.1007/978-3-031-86512-1_3

In this series, Kuhn intended to show the genesis of scientific objects from the logical and psychological operations whose characteristics he had come to understand more deeply during the previous year. But, in general terms, we are only facing an elaborated version of the 1947 scheme, in which the fundamental elements of physical theory were a series of scientific objects that arose from the classification of essential features obtained from perceptual objects. Creativity was of the utmost importance in such a process.[5] In the lectures, he aimed to describe how that process used to take place on the basis of his newly acquired knowledge and perspective. He developed this against a background of historical cases that served as significant evidence for his philosophical arguments. Thanks to such historical cases, the parallelism that he established theoretically was also empirically likely. The examples exhibited historical evidence that required explanation—a better explanation than the one already available in empiricist philosophies of science. Analogies and ideas coming from contemporary perspectives on developmental and experimental psychology, comparative linguistics, philosophy of language, and epistemology were going to play an important role in his own explanation.

Kuhn finally changed the title of the cycle to the nowadays better known, "The Quest for Physical Theory: Problems in the Methodology of Scientific Research," although he did not allude at any time to the reasons for the change. The list of lectures resembles the list of chapters in the third scheme for a book that he had prepared shortly before, as I mentioned before.[6] The new list, as it appeared in the 1951 program, is as follows—dates follow the titles[7]:

1. Introduction: Textbook Science and Creative Science. March 2nd, Friday.
2. The Foundations of Dynamics. March 6th, Tuesday.
3. The Prevalence of Atoms. March 9th, Friday.
4. "The Principle of Plenitude": Subtle Fluids and Physical Fields. March 13th, Tuesday.
5. Evidence and Explanation. March 16th, Friday.
6. Coherence and Scientific Vision. March 20th, Tuesday.
7. The Role of Formalism. March 27th, Tuesday.
8. Canons of Constructive Research. March 30th, Friday.

The way the lectures were promoted was the subject of a small controversy between Kuhn and Lowell and some of his collaborators. The promotional leaflet, with the program on its back, had on its front a question (followed by a large question mark): "What are the problems of scientific research today?" The question was followed by this message: "In a world in which science's quest for physical theory has already had results that promise to change the course of history, the fate of mankind

[5] See our Chap. 2.

[6] See our Chap. 2, esp. Section 2.14.

[7] See Kuhn to W. H. Lawrence (Curator, Lowell Institute), 13 April 1950, and Lawrence to Kuhn, 2 September 1950, both in TSKP 3.10 (the second one has an attached copy of the program). My arrangement of titles and dates is different from the one that appears in the program itself. On these lectures, see Andresen (1999, S61–S63); Marcum (2015a, 32–36); Hufbauer (2012, 434–436); Mayoral (2013); Galison (2016, 59–63); Reisch (2019, Chap. 9).

may depend upon solving the problems of research."[8] The advertising seemed to suggest that Kuhn would enlighten his audience with the nature of the overwhelming and transformative current scientific research and discuss certain aspects of it.

Of course, Kuhn did not intend to do anything like that. His intention was to report on and illustrate the historically variable nature of the scientific enterprise, about theoretical change, about the nature of scientific method, and about a new understanding of science, and this had little to do with the apparent promises of the pamphlet. The advertisement appeared in the *Boston Globe* and in the *Boston Herald*, too. After seeing the announcement of his lectures in that format, Kuhn was infuriated. He called Lowell on the morning of February 17, 1951 (there were only a few days left before the start of the lectures) and reiterated his complaints three days later. He was assured that this specific advertising would be suppressed as much as possible.[9]

The episode shows us to what extent Kuhn wanted his message to be clearly conveyed, something that was not unusual for him. He was not trying to offer a popular portrait of the kind of science that many people had in mind, and that had both amazed and terrified the average individual. He wished to prove that science was much more complex than philosophers had contemplated in it—a thesis that was not unusual in his academic group, as it was explicitly shared by Conant, for example—and that, at the same time, it was closer and more understandable than it might initially seem on that basis. He was proprietorial of his thinking and of the message that he intended to deliver.

3.2 Kuhn Versus Pearson

In the first lecture, "Introduction: Textbook Science and Creative Science," Kuhn sets out to show the limitations of the empiricist (broadly speaking) version of the scientific method. Like Conant, Kuhn takes *The Grammar of Science* by Karl Pearson as his main example of this methodology (and of its weaknesses). As a central element of this critique, Kuhn sees in Pearson an elimination of subjectivity when ordering empirical evidence and, as Kuhn might have said, when selecting the essential features of scientific objects. For him, dispensing with this human element eliminates the very key to productive research.[10]

Let us remember that Conant denied the existence of a pure classificatory phase as described by Pearson, prior to the development of laws.[11] For Kuhn, classification is an integral part of science, but it is not exactly the mechanical process that Pearson

[8] A copy of that advertising is in TSKP 3.10. See GR viii–ix for pictures of the leaflet.

[9] See Kuhn to Lowell, 17 February [1951] and February 20, 1951, both in TSKP 3.10. See Marcum (2015a, 12) for a good description of all this controversy. Reisch (2019, 171–173) interprets it in a political sense.

[10] Lowell 1, 1–2, 4; GR 1–3.

[11] Conant (1947a, 5–6, 9–11, 14–15).

talks about, in which the human mind only receives the features of reality and adds them up in a complete snapshot of it, so to speak. The essential features of objects are not given. They are subjectively selected on the basis of our goals. Among such goals we can cite as an example our desire to establish certain laws that we believe can be obtained from those phenomena with whose repetition we are familiar.[12] Scientific objects are created. Obtaining laws is also a creative process, but it is also linked to the classification and qualification of the objects that such laws govern, which is, in turn, based on a creative (or let's call it *intentional*) selection of its essential features. Let's remember that in 1945, in his graduate essay on metaphysics for Donald Williams, Kuhn described the close logical link between Newton's laws of motion and the scientific conception of objects obeying such laws. In this first Lowell lecture, we see a general perspective that would have been useful as a background for this old example.

In this first lecture, Kuhn also presents a difference between two visions of science, each with a function that is not to be confused with that of the other. First of all, we find a pedagogically useful vision of science, which appears in the textbooks of many disciplines and is characterized by the separation of the evidence available for the theoretical part of a science from its theoretical statements and the training with them. Secondly, we also find a perspective on science that comes from the observation of the way scientists carry out the aforementioned process of selecting essential features in perceptual objects and obtaining the discipline's laws and other central statements. This second view is not very useful if we wish to introduce students to their future work, but if we want to build the best possible version of the scientific method, then we must pay attention to this alternative. The former view produces examples of "textbook science"; the second leads to what Kuhn calls "creative science." His lectures are dedicated to showing the main elements of this second perspective on science, with a view to developing a renewed version of the scientific method, one that is also closer to historical facts. From these lectures, we would also be provided with an explanation of the role played by the first vision, that of textbook science, in the common work of the scientist and in the history of science. Kuhn would also provide us with a comparison on the linguistic perspective involved in both views of science, which helps us to see clearly the functions they play, as noted above (see the last Lowell lecture).[13] We must (and shall) study that part carefully, because it would survive in Kuhn's mind for a long time. When he returned to some themes from *Structure* shortly after its publication, some ideas from the latter parts of the Lowell Lectures re-emerged as a part of his criticism against logical empiricism and of his own perspective on scientific language.[14]

---

[12] Lowell 1, 22; GR 15.

[13] See Lowell 1, 5–7, 9–12 (GR 4–6, 6–9) for the description in this paragraph and the expressions in quotation marks.

[14] See our Chaps. 6 and 7 for further details.

3.3 Lectures with Historical Content

Lectures 2, 3, 4 are a presentation of historical cases. In the first one, "The Foundations of Dynamics," Kuhn shows the history of the problem of motion, from Aristotle to Galileo. He highlights that we are not facing the same problem in one historical period as in another. To this end, Kuhn describes the different points of view or orientations that are transforming not only the way of dealing with the phenomena of motion but phenomena themselves—our conception of them. In this way, he provides a key to seeing how the problem itself is transformed.[15]

Lectures 3 and 4 are, respectively, "The Prevalence of Atoms" and "'The Principle of Plenitude': Subtle Fluids and Physical Fields." In these lectures, he reverses the use of historical examples. Instead of seeing how a problem changes depending on the different approaches, or orientations (I will talk about this term and its use in these lectures later), he explores two such orientations as well as the way each of them affects various problems upon which they are used. He highlights how such orientations adapt to the problem they deal with and change with it, sometimes significantly.

In the first lecture, he deals with atomism and shows the ontological and cosmological transformations that its adoption brings with it, such as a focus on primary qualities (according to Galileo's distinction) that are outside the scope of an immediate observation of objects. A new, directly related idea is that of an infinite and mechanical universe in which the cause of phenomena is the interaction of minimal and indivisible parts, whose ontology is composed of matter in motion. These features admit a wide range of qualifications, however. Orientations serve objectives that vary over time and with the researcher who applies them. That's why he makes a journey through the concrete applications of this vague scheme from Leucippus and Epicurus to John Dalton, passing through Robert Boyle's mechanical philosophy (whose chemical research was the subject of a well-known paper that Kuhn published in *Isis* shortly after), Descartes' atomist *plenum*, and Newton's atomism with forces of attraction at a distance. The journey is important, as it exemplifies the function of an ontological resource that is not directly verifiable (a scientific orientation) in the organization of scientific research.[16]

In the second of these two lectures, the one dedicated to the "principle of plenitude" (Lecture 4), Kuhn talks about other orientations—about alternatives to atomism; for example, the notion of "subtle fluid." A subtle fluid, he says, is not so much a *general* visualization of nature (that is, a total ontological perspective) as a *part* of one. It is a theoretical, explanatory resource that can be used in a specific field of phenomena, for example, Kuhn says, in order to talk about observable qualities.[17] He begins his narrative in antiquity again, starting with the Platonic *anima mundi* and the Stoic *pneuma*. He then goes through the Middle Ages and the "substantial forms," talks about the role of subtle fluids in the chemistry of the seventeenth and eighteenth

[15] Lowell 2, 1; GR 21.

[16] See Lowell 3, 1–3, 7; GR 153–157, 161–162.

[17] Lowell 4, 1–4; GR 57–58.

centuries, and gets to the phlogiston theory, to which he dedicates quite a bit of space (once again, it is a classic example for him as well as for Conant), and then to the caloric theory. Finally, he deals with the application of the notion of subtle fluid to the study of electrical phenomena.

3.4 Lectures with Philosophical Content

Lectures 5–8 bring us back to a philosophical, theoretical plane, although with continuous references to other disciplines—psychology, linguistics and logic—that allow us to understand what mental processes underlie creative science and see their effects in the historical examples we are already familiar with.

In Lecture 5, "Evidence and Explanation," Kuhn describes the general characteristics of scientific "orientations," also cited as "points of view," "prejudices" or "preconceptions." He shows that they are the elements of judgment that, according to Pearson, must disappear from every perspective on scientific method.[18] Kuhn shows that these orientations can be distinguished by their functions, and deals with three main ones: cosmological, metaphorical and normative.[19] Finally, after this description, he goes on to study their development and modes of change in the part he calls "dynamics of scientific ideas,"[20] a section that constitutes the bulk of the lecture and which articulates the historical examples and the theoretical and methodological notions.[21]

Lectures 6 and 7 (respectively "Coherence and Scientific Vision" and "The Role of Formalism") delve into psychology, linguistics and logic. In the first of them, Kuhn shows how our psychological world simplifies the complex flow of stimuli that reaches our senses. We are the ones who select the features of that world and form objects in an unconscious, vague and pragmatic way. Citing William James (as he will also do later in *Structure*), Kuhn says that in the absence of this selection process, the world would be "a bloomin', buzzin' confusion."[22] With such features and objects, we configure a "behavioral world," which helps us to face future experiences and, moreover, to advance from them—to learn, in short.[23] Therefore, it is not stimuli but behavioral needs that determine the characteristics of perception. There is no rigidity in the source of stimuli that does not allow us to adapt perception to the requirements of the world in which we deploy our behavior.[24]

In this lecture, Kuhn cites the psychological experimentation of Bruner and Leo Postman, that of the Hanover Institute, Piaget's genetic epistemology, and Whorf's

---

[18] Lowell 5, 1; GR 77; Lowell 1, 25; GR, 16–17.

[19] Lowell 5, 6–18; GR 79–84.

[20] Lowell 5, 18; GR 84.

[21] Lowell 5, 18–44; GR 84–94.

[22] Lowell 6, 22; GR 104–105. See *Structure*, 113.

[23] Lowell 6, 23; GR 105.

[24] Lowell 6, 31–32; GR 108.

research in comparative linguistics, all of which he knew well from his period in the Society of Fellows. Taking these works as his background, Kuhn talks about various fundamental features of his new epistemological organization. He argues, for example, that language and perception are determined by the needs of behavior; or that all difficulties find a solution within it, and pragmatic efficiency, not empirical correspondence, is the basis for evaluating this world and its attached language; or that it is the failure to face behavioral needs efficiently that triggers change in this world.[25] He finds a highly fruitful parallelism between the behavioral world and scientific orientation by virtue of these features. Both are appropriate vehicles to turn stimuli into the sensed objects, thereby enabling the subject's activity—everyday activity and research activity, respectively. In functional terms, they are equivalent. He argues in this lecture that the scientist deploys his or her activity in a kind of behavioral world typical of his or her expert knowledge, which is not that of the usual behavior of those who do not belong to that profession. Beyond that, the idea of a behavioral world does not differ significantly when applied to the case of the scientist.[26] As he says at the end of lecture 6,

> I suggest we equate the notion of scientific orientation with that of a behavioral world. [... I]t is in part the psychological necessity of some behavioral world as a mediator and organizer of the totality of perceptual stimuli that I suggest we will never be able to eliminate from the scientific process orientations which originate in experience but which subsequently transcend it and legislate for it.[27]

It might be argued that Kuhn spent a great part of his career trying to find out the details for such a view; more precisely, developing an explanation for that position about the preconditions for scientific experience. As we shall see in our Epilogue, Chap. 4 of his draft *Plurality of Worlds* testifies to the importance he still granted to describing in detail this kind of "mediator" more than forty years later.

Lecture 7 deals with the role of logic and mathematics in the physical sciences. Kuhn talks about the way in which logic and mathematics have become autonomous with regard to empirical data in the preceding century.[28] Physics has benefited from this autonomy. However, he has reservations about the role of logic in science and, especially, about the formalization of scientific language, which are related to what we see in these lectures. In particular, the clear possibility of formalizing mathematics does not seem to make sense in scientific language except for very specific parts. He points to the application of geometry in the representation of physical space. Such application to physics involves the reduction of some basic characteristics of our perceptual world to a completely coherent and predetermined conventional system. With this, we fix the meaning of the terms used to refer to the phenomena of physical space.[29] This is fruitful in this and some other cases in which the application of logic

---

[25] Lowell 6, 16–20, 28–31, 34–43, 44–46; GR 102–104, 107–108, 109–114.

[26] Lowell 6, 42–43; GR 112–113.

[27] Lowell 6, 47; GR 114. Reisch's version is edited slightly as compared to the original. In my case, I have only eliminated the plural of "organizers" from the original.

[28] Lowell 7, 1–17; GR 117–124.

[29] Lowell 7, 32–34; GR 129–130.

has been useful as a research tool. This is the case for the logical analysis of the problem of motion in medieval physics.[30] However, Kuhn considers that the plan of a full formalization of scientific language is a program whose future is doubtful. He uses the example of operational analysis—of which he is aware, as we have already seen—as a project for clarifying the content of physics that relies on that kind of plan.[31] For him, the main problem of this plan is its unfeasibility, a contention he had already had in mind years before, as is evident from his 1949 notes on the fringe of vague meaning.[32] Scientific language is not a language given in advance. A scientific orientation, that is, the behavioral world in which the scientist lives and practices his or her profession, accommodates external information to the needs of his or her behavior, as Kuhn explained in the previous lecture. This adaptation is also liable to change with the same goal—efficient activity in the behavioral world. Something like this would be impossible in a language in which changes in the criteria of application of terms are not admitted; that is, in a formalized language.[33] The idea of a natural language that he is pursuing—scientific language is nothing but a specialized version of a natural language—is incompatible with the alternative idea of a full formalization, because the latter involves a conventional fixation that hampers the exploration of the world of our experience.

The last of Kuhn's Lowell Lectures, "Canons of Constructive Research" (Lecture 8), talks again about the fringe of vague meaning. Kuhn describes natural language as the starting point to reach scientific language. He talks about natural language as a vehicle of communication in which the meanings of words are established through use.[34] He also returns to the idea of meaning as something that depends on our connotative abstraction from our experience of past denotative applications of the same term. In scientific language this use is established scrupulously, and if something like this is not possible, expectations are generated regarding the future application of terms. Such expectations belong to the fringe of vague meaning, a realm that reflects nothing more than an incomplete understanding of the perceptual flow, an uncommon use, or one only anticipated. In such a margin we can be wrong when applying the terms of language and we may not coincide with others as regards common use and the meaning of terms.[35] Despite that, that fringe is the basis of scientific research. This latter takes place within it when trying to clarify the language or to modify it from its bases.[36] Kuhn returns here to the relationship between the scientific textbook and creative science, as their field of activity is seen in a new light; specifically, the textbook is seen as a good reference tool for creative science, but not as a sample of scientific language.[37]

---

[30] Lowell 7, 38–39; GR 132.

[31] Lowell 7, 40; GR 132–133.

[32] See our previous chapter.

[33] Lowell 7, 44–45; GR 134–135.

[34] Lowell 8, 10; GR.

[35] Lowell 8, 14–20; GR 141.

[36] Lowell 8, 35; GR 150.

[37] Lowell 8, 31–33; GR 149–150.

The end of the lecture includes an interesting comment. What we have studied in these lectures—from historical cases to psychological experiment and theses about perception, and a critical review of the relationships between formal and natural languages—reveals a similar organization, Kuhn says, a very *similar pattern*, which can be understood if we consider that the three areas studied here are, though seemingly disparate, just *aspects* of the same cognitive activity that revolves around the adaptation of incoming information into our experience so that it is understandable, usable, suitable for our behavior and also modifiable to improve its utility.[38] This plural, naturalized, and developmental idea that Kuhn applies in different areas to arrive at a common vision is a good example of what he tries to show in *Structure* and further work.

3.5 The Renewal of Epistemology

The Lowell Lectures have a dual purpose. The first of them is purely informative, descriptive. Kuhn's intention is to offer an image of how scientific theories, concepts, laws and objects are generated. However, the way in which Kuhn explains this differs from the usual description. In addition, he does not avoid controversy. Many of his philosophical arguments are presented as alternatives to the empiricist methodology of science. From the very outset, Kuhn is openly critical of a widely disseminated point of view of this methodology, *The Grammar of Science* by Karl Pearson—which, as we saw, had been the target of Conant's criticism too. Kuhn is especially critical about the epistemological foundations of the methodology that such work represents. Later his criticisms were directed at other aspects of this methodology, which were difficult to sustain after the first kind of criticism. For example, he deals with the conception of scientific language that was characteristic of a logico-positivistic methodology. This critical plan was the second objective of the Lowell Lectures.

In principle, Pearson's general perspective is the target of his criticisms. Like Pearson, Kuhn still believes that science is cumulative. However, there is a fundamental difference between them: for Kuhn, science accumulates "understanding of nature," as he indicates at the beginning of his first lecture:

> I hope that these lectures may also serve a more constructive function. For I was trained as a physicist; I have done research; and I am committed to the belief that scientific knowledge is good knowledge, that it is useful knowledge, and above all that it is cumulative knowledge. I believe that the work of scientists has resulted in an increasingly detailed and an increasingly far reaching understanding of the operations of nature, and that this progress can be objectively described so that a seventeenth, eighteenth, or nineteenth century scientist, confronted with a textbook containing contemporary scientific theories would, without reluctance, admit that science had proceeded a long way since his own day.[39]

---

[38] Lowell 8, 36–37; GR 150–151.

[39] Lowell 1, pp. 4–5; GR 3–4. Conant had already expressed something similar; see, for example, his paper "Science and the National Welfare" (Conant 1944): "there can be no doubt that the pioneers

The sense in which Kuhn alters the meaning of "cumulative knowledge" has to do with his refusal to accept Pearson's idea that there is no place in the scientific method for the subject and his or her inclinations. Cumulative development, from this perspective, would involve a disconnection from these subjective elements and a focus on obtaining empirical knowledge and the laws of nature. Only in this way is this knowledge and control of nature reliable enough.[40] This is the idea of an "understanding of nature" that the traditional view of the method supports, which might not be completely unreasonable. For Kuhn, however, it is an idea that must face counterexamples from the history of science and, to do so successfully, it must be corrected and the result of its correction is a new understanding of science that has little relation to (and tends to displace) the one underlying the traditional method, ending by replacing the latter.[41]

In the previous sections we have seen the pedagogical organization of Kuhn's lectures at the Lowell Institute. Now we are going to alter that arrangement, which was convenient for his audience, in favor of a more thematic organization. For the critique of empiricist methodology, he first addresses its epistemological roots. As in *Structure*, he shows that such roots do not help us to face some recent results of the psychology of perception and of linguistics. He has an alternative viewpoint on the mechanism of human knowledge that is more plausible than the empiricist one, and that is more consistent with the historical development of science, as he shows in these lectures.

Kuhn exhibits the traditional epistemological viewpoint in the sixth conference.[42] This description aims to present a conclusion of this epistemology (which we will see next) that is then rejected on the basis of arguments from psychological research by Piaget, Bruner, the Hanover Institute, and the Gestalt school, and from Edward Sapir's and Whorf's linguistics. First of all, it is important to underline that Kuhn does not discuss the empiricist epistemology from within, nor does he attack specific versions while simultaneously developing his own philosophical arguments. He presents an alternative that fits the data from the psychological and linguistic currents already cited (to which he does not adhere in interpretive terms). The alternative that he seeks has a pragmatist outlook—in some aspects it is very similar; for example there are similarities to Dewey's functionalism.[43] Yet, Kuhn does not present it as an outcome of his own work, nor does he discuss it in philosophical terms. He simply subscribes to it and defends it thanks to evidence coming from experiments in psychology.[44]

---

in chemistry, physics, mathematics, geology, and biology, if brought to life today, would exclaim with delight and wonder at the progress which has been made" (Conant 1944, 400).

[40] Pearson (1900, 6–9); cited by Kuhn, Lowell 1, 1; GR 1.

[41] Lowell 1, 19–21; 13–15.

[42] Lowell 6, 12–15; GR 100–102.

[43] On the parallelism between Kuhn's mature views and Dewey's arguments in *Logic*, his main work on scientific method, see Mayoral (2022). On the connections of Kuhn's mature views with pragmatism at large (and with Dewey's views in particular), see Mladenovic (2017, Part III).

[44] On this current of epistemological reform (especially around perception), see Firth (1949, 1950).

As Kuhn talks about it, traditional epistemology (that is, the one that serves as a basis for Pearson's argument) shows that the objects we handle in ordinary language, to which we refer when we say that we see something typical of our everyday environment, are the result of a simple sum of qualities in our mind. An image is formed in it, whose identification through trained judgment allows us to assign a name to that object. Kuhn draws a conclusion from this organization of human cognition against which he wants to present an alternative. For him, in this view (which is about three centuries old at this point), rather than *seeing* the object, we interpret that it is there by virtue of some indicators—qualities or appearances. If I look at a sample of coal before me, I do not actually *see* the coal, but I observe the irregular profile of a black rock, often matte, perhaps a rough surface, and so on. I do not *see* the coal as such, but I end up taking it as a sample of that kind; *interpreting*, if you will, that it is that kind of rock.[45] From this perspective, if such traits are the only elements that lead us to identify the object as a member of a certain kind, we have the path prepared for an empiricist methodology like Pearson's.[46]

After introducing this perspective on perception, Kuhn presents an alternative. For him, there is not any process that, on the basis of its qualities, leads us to see the object; we just *see* an object (*that* object), most probably (though not necessarily) as soon as we *identify* it as an instance of its kind.[47] The fundamental difference with the previous view, according to Kuhn, is that we lack an alternative in cognitive terms. As in the previous situation, there is a certain variety of traits from which we *form* some image (we represent the object) in our mind. But "we form" has a different meaning than it would have if we had used the same expression in the previous view. "We form" should not be replaced by "we interpret," but by something more similar to "we react"; and the product of that reaction, which is *inevitable* for us, would be the visual representation of the object. Kuhn shows that the number of stimuli is different in each situation in which we come across common objects. However, despite this variety, we *perceive* these objects. We start with them, so to speak, not with their apparent qualities.[48]

The main consequence of this perspective, which directly affects the methodology criticized by Kuhn here, is that the *subject's* active role, which somebody like Pearson preferred to exclude from the basic steps in scientific method, is after all required for

---

45 Lowell 6, 14–16; GR 101–102.

46 Kuhn's argument in Section X of *Structure* is not different from this one. And, as we shall see in Chap. 7, his Postscript—1969 (see p. 194) persists in using that line of reasoning. See our Chap. 7.

47 Crane (1992, 175–176) shows an argument against perceptual relativity as expounded in *Structure*. What happens to those that do not have the classification of rocks necessary to say "I see a sample of coal"? According to Kuhn, do they perceive (see) anything or nothing at all? His likely answer would be that, evidently, in normal conditions, they are aware that they are looking at something. His main point, in any case, is not against that possibility, whose main difficulty would be that—leaving aside the exceptional case of something that completely alters the order of things—there is an object pending identification. In that case, once our classification gets enriched, the object would be identified not as an unknown though not necessarily anomalous item, but as a sample of a certain kind. Kuhn's argument is rather against the *way* the object, identified or not, is first perceived. On Kuhn and perception, see also Bird (2000, Chap. 4) and Hoyningen-Huene (1993, 83–90).

48 Lowell 6, 15–16; GR 102.

the *objects* to appear in the perceptual experience, and so in what he saw as the *basic* steps. This consequence justifies Kuhn's attention to this alternative point of view in epistemology. However, the question that immediately arises from this alternative approach is: where do the objects we see come from, if not from their observable features? An obvious way for these objects to enter our mind is through education. Now, the question would be, *how* are such objects introduced into the mind in this way? *What* process is followed in education, during the training? Kuhn's answer is based on the aforementioned psychological perspective (partly pragmatist and functionalist and partly Piagetian). Thanks to that perspective, we see the fruit of his investigations while in the Society of Fellows.[49]

3.6 Behavioral Worlds

According to Kuhn, the objects we perceive are the result of our reduction of the chaotic complexity of information that external stimuli bring to our sensory apparatus. The objects thus offer a functional and recognizable sketch of the world, constituted by an ontology of objects and relationships with respect to which we develop our behavior, our daily activity. Such objects and relationships shape what Kuhn calls a "behavioral world," an expression that can be observed in Kurt Koffka, for instance.[50] This metaphor helps us to understand, but does not fully explain things. If these objects do not come from a priori categories of a Kantian type and with a transcendental nature, what is their origin then? In the sixth lecture, he explains that the behavioral world is the environment in which we live and act. Our actions contribute to generating this environment of our experience.[51] Thus, both this channeling of the information coming from the outside and its enrichment and correction pursue the same objective: efficiency in our behavior.

The objects and relationships of our behavioral world originate in the society in which we live, as well as in the role that we play in it. (Needless to say, they are also based on the stimuli flow input from the outside that I mentioned earlier.) There is, therefore, no single or uniform origin for the mental objects of all members of a society, although there is seamless communication between them that involves some (and frequently a high) degree of ontological uniformity. The world in which we live from childhood alters our initial cognitive categories—those with which we face the world from the beginning—but our own experience serves as a basis; it is the material

[49] In fact, Kuhn shall devote much effort to answer those questions properly. How education in a science is possible without resource to rules and definitions, and so without the kind of arrangement that characterizes the empiricist construal of language-acquisition, is a main concern now and later, after *Structure*, for instance. This persistent interest in that problem shall be evident again in our Chap. 7.

[50] Lowell 6, 21 ff.; GR 104 ff.; see Koffka (1935, 31); see also our Chap. 2, Sect. 2.12. However, as noted in our previous chapter (Sect. 2.10), a similar term, "thought world," is also present in Whorf; see Whorf (1956, 147–148).

[51] Lowell 6, 21; GR 104.

with which our personal ontology is molded, with forms that enable us to have such seamless communication with other individuals in the group. As Kuhn expresses it, in a clear Piagetian tone:

> As the child grows up [his] modes of thought are no longer adequate to deal with the developing complexity of the problems which he meets. The thought process and the concomitant behavioral world are changed, and the behavioral world at which he arrives is a joint product of the older society in which he is raised and of the new problems which he, as a member of that society, may have to face. It thus need not and usually will not be quite the same as the behavioral world of the preceding generation.[52]

In summary, as Kuhn says in the sixth lecture, perceptual experience is shaped in us according to our objectives, in harmony with what our behavior demands.[53] In this lecture, he then speaks more deeply about it. There are differences in the behavioral world that, he says, respond to various kinds of demand. He speaks, specifically, of the demands of the environment and of the way of life of the human group in which the individual is integrated. He reminds us that, for an American like him, the sight of a reindeer involves few nuances. For certain Eskimo groups, however, such a sight adds some more details. These groups, he says, see and name more than twenty different types of reindeer skin. Of course, unless we are very interested in reindeer for professional or amateur dedication, this classification will hardly be useful and, for that reason, they do not appear in our behavioral world. In contrast, Eskimos live together with reindeer and the latter are particularly important to their way of life, so such differences have immediate utility. Conversely, these Eskimos, Kuhn also says, have similar difficulties with color discrimination forms that for an American like him are usual and easily identifiable. For Kuhn, we learn to discriminate what plays an important role in our behavior.[54] He also mentions those demands that are specific to purely individual activity. As he says, an expert's behavioral world is usually very different from that of the average individual, even though both share a common language and culture.[55]

Kuhn's behavioral worlds also have an instrumental nature summarized in the maxim that they are not susceptible to evaluation in terms of the usual values of truth and falsehood, but by virtue of their efficiency in leading our behavior, and in helping us adapt ourselves to the surrounding environment, social or natural, and to the challenges of the novelties that often emerge in it. This proximity to pragmatism, somewhat similar to Dewey's instrumentalism, is striking and would be a constant in Kuhn's thinking, even in his last phase, when he elaborated his vision of conceptual schemes until he related them to his lexicons.[56] In this latter theory, only statements

52 Lowell 6, 38–39; GR 111.

53 Lowell 6, 31; GR 108.

54 See Lowell 6, 25–27; GR 105–107, for this example.

55 See Lowell 6, 24, 27–28; GR 105, 106–107.

56 See Kuhn (1991a, 94). An improved version is available in his Shearman Memorial Lectures, delivered at the University College London in 1987, a version of which is published in *The Last Writings*; also see our Chap. 8 for more details. On the connection with pragmatism, see again Mladenovic (2017, Part III).

(and thus, specific beliefs) are evaluable in terms of truth and falsehood *within* a lexicon (or conceptual scheme), from which they acquire meaning and through which the reference of the terms in use is specified. The lexicons themselves (and again, the conceptual schemes) cannot be either true or false, but pragmatically efficient or inefficient. The same perspective for the lexicon appears here in this environment for experience that he calls the behavioral world.[57]

Piaget again offers a good example of the kind of dynamic change in the behavioral world that is caused by behavioral demands that go beyond the world's resources. Piaget's children are not miniature adults, with only a partial development of the adult ontology, Kuhn reminds us. Quite the contrary, as we have already seen, children have systems of objects and relationships moderately valid to face the world they see. The interaction with their older peers and the needs that gradually arise as they integrate into the adult world force them to seek greater coherence in their world, in their own experience. This is a process in which the child gradually finds greater autonomy from the world that requires him or her to resolve many of the incongruities of his or her egocentric stage. Children's learning happens as a result of vital needs like the ones we have seen here.[58]

Some of the perspectives on human cognition that, in recent years, were drawn from the psychology of perception and of education (and from other disciplines like comparative linguistics) advocated for an epistemology more inclined towards Kuhn's alternative than to its traditional empiricist predecessor. It is not exactly the simple correspondence with the physical world that validates our mental images. Although this link is necessary in the formation and correction of our mental categories, we can only verify beliefs that are already in operation, that is, which are a functional part of the subject's psychological world. What is checked is the *efficiency* of this psychological world, because there is no other way of making the comparison than by means of beliefs themselves. The correspondence with the physical environment is *part* of the information with which we evaluate the coherence of our world. This point of view is identical to the one already seen in Kuhn's 1949 *Notebook* that we examined in our previous chapter. In support of his view, he adds some detail that is crucial for the later application of this epistemological perspective to his methodological vision.

3.7 Psychology of Perception

For Kuhn, every interaction of the subject with external stimuli has a response in terms of the subject's mental categories—kinds of objects, relationships, and so forth. As he emphasizes, there is always a response available, formulated in accord

[57] See Kuhn (1993, 244–245); Lowell 6, 28; GR 107.

[58] On Piaget, see Lowell 6, 36–40; GR 110–112.

with the behavioral world's ontology and vocabulary.[59] To illustrate this thesis, he refers to the same experiment with a deck of anomalous playing cards carried out by Harvard psychologists Bruner and Postman that he will later describe and comment on in *Structure*.[60]

In this experiment, the subjects had to identify cards similar to the usual ones in which a variation had been introduced; for example, a six of spades can be red (as in *Structure*).[61] One of the factors to observe was the time required to perceive the changes introduced in the cards, which was sometimes very prolonged for the first time (and also shorter for subsequent tries). For him, however, the most significant result was that even an altered card had been perceived as a usual card, without any variation at all. In experience, he shows, an object has the features, the form, of a regular inhabitant of the subject's behavioral world. As he says in *Structure*, "novelty emerges only with difficulty, manifested by resistance, against a background provided by expectation."[62] Obtaining an experience consistent with the behavioral world prevails over the perception of its counterexamples.

If we can only evaluate our psychological world from the psychological world itself, then any alteration will involve a previous stage of confusion, of uncertainty. Bruner and Postman's example also illustrates this stage that precedes a new classification. Observing the counterexample, perceiving it for the first time, changing the scenario of our experience so that the new object fits, is something that requires a longer duration than common perception.[63] Kuhn illustrates this with two more experiments, in this case from the Hanover Institute and the Dartmouth Eye Institute.[64] Both experiments illustrate a similar idea: an individual is subjected to modifications of the visual environment through glasses that alter the subject's vision. The subject may be watching an ordinary or a modified environment. The subject must interact with the environment under such conditions. The result is that more or less normal behavior is eventually visible, although it usually takes a certain period of time for adaptation to the new environmental conditions.[65]

Kuhn concludes his reflections in Lecture 6, arguing that the solution to problems of visual paradox, such as those artificially created in the aforementioned experiments, involves a new psychological, phenomenal environment with new possibilities for perception. He adds to this a consequence that we also find in his better-known work—that when the new world appears, the previous one *disappears*. Let us recall that, according to Kuhn, our behavioral world *always* provides room, classification,

---

[59] See Lowell 6, 43; GR 112–113. In Reisch's edition, the original word "complete" in the sentence "all these behavioral worlds are complete" at the top of that page (Lowell 6, 43) is rendered "incomplete" (GR 112).

[60] See Bruner and Postman (1949); *Structure*, 62–65, 113.

[61] *Structure*, 63. See Lowell 6, 18; GR 103.

[62] *Structure*, 64. See Lowell 6, 18–20, 43 (GR 103–104, 112–113) for this same experiment and its corresponding comments.

[63] Lowell 6, pp. 44–45; GR 113–114.

[64] Kuhn speaks of the first institution in *Structure*, 112.

[65] See Lowell 6, 29–31, 45; GR 107–108.

for incoming information. A variation in that world that leads to essential alterations of it—that is, in essence, to another, different one—also brings about the elimination of the previous one. In other words, one comes after the other *and* one *suppresses* the other.[66]

This new perspective on the nature of the perceptual experience affects, of course, the vision of the scientific method as soon as it shows that experience in the scientific domain is not a very different scheme from everyday experience. Kuhn's aim is to make this clear, as he does on the last page of the sixth lecture and as he continues to do in the following lecture.[67] Thus, while the empiricist methodologist considers the action of metaphysical beliefs a hardly admissible intervention of extra-scientific, subjective, and even irrational components in the scientific domain, Kuhn invites the new philosopher of science to explore such components and to overcome the empiricist point of view. Only in this way, he tries to suggest, will rational research delve into the *psychological* medium—and then into the, for him identically important, *social* medium—in which real science is practiced and built.

3.8 Scientific Orientations

The scientist is trained for a working practice. This involves a language based on the reference to those objects and relationships that, in turn, make up the scientific world, which follows the model of the individual's behavioral world as seen before. If we follow the same line initiated by Kuhn in 1947, we can call these referents "scientific objects".[68] These new objects are referents for our mode of communication.

What specific characteristics mark orientations? We must consider that Kuhn treats them as a specialization of the individual behavioral worlds. The meaning of the terms of the language that rests on these orientations is much more specific than in common speech, but, he also argues, this specification should not be taken to the extreme of completely eliminating the pragmatic background of its application. Some of the application criteria, which have such a pragmatic nature, would surely not withstand scrutiny in the search for an appropriate definition of the components of theoretical vocabulary, but, for him, these criteria nonetheless serve a function that should not be blocked. The ultimate goal of such a specification is to facilitate a common framework of communication that, at the same time, contributes to enhancing the virtues of the theory—although Kuhn is not yet very explicit about these latter and about this point. The virtues that we can observe here range from the most obvious, the degree to which a theory is confirmed (although its particular application depends on the orientation), to the unified vision that the theory (again based on an orientation) provides about the phenomena, which contributes to its explanatory power; and we must not forget the manageability of the theory: the fact that its theoretical and

---

66 Lowell 6, 46; GR 114.

67 See Lowell 6, 47 (GR 114), and our next section.

68 See our Chap. 2, Sect. 2.2; see also Lowell 5, 11 (GR 80–81), for a similar use.

conceptual structure is clear enough for a generalized practice on the same problems. Conceptual clarification is thus a consequence of the adoption of an orientation that must be worked out, since it is nothing more than a cognitive precursor of a theory, or several of them, which, although Kuhn does not explicitly say so, ends up satisfying certain purposes such as those already indicated.[69]

Orientations like atomism, for example, or the one based on subtle fluids, are originally vague, come and go throughout the history of science, and are adapted to specific problems and problem-solutions in each context. Some orientations—and atomism is a good example—promote an overall vision of nature; even an entire cosmology. We may also consider that Kuhn's idea of the "two-sphere universe," as he introduces it in *The Copernican Revolution* only six years later, is an orientation of this kind.[70] Kuhn chooses atomism to show that the way our orientation selects for us the basic ontological ingredients that make up reality—from ideas about the limits of the universe to the continuity or discontinuity of matter, or the privileged points or shapes in it—conditions the kinds of problems that will arise (that is, the way we will understand phenomena and question our initial explanations of them), and the methods and material resources with which we are going to tackle them. A natural philosopher after Newton did not see the world like an Aristotelian philosopher. The problems discovered in those worlds, and their respective methods to face them and find solutions, were completely different. Science as practiced under an orientation of this kind (like atomism) easily provides a unified and considerably general point of view—an all-encompassing vision that contributes to the agglutination of the problems, methods, and solutions of its time.[71]

Problem-solving is a central goal in science, and especially the pursuit of solutions to difficult problems. In his historically-inclined lectures, Kuhn showed that new evidence or inspiration for research often comes from altering the point of view on the intrinsic nature of phenomena—on the composition, if you will, of the objects or processes involved in them. Some orientations predominantly deal with this, thus becoming heuristic guides for research. This role is metaphorical and leads to significant reunifications. Their function has to do with usual problem-solving, but sometimes also with extraordinary pathways to a solution to a stubborn problem, that is, with that which requires a certain change in perspective. Less general in action than orientations with the previous cosmological function, these others enable a new perspective on familiar phenomena. He mentions the Galilean regrouping of pendulum oscillation with other types of movement, which contributed to a more precise physical description of such phenomena.[72]

The last mode of orientation that Kuhn mentions can be interpreted as something that does not have so much to do with the satisfaction of certain ends as with the

---

69 See Lowell 5, 13; GR 81–82; Mayoral (2013, 464).

70 See *Copernican Revolution*, 25 ff., 104; on 104, he uses the term "orientation" and "cosmological" in the sense that I have just shown. The Kuhnian description of the "two-sphere universe" occupies a good part of the book's early chapters. See our Chap. 4, Sect. 4.6, for a brief study on that book.

71 See Lowell 5, 6–8; GR 79–80.

72 See Lowell 5, 9–10 (GR 80), for this new kind of orientation.

more general idea that any ontological disposition born from an orientation prompts the fulfillment of some research objectives and, most importantly, the *suppression* of others. His basic idea, which we saw in his work on causality a few years before, that every point of view does not seek to unify only by agglutinating common aspects but also *excluding* the inadmissible, whether ontologically or epistemically, reappears here.[73] For him, orientations of any kind tend to *prescribe* a series of procedures and to admit only certain types of solutions, while *suppressing* certain others. The mechanistic viewpoint mentioned earlier, of pure inert matter and movement, promotes answers that have nothing to do with active principles. This idea constituted a source of opposition to Newton's work, for example. This function, or emphasis, of an orientation is, therefore, of a *normative* nature.[74]

3.9 The Dynamics of Scientific Ideas, I: Classical and Crisis Stages

The Lowell Lectures present the first complete version of Kuhn's cycle of scientific revolutions, that idea for which he is best known. Let me repeat this idea once more as it is usually understood after a first reading of *Structure*. Our usual vision of science typically starts from what Kuhn identifies as "normal science," a state of research whose widespread image non-experts obtain through textbooks, popular histories of science, or books on recent science, and whose true nature requires a practice and a language that are incomprehensible to the layperson. It is the image of science typical of disciplines that explore the nature we live in. In *Structure* Kuhn shows that this image is only of a specific stage of scientific development—one in which the development and application of a certain series of theoretical and technical resources and of certain kinds of (paradigmatic) solutions to problems is continuous, as there are no variations in any element of that group. Its learning is complex and involves immersion in a group of experts until developing a professional career as one of them. Hence, there is a barrier of incomprehension between professional science and the lay public, only partially demolished through the aforesaid popular science books—historical or otherwise. These, however, hide a more agitated and insecure process than it seems, and above all, behind the appearance of uniformity, they hide previous stages in which the social group dealt with other problems and with other resources. There are two relevant stages here: one of "crisis," in which normal science has difficulties in going on with its continued development; and one of "revolutionary science," in which old practices and projects are replaced by new ones.

In an excessively synthetic way, this is the idea of the successive stages of scientific development that Kuhn explains vividly and convincingly in *Structure*. In this section and the following we examine the very similar version that he offered a decade before,

[73] See our Chap. 1, esp. Section 1.12.

[74] See Lowell 5, 10–13 (GR 80–82), for this new form. Kuhn gives other examples there.

in the Lowell Lectures. Let us start, as we just did, with the predecessor of his normal science.

In Kuhn's dynamics, following Fleck, we must expect "classical" stages in the development of a science.[75] The "classical" nature of such periods is a qualification that he later converts into his normal science. The defining features of this stage follow on from what has already been seen. If a theory and its finished, organized structure—arranged as a logical and nomological structure, which is also well confirmed and incorporated into manuals that are central to the training of new scientists—is the more precise manifestation of a scientific orientation, the historical process through which that theory is optimized is the developmental stage in which such "classicism" is achieved. The theory in this phase is a powerful, effective, precise, exact instrument for understanding nature. In that phase, the theory manages to include some kinds of phenomena it initially did not start from. The orientation, its cognitive and conceptual predecessor, and its functions—cosmological, metaphorical, normative—act in all fullness. This period is, therefore, of optimal operation of a scientific orientation through the theory that is developed from it.[76]

There is a consequence derived from the very idea of orientation that is now important to highlight. The experiment does not invalidate this operation. Something like this only comes from the irruption in the scientists' minds, especially in those of the least experienced, of a new scientific orientation. This is a process of revolutionary change. And before the entry on the scene of this new orientation, there is usually a period of crisis.[77] In a crisis, a theoretical weakening occurs that can be explained as follows. In the classical period the scientist always knows what he or she is working on; that is, he or she knows and agrees with other peers on the particularities, features, scientific objects, relationships, generalizations, that make up the behavioral world of the scientist, and on the details of the theory that tells them about it, that specifies it. In the crisis stage, divergent results regarding the expected behavior of nature begin to be noticed. Their origin is usually the attention to results of experiment or measurement, whether old or new, that show a clear departure from the nomologically described behavior.[78]

In terms already explored in relation to the fringe of vague meaning, we can describe this last phase by saying that there are new phenomena and scientific objects that are not easily described by our language and that do not conform to the pre-existing conceptual pattern, which seems to indicate that there should be some kind of modification in the meaning of the terms we use. The new objects do not fit well into our initial classifications. They may even fit well into more than one category, leaving us with the question of which one we should assign them to. The questioning of the current classification, the doubt about whether there are now terms that were

---

[75] Lowell 5, 18–19; GR 83–84; see Fleck (1979, 9, 28). Kuhn does not mention Fleck explicitly in these lecture notes, but he says that this stage "has recently been called its classical period" (Lowell 5, 19; GR 84).

[76] Lowell 5, 19–20 GR 84–85.

[77] Lowell 5, 25, 31; GR 86, 88–89.

[78] Lowell 5, 21–22; GR 85.

previously considered synonyms that should now cease to be so, appear at these times.[79]

In a crisis phase, Kuhn tells us, indecision prevails. He is not only talking about problems in deciding the correct procedure to resolve this ambiguity, uncertainty or misunderstanding, but rather about the root of these problems. Facing the difficulties is now the main objective, but the search for a solution can only make use of the resources that are available to the scientist, that is, those provided by the orientation. Thus, indecision is not relative to the procedure to follow, since there is only one way to practice science that the scientist knows from education and research. Indecision rather occurs with respect to the theory that is being handled, which is difficult to define in every aspect of it in accord with its many available varieties. In this phase, Kuhn indicates, the widespread belief in the theory has not broken, but the question about the *specific* theory that is being handled, about its details, is a very different matter, on which there is no close agreement.[80] How is this state of indecision reached?

This state of indecision is a state of uncertainty. It is not created by inactivity in the face of certain problems, but by a certain excess of it. Perhaps it is worth emphasizing at this point that the behavior of scientists in a period of crisis, even though it leads to a generalized discrepancy (or rather, to a series of divergent behaviors), does not constitute a less scientific behavior. Why do we say this? Kuhn shows us that the state of crisis arises in the face of problematic observations or results, which are difficult to fit into the current theory and, at the same time, not easy to avoid. The responses to these problems are diverse. Some involve questioning these results and wondering whether the divergence is real. Others lead to slight modifications of the theory, introducing small changes intended to alter the theory only in those aspects that have been questioned by such adverse results. Often, the only intention behind such modification is to avoid the counterexample, a procedure openly criticized by Popper.[81] Kuhn, however, already diverges from Popper regarding the function of ad hoc solutions. They give rise to one of the stages a science goes through in its development. In that stage, the theory loses manageability due to multiple solutions, not all of which contribute to the old, unified image that appeared in its orientation. At such a moment there are not yet secessions in the group, but there is a lack of unanimity and a certain threat of division. This occurs with the appearance of a new unifying point of view, a new orientation.[82] Kuhn presents his idea of the crisis in these lectures as follows:

> It is a stage during which all professional scientists continue to adhere to the point of view and to the theories derived from it; it is a period in which the bulk of research is directed by the old orientation. But it is equally a period in which the theory itself has become cumbersome due to the accretion of numerous ad hoc hypotheses designed to save it, and it is characterized by a lack of unanimity among practicing scientists as to which ad hoc hypothesis should be

[79] Lowell 8, 36; GR 150–151.

[80] Lowell 5, 26; GR 86–87.

[81] See Popper (1959, 19–20).

[82] See Lowell 5, 24–26 (GR 86–87), for this perspective on the crises summarized up to this point.

> adopted to explain the results of particular experiments. Every one believes the theory, but no one is quite sure what it is.[83]

These two ideas—the classical and crisis stages—are well illustrated with an example developed by him in his fourth lecture, "'The Principle of Plenitude': Subtle Fluids and Physical Fields." This example refers to a classical historical event that Conant's writings and lessons studied in depth—Lavoisier's Chemical Revolution.[84] Kuhn, in this case, paid particular attention to the main theory that preceded the revolution: the phlogiston theory.

Kuhn shows that the phlogiston theory meets that requirement—an orientation that gives unity to different processes. As a subtle fluid, Kuhn says, phlogiston fits better among those orientations of a metaphorical nature than in the cosmological ones, since, as he says at the beginning of the fourth lecture, these ontological elements are rarely points of view of the second kind (unlike atomism, for example).[85] Phlogiston, for him, certainly opened the door to a classical stage thanks to its unifying virtues. Experimentation developed well thanks to it.[86] Phlogiston thus had explanatory and operational virtues. There are not many differences, in this respect, between Kuhn's version here and Conant's previous one. In *On Understanding Science*, Conant recounts the function of phlogiston in a similar vein:

> The phlogiston theory in its day was, we must first realize, a long step forward. In the sixteenth and seventeenth centuries those who were interested in making some sense out of what we now call chemistry were wandering in a bewildering forest. From the alchemists and the practical men, particularly the metal makers, they had acquired a mass of apparently unrelated facts and strange ideas about "elements." [...]
>
> How were all these facts, inherited from the Middle Ages and before, to be fitted together? By the introduction of a principle called phlogiston, closely related to Aristotle's old element, fire—closely related, yet the relationship was never clear.[87]

For Conant, just like for Kuhn, phlogiston promoted research in chemistry during a long and productive period.[88] Kuhn emphasizes, on the other hand, that the process of abandoning the phlogiston theory was not simple and was marked by problems that directly affected it and by the appearance of ad hoc solutions that aimed to eliminate those problems. The case of the weight gain of metals after their calcination (despite the phlogiston loss) is one of them. Reactions to this anomaly included downplaying its importance and suggesting ad hoc explanations of the phenomenon. This illustrates one of the causes of the crisis phase, according to Kuhn. That cause is not the ad-hocness of the hypotheses offered, as Popper would have pointed out, but their proliferation and the resulting weakening of the agreement around the theory itself as a clearly identical tool for all practitioners of the current chemistry to continue

---

83 Lowell 5, 25–26; GR 86–87.

84 See our Chap. 2, Sect. 2.4, for some details about Conant's account of this episode.

85 Lowell 4, 3, 17; GR 58, 64.

86 Lowell 4, 25; GR 67.

87 Conant (1947a, 80–81).

88 Conant (1947a, 103).

with the ongoing research. Once in this state, a theory begins to lose its virtues as an orientation, and science is not practiced on a unified ground, but from a much more fragmented conceptual base.[89] Kuhn speaks of it in the following way:

> New experiments called for new modifications and distortions of the theory, and though individual chemists adopted appropriate distortions to fit the theories to the new facts, there was a complete lack of agreement as to the particular manner in which the theory ought to be modified for a particular application. As a result, by 1775, when the theory was first publicly attacked, there were really a number of phlogiston theories. Everyone believed the phlogiston theory, everyone used it, but no one could agree on just what it was.[90]

In summary, the phlogiston theory is a good example of the nature of an orientation and of the two stages that it goes through. The third phase to study is the well-known period of scientific revolution. The phlogiston theory also suffered the consequences of that new stage of change. Let us see next what this stage consists of for Kuhn.

3.10 The Dynamics of Scientific Ideas, II: Scientific Revolutions

The last third of Kuhn's fifth lecture is dedicated to the concept of scientific revolution as a stage of his dynamics. One orientation is replaced by another in this phase. It is not, as he says, simply that evidence contributes to show the need to choose between two alternatives that respond to the same facts. Everything is, according to him, more complicated than that. The behavioral worlds and the scientific objects to which two different theories—also from different scientific orientations—respond are unalike. Choosing between them involves losing the cognitive and theoretical resources that have, thus far, been useful in dealing with phenomena, and also losing the phenomenal world itself, so to speak, and going through the difficult period of learning to behave in a very different cognitive scenario. We can rely on the examples of experimental psychology already mentioned in order to understand what kind of adaptation process Kuhn has in mind here.[91] The process is even more complicated in science, because it is not only about executing simple, everyday activities in a slightly modified scenario, but about practicing the complex theoretical and experimental activity to which the mature scientist has become accustomed over a long period of time. It is difficult to abandon our old cognitive scenario, the conceptual world we share with professional colleagues. For him, the classical values of rational theory choice are not the key to make the scientist shift from one orientation to another. Although some elements from a prior theory remain in its successor—like, for instance, the statements that

---

[89] For the entire case of phlogiston according to Kuhn, on which we have based this paragraph, see Lowell 4, 17–30; GR 64–70.

[90] Lowell 4, 30 (GR 69–70). Kuhn says something similar in Lowell 5, 25–26 (GR 86–87).

[91] See *Structure*, 112 ff. Kuhn adapts the metaphor of Gestalt change in these pages (of Section X) with some qualifications.

(without a deep ontological commitment) have passed tests successfully—the choice is not easy.[92]

There is no reference here yet to the "new-world problem," as Ian Hacking called it, which has nourished debate among philosophers of science.[93] This problem is well represented by some passages from *Structure* that have been repeatedly reproduced. A good example can be found in Section IX of the book; Kuhn says: "I have so far argued only that paradigms are constitutive of science. Now I wish to display a sense in which they are constitutive of nature as well."[94] Then, in Section X he adds: "though the world does not change with a change of paradigm, the scientist afterward works in a different world."[95] He also clarifies later that: "Rather than being an interpreter, the scientist who embraces a new paradigm is like the man wearing inverting lenses [i.e., a well-known experiment performed at the Hanover Institute that Kuhn cites shortly before in the book]. Confronting the same constellation of objects as before and knowing that he does so, he nevertheless finds them transformed through and through in many of their details."[96] This last clarification reveals that the previous cognitive explanation of the orientations that we have seen before, the idea of a behavioral world, is the basis for properly explaining what Kuhn has in mind in these passages from *Structure*. For this reason, when Paul Hoyningen-Huene promoted the differentiation between a "world-in-itself" and various "phenomenal worlds" to make sense of this Kuhnian idea about world-change, he was reviving an idea previously exposed in the Lowell Lectures, now more explicitly from a neo-Kantian theory that fit well with Kuhn's perspective.[97]

The "new-world problem" arises from the need, exposed in these lectures, to talk about the research *behavior* that originates when postulating the kind of cognitive scenario (the behavioral world) that Kuhn presumes necessary for a correct view of scientific activity. In this behavior, the scientist does not seem to be simply *interpreting* a set of universally accessible and accepted observations, but constructing a hypothesis or solution to a certain problem, all of which requires a complete ontological scenario whose description only that scientist and the rest of members of his or her community would find coherent. Perception, as Hanson would later say, is "theory-laden" and Kuhn's idea of the behavioral world explains this point.[98] Obviously, there is a difficulty here—a problem, as Hacking says. If there is incompatibility between the ontological scenarios of two different communities, how do we avoid a skeptical attitude about whether such scenarios offer us contact with the unaltered world that underlies the behavioral (or phenomenal) world? Hacking

---

[92] See Lowell 5, 28–31 (GR 87–89), for these ideas on the scientific revolution in Kuhn.

[93] Hacking (1993, 276).

[94] *Structure*, 110. This is a statement often reproduced in studies on Kuhn.

[95] *Structure*, 121; cited by Hacking (1993, 276) and by many other authors. For similar passages, see *Structure*, 111, 117, 134. Hacking (1993, 275–276) gives a representative list of such statements.

[96] *Structure*, 121–122.

[97] See Hoyningen-Huene (1993, 31–36 ff.).

[98] See *Structure*, 120–129, for his critique of interpretation. For the expression "theory-laden," see Hanson (1958, 19).

and Hoyningen-Huene offered answers to this problem.[99] Even in his later years, Kuhn believed that this little philosophical puzzle could be explained by his lexical theory—a development of the views expounded in these early Lowell Lectures.[100]

Leaving this difficulty aside, Kuhn believes that a scientific revolution is, in the development of a science, an episode which—although it may overthrow a theoretical building that for its architects and inhabitants was sufficiently well constructed—is also an opportunity to solve old problems. This is, in my view, a central point of these lectures. At the end of the fifth lecture, he proposed a small puzzle to his audience. It is in these lectures that he uses the notion of a paradigm for the first time, even if it is only to refer to the example in question, that is, without yet occupying the position it will later hold in *Structure*.[101] Kuhn leaves no written clues in the fifth lecture about the original setup; he seems to have left this to improvisation. In the sixth lecture, however, he shows that everything is based on a blank eight-by-eight grid that lacks two squares in two opposite corners. All the extant squares in the grid must be covered in pairs by something such as domino pieces (these are also mentioned in the sixth lecture). In this new lecture, he says that he has already converted the board into a chess board, with the squares missing in the corners; of course, he was alternating colors for each space. He shows that we now have an intuitive idea that reveals that we will not be able to fill such a 62-space grid with such pieces and tries to show two things with this demonstration: first, that we find out whether the problem can be solved or not (in this case, it cannot) from the moment we adopt a new point of view about it; and second, that perhaps the new approach suggests a likely solution or at least a new path to obtain it, such as, for example, distinguishing the spaces of the grid in another way, or composing the pieces so that they cover more spaces (three, for example) and with a different spatial order.[102]

Kuhn emphasizes that we gain a lot with the rearrangement of what we already know, without additional empirical information, as soon as we set up new connections within our previous knowledge. Grids and chessboards are not new perceptual objects for us. The orientation in question has simply offered us a new heuristic, metaphorical path, which seems to be fruitful. Historical cases of scientific change examined in the Kuhnian way do not differ from this example.[103] The case about Galileo

---

[99] See Hacking (1993); Hoyningen-Huene (1993, Chap. 2).

[100] See Kuhn, *The Presence of Past Science*, The Shearman Memorial Lectures, University College London, TSKP 23.32, Lecture III, 79–83 (*The Last Writings*, 80–83), for his later treatment of what Hacking called a "new-world problem." Also, see Kuhn, *The Road*, 101–104, 206–207, 228–233, 245, 249, among other relevant passages. Hacking (1993) discusses Kuhn's stance on this matter in *The Presence of Past Science* and other texts. Other discussions of the same problem worth mentioning are those of Fuller (2002, 87–89); Sankey (1994, 152–153, 179–180, 187–190); Bird (2000, 123 ff.); Read (2012, 34–40); and, of course, the aforementioned classic book by Hoyningen-Huene (1993). See also Hoyningen-Huene (2021).

[101] See Lowell 5, 46; GR 94; Lowell 6, 1; GR 97. See also our Chap. 5 on this point.

[102] See Lowell 6, 1–5; GR 97–99. See Galison (2016, 61–62), who recreates (in his words, "as best [he] can") this little mental challenge. My reconstruction is indebted to Galison's description, and also to Reisch's (2019, 240, 410, n. 3; GR 95, n. 19), who traces its origins to Max Black (1946).

[103] See Lowell 6, 4; GR 98.

and the scientific method that, inspired by Alexandre Koyré, he expounds in these lectures, has much to do with the metaphorical function of orientation in theory change that this small abstract example helped him to illustrate. Kuhn says that Galileo's experimentation depends on a new problem and a new perspective on motion. Experimentation is thus submitted to intellectual elaboration. As Koyré said in his *Galileo Studies*, "good physics is done a priori."[104] For Kuhn, many of the observations on which Galileo based his research work were previously available (for centuries, we can say). From this point of view, Galileo is seeing the world from a new orientation that affects observation, helping him to grasp new details in old phenomena—or new phenomena in old situations. This is truly unacceptable for Pearson, but pretty convincing for Kuhn (and for Koyré, of course).[105]

Kuhn also comments here on a question relevant to scientific revolutions: the relationship between youth and the innovative processes of theory revision that lead to such kinds of change. This is also related to the so-called "Planck's Principle," according to which innovative theories are finally accepted by a scientific group because its older members eventually disappear; this is a relationship that Kuhn seems able to explain in the terms expounded here. For him, the apparent role of youth in innovation—just as in the case of older outsiders that became innovators in a field that is new for them—has to do with the still untrained human mind during its early steps in a given field that remains, as yet, pretty new for that mind. In comparison, the older scientist, especially one who can be considered a well-established insider, has his or her view of nature already completely arranged according to a certain, usually deep-rooted, orientation. So, the youth–revolution correlation (and the contrary one, mature-expert–conservativeness) is thus explained by virtue of the theory expounded here.[106]

3.11 Formal Languages and the Structure of the Scientific Textbook

As we have seen in this chapter, there are no qualitative differences, in cognitive terms, between the behavioral world of the common individual and the mental world of the scientist, forged by means of an orientation. From both, the same type of language arises: the natural in the first case and the scientific in the second. The difference between both is the degree of precision that is achieved in the second case. The goal of that precision is that terms do not refer to scientific objects ambiguously. Terms

[104] Koyré (1978, 166).

[105] See Lowell 1, 20–21 (GR 13–15); see also 14 (GR 9–10). For a more extended account of Galileo's work and its sources, see also Lowell 2, 1, 8–15, 19–23 (GR 21, 25–31, 32–36); Lowell 3, 2–5 (GR 154–159).

[106] See Lowell 5, 35–43; GR 90–93; see also Zuckerman and Merton (1972), Wray (2003), Hull (2001) on this issue; see also Hull et al. (1978) on "Planck's Principle." See *Structure*, 149–150.

must have clear applications and give rise to the least possible confusion when used in our scientific expressions such as generalizations and laws.

The appearance of the scientific textbook is that of a scheme free from ambiguities. The scientific textbook seems to be written in a language whose logical structure is thoroughly respected, and in which the theoretical part is clearly separated from the applied part. When we study the theoretical part of a nomologically-structured discipline (like any of the physical sciences) we can grasp the conceptual meanings in the relationships that the generalizations or laws exhibit between terms. Explicit reference to entities and concepts is usually not added. The results of observation directly or indirectly support the predictions made from the theory. Thus, fundamental entities are rarely seen, but their place in the theory is respected because both the features of these basic entities and the concepts by which we talk about them are part of a theoretical framework that successfully confronts the natural world.

This organization of the scientific textbook leaves room for the scientist's work. The scientist can focus on making scientific objects more precise, as there is room for ambiguity in the reduced explicit reference to basic entities and concepts. Up to this point, Kuhn could agree (and in fact did agree) with the organization of scientific education. As a freshman student in a particular field of physics, for example, a young individual could delve into the vague meaning system of the scientific textbook and then develop his or her career trying to better specify the various relationships between scientific objects, or trying to understand those objects in and of themselves; normally all at once.

However, the view from empiricist methodology is that the scientist does not behave like this. For the empiricist methodologist, from Pearson to Bridgman, the process to follow is very similar to the scheme in the scientific textbook. Since there is no reference to the non-explicit content in the formalization, that content *should not* be part of the scientific language. Phenomena should speak for themselves and we should generalize from their obvious features. Kuhn, however, argues that we should make use of all the features that, although they are not fixed in our view of the world (and perhaps cannot be, according to a principle of verifiability), can help us clarify the features of objects, so that our identification is exact, precise, and our world logically and empirically coherent. What he says in his last two lectures is that empiricist methodology leaves that baggage of meanings that *is an integral part of science* inactive. In those lectures, he tries to overcome the barrier that empiricist methodology establishes between natural language, in which that margin of semantic vagueness is handled with ease and without fear, and scientific language. For him, once the empiricist methodology is corrected in this sense, thanks to all that we have just seen, it seems to be an obsolete monument to a sterile formalist vision of science. Let us start our exploration of these two last lectures by getting to know a little better what this formalist plan consists of and what its scope of validity is in scientific research, according to Kuhn.

As he points out at the beginning of the seventh lecture ("The Role of Formalism"), one of the most widespread characteristics of the physical sciences, one of their essential attributes, is their mathematical and logical structure, their deductive structure. With it, we face an understanding of logic as a conventional system, independent of empirical reality, to which mathematics can be reduced. The logical structure of mathematics and, in particular, of geometry stems from this disconnection, especially following the axiomatization of Euclidean geometry by David Hilbert, an example that Kuhn uses in order to illustrate this idea.[107] There is no need to relate this language to its possible interpretations to carry out research in mathematics. We deal with symbols and principles of a conventional nature and obtain new statements through pure logical manipulation. Conventionalism is an appropriate approach for mathematics; intuition does not play an essential role in it. This, he shows, leads to the reduction of mathematical truth to pure logical truth, thus completely detaching mathematics from the perception of the natural world.[108] He is aware of the problems of this conventionalist stance, although it fits well with his view of the social nature of language, as he also shows here. He says:

> The rules of formal logic or of formal language are conventions. They are the rules of the game which we play with other human beings when we communicate. In themselves they are no more necessary than any other adequate set of rules, but without some such set no communication or very little communication would be possible. We are not then obliged to admit the necessity of the syllogism, but an announcement that we will not do so is an announcement that we will not abide by the rules of the game, that we will not play. It is therefore in the most literal sense anti-social, and it carries the penalty of other anti-social acts, it deprives us of the privilege of learning certain things from the experience of others.
>
> I confess that this view that the truths of logic are products of linguistic convention is not by any means free of difficulty. It is probably true that it raises as many questions as it answers, and the problems raised have led a number of philosophers and logicians to reject it. I adhere to it at this time in spite of my inability to resolve all the difficulties simply because the difficulties seem no more severe than those which arise in the attempt to root logical necessity in the external world or in innate categories of the mind.[109]

However, the independence of mathematics thus expounded does not exempt the field from a potential and fertile application to the study of empirical experience. Moreover, the wide variety of postulate systems corresponding to different geometries and existing algebraic systems find very useful interpretations in physics, such as general relativity and quantum mechanics, for example. The same happens with logic. Logic is a system of symbols, axioms, theorems, and rules that is built and developed independently of the observation of the real world, but it is very useful as a foundation for the study and analysis of all kinds of inference; in particular, scientific ones. Furthermore, a formal language enables the study of recurring modes of argumentation visible in ordinary speech and the representation of these in its own vocabulary. The reconstruction of non-formalized languages through logic, their reduction to interpreted logical calculi, is a project that, Kuhn shows, has seduced

[107] See Lowell 7, 5–7; GR 118–120; also see Lowell 7, 1–4, 8–10 (GR 117–118, 120–121), for what was indicated earlier in this paragraph. Kuhn does not explicitly talk about David Hilbert, however.

[108] See Lowell 7, 1, 8–13, 27–28; GR 1, 120–122, 127–128.

[109] Lowell 7, 27–28; GR 127–128.

many philosophers for this reason. He seems to have in mind, obviously, logical positivism and its program of rational reconstruction of language.[110]

The typical scheme of the scientific textbook, Kuhn says, fits well with this formalist and reductive path. Physics, for example, has laws, scientific objects, and operations that are carried out in the laboratory and that anyone with the right equipment can repeat. In laws, we summarize the recurring behavior of scientific objects. The investigation of laws consists of showing what there is in the objects that leads us to deduce that this particular behavior should be considered a law of nature and to exclude any other behavior. This is carried out through the application and development of laws themselves. Through them, we explore the lesser-known areas of the behavior of scientific objects. This process becomes quantitatively precise in the mathematical formulation of laws. Once this step (directly related to the mentioned use of formal languages) has been taken, laboratory operations allow us to contrast the scheme of scientific objects and laws with empirical reality. Through experimental instrumentation, we carry out the corresponding measurements. With this, we calibrate the system and provide empirical content for the scientific objects and the features with which we define them, which we usually summarize in concepts such as distance, mass, force, etc.[111] The formalist tendency of the philosophy of logical empiricism easily coordinates with the point of view of operational analysis in the Kuhnian idea of the scientific textbook.

For Kuhn, however, this clear separation between the formal and the empirical does not represent real scientific research. To begin with, only on certain occasions, in reduced areas of science, and always in well-established theories, have both works been separated in this way. The formalization of geometry and the development of non-Euclidean geometries seems to be a good example of this, but, in this case, we are talking about a mathematical theory, a science whose research develops in a purely formal terrain. As Kuhn reminds us, however, Newton's mechanics benefited from this method in the nineteenth century, and so did Aristotle's physics in the Middle Ages. In both cases, new opportunities for research were created. However, Kuhn warns us, this is a special case within creative science. The formalist arrangement of theory and observation (or experiment) in the scientific textbook is, above all, a good way to transmit scientific knowledge when educating individuals. Leaving aside these two uses (that is, specific research and teaching methods), it is a mistake to consider that this resource of a fundamentally pedagogical nature is a good representation of real scientific research.[112]

Kuhn began the Lowell Lectures by distinguishing between textbook science and creative science. One of the pending tasks for methodology is, for him, to overcome the empiricist fallacy that subsumes the second into the first. A theory of scientific

---

[110] See Lowell 7, 13–17, 23–27, 31–34 (GR 122–124, 126–127, 129–130), for Kuhn's view as expounded in this paragraph. On the notion of rational reconstruction, see Carus (2007, 14–19, Chap. 6). On Kuhn's criticism of this program, see our Chap. 7. It is relevant Reisch's (2019, 192) comment on the plans of full formalization of language.

[111] See Lowell 7, 40; GR 132–133; Lowell 8, 31–33; GR 149–150.

[112] See Lowell 7, 36–39, 42 (GR 131–132, 133), and Lowell 1, 10–13 (GR 7–9), for this perspective of Kuhn.

method, he says in his first lecture, uses a teaching resource as a guide for the representation of all science, including the one practiced at the frontiers, or margins, of current knowledge. That theory, the traditional one, the empiricist view, is the one he tries to correct by going deeper into the genuine nature of creative science.[113]

3.12 The Real Language of Science

The idea of language that Kuhn tries to communicate has more to do with three of his influences from philosophy and linguistics than with the logico-empiricist tradition that has addressed the study of science from the resources of the linguistic turn. These three influences are well known: Whorf from linguistics and Quine and Wittgenstein from philosophy.[114] The vision of language from which Kuhn starts is well represented by the viewpoint on language and knowledge that Quine outlines in "Two Dogmas of Empiricism." For Quine, language forms a network of statements from which each of them extracts its meaning; that is, if that meaning is empirical, a statement does not obtain it from its comparison in isolation with experience.[115] Language helps us, as a holistic set, to deal with the world we experience without there being a statement that is not susceptible to revision when language, and the knowledge it carries, proves ineffective to guide our behavior in our environment.[116] In this way, Quine says, "total science is like a field of force whose boundary conditions are experience. A conflict with experience at the periphery occasions readjustments in the interior of the field."[117] This holistic point of view in Quine involves that the "readjustments" he talks about may affect statements that we previously could consider immutable, including its logical foundations. Quine says, in this sense, that: "Even a statement very close to the periphery can be held true in the face of recalcitrant experience by pleading hallucination or by amending certain statements of the kind called logical laws."[118] Quine's attack on analyticity and empiricist reductionism allows him to offer a holistic alternative (which he later qualified) to the project of rational reconstruction of language in the manner of Carnap and other empiricist thinkers.[119]

---

[113] Lowell 1, 12; GR 8–9.

[114] See our Chap. 2 for details on these influences.

[115] See Quine (1951, 41–43).

[116] Quine thus develops what we know as the Duhem–Quine thesis, adding the ideas of the French physicist and philosopher Pierre Duhem, who advanced a similar thesis. See Quine (1951, 41–42). The Duhem–Quine thesis has been studied and discussed on countless occasions. For a recent historical review, see Zammito (2004, 15–25). We shall return to some consequences of this thesis according to Kuhn in Chap. 8.

[117] Quine (1951, 42).

[118] Quine (1951, 43).

[119] For his critique of Carnap, see Quine (1951, 45–46). For further qualifications, see Quine (2008). For a study of his versions of holism, see Zammito (2004, 25); Gibson (1996, 80–89).

Kuhn's viewpoint on language in the Lowell Lectures (and afterwards) is identically holistic and qualifies something that also coincides with Quine's view: the idea that our rules for applying names are neatly divided into definitions (stipulations) and empirical generalizations does not make sense in the real use of language.[120] As in the case of the epistemological study of orientations through behavioral worlds, Kuhn uses what we know about ordinary language to approach scientific language.[121] He shows that, just as for Quine and for Wittgenstein, our use of language does not depend on an organized set of rules, but on learning to consistently apply grammar and a series of terms in such a way that our adaptive behavior to the social and natural environment is consistent.[122] The price of not doing it right is not as simple as contravening a set of rules; sometimes it entails problems for our public life; sometimes, for our life, period. Not knowing how to recognize, for example, that the animal before us fits into the category of rattlesnakes has dire consequences for more than just linguistic behavior.[123] The cost of misuse of language and the concepts it contains is, therefore, high. For Kuhn, as for Quine, language has an inalienable *pragmatic* component, even though its formalized version occasionally contributes to clarify the foundation of the concepts transmitted by it, or an operational analysis manages to clarify the limits that observation establishes for such concepts. In this lecture and in the previous one, Kuhn tries to convey the critical thesis that if the rational reconstruction of scientific language (key to logical empiricism's criterion of cognitive meaning) affects this pragmatic, adaptive function of language, we lose in terms of functionality what we gain in conceptual clarity, thereby subtracting *utility* from language—the end that it satisfies.[124]

The actual scientific language in Kuhn is, therefore, a vehicle very similar to common language and shows those traits of holism and practical virtues that we see in Quine. Like its mental counterpart, the behavioral world (the orientation, in the case of scientific knowledge), language changes according to the practical needs it has to satisfy. Polishing it, therefore, is a specific function with a particular objective, not a total renovation plan. Kuhn would agree with Wittgenstein (even if he is not actually *following* him) when he states that "ordinary language is all right. Whenever we make up 'ideal languages' it is not in order to replace our ordinary language by them; but just to remove some trouble caused in someone's mind by

---

[120] Lowell 8, 15. See *The Road*, 71–73, for a later example of this idea.

[121] See Lowell 8, 37; GR 151. This perspective in which science is a continuation of our usual forms of cognition and communication does not distance him much from Quine, who says in "Two Dogmas of Empiricism": "Science is a continuation of common sense, and it continues the common-sense expedient of swelling ontology to simplify the theory" (Quine 1951, 45).

[122] See Lowell 8, 9–21; GR 140–145. We will address this idea later. There are parallels with some passages from *The Blue and Brown Books* by Wittgenstein, which Kuhn says he read around 1950 (see our Chap. 2). See also Mayoral (2015, 196–197).

[123] Lowell 8, 20; GR 144–145. Also remember, in this same sense, a paragraph reproduced pages back, in which Kuhn warned of the social implications of the anomalous use of one's own logic; see Lowell 7, 27; GR 127.

[124] See Lowell 7, 40–45 (GR 132–135), Lowell 8, 13, 27–38, (GR 142, 147–152) for this critique of Kuhn.

thinking that he has got hold of the exact use of a common word."[125] In short, the same instrumentalism that prevails in Kuhn's reconstruction of the psychological world in these lectures is reflected in its linguistic counterpart. Both psychological world and language are two sides of the same coin, and have a similar instrumental nature, although language is publicly accessible, while the behavioral world (and the orientation, of course) is part of our private experience.

The evolution of the behavioral world (and of an orientation) and of its corresponding language follow a similar scheme according to the pattern established in the vision of education that Kuhn obtained from Piaget. Changes are motivated by the new ingredients that emerge in experience, and language, just like the behavioral world that underlies it, must be flexible enough in order to face them. Again, the formalization of language is an obstacle. This additional consideration is important for Kuhn. A fully formalized scientific language, in which perceptual contact with the world constitutes a firm foundation for the rest of the theoretical building, would be useful if there were no (minor or major) novelties in experience. However, ordinary and scientific experience reveals that we cannot trust in such absence of variation.[126] Let us remember an even younger Kuhn who wrote the essay "The Metaphysical Possibilities of Physics" and who speculated in it about the plausibility of a state of finitude, and so of stability, of information.[127] Now, a decade later, this older Kuhn no longer considers that possibility, or at least assumes that the contrary one is likely (in *Structure* he seems to make a similar assumption as regards the emergence of the anomaly).[128] Novelty often emerges, so a formalized language such as that fostered by the logico-positivistic program is unable to deal efficiently with it; it is, rather, an obstacle to the advancement of science, he says there.[129] Only a naturally evolved language like ordinary language, or the real scientific language, can deal with that novelty; or, at least, only a language such as Kuhn considers these last two to be (or should be) can.

To understand why Kuhn defends this position, it is advisable to study the image of language, of its learning and functioning, that he is already demonstrating here, in the Lowell Lectures. We have said that language, for him, has holistic and pragmatic characteristics. Language evolves as a whole and is altered both individually (in education) and collectively (in scientific research, for example) due to the needs of behavior. So, he assumes a *dynamic* vision of language, identical to the one that affects orientations, and we should study it in that field of, so to speak, permanent evolution.

Language-learning shows us how the meaning of terms—whether they are proper names or kind-terms (he does not mention other words)—is assigned. He shows that our main resource for assigning and applying them is a combination of rules and ostension. We need contact with representatives of a kind and some rule to refer to the

[125] Wittgenstein (1958, 28).

[126] Lowell 8, 6; GR 139.

[127] See our Chap. 1, Sect. 1.10.

[128] See Lowell 8, 14, 28; GR 142–143, 147–148. See also *Structure*, Section VI.

[129] *Structure*, 98–100.

class through its meaning. Within this last method, we may divide the work between the use of stipulative statements and that of empirical generalizations—although for Kuhn this division is more functional than logical (surely a notion inspired by Quine). Our ability to assign a proper name to a subject, or to apply a kind-term to a member or sample of a kind does not depend exclusively on a description, or series of them, that allows us to identify them without ambiguities. However, that ability is usually as successful as behavior demands. When we assign a term, we use a combination of empirical generalization, definition, and ostension, he shows. In this way, we discriminate the objects that appear in our behavioral world for the first time, and keep that world ordered. Around our individual, property, or substance term, we can draw, metaphorically speaking, a certain sector of meaning formed by the criteria that allow us to apply it. Kuhn shows no intention to affirm that necessary and sufficient conditions are needed to identify an individual and apply a term to it. We are conditioned by the pragmatic, behavioral nature of language usage, and we select the term after satisfying some criteria that are sufficient for the particular behavioral situation we face. I will not try to satisfy too many criteria to identify a rattlesnake before moving away from it in its presence, but I may do so in order to assure myself that it is a rattlesnake when examining a picture book on snakes, or when answering a question about its species in an exam.[130]

In this account of language, Kuhn also gives some features of the kind of ontology it requires—that is about what groupings of objects, substances, and relationships are needed. In doing so, he makes a statement whose content would remain roughly unchanged in his later work—from *Structure* to his Shearman Lectures in 1987 and later on in *The Plurality of Worlds*.[131] In *Structure*, he refers to this statement when he mentions what he adds to Wittgenstein's view. Kuhn says: "Wittgenstein, however, says almost nothing about the sort of world necessary to support the sort of naming procedure he outlines."[132] In *Structure* and later, Kuhn argues that the world has its own distribution in "natural families," among which there are unpopulated spaces, which makes his model of language-learning appropriate from a pragmatic, behavioral point of view.[133] If Kuhn's aim is to show that the use and application of a theoretical vocabulary may be governed by the pragmatic mode surrounded by vagueness that he defends here, the distribution in "discrete natural families," as Hoyningen-Huene calls them,[134] must be part of our experience of nature. Looking

---

[130] The perspective of this paragraph is a reconstruction from Lowell 8, 3–18 (GR 138–144). What is said next is also based on those (and other) pages of the eighth lecture, which constitute Kuhn's version of language-learning and meaning at this time.

[131] See our Chap. 8 and the Epilogue for this book.

[132] *Structure*, 45, fn. 2.

[133] See *Structure*, 45–46; Postscript—1969, 196, n. 14; *Essential Tension*, 312–318; *The Presence of Past Science*, The Shearman Memorial Lectures, University College London, TSKP 23.32, Lecture II, 45–46; *The Last Writings*, 58–59. Kuhn would also theorize about other kinds of terms and their relationship with the phenomenal world and the learning process, of course, especially in writings of the 1990s. See our Epilogue for further details.

[134] Hoyningen-Huene (1993, 95); see also its Chap. 3 ("The Constitution of a Phenomenal World") for more details.

at it in this way, in the Lowell Lectures, Kuhn is already building his perspective of "family resemblances" on which he relies both in *Structure* and in post-*Structure* writings in order to show how the natural and social categories are arranged and how kind-terms are applied. In fact, he would include a more finished vision of this in the revised manuscript of *Structure* in 1962—its current Section V—and its further evolution would lead to a synthetic presentation in "Second Thoughts on Paradigms" and in further writings. In that sense, the seemingly fresh renewal of his views in *Structure* years after the book was published has recourse to much older notions.[135]

One may reasonably wonder whether language can be improved according to this model. The answer is of course affirmative. For him, in fact, it *is* always improved. But the improvement only solves specific problems of behavior according to goals such as those set by scientific research. Later, we will be able to understand that, for him, such improvements are positive just as they are also sometimes (at least, potentially) negative. So, in Kuhn's view, a general plan of clarification as envisaged from logical empiricism or the reconstruction of language according to formal criteria of whatever kind constitutes, even in 1951 (and in a similar way afterwards), a serious mistake.

We will return to this later. Let us go back to language now. Our contact with nature in our behavioral world sometimes presents novelties in experience, which sometimes involves an adjustment of our behavioral world in order to put behavior in line with the old and the newly discovered elements of our natural and social environment. This is Kuhn's view of learning as inspired by Piaget. It means that our resources for naming individuals and for applying kind-terms largely depend on our education. Kuhn's idea here is not very different from the one he would demonstrate many years later in "Possible Worlds in the History of Science," when talking about the alternative ways in which different individuals within a community learn the same vocabulary, such as that of Newtonian mechanics. Other later works develop that view, too.[136] He shows in them that, despite that variation, these methods allow the use of vocabulary in such a way that the discrepancies are merely complementary; if they were not, there would have been a rupture between those two individuals like the one that results from a scientific revolution. He says something very similar here. Given the combined methods with which we learn to use the terms of our vocabulary, there may be discrepancies from one individual to another within the same community as regards the concepts behind them, which are marked by the different generalizations, definitions, and experiences that we have accumulated for their reference. This variety, in Kuhn's view, helps an individual draw a certain "map" for the distribution of meaning around a term. Let us remember that I previously talked about "a certain sector of meaning formed by the criteria that allow us to apply it." That sector is not uniform, in Kuhn's view, and its lack of uniformity, and a remnant

[135] See our Chaps. 5 (for the detail about the draft of *Structure*) and 6–7 for the further evolution of this important matter in Kuhn's thought. This view of a recovery of old thinking is also reflected in Reisch (2019, 355), Mayoral (2023b, Sect, 5.4), and Melogno and Giri (2023). For this perspective in the Lowell Lectures, see Lowell 8, 19–20 (GR 144–145).

[136] See our Chap. 8 for more details on this view.

of vagueness that we already discussed in the previous chapter, stimulates research in a classical phase.

Let me describe this "map." For Kuhn, almost all speakers of a language use similar criteria and have similar experiences with certain aspects of their respective behavioral worlds. If we talk about scientific research, learning through textbooks—and, in many sciences, with the help of similar laboratory resources—contributes to further unify these common resources. So, most speakers in the community share a part of the meaning of each term. That part—the senses included—form what Kuhn calls here the "hard central core of meaning" for each term; it is a first, central sector within the full area of meaning.[137] However, not everything we know about an object or situation to which we apply a term or expression is known by everyone. The term "force," for example, does not mean the same for a theoretical physicist as for a lay person. Even in each of these cases there may be discrepancies around the use of the same term from one group to another. Kuhn gives an example of this kind in *Structure*.[138] Therefore, within the sector of meaning we are talking about, some individuals accumulate criteria and experiences that differ from those that other speakers of the language have at their disposal. In the lexical theory of the later Kuhn, in which these discrepancies are also present, he redirects his idea of common meaning to the stability of the structure of relations between the terms of the vocabulary; the lexical *structure* takes full responsibility for the declaration of incommensurability and its explanation. Here, in the Lowell Lectures, Kuhn assumes this viewpoint for the first time (these early considerations are the basis from which, more fully, the later, lexical viewpoint evolves). For Kuhn, the important point is that communication between speakers of the same language does not break down. He also adds for the first time the *variation* of meaning in such linguistic (and experience) communities, and locates it in a zone *further away* from the core, what he calls the "vaguer fringe of meaning," whose first notions we examined in our previous chapter.[139]

This gives us an idea of the areas of meaning that concentrate around a term, based on the knowledge about the object or property we name on the basis of pragmatic reasons. This view gives us an idea of the nature of scientific research. For Kuhn, this latter takes place in the area of meaning where there is not a clear coincidence of criteria—that is, where there are *divergences*. If the divergences go beyond the subsequent agreement, we enter a phase of crisis and subsequent scientific revolution, while, if there is a full coincidence between the uses of all speakers of the language, progress will be typical of a classical phase. As we can see, the research process is very similar to the path that any individual follows in his or her learning, with the difference that in the former a better orientation is enabled—rather than a behavioral

---

137 Lowell 8, 17; GR 143–144.

138 See *Structure*, 50–51. The well-known example shows the disagreement of a chemist and a physicist in their answer to the question whether helium is a molecule.

139 Lowell 8, 16–17 (GR 143–144); see also 32 (GR 149): "vague fringe of meaning." See also our Chap. 2; Kuhn called it in his 1949 notes "fringe of vague meaning," as we know.

world—and that the use of the vocabulary is clarified for its collective use, not only for individual understanding. The explanatory basis is, however, the same.

A question that arises at this point is how this research process varies with regard to the one suggested by the perspectives that suggest a logical reconstruction of language—which Kuhn criticizes here. We will see this in the next section, which complements this one. Now, however, we must add one more aspect of this vision of language: its relationship with the idea of *knowledge* in Kuhn.

On this last connection, he shows that these sectors of *meaning* around a term have a similar function in *epistemological* terms. After all, the meaning of a term depends on the beliefs that we have and use as criteria for the development of categories and the application of terms. We know, for example, that certain kinds of small animals, of different colors, surrounded by a certain number of legs, that can sting us and inoculate us with venom and that often weave webs of a substance similar to thread, but with greater adhesion capacity, are normally called "spiders." The use of this last term depends on a series of beliefs, some of which (the ones I just mentioned, for example) are known by almost all speakers. They are justified in holding them and they count as true. Our coherent learning and use of language bring with them the possession of certain beliefs that, according to the classical canons, we can call knowledge. Kuhn says, in this regard, that the closer we get to shared beliefs, the closer we approach the sector of meaning that is also frequently identified with the most secure knowledge we possess. In accordance with this idea, in the eighth lecture, he says that: "The area of stable meaning is an area of what we take to be certain knowledge."[140] This relationship between the acquisition of language and the acquisition of knowledge about nature is recurrent in his thinking. In the 1960s, right after *Structure*, and then in later work, it re-emerges in his writings every so often, and in "Does Knowledge 'Grow'?", his 1976 Foerster Lecture, which he would repeat on some occasions in the late 1970s, he returns to the question about knowledge from this perspective. We shall see all of this in later chapters.[141] Now, let us see what to do, according to him, with the plans of limiting meaning that try to suppress vagueness in the fringes (another recurring argument, by the way).

3.13 A Difficult Equilibrium

For Kuhn, language is a collective and tacitly created tool, which includes the principles of logic that govern our reasoning, and which makes sense within the behavioral worlds of each speaker. Language and experience advance hand in hand, therefore. As Hoyningen-Huene says for the perspective of *Structure*, "The relationship between language and perception thus shouldn't be thought of as either sort of unidirectional dependence but should rather be construed as a dialectical process of differentiation

[140] Lowell 8, 35; GR 150.

[141] See our Chaps. 6–8.

and transformation."[142] We might think that, in this process of mutual improvement, a formalization plan would be correct. Indeed, for Kuhn it could be, in particular applications, but he distrusted plans for a complete clarification. If, as Bridgman indicated, operational analysis was originally designed to keep the conceptual systems stable after revolutionary episodes like the irruption of the theory of relativity, Kuhn was, of course, against a project like Bridgman's.

For Kuhn, "the vague and behaviorally determined meaning systems of natural language are one of the most important vehicles for what we have previously called scientific orientations."[143] He seems to be inspired by Wittgenstein, once again, when the latter says that "[m]any words in this sense then don't have a strict meaning. But this is not a defect. To think it is would be like saying that the light of my reading lamp is no real light at all because it has no sharp boundary."[144] Those systems of meaning that Kuhn talks about, marked by vagueness and practical use (by behavior), involve an approximate meaning for terms (related to an also approximate knowledge) that is nonetheless useful for predicting the behavior of objects in experience. In a significant passage from this eighth lecture, Kuhn states:

> For names, let me repeat, are not simply artificial and arbitrary labels. They are focal points about which crystallize our expectations about the perceptual worlds, and the act of naming a particular perceptual complex is a positive act. It is a statement of belief about the future history of the named complex.[145]

Kuhn's claim for the preservation of the fringe of semantic vagueness around language terms is caused by the conviction that these predictive, practical functions, marked by effectiveness in behavior, involve beliefs and convictions about objects whose "future history," as he says, we want to predict, which are not present in the minds of all speakers of the language. If we were asked about the meaning of a term by which we identify a certain object or kind, we would find the sort of discrepancy in the fringe of vague meaning that we have already discussed. For Kuhn, thus, the meaning of a term is not exhausted by what *all* speakers have in mind regarding it—it also includes what *various subgroups* consider the named object or kind to be, which would be lost in a process of strict definition of the term in question.[146]

We should not take this perspective too far, however. Kuhn is aware that ordinary-language terms are sometimes pejoratively laden, which sometimes affects individuals to whom they are applied. This unwelcome consequence could be avoided with greater accuracy of definition, and thus by stricter criteria for application. He mentions as an example here political vocabulary, by which we ascribe certain labels to some individuals just because they sometimes behave according to a vague and maybe even circumstantial generalization, typical of the fringe of vague meaning.[147]

---

142 Hoyningen-Huene (1993, 95).

143 Lowell 8, 34–35; GR 150.

144 Wittgenstein (1958, 27).

145 Lowell 8, 30–31; GR 148–149.

146 Lowell 8, 13; GR 142.

147 Lowell 8, 21–22; GR 145–146. See Reisch (2019, 191–196) for a reading of these passages of the Lowell Lectures in political key.

However, he also considers that an increase in semantic accuracy, the search for definitions, is something that should be achieved only *to a certain degree* and is part of a difficult equilibrium that we should maintain. In his eighth lecture, Kuhn presents a position on the reduced desirability of drawing precise limits for the meaning of a term and on the dangers that the extremes of that strategy create. That is an argument that he recovers and elaborates on in papers like "Second Thoughts on Paradigms" and in his preparatory work for such publications.[148]

For Kuhn, there is a difficult balance between *accuracy*—typical of exact definitions—and *effectiveness* when facing new experience of objects. The domain of objects and kinds that we face is categorized, but the margins of categories are not drawn by completely sharp borders in all cases. The exceptions to this vagueness are those aspects of the world in which definitions are needed in order to distinguish between categories. In the rest of the cases, like those in which natural families are involved, the limits are not wholly precise. An increase in accuracy, which is one of the possible ends to which language evolution can tend, would contribute to specifying these limits. However, Kuhn says, this is also a step taken in favor of arbitrariness. As we saw previously, for him there is no arbitrariness in vocabulary, but an attempt to preserve its predictive functions—to adapt ourselves to behavior. What we do by being more precise is to select which beliefs we want to guide the application of our vocabulary and which others we do not. One extreme of this process would be to restrict ourselves to what we only know for sure—the content that is part of the core of meaning, exclusively. A second potential end is to expand the meaning of a term until it includes everything that can be taken into account about it, and thus, to draw clear boundaries for the entire domain of objects, actual and possible. We should avoid the intersections of two categories that are supposed to be mutually exclusive or even complementary. The entire domain of inhabitants of the experience (actual and possible) would be completely covered by our classification. But this, again, is a step also taken in favor of pure arbitrariness.[149]

Semantic accuracy (of whatever type) that allows us to draw those clear boundaries for the meaning of terms, suppresses one of the fundamental goals of Kuhn's vision of language, as already pointed out: the flexibility that gives it a great adaptive effectiveness, the ability to face novelty in our experience efficiently. We already know that, for him, there are two steps to take in this regard. One of them is to group—sometimes in a somewhat artificial and ad hoc way—the variations in the classification that is already in place. Fringes of vagueness often assume such exceptions. A second step, in another direction, is the alteration of the categories already established in the processes that we know as revolutionary science.[150] Whatever the step to take in the face of novelty, and especially if it is the first step, the first action

---

[148] See Kuhn (1974, 313–318), for the elaborated version of what we are going to discuss in the rest of this section. See also my discussion of that work in the 1960s, which is connected to the argument here, in Chap. 7.

[149] See Lowell 8, 25–31 (GR 146–149), for this perspective of what I have called "balance" (or "equilibrium") in Kuhn. See also our Chap. 2, Sect. 2.15.

[150] See Lowell 8, 35–38 (GR 150–152), for Kuhn's discussion of these two modes of research in relation to language.

is to try to integrate the experience within the existing categories and apply the same terms that we already use. To do so, we cannot depend on so exact or so ample criteria that their application deprives the term and the corresponding concept of any value for behavior—a form of arbitrariness. Our language must be a means to discriminate what is experienced and act accordingly. If we restrict it to what we have already perceived and classified, or to anything that could fall within the current classification, we reduce our capacity for an effective subsequent action. Either of the two steps previously considered—a classical behavior or a revolutionary one—would be hindered by the establishment of such precise limits.[151]

For Kuhn, therefore, the improvement of language is linked to the preservation of its adaptive function, of its ability to promote effective behavior, which has consequences for the evolution of language itself and, of course, of the behavioral world or scientific orientation in which its vocabulary is applied. Science, he shows, is characterized by increasing precision. The textbook, along with all the educational function that unfolds around it, bases its structure on the extreme satisfaction of that end. But Kuhn, as he began to assert in the first conference, shows that science is not reduced to that model of artificial and somewhat arbitrary semantic and categorical accuracy. The world of experience in his view is full of novelties and our cognitive and linguistic resources must be able to face them effectively. A plan for formalizing language leads, in his opinion, to the elimination of precisely everything that makes the scientific view of the world an effective cognitive and communication tool.

3.14 Meaningful Absences

The Lowell Lectures of 1951 are a synthesis of all the ideas that Kuhn had accumulated over the previous years, from those visible in his works of 1945 to those of the *Handy Notebook 1949* and later. He took a further step in the following decade that led him to the arguments in *Structure*. References to the linguistic dimension of these lectures were, however, minimal in that book. As he pointed out shortly after the Lowell Lectures, he hoped to explore such an aspect of his view more exhaustively later, but not in *Structure*.[152] Indeed, we have to wait until after *Structure* to see how he reopened old questions and viewpoints that he would never disallow as central.[153]

However, as early as the first half of the 1950s, there was already an important dimension of scientific orientations, especially for *Structure*, to which Kuhn paid renewed attention: the *social* dimension of knowledge. Just as the epistemic and semantic aspects of his view are only passingly examined in *Structure*, attention to the specifics of the scientific community, and to its epistemic significance, are not

[151] See Lowell 8, 23–31 (GR 146–149), for this critique of the formalist program by Kuhn.

[152] Kuhn, "Plans for Research," 22 October 1953, UCBTK, 3–4; edited in Hufbauer (2012, 459–460).

[153] Our Chaps. 6–8 and the Epilogue shall be devoted to that post-*Structure* work.

prominent features of his Lowell Lectures. However, in its fusion with the psychology of research, with the study of the origins of scientific languages, and with inquiry into the cohesion and disintegration of research groups, the sociological study of science would be crucial for him. Our next two chapters show how he combined the two key ingredients of the history of science and the sociology of science in his book.

Chapter 4
The History and the Sociology of Science

4.1 Kuhn as an Assistant Professor of the History of Science

In April 1951, after delivering the Lowell Lectures in March, Kuhn was hired by Harvard University as an instructor in General Education and also as a tutor in the history of science.[1] It would be another year before he reached the position of assistant professor which he would hold for the next half decade, a period that would end a year earlier than expected owing to his departure for the University of California, Berkeley in the fall of 1956.[2]

During his last period at Harvard (in 1951 he had already spent more than a decade at that university, from student to assistant professor, only leaving it during the war), Kuhn dedicated himself to teaching, with only occasional forays into publication—mostly very brief papers on the history of science. In fact, except for his work on the impact of atomism on seventeenth-century chemistry,[3] his publications before 1955 are notes and reviews, mainly published in *Isis*—George Sarton's and the History of Science Society's main publication—although he also published a couple of reviews in the journals *Science* and *Speculum*.[4] During those years, Kuhn was known above all as a meticulous professor of the history of science, quite serious (especially compared to his colleague L. K. Nash), tied to his notes and rigorous in the details of facts. However, he soon revealed himself to be an enthusiastic historian with philosophical inclinations as uncommon as they were accurate in their judgments. In 1955, Quine wrote the following about him on the occasion of his potential move to Berkeley:

> I think well of Tom Kuhn, and am sorry he is leaving. His is the only philosophical mind in Harvard history of science. Moreover the chairman in that field, though amiable and not

[1] See Kuhn, "Vita," 1 October 1955, UCBTK, 2.

[2] See the short vita for the Department of History at UCBTK, Fall 1957, 1.

[3] Kuhn (1952a).

[4] Hufbauer (2012, 437) already highlights these data on Kuhn's publications up to those dates with the intention of showing that, thanks to them, our budding historian and philosopher promoted himself in the field of the history of science in the United States.

J. V. Mayoral, *Thomas Kuhn*, Springer Biographies,
https://doi.org/10.1007/978-3-031-86512-1_4

> without gifts, is so impervious to philosophy that he is even unaware of this imperviousness; so Tom's excision won't soon heal over.
>
> Tom's philosophical slant is towards positivism without its frequent sanguine over-simplifications. Cautions on the latter score cause him to hedge his generalizations a good deal, and to be far readier to question philosophical positions than to propound them. But he does have philosophical sense, and his studies in the history of science are epistemologically motivated.
>
> I think his appointment, though not squarely in your department [of philosophy], should be welcome by your department. I should want to see further developments in Tom's thought before urging him as a squarely philosophical appointment; though it is not clear to me that even this would be a mistake.[5]

Around 1967, Kuhn gratefully recalled that historian Richard Shryock, who had been president of the History of Science Society, had first treated him as part of his profession back in 1948, during a break at a conference at Harvard on general education.[6] From then on, his relationship with the community of historians was close, although, as we will see, also selective. Alongside them, he cultivated good friendships with philosophers on both sides of the Atlantic. In England, for example, he had a close colleague in Mary Hesse.[7]

About his qualities as a teacher in his early years there were, as usual, varied opinions. Hufbauer highlights that the *Confidential Guide* (prepared by the student newspaper *The Crimson*) informed freshman students from 1952 to 1954 that Kuhn was clear, orderly, offered many facts to support what he said, but also that he was tedious—unlike Nash, who aroused enthusiasm among the students with a more lively and "witty style of delivery."[8] Certainly, Kuhn used to set high standards for students, probably the same standards he set for himself, and not everyone was prepared or willing to meet them. Some, however, were encouraged by his rigorous teaching. This is the case with J. R. Levenson's brother, a General Education student at Harvard, who thought Kuhn was a skilled teacher. According to Levenson, his brother had been especially interested in how Kuhn examined *Gulliver's Travels* in the context of the science at the time the novel was published.[9] We know, from his old literature works, that Kuhn liked this type of contextualized approach, and it is not strange that he used it in his classes, surely for the benefit of part of his audience. This image of a young Kuhn who liked teaching and research is remembered for example by his former student Joy Harvey, in a study about the history of science at Harvard:

> In my junior year, I had the pleasure of being tutored by Thomas Kuhn, who first introduced me to the field [of the history of science] as a discipline. I recall his brilliance and erudition, his love of his subject, and his concern for his students. He went carefully and critically over

[5] Quine to W. R. Dennes, 4 October 1955, UCBTK. Reproduced by permission of Douglas B. Quine, PhD.

[6] Kuhn to W. J. Bell, 3 February 1967, TSKP 6.9.

[7] Interview 1995, 285–286.

[8] Hufbauer (2012, 436); the quoted words come from the 1953 *Confidential Guide*. On Nash's thinking, see Jacobs (2010).

[9] J. R. Levenson to G. H. Guttridge, 11 October 1955, UCBTK.

> the short papers he required in the ungraded junior tutorial, displaying his concern that his students express their ideas clearly and directly. Some of points he raised as a tutor are stated in a more developed manner in the introduction to his later book of essays, *The Essential Tension*, emphasizing his concern for encountering historical material in its own terms.[10]

Historian of science Jed Buchwald studied with Kuhn at Princeton University towards the end of the 1960s and remembers that the history of science course he followed with him and with Michael Mahoney "proved to be peculiarly intense," as Kuhn supplied them weekly with "hundreds of pages of *arcana* from the distant past" to work with them to find the meaning of the text.[11]

When he replaced Conant in General Education classes in 1951, the experience was very different compared to that of a few years before, in 1947, with the president leading them. As Kuhn admits, most of the students in the audience remained silent. He wanted to see them involved in the subject matter—young people stimulated by it and by his teaching. But such a thing only happened with a small group of the class, of whom he had pleasant memories and with whom he stayed in contact. However, that group was small, and the size of the greater class diminished due to the fact that Conant was not involved with the course or even with the university itself. For Kuhn and Nash, it was Conant and not the course itself that attracted large groups of students to the "Natural Sciences 4" classes, as the course was called (with the subtitle "Research Patterns in Physical Sciences").[12] Kuhn combined this unsatisfied desire for student empathy with progressively well-prepared material—which was also, unfortunately, increasingly difficult to transmit. As he admitted:

> Now I started spending too much time preparing, getting very nervous in advance, and I've never altogether gotten over that. I mean I've never regained the original freedom to just go in with rough notes—knowing I knew the stuff—and start talking.[13]

Harvard allowed him to teach his own history of science courses for more advanced students (upper division undergraduates and also graduates). This happened from 1953, his third year as a professor. He taught, for example, a seminar on nineteenth-century physics, which that year was devoted to the history of thermodynamics, starting from Lavoisier's time, during the spring semester of 1953–1954 ("History

---

[10] Harvey (1999, S281–S282).

[11] Buchwald (2010, 279).

[12] See the *Final Announcement of the Courses of Instruction Offered by the Faculty of Arts and Sciences During 1951–52, Official Register of Harvard University*, vol. XLVIII, No. 21, 10 September 1951, 22. (Available at <https://iiif.lib.harvard.edu/manifests/view/drs:467242686 $24i>.) In the same page, Kuhn appears as part of the staff for "Natural Science 3" as well, in that case devoted to "The Nature and Growth of the Physical Sciences," and shared teaching duties with Nash and Holton in the spring semester, while I. B. Cohen alone directed the course in the fall semester. Kuhn also appeared as part of the staff of "Natural Science 4" (with Nash), though as ascribed to the Society of Fellows as yet. See the *Final Announcement of the Courses of Instruction Offered by the Faculty of Arts and Sciences During 1950–51, Official Register of Harvard University*, vol. XLVII, No. 23, September 1950, 22. (Available at <https://iiif.lib.harvard.edu/manifests/view/drs:467242132$24i>.)

[13] Interview 1995, 284.

of Science 212: Seminar in the Development of Modern Physics").[14] In this seminar, Kuhn tried a finer and more personalized style of teaching, which mixed teaching with research and even with the possibility of publishing the results. Naturally, that involved handling original documentation and, of course, the same dedication that Buchwald talks about.[15]

Among his students in that seminar were S. G. Brush, S. Nakayama, E. Hiebert and R. Hahn, who would end up becoming renowned historians of science. Brush, for example, remembered that this seminar had been his own initiation as a historian of science and the first contact with the problems to which he would dedicate much of the next twenty years of his historical research. He outlines the principles of the methodological doctrine that Kuhn instilled in him and his seminar colleagues as follows:

> In Kuhn's seminar we learned that the history of science must be studied by careful reading of original sources. That means reading them in the original language, not relying on translations; it also means becoming aware of the precise meanings of technical terms by reading other works by the same author and works by other authors on the same subject at that time. One must be careful not to read modern meanings into older writings.[16]

Kuhn's recommendations are common in the historiographic method and are not at all exclusive to his specific way of teaching. Using the original sources is a fundamental rule for any historian, as well as not trusting translations or not projecting present points of view onto the past. However, beyond Brush's description, we can find something that is indeed novel in Kuhn's research method in the history of science: the application of his views on the nature of science that we have seen develop in previous chapters. This application provides a perspective on the evolution of science that was a new contribution and that would leave its mark on *Structure*.

4.2 Kuhn and the Historiographical Method

In the early years of his career in the history of science, Kuhn was part of a group of historians who criticized a positivist-style history—that is, one based on the collection of significant facts, and with a triumphalist tone, which in turn means that the achievements still valid in our present time and the figures to whom such scientific discoveries were attributed constituted the criterion of relevance for the selection of facts.

In the local context of Cambridge, Massachusetts, Kuhn found a few historians with a similar approach to his own. One of these was Charles Coulston Gillespie,

---

[14] See the *Announcement of the Courses of Instruction Offered by the Faculty of Arts and Sciences During 1953–54, Official Register of Harvard University*, vol. L, No. 21, 28 August 1953, 179. (Available at <https://iiif.lib.harvard.edu/manifests/view/drs:467248356$1i>.)

[15] [Kuhn], "History of Science 212," 4 February 1954–15 November 1955, TSKP 1.22, esp. "Meeting 1," 4 February 1954, "Meeting 2," 10 February 1954, and "Meeting 3," 17 February 1954, 1.

[16] Brush (2000, 40).

his colleague as a tutor at Kirkland House and later founder of the program of History and Philosophy of Science (HPS) at Princeton, which Kuhn himself joined and later chaired. He was part of this group until his departure for Princeton.[17] Other members of this group included Marie Boas, William Stahlman (a student of Otto Neugebauer), A. Hunter Dupree (Kuhn's future colleague at Berkeley), and Edward Lurie. The group met periodically at MIT during the first half of the 1950s to discuss the history of science.[18] It seems that they didn't share identical convictions about the historiographic method, but they did agree that the triumphalist, positivist, Whiggish, or "ethnocentric" viewpoint, as Kuhn also called it,[19] required revision. There was a goal beyond the exaltation of scientific achievements, but it seems that Kuhn was yet to discover this goal; later in this section we will see what that was. Gillispie recalled that "Tom and I talked shop from the outset, he feeling his way to history of science from physics, I from history, neither of us with a clear notion of what the subject might become."[20] The notion was, of course, much clearer when they collaborated on Princeton's HPS program from 1964 to 1979, already as senior professors and historians of science, and renowned authors in their fields.[21]

The positivist approach they were confronting was represented at Harvard by the historian of science George Sarton. This historian, born in Belgium, had arrived in the country in 1915 and a few years later had ended up teaching at Harvard, where he was well respected due to his great erudition. Until 1940, when he was appointed professor at that university shortly before retiring, his fees actually came from the Carnegie Institution in Washington, of whose president, R. S. Woodward, he was a good friend. Harvard, in fact, paid him very little—not enough to live on. His historiographic method did not convince everyone, including Conant, who supported his stay at the university along with Henderson. Sarton's historiography of science is a description of the progress of the history of science's great geniuses and theories in a way that contrasts Herbert Butterfield's, Alexandre Koyré's and later Kuhn's.[22] I. Bernard Cohen, one of his main disciples, provides a very illustrative image in the following passage:

> George Sarton [...] spoke of many men and their writings. Each lecture would begin with Sarton writing a list of names on the blackboard, together with dates, of the individuals he would be discussing. Much of the presentation was anecdotal, describing the lives and achievements of the men in question. He spoke with great enthusiasm, clapping his hands, hitting the table, bounding about the lecture platform like a young deer. [...] Sarton's approach in these lectures [...] was to show in as complete a form as possible the great panorama of scientific development, concentrating on the lives of the scientists in the manner that a

17 Gillispie (1999, S86).

18 Boas Hall (1999, S78).

19 Kuhn (1984, 364).

20 Gillispie (1999, S86).

21 See our Chaps. 6–7 for details.

22 On Sarton, see Thackray and Merton (1972; 1976); Harvey (1999, S272–S274); Conant (1957, 302, 304). Also see Sarton (1927–1948; 1936; 1950; 1954) to understand his historiographic perspective. See also the issues from the first decades of *Isis* and *Osiris*, the journals he created as a means of disseminating the history of science.

hagiographic account would present the lives of the saints. The scientists presented would be grouped according to ideas, for instance, wave theory of light or electromagnetism. But it must be stressed that Sarton did not lecture on the history of ideas as the history of science, but rather on the history and lives of scientists, together with their achievements.[23]

With the help of Koyré, Henry Guerlac, Arthur O. Lovejoy and others, the history of science was advancing along paths closer to the history of ideas than to that kind of scientific hagiography. As Kuhn indicates, the intention of this renewing current was not to make great reviews of the history of science, especially with a triumphalist intention, in which the successes and mistakes are always distinguished against the backdrop of current science, which serves as a standard. Rather, the intention was to understand historical episodes or scientists in their own context of ideas and concepts. In order to observe the "creative scientist," as Kuhn would say, it is necessary to know all his or her work well, including the unpublished sources, if we want to arrive at a genuine description.[24] The goal is less grandiose than a great historical tour through the great scientific achievements, but at the same time it is more ambitious, as it is about delving into real scientific research in a field that is completely foreign to us, for which current ideas, concepts and problem-solutions do not quite prepare us. In *The Astronomical Revolution*, Koyré expounds this perspective as follows:

[N]othing can take the place of original sources and texts. They alone enable us to catch the spiritual and intellectual atmosphere of the period under study; they alone enable us to appreciate at their true value the motives and incentives which guided and impelled the authors of them; they alone enable us to understand the powerful nature of the obstacles that were erected on the difficult, tortuous, uncertain path which had led them to abandon the ancient truths in order to discover and proclaim fresh truths.

Itinerarium mentis in veritatem is not a straight line. The road must be traversed, no matter how circuitous or mazelike; blind alleys must be negotiated; wrong paths must be retraced in order to discover the facts of the quest and hence the truth. Then, with Kepler, we can acknowledge that the ways by which the mind attains the truth are even more wonderful than the achievement itself.[25]

The tendency to overcome triumphalist positivism became manifest in that local context of Harvard as Sarton advanced towards his retirement. This was especially evident in the new generation to which Kuhn belonged. Gillispie recalled that Guerlac had recommended that he work on his doctoral thesis in the Department of History at Harvard, thereby evading the Belgian historian.[26] The gloss of the achievements

[23] Cohen (1984, 13–14).

[24] See Kuhn (1979a, 68); see also Kuhn (1958a, 29), from which the expression "creative scientist" is taken. Marie Boas (1999, S76) shows how the historians of science of her generation began to make more frequent use of archival sources. Buchwald (2010, 281) states that this was not a common practice in Kuhn. However, Kuhn emphasizes the importance of using more than just the "official" scientific communications (which leads us to unpublished material) in cases like Newton's, because, in this way, that "creative scientist" is uncovered, which sometimes significantly changes our perception of him. See Kuhn (1958a, 29).

[25] Koyré (1973b, 10–11). Beltrán (1995, 19–20) points to a similar passage at the end of Koyré (1973a).

[26] Gillispie (1999, S85).

of great men no longer seemed to make sense. As Kuhn noted, the goal was to make the real, internal nature of science visible. After World War II, in an era in which science overwhelmed the general public with its powers and, at the same time, with its opacity, the history of science was a suitable bridge between both. Conant's General Education plan followed such a path, Kuhn adds.[27] He already saw, at that time, that

> there was a sort of history of science to do that Sarton wasn't doing. [... H]e certainly was a Whig historian and he certainly saw science as the greatest human achievement and the model for everything else. [...] I could have learned a lot of data from Sarton but I wouldn't have learned any of the sorts of things I wanted to explore.[28]

One of the central influences on Kuhn is, as he himself acknowledged (and as is clear from the way he discusses Galileo in the Lowell Lectures),[29] Koyré.[30] Koyré's approach is focused on ideas, not on social contexts and their explanatory powers.[31] "Internal" has often been considered an appropriate term to name his viewpoint. This approach focuses on ideas rather than on the influence of social, technical, and economic factors to find the origins of scientific discovery and development. That alternative point of view was more typical of Merton's sociological approach, and is usually called "externalist."[32] For Kuhn, Koyré's approach was not exactly internal since, though external factors of a social, political or economic nature had no place in his historiographical inquiry, those of a philosophical or religious nature, no less external for Kuhn, *did* have a place.[33]

In Kuhn's (also) internalist approach, the main focus is the set of linguistic, conceptual and material resources that allow us to reconstruct the world of phenomena that (through their linguistic expression) we consider shared by the members of a certain past community.[34] This attention to communities focused Kuhn on sources that could be considered extra-scientific, and to the material resources of the scientific traditions. The case of the Baconian tradition is a good example. Koyré's perspective on "the Baconian movement as a fraud," as Kuhn put it, marks the difference between their respective modes of internalism.[35] Koyré, for his part, ended up underscoring the merits of Kuhn's point of view in a 1963 letter to Kuhn: "Your concept of 'paradigm'

---

[27] Kuhn (1986, 30–31).

[28] Interview 1995, 282. Kuhn does not minimize the merits of Sarton, nor the idea that science is "a great human achievement," but for him it is not the only one (282).

[29] See our previous chapter.

[30] See Kuhn (1984, 361).

[31] See Murdoch (1987, 77–78), who shows why for Koyré social explanations (or social history) did not go as far as he intended.

[32] See Kuhn (1979a, 67, 69), for a distinction between both perspectives (without that assignment of names) and for Koyré's fundamental role in the first of them (internalism).

[33] Kuhn (1983b, 27). Something similar is defended by Beltrán (1995, 184). See Murdoch (1987, 77).

[34] For a recent discussion of Kuhn's historiographical approach, especially focused on *Black-Body Theory*, see Wray (2021, 128–132).

[35] Kuhn, *Essential Tension*, 46; Kuhn (1983b, 27). Kuhn provides more details of this opposition to Koyré in a long letter to Merton about the new attitude towards experimentation in the seventeenth century (Kuhn to Merton, 7 December 1959, TSKP 23.6, 4–5). On the relationships between Koyré's

and your socio-psychological considerations about the behavior of the 'profession' seem to me to be really illuminating and to fill the gap between the history of science as such and the social history that till now were miles apart."[36] It should not surprise us that Kuhn emphasized the importance of experimentation. From his own experience in physics to his study of the history of science under the aegis of Conant's plan, the operational view of the meaning and importance of the experiment was always present. It is a contextual cause of a part of his internalist viewpoint that is not easy to overlook. In Kuhn's view, as we have already seen, the experiment was, from the beginning, at the service of a certain conceptual background, of the interaction between ideas and worldviews, but that role did not diminish its importance.

This aspect of Kuhn's internalist approach shows the first characteristic element of the idea of well-conducted historiographical research that is paramount for him: the *reconstruction* of the scientific vision of the community that is the object of historiographical inspection.[37] For him, showing familiarity with the kind of world explored by a past community, with its scientific objects, with its conceptual, linguistic (symbolic, formal, mathematical) and material resources, and being able to describe nature according to that way of understanding it constitutes the first mandatory step of the historian.[38] Buchwald notes what, prima facie, looks like a *realistic* approach to historiography by Kuhn; it has also been highlighted by Hoyningen-Huene.[39] Thus, Buchwald points out that:

> His need, even compulsion, to find *the*—not *a*—core of meaning that unites a disparate series of texts, to extract that largely-implicit structure and to display how it governed and connected to a set of canonical problems, powerfully directed his historical research.[40]

This core of meaning, an implicit structure, is well represented in his more mature philosophical theory by the idea of the structured lexicon, or kind set, of a community.[41] And it is this lexicon that his historiographical research adopts as a basis for formulating hypotheses and continuing to investigate the research problems generated by his immersion in a science of the past. We will see an example of this below, in Kuhn's work on Boyle.[42]

---

approach and Kuhn's, see Beltrán (1995, 167–184); Solís (1994, 15–16; 1997); Westfall (2014, 31–32).

36 Koyré to Kuhn, 1963, partially reproduced in Kuhn (1983b, 27).

37 See Kuhn (1984, 361–364).

38 Kuhn, *The Road*, 112, 233–234. Kuhn presents the same perspective in *The Presence of Past Science*, Shearman Lectures, University College, London, November 23, 24, and 25, 1987, TSKP 23.32, Lecture I, 3–4, 29–30; see *The Last Writings*, 29–30, 47. On these lectures, see our Chap. 8.

39 See Hoyningen-Huene (2012, 289–290).

40 Buchwald (2010, 280). Something similar is described by Murdoch (1987, 76) as the central objective of Koyré's historiography.

41 On lexicons, see *The Presence of Past Science*, November 23, 24, and 25, 1987, TSKP 23.32, esp. Lecture I, 32–33; Lecture II (also in Mladenovic's edition, of course, *The Last Writings*, Lecture I, 48–49; Lecture II). On the idea of "structured kind set," or simply "kind set," see his last book, *The Plurality of Worlds*, in Mladenovic's edition, *The Last Writings*, esp. 113–114, 168, 181–182, 213, 224, 247.

42 On historical knowledge, see his reflections in the 1960s in our Chap. 7, Sect. 7.7.

Buchwald notes that, in certain discussions with him, he could see how Kuhn paid more attention to linguistic aspects than to material ones (for example, to measurement operations), which give rise to the language of a paradigmatic phase of history, although he also shows that Kuhn knew that such material aspects were not negligible.[43] This attitude, seemingly occasional, does not look complete enough to represent his overall stance as a historian and philosopher, as Kuhn did not overlook the importance of the interaction of the subjects with the language they learn and also with the material resources that allow them to confer a specific meaning to terms. An example of this is the way Kuhn recounts the learning paths that can be followed in order to learn the meaning of some central terms of Newtonian physics, where a device like the "spring balance" is an integral part of this process.[44] For the same reason, his internalist approach is not as intellectualist as Koyré's.

If this kind of reconstruction is the first target of his research work as a historian of science, Kuhn also understood the essential role of his abilities in achieving it. As he said on more than one occasion, "[p]robably the thing I do best and certainly the one to which I have devoted most time is climbing from the writings into the minds of dead scientists, figuring out how they thought, why they believed what they did, and how they came to change their minds."[45] Similarly, he told Buchwald in a letter:

> I think of my primary talent as a hard-earned ability to read a text, find a way to make it make sense by discovering the conceptual structure that lies behind it. It's the experience of finding hidden structures that underlies *The Structure of Scientific Revolutions* and that I'm now back trying to analyse again.[46]

With this we come to the second important aspect of his historiographical approach: the *ultimate goal* of the historian. Kuhn, as we will see in his account of Boyle a little later, focuses particularly on the details of linguistic and conceptual change—details that lead to (and clarify) the major transformations in the history of science, the processes of discovery and revolution. In his well-known article "What Are Scientific Revolutions?" (1987) such processes are exposed through a fine examination of conceptual change; he exhibits three cases of scientific revolution based on the study of the transformations of the language-structure with which scientific phenomena and objects are categorized.[47]

Kuhn's research method is also considerably demanding at this point in normative terms. Buchwald recounts that Kuhn invested a lot of energy in the strict understanding of documents.[48] When Kuhn himself shows how historiographical research should be, he emphasizes caution when talking about scientific discovery. There are

---

[43] Buchwald (2010, 282).

[44] See, for example, Kuhn, *The Road*, 69.

[45] Kuhn (1983b, 27). He says something similar in Interview 1995, 276.

[46] Kuhn to Buchwald, 1986, cit. by Buchwald (2010, 281).

[47] Kuhn, *The Road*, Chap. 1; see also *The Last Writings*, 30–46, 135–153.

[48] Buchwald (2010, 280–281).

two important features in this caution. The first one consists of considering the influence of the post-revolutionary rereading of the scientific past.[49] This occurs immediately after the establishment of the new theory (or paradigm) and hides previous incongruities of the discoverer in question. This, Kuhn shows in *The Copernican Revolution*, mixes two worldviews.[50] If this happens, texts show an incongruent combination of conceptual systems. For him, the historian must be careful not to limit himself to "explaining" such incongruities as "confusions" typical of those who intuit a new vision without fully seeing it—what for Kuhn is a pseudo-explanation.[51] This inclination toward what looks like a *realistic* reading of the discovery process, often present in the kind of post-revolutionary reading of the past that I mentioned earlier, does a disservice to historical explanation, as it adopts a Whiggish tone, in which the discovered object and the conceptual system that explains it are projected onto the real situation that generated the discovery.[52]

The second feature affects the internal coherence of the historical description, which is directly related to his alleged historiographic realism, previously mentioned. As a historian, Kuhn says, he seeks to "get the facts straight," and that goal should be a requisite for everyone devoted to such an enterprise.[53] The clarification of the facts involves that, once the starting points—that is, the vocabulary and the basic conceptual scheme—have been established, the discovery process and the steps attributed to the discoverer must be accurate. He advocates for great accuracy in the attribution of roles to historical figures. This is in line with his idea that discoveries and revolutions are extensive processes over time, and with the related idea that the language and the worldview that eventually emerge from these processes project their coherence onto the past and, in some way, *create* their own history.[54]

A good summary of these two main objectives of the historian can be found in *The Road since Structure*, in his Rothschild Lecture at Harvard, in 1991:

> The characteristic concern of the historian is development over time, and the typical result of his or her activity is embodied in narrative. Whatever its subject, the narrative must always open by setting the stage [...]. If that narrative deals with beliefs about nature, then it must open with a description of what people believed at the time when it began. [...] With the stage thus set, the narrative proper begins, and it tells the story of change of belief over time and of the changing context within which those alterations occurred. By the end of the narrative those changes may be considerable, but they have occurred in small increments, each stage historically situated in a climate somewhat different from that of the one before. And at each of those stages except the first, the historian's problem is to understand, not why people held the beliefs that they did, but why they elected to change them, why the incremental change took place.[55]

---

49 *Structure*, Section XI.

50 *Copernican Revolution*, 182–184. More recent examples emerged when studying Niels Bohr's work in deep in the 1960s, for instance. See the last section of our Chap. 7.

51 Kuhn (1984, 367–368). "Explain" and "confusion" are used by Kuhn in such pages. See his exchange with Martin Klein as examined at the end of Chap. 7.

52 Kuhn (1984, 367–368).

53 Kuhn (1984, 370).

54 See Kuhn (1962) on the idea of discovery as a lengthy process over time.

55 Kuhn (1992, 112).

This paragraph shows Kuhn's vision of the historiographic description very clearly. In his case, as he says in "Revisiting Planck," it has philosophical significance—it is relevant in building the kind of developmental view of philosophy of science that he puts forward.[56] We return to this in Chap. 8. However, in Chap. 7 we discuss reasons for considering that his view on the historical reconstruction of the past is not too far from his account of the scientific reconstruction of natural phenomena, as in chemistry or physics. As in the latter case, reading Kuhn's position from a realistic viewpoint is not completely convincing. Moreover, I am not even sure that the opposition realism/anti-realism, understood from the traditional epistemology he usually attacks, works well for asking the correct questions about Kuhn in that respect.[57]

4.3 Teaching the History of Science at Harvard

Even before joining the faculty at Harvard, Kuhn strove to make the practice of history and its philosophical foundation his teaching work, not just his research. While Kuhn was still in the Society of Fellows, the chairman of the General Education Committee, David Owen, asked him for any subject he wished to teach. Kuhn, of course, sent Owen a proposal and explained his underlying intentions. For Kuhn, the methodology of the natural sciences, physics in particular, could benefit from a historical study that went beyond the known results of scientific research, advancing to its origins in perception, thought, and experimentation (in this latter sense, he went beyond Koyré). Kuhn refers to Pearson in that letter as a representative of the empiricist tradition in the study of the method that has its roots in the work of Descartes and Bacon. As we know, he would also refer to Pearson a few months later in the Lowell Lectures.[58] The conviction concerning the role the history of science may play in the philosophy of science reappears throughout Kuhn's work and is well known, especially because of its presentation in the introduction to *Structure*, "A Role for History," where he makes a famous statement: "History, if viewed as a repository for more than anecdote or chronology, could produce a decisive transformation in the image of science by which we are now possessed." Several pages later, at the end of the section, he says: "History, we too often say, is a purely descriptive discipline. The theses suggested above are, however, often interpretive and sometimes normative." He ends the section by saying: "How could history of science fail to be a source of phenomena to which theories about knowledge may legitimately be asked to apply?".[59]

The courses that Kuhn presented to the committee, thanks to Owen's petition, selected some conceptual systems and then proceeded to determine their sources. He put forward two courses, which arose from the material also included in the Lowell

[56] See Kuhn (1984, 370). See our Chap. 8 for that philosophical view.

[57] See our Chap. 7, esp. Sect. 7.7.

[58] Kuhn to Owen, 6 January 1951, TSKP 5.84, 1–2.

[59] *Structure*, 1, 8, 9, respectively.

Lectures—both its underlying philosophy and the examples. In one of the courses, Kuhn was going to explain how modern dynamics since Galileo involved a problem change. Newtonian dynamics was not a response to a change in the kind of question that scientists used to formulate before, as had commonly been assumed. For Kuhn, it was not, as it was often said, that Galileo had gone from questioning the *why* of movement to asking about the *how*. Kuhn criticized this interpretation of the Galilean change.[60] For him, the dynamics in which Newton had worked had a new class of facts and the shift towards new questions was not something as simple as the step from the "why" to the "how." The new class of facts responded to a new orientation towards experience, to a new selection of objects, to new criteria—as we can see in his (at this point, still future) Lowell Lectures. This is the truly crucial problem in methodology, Kuhn told Owen, and solving it allows us to better understand both Newton and scientific thought in general.[61]

The second course dealt with mathematics and formal languages and its contents were those of the forthcoming seventh Lowell lecture. In it, Kuhn planned to address formal systems, their origin in operations with abstract entities devoid of any connection with the physical world, and the proliferation of alternative series of such systems. For him, the discovery of this disconnection of the formal system from physical experience and the consequent proliferation of such abstract systems constituted the main intellectual revolution of the last hundred years. As in the corresponding Lowell lecture, he emphasized in this subject the impact of this revolution on the physical sciences. This impact was based on the application of the mathematical tools thus developed for the exploration of the physical world, which was carried out through the interpretation of abstract mathematical systems by means of experimental results and following pre-established correspondence rules.[62] Since all this would help us to understand contemporary science and mathematics a little better, Kuhn anticipated that the committee would probably be more interested in this second course than in the first. However, he was much more inclined towards the former course, as it accorded better with his own research.[63]

In the end, the committee did not choose either of them. Kuhn was appointed General Education instructor in order to replace Conant alongside Nash in "Natural Science 4," and the committee planned to reconsider his appointment in a year. After that reconsideration, he was finally appointed assistant professor. It seems that his course on the revolutionary change in the seventeenth century was eventually accepted, as it was thematically and chronologically close to a course he began to teach in the 1953–1954 academic year, "The Rise of Scientific Cosmology: Aristotle to Newton" ("History of Science 105").[64]

---

60 Kuhn to Owen, 6 January 1951, TSKP 5.84, 4–5.

61 Kuhn to Owen, 6 January 1951, TSKP 5.84, 5.

62 Kuhn to Owen, 6 January 1951, TSKP 5.84, 6.

63 Kuhn to Owen, 6 January 1951, TSKP 5.84, 6–7.

64 See the *Announcement of the Courses of Instruction Offered by the Faculty of Arts and Sciences During 1953–54*, *Official Register of Harvard University*, vol. L, No. 21, 28 August 1953, 178. (Available at <https://iiif.lib.harvard.edu/manifests/view/drs:467248356$1i>.)

In any case, Kuhn's appointment to the teaching staff at Harvard was not an easy process. The first proposal presented in his favor (November 28, 1950) did not result in a decision on the matter, and on the second and third tries (January 9 and February 20, 1951) the committee finally decided to grant him the aforementioned instructor position only. This was due to his lack of published research in the field in which he claimed to work, as noted at the beginning of this chapter. He surely deserved an appointment, but at that time he still had to prove his worth in the field he had chosen.[65] Actually, Kuhn was coming out of a short period of field change within the Society of Fellows, an environment that, as we know, privileged long-term research and freed its fellows from other obligations. In practice, however, things were different. From that same society other young talent had emerged with a good number of articles published, not just prestige. Other fellows had managed to get the necessary background to become part of the faculty as soon as they left the society, as was the case with F. X. Sutton.[66] This was not yet the case with Kuhn, though.

4.4 Kuhn's Historiographical Method in Practice

In 1951 and 1952, Kuhn's first publications on the history of science appeared. He also gave his lectures at the Lowell Institute and taught his first General Education lessons at Harvard. In this latter area, as early as May 1949, he had represented Conant's plan and group at the University of Washington, invited by A. H. Compton and T. S. Hall for a conference on general education, where he presented a paper.[67] His activity as a member of the General Education group increased from 1951 onwards, not just by teaching classes. In June of that year, for a whole week (17–23 June), he attended a meeting of general education teachers at the Teachers College, Cortland, State University of New York. He gave a talk there and participated in a workshop on general education in the sciences that met from 9 to 12 a.m. and from 1:30 to 4:30 p.m. daily from Tuesday to Friday.[68]

---

[65] See *Minutes of the Committee on General Education*, meetings of 28 November 1950, 1–2, 9 January 1951, 1, 20 February 1951, HUA, Ref.: UA III 10199.2, folder "1946–1951 NOV"; Kuhn, "Vita," 25 October 1955, HUA, Ref.: HUG 300, 1–2; Hufbauer (2012, 434).

[66] See *Minutes of the Committee on General Education*, meeting of 18 January 1949, 2, HUA, Ref.: UA III 10199.2, folder "1946–1951 NOV," for the appointment of Sutton.

[67] See Kuhn, [Untitled Notes], Washington University Conference, 12 May 1949; A. H. Compton to Kuhn, 2 May 1949; both in TSKP 12.33.

[68] Kuhn, "The Sciences in General Education," Conference at Brockway Hall, Teachers College (Cortland), SUNY, 18 June 1951, TSKP 12.33. Still in the 1990s, there was a rumor related to Kuhn's activity in general education. According to it, J. R. Oppenheimer had thought about Kuhn as the new leader of a general-education movement in the United States. Unbeknownst to Kuhn, the story goes on, Oppenheimer's opinion about Kuhn helped to boost his career. Paul Hoyningen-Huene reported to Kuhn about this rumor in 1990, but Kuhn was unable to confirm or refute it, although he was skeptical about it as a sort of Oppenheimer's long-term plan for him. Probably, he thought, Oppenheimer would have heard of him and would have a positive opinion about his work

We have already seen that Kuhn's first historiographical publications were first published in *Isis*, tightly grouped, and that they were contributions that largely arose from the same research that also led to the Lowell Lectures. Thus, although later in appearance, the primitive idea of "Robert Boyle and Structural Chemistry in the Seventeenth Century" was among the historical cases illustrated in the Lowell Lectures (specifically, in the third and fourth conference) and, in inception, it was even prior to the lectures. Nine months after the Lowell Lectures, on December 30, 1951, Kuhn presented a paper before Section L (History) of the American Association for the Advancement of Science titled "The Structural Chemistry of Robert Boyle," in which he summarized that article.[69] This appeared only four months later, in the April issue of *Isis*. It had been preceded by a brief note on what he believed was a mistranscription of a passage from Newton's *Opticks*, and was followed by two more notes in response to the criticisms of his friend Marie Boas.[70]

These articles and shorter notes illustrate Kuhn's research method well. As noted, the first time he spoke extensively about Boyle is in the third and fourth Lowell lectures—those dedicated to atoms and subtle fluids, respectively, as scientific orientations. In the first of both, he pointed out that Boyle had applied atomism, which had been successful in its applications to dynamics, to chemistry.[71] In the fourth lecture, he added that the Boylean application of atomism had had very little resonance in the course of the historical development of chemistry, which in Kuhn's opinion had not been a problem. Boyle's structural chemistry considered certain concepts such as that of *element* from a new point of view, which involved a new approach to chemistry.[72]

Kuhn showed that Boyle's new approach had questioned whether there was anything that could really be called "element" in the terms in which the notion was then understood; that is, a substance that is the final *result* of all kinds of analysis (chemical or physical). The chemistry prior to Boyle—that of Étienne de Clave, Daniel Sennert, or Joachim Jung—was no less corpuscular than his, but the corpuscles of these chemists were distributed in elemental classes while those of Boyle were undifferentiated matter. Boyle sought to establish an ontology of matter in motion for chemistry.[73] Although he understood the obstacles to transmuting a substance into any other, that ontology made such a type of reaction possible. In this way, he opposed peripatetic chemistry and iatrochemistry. Not even Newton shared such a perspective, despite the fact that his vision of chemistry was the closest thing that could then be found to Boyle's vision. It was similar, but not exactly identical. Newton adds to the corpuscles the forces of action at a distance; inverse forces of powers of distance

---

in general education, but that would be all. See Hoyningen-Huene to Kuhn, 24 January 1990, and Kuhn to Hoyningen-Huene, 2 February 1990, both in PHA.

69 Kuhn, "The Structural Chemistry of Robert Boyle," Section L, AAAS, 30 December 1951, TSKP 12.33.

70 Kuhn (1951b; 1952b; 1952c); Boas Hall (1952).

71 Lowell 3, 4; GR 157–158.

72 Lowell 4, 9–15; GR 60–63.

73 Kuhn (1952a, 16–18).

greater than the square. On the other hand, Newton's corpuscles are indivisible and eternal, unlike Boyle's.[74]

Kuhn emphasized that Boyle's corpuscular philosophy had not transformed chemistry, and that it was not the imperfect precedent of the perspective of Lavoisier, either. Chemistry right after Boyle (from 1670 onwards, specifically) was faithful to the image of the chemist as a separator of "elementary" substances. John Mayow and Nicolas Lémery, for example, two of Boyle's contemporaries, had focused the chemical exercise that way. Kuhn also warned that Lavoisier did not show features that could refer to both structural chemistry and the chemical tradition of his time. Boyle had intended to take an atomistic approach that had been effective in dynamics and apply it to chemistry, but this had only ended up being his own delusion.[75]

Kuhn's article presented this historical case with a great bibliographic apparatus and a notable abundance of data. The article still seemed very good to him in 1995, but he admitted that "it's totally unreadable because I thought I had to persuade a very learned group of historians of chemistry out there." He then added that "what I gradually discovered was that nobody knew nearly as much about this problem as I did."[76] It is not surprising that Kuhn became an authority on the problem, because the starting point to reach it belonged to the methodological approach that we have already seen. In fact, the starting point, as he talks about it, is an instance of the scientific orientations he spoke about in the Lowell Lectures. As he recounted in his article, Boyle's corpuscles were pure matter without those qualities that helped to distinguish them from each other, nor to make them attract each other (with different forms of affinity, for example, as for Newton). Boyle's atomism, Kuhn said, "was a matter and motion atomism. It suddenly occurred to me that if you believed *that*, you would believe that you can make anything out of anything—it's a natural basis for transmutation."[77] For him, Boyle's atomistic orientation introduced the existence of fundamental entities of matter that did not exclude a general phenomenon of transmutation, which undermined the chemical tradition based on *elements* as fundamental, essential, basic, permanent substances.

At Nash's suggestion, Kuhn rushed to read *The Sceptical Chymist* to verify his conjecture and found that there was a frequent ellipsis in every customary reproduction of the paragraph in which Boyle talks about the elements, and whose inclusion (along with Kuhn's point of view) completely changed the common understanding of Boyle:

> Very early there is a remark [in *The Sceptical Chymist*] in which one of the interlocutors says to the major figure who represents Boyle, "It sounds very much to me as though you don't believe in the elements," or something like this. And Boyle says, "That's a very good question. I'm glad you asked me." And then he proceeds to say, "I mean by element those things out of which all things are made, and into which they can be divided." Now, that is taken to be, and it isn't quite, the definition of an element. And Boyle is given credit for the

[74] See Kuhn (1952a, 24–25, 33, 35); Newton (1730, 400, Query 31).

[75] Kuhn (1952a, 15, 31–36).

[76] Interview 1995, 291.

[77] Interview 1995, 290.

> first definition of an element, but what he's doing at this point... he says, "I mean by element, *as I take it all chemists do*"—[but that phrase is] replaced by dot dot dot, when that definition is quoted! And he says, "I'm going to give you reasons for believing that there are no such things." And that was almost my first article.[78]

In his paper, Kuhn considered what kind of semantic alterations certain chemical notions such as "element" or "transmutation" could be subjected to when viewed from a different orientation that turned the fundamental entities of that science into a different type of objects. Kuhn did not delve into the respective meanings of these terms in structural chemistry, in peripatetic chemistry, and in iatrochemistry; he did not establish a complete comparison. He did not fully reveal the fate of Boyle's structural chimera, what research problems it could generate, and why they were not cultivated, either. Instead, he displayed the chemical universe that such a transformation gave rise to, and he showed that something like that would not have been acceptable to a chemist of the time. So, he showed that the fate of a conceptual transformation depends on a collective, generalized practice, which may not employ a new solution—a new approach.

As Kuhn's professional career progressed, the articles and longer texts that resulted from his investigations into diverse episodes of change in the history of science, which include a book like *Black-Body Theory*, covered a wider spectrum of scientific problem-solving, always within the physical sciences (broadly understood), and examined a greater number of consequences of such problem-solving processes in science. Yet, they started from the same point of view as these initial works. To understand the solution we are talking about, and why (and how) it was or was not accepted (that is, what made it a genuine solution), Kuhn used to start with the type of entities that scientists of a certain era saw—about which they thought in order to qualify them better. *Structure* provides us with a theory that supports this image of science from a philosophical point of view.

4.5 Kuhn's Seminar on Thermodynamics at Harvard

A second historical episode that Kuhn liked and pursued, and a more enduring one in his career, was the origins and subsequent fate of thermodynamics. His research on this key theory resulted in some texts that are worth commenting on, as they reveal new aspects about the complexity of scientific language that Kuhn would later incorporate into his philosophy of science.

Kuhn was always attracted to thermodynamics, he says. He studied it with Bridgman at the post-graduate level, and he dealt with its history for the classes with Conant in "Natural Science 4."[79] While preparing the classes for Conant's

[78] Interview 1995, 290–291. Square brackets and ellipsis are in the original.

[79] Interview 1995, 267–268. Indeed, Bridgman gave a course on "Thermodynamics," "Physics 41a," in the academic year 1945–1946. See the *Announcement of the Courses of Instruction Offered by the Faculty of Arts and Sciences During 1945–46, Official Register of Harvard University*, vol.

course as a fellow of the Society of Fellows in the fall of 1949, Kuhn exhibited the intention to offer an independent point of view, which turned him against Duane Roller, who was collaborating with Conant in the preparation of the *Harvard Case Histories in Experimental Science*. Roller planned to offer comments and questions in his historical study on the transition from the caloric theory to thermodynamics, but Kuhn suggested (and preferred) that such interpretative and pedagogical ideas be considered aside from the historical description. In this way, the instructor—such as himself—would be free to handle the historical case according to his or her plan.[80] By 1953, Kuhn was already beginning to delve deeper into the origins of thermodynamics, especially into the scientists who promoted a revolutionary change: Sadi Carnot and James Joule. In the courses from that year onwards, Kuhn's notes show an increasingly deep understanding and analysis of the episode in question.[81]

It is difficult to pinpoint the source of Kuhn's interest in this episode of the history of science. We can speculate that the complexity of the case itself was a factor for him to consider. The historical role Joule and Carnot played, with their respective biographical itineraries, must have been another significant factor. Carnot's biography, his sudden death, conceptual changes of great significance both for his work and for the development of his field that, in addition, did not come to light at the time—all of these facts made this scientist an attractive challenge for Kuhn's practice of the history of science. It was a good case for him to study with his historiographical point of view, anyhow, which in this particular case included the specific caveat that one had to avoid projecting a modern conception of thermodynamics onto the development of Carnot's own ideas. Another interpreter of Carnot, Victor Kuhn La Mer, had fallen prey to this problem of Whiggishness, in Kuhn's opinion, and, as we will discuss shortly, Kuhn himself made this clear in his publications.

Whatever his interest in this historical episode, thanks to the appearance of his seminar on thermodynamics ("History of Science 212") in the course offerings at Harvard, Kuhn found the environment in which to develop his own in-depth research on this subject. In his seminar, he left Conant's pedagogical objectives behind and focused on teaching his method of historical research through its application to the origins of thermodynamics. He began to detail his vision of this historical issue, with his students as spectators and participants.[82] In later years, it would be this specialized practice of the history of science, not its pedagogical adaptation to general education, which Kuhn pursued in his classes. Right after Harvard and *The Copernican*

XLII, No. 8, 31 March 1945, 106. (Available at <https://iiif.lib.harvard.edu/manifests/view/drs:467240026$108i>.)

[80] See Roller to Conant, 12 December 1949; Conant to Roller, 26 January 1950; Roller to Conant, 28 February 1950; Conant to Roller, 10 April 1950; all in JBCP 23, folder "Case Histories: Harvard 1949–50.".

[81] See [Kuhn], "NATURAL SCIENCES 4. Technology," Lecture 4 (24 February 1956)–Lecture 7 (2 March 1956), and Lecture 10 (9 March 1956)–Lecture 11 (12 March 1956), all of them in TSKP 1.20; see also [Kuhn], "NATURAL SCIENCES 4—El[ectric] P[o]w[e]r," Lecture 6 (28 April 1954)–Lecture 7 (30 April 1954), both in TSKP 1.19.

[82] See Brush (2000, 40).

Revolution, his objectives were those of the historian or philosopher of science who aims to understand a past scientific theory as a quasi-practitioner.

Kuhn's published papers reveal that most of his research on the history of science up to *Structure* originated in that seminar and, especially, in its subject matter.[83] Thus, he devoted two papers to Carnot's relationship with the engineering tradition in France: "Sadi Carnot and the Cagnard Engine" and "Engineering Precedent for the Work of Sadi Carnot."[84] Another paper dealt with the investigation of the adiabatic compression, as examined from the point of view of the caloric theory, "The Caloric Theory of Adiabatic Compression."[85] Similarly, "Energy Conservation as an Example of Simultaneous Discovery," one of the best-known examples of his historiography of science, also emerged from this material.[86] Finally, his first paper on the subject led to a controversy in the *American Journal of Physics* with La Mer and led to an additional reply by Kuhn.[87] In this exchange, the root of the dispute is Kuhn's consideration that La Mer had done exactly what he believed someone who wanted to get to know a historical episode genuinely should not do: projecting onto the early stages of an investigation concepts that were not yet there. It is the usual charge by Kuhn, based on the methodology of the history of science that he defended, as shown above. Many controversies with other historians (or scientists, as in La Mer's case) had their origins in that anti-Whig assumption.[88] He certainly took this subject-matter very seriously. In October 1953, when taking stock of his projects (including the forthcoming publication of *The Copernican Revolution*), he said that he had plans to travel to England for a lengthy period of time to write a book about Joule—something that in the end, as we well know, he never did.[89] And there were also testimonies to that expertise. In 1961, for instance, Marshall Clagett

[83] There are exceptions, of course, like his "Newton's Optical Papers" (Kuhn 1958a). Kuhn, however, did not focus much more on such a subject afterwards. In 1976, the general editor of that volume, I. B. Cohen, told him that the general introduction to the volume was going to have an updated version, but that it was not going to happen with the introductions to every section—as Kuhn's own introduction. Kuhn confessed himself "somewhat embarrassed at having my own contribution reprinted" (Kuhn to Cohen, 21 July 1976, TSKP 21.18, 1). The field had advanced at a pace that Kuhn had been unable to keep, and he would have suggested, if asked, to have a new introduction prepared by some Newton scholar instead of himself. He asked Cohen to leave clear this—which he did in the "Preface to the Second Edition"—and to show his bibliographic interventions in Kuhn's text within square brackets, or in any other way that showed them as additions. Cohen satisfied this petition by two references to the new "Supplement" with an updated bibliography at the end of the book. See Kuhn to Cohen, 21 July 1976, TSKP 21.18, 1. Compare both versions of Kuhn's paper (1958a, 33, n. 11, 42, n. 24; 1978, 33, n. 11, 42, n. 24). See also Cohen and Schofield (eds.) (1978, v–vi, 498–501).

[84] Kuhn (1960; 1961a).

[85] Kuhn (1958b).

[86] Kuhn (1959b).

[87] Kuhn (1955a). The paper was a correction of La Mer's (1949; 1954) arguments about the meaning of the Carnot cycle. La Mer (1955) was a reply to Kuhn, and he also responded in turn (Kuhn 1955b).

[88] See the end of our Chap. 7 for further examples from the history of recent physics and his exchange with the historian of physics, Martin Klein.

[89] Kuhn, "Plans for Research," 22 October 1953, TSKP 25.53, 4. See an edited version in Hufbauer (2012, 457, 460). On *The Copernican Revolution*, see the next section.

considered his historical research on nineteenth-century physics to be among the best in the country.[90]

"Energy Conservation as an Example of Simultaneous Discovery" synthesizes his most complete vision of the revolutionary transition that led to the relationship between diverse subfields within the physical sciences through the concept of energy. His account dealt with the emergence of a new vocabulary, or new common language, and of a brand-new unified vision of physical phenomena from numerous partial works in dispersed contexts, with geographically and culturally distant approaches, where the available information was usually partial. Therefore, it was necessary to cover a wide range of topics and historical and theoretical resources—cultural and philosophical, too. He tried to communicate to his students the dimensions of this difficult, but at the same time exciting, historiographical research in which they could also play a role.[91] He tried to disclose the basic structure of the entire complex historical episode in collaboration with the seminar members. However, this did not always produce the enthusiasm and interaction he desired.[92] It also seems likely that his investigative zeal was not easy to match for many of his students.

Kuhn was interested in the process of revolutionary change that had resulted in the principle of energy conservation.[93] The change had mainly occurred between two main theories, the material theory of heat, whose main theoretical concept was known as caloric, and the dynamic theory of heat—or thermodynamics, as it is better known today—in which the new theoretical notion, energy, played a central role. The main problem to be addressed had been, towards the beginning of the nineteenth century, the conversion of heat into mechanical work and, vice versa, from this latter back into heat. The practical context in which this problem was visible was that of machines like steam engines, in which a heat source could generate a certain amount of mechanical work. Some engineers of the time, such as Carnot in France, had worked on the production of mainly theoretical studies on the efficiency of such machines. Others, such as Joule in England, had done the same from a slightly different approach. Here, as we can guess, the national traditions of scientific education in physics, mathematics, and engineering play a central role in obtaining an adequate historical description.

The main source of the pre-revolutionary disagreement that interested Kuhn had to do with the prevailing conceptions of heat at each specific moment in time. Carnot, at least in his *Réflexions sur la puissance motrice del feu* (*Reflections on the Motive Power of Fire*, 1824), had used the aforementioned theoretical concept, the caloric, which was a form of fluid conducted by other substances such as, for example, steam. The caloric was the causal agent of work production and remained within the

[90] Letter from Clagett to D. M. Brown, 6 March 1961, UCBTK.

[91] See [Kuhn], "History of Science 212," "Meeting 1," 4 February 1954, TSKP 1.22.

[92] See, e.g., [Kuhn], "History of Science 212," "Meeting 9," 31 March 1954, TSKP 1.22.

[93] My description of this historical episode in this paragraph and the following two is based on Harman (1982, 45–58); Nye (1999, 90–95); Smith (2003); Solís and Sellés (2005, 844–860). On the role of H. v. Helmholtz, not commented here, see Solís and Sellés (2005, 854–855).

machine, in which work was obtained by virtue of its activity.[94] Conversely, Joule defended a view that was incompatible with this caloric, as for him, heat was the product of the motion of molecules of a gas, of steam, or even of a rigid body, for example a metal, and the production of work implied the gradual mitigation of such motion. From Joule's point of view, heat was not any fluid, but an effect of the motion of matter, which did not entail its conservation, but its dissipation.[95]

These two ontological perspectives were difficult, but not impossible, to reconcile. The path to their reconciliation was mainly produced by the works of three scientists: the physicists William Thomson (future Lord Kelvin), William J. M. Rankine, and Rudolf Clausius. Thomson began to explore both perspectives, intrigued as he was by Carnot's theory but also by Joule's works, about which he also had doubts.[96] Clausius (and also partially Rankine) then arrived at a synthetical view in which both perspectives were appropriately conjoined, and set up the basis for the discipline which is now called thermodynamics, including the statements of its two laws inspired by and born from the works of the previously noted pioneers.[97] Thomson followed that work, and coined the term "energy," which would give rise to a brand-new worldview. The explanation of the new dynamic vision didn't have a material background as in Carnot's earlier theory, but a mechanical background as in Joule's perspective.[98] Kuhn's seminar was arranged to explore and explain this process of scientific revolution: the transition from the material theory to the dynamic theory of heat, and the emergence of the first two laws of thermodynamics. "[The study of f]irst and second laws of thermo[dynamics]," Kuhn wrote in his notes, "[…] is not just the recognition of certain restrictions on nature. It's [the] introduction of a new unifying principle, ENERGY, and of a new notion as to what science is and how it should be conducted. An immense *reorientation*."[99]

In his research, Kuhn aimed to reconstruct the reigning conceptual scheme, the one that played the role of a scientific orientation, and according to which researchers worked.[100] Among the key concepts Kuhn's research had to talk about, the aforementioned ones—those to which terms such as "caloric" or "energy" refer to—are central; and there are other equally important ones, such as the notion of "work," or those of "conservation" and of "conversion" of forces. Expressing it in terms of Kuhn's mature theory in the 1980s, in that seminar and in his classical work on the history of the principle of conservation of energy, he aimed to reconstruct the lexicon

94 Precisely this usual interpretation of Carnot's work had been questioned by La Mer, who argued that Carnot already had a more modern vision similar to that of Rudolf Clausius (see below) at the time of his *Réflexions*. It is true that, shortly before his death, Carnot changed his mind about his view of heat as an imponderable fluid, but this does not affect his *Réflexions*, Kuhn showed. See the references to this dispute previously cited in footnote 87. On Carnot and the caloric, see Harman (1982, 48–49).

95 See Harman (1982, 49, 51, 53); Smith (2003, 293–295).

96 See Smith (2003, 295–298); Solís and Sellés (2005, 856–858).

97 For this perspective on Clausius, see Harman (1982, 53); Solís and Sellés (2005, 856–857).

98 See Harman (1982, 56–58); Smith (2003, 298–300).

99 [Kuhn], "History of Science 212," "Meeting 1," 4 February 1954, TSKP 1.22. My italics.

100 Kuhn (1959b, 72–73).

or lexicons that were replaced by a new unified (we would now say "paradigmatic") view of phenomena and the solutions to the problems they posed. As to the birth of thermodynamics, Kuhn was aware of the diverse map from which he had to start. Except for the philosophical vocabulary that we are extemporaneously applying to the historical method he followed at that as yet early time, his interpretive method can be approached from that point of view. It would not be incorrect to say that such vocabulary finds support in that practice, as Kuhn himself said.[101]

In my previous brief presentation of the case, I have mentioned only a few basic figures in the birth of thermodynamics. In contrast, Kuhn's main article, which only referred to the principle of conservation of energy, is much more exhaustive and, of course, more representative of the kind of work that he argued that a historian should do. First and foremost, he explores the work of a dozen scientists.[102] In addition, a key concept, energy, is traced back to its origins through sources of a diverse nature, not only conceptual. He matches that concept, energy, with experimental, conceptual, and philosophical sources, and few of them are identically distributed in the scientific geography of nineteenth-century Europe. The experimental sources mainly refer us to the so-called "conversion processes," which establish relationships between the phenomenal domains of various physical and chemical sciences explored in the laboratory, from the relationships between magnetism and electricity to those of chemistry with heat. Obviously, not all scientists of the time had first-hand knowledge of each of these processes.[103] Something similar happens in the conceptual field, where the concept of work (and its close link with the most practical research in engineering, not in pure science), which is key for Kuhn in the development of the conservation principle, is not treated in the same way by all the scientists involved in the extended process of collective discovery.[104] Finally, there are philosophical sources as well, as I have noted, especially the *Naturphilosophie* in Germany, which creates a metaphysical viewpoint that is identically influential in the discovery process for some of those involved in it.[105]

In view of such a complex map, constructing the history of the discovery of this principle involves delving into a branched arrangement that extends over an intricate practical, conceptual, and philosophical terrain, and that implies highly diverse geographical, linguistic, and cultural origins. Kuhn's idea was that close collaboration with his students in the seminar would help him to reconstruct this complex map, dispersed across time and among the European national cultures, which would end with the most modern theory of thermodynamics. The conditions

[101] See Kuhn (1984, 361–364).

[102] See Boyer (1959, 384); Smith (2003, 289).

[103] See Kuhn (1959b, 73–82).

[104] See Kuhn (1959b, 82–90).

[105] See Kuhn (1959b, 94–100). See Boyer (1959, 384 ff.) for a reconstruction of Kuhn's argument and a critique of these three points.

are those of a revolutionary process in which different visions, objectives and results are shared after having started from dispersed and diverse origins.[106]

Despite the variations with respect to Conant's historiographic approach that occur in other works, such as *The Copernican Revolution*, in this one we can observe a certain connection with them. Laboratory work serves as the empirical basis for which the concept of energy and the orientation it generates—as we saw he told his students—are the key to an entire conceptual scheme. Thus, he says in his article:

> In short, just because the new nineteenth-century discoveries [of conversion processes] formed a network of connections between previously distinct parts of science, they could be grasped either individually or whole in a large variety of ways and still lead to the same ultimate result. That, I think, explains why they could enter the pioneers' research in so many different ways. More important, it explains why the researches of the pioneers, despite the variety of their starting points, ultimately converged to a common outcome. What Mrs. Sommerville had called the new connections between the sciences often proved to be the links that joined disparate approaches and enunciations into a single discovery.[107]

However, beyond this experimental basis that plays a key role for unification under a single point of view, Kuhn goes beyond Conantian historiography and tries to adopt a vision of revolutionary transitions of a more abstract and, at the same time, social nature. The vision is more abstract because it surpasses the psychological point of view of the innovator to focus on the simultaneous creation of an orientation, a theory and a practice that enable the development of scientific research. At the same time, it has a *social* character because, more than in other cases, Kuhn traces the conceptual, linguistic, and observational links that are created and that advance almost at once for an entire community—a geographically dispersed group—of scientists over a certain chronological parenthesis.[108]

The social aspect of scientific research that is already visible in this kind of work in the history of science emerged through a key piece in *Structure*, the paradigm, and turned that work into an exploration of the socially extended process of research and discovery and a social account of the nature of science with philosophical consequences for our view of scientific method. We shall see later how the clarification of that social dimension was a crucial part of *Structure* and how those who invited Kuhn to write that monograph—the members and collaborators of the logico-empiricist Institute for the Unity of Science, at that time with Philipp Frank in charge—stimulated that pursuit. Now, however, we must take a look at another product of his practice as a historian of science and as a science educator—more precisely, a by-product of his practice in the latter role—namely, *The Copernican Revolution*. The book on Copernicus was the first step in his acquisition of some academic notoriety. In purely practical terms, it was an opportunity to use the material from his lectures

[106] See [Kuhn], "History of Science 212," "Meeting 1," 4 February 1954, and "Meeting 7," 17 March 1954, TSKP 1.22.

[107] Kuhn (1959b, 76).

[108] There are relations between the historical case that he presents here and his overall vision of discovery in "The Historical Structure of Scientific Discovery," a paper that, as he says, reflects ideas that were previous to those in *Structure*. See *Essential Tension*, xvi–xvii, and that paper reprinted as the Chap. 7 of that book (Kuhn 1962).

in General Education to publish a book. However, he could not help but give them the kind of theoretical perspective that he had exhibited in previous writings such as the Lowell Lectures. Owing to such practical and theoretical significance, we have to examine Kuhn's well-known first book, *The Copernican Revolution*, and try to establish what kind of monograph it was.

4.6 The Pedagogical Nature of *The Copernican Revolution*

In October 1953, pretty early in his career, Kuhn optimistically assumed that his first book, based on his lectures for "Natural Sciences 4" at Harvard, would be published, sooner or later, by Harvard University Press. He had a title for the book, *The Copernican Revolution*, and also a subtitle that was a little different from the one that now appears on its cover—*A Historical Introduction to Planetary Astronomy*.[109] However, *The Copernican Revolution*—as happened with the rest of monographs that Kuhn wrote in his lifetime—took him much longer to produce than he had planned. *The Structure of Scientific Revolutions* was expected to be completed during more or less the same time period, and *The Copernican Revolution* delayed the latter in turn. I shall examine this interaction in the next chapter, but I would like to examine the contents and structure of *The Copernican Revolution* in the current one, as they are related to his practice as a General Education instructor and historian of science.[110]

The book was the result of his work in General Education and in the history of science. It was related to Conant and Nash's *Harvard Case Histories in Experimental Science* as a somewhat independent outgrowth of the original program. *The Copernican Revolution* was dedicated to his colleague Nash, who gratefully accepted—although he also added, politely, that he could think of better candidates for such an honor—and it also included a foreword by Conant.[111] In that foreword, Conant makes a comparison between Kuhn's book and his and Nash's *Case Histories*. He said that this book was rather different. In the questionnaire that Kuhn submitted to Harvard University Press in November 1956, he also wrote that it had some relationship with the *Harvard Case Histories*, but that it did not belong to them.[112]

I have said that Kuhn made his plans somewhat optimistically because—judging from his exchange in June 1956 with his close friend in the Harvard Department of

[109] See Kuhn, "Plans for Research," 22 October 1953, UCBTK, 1–2; edited in Hufbauer (2012, 457–458).

[110] See our Chap. 5, esp. Sect. 5.2. On Kuhn's book, see Swerdlow (2004), who connects *The Copernican Revolution* with Conant's educational plan (Swerdlow 2004, 64–76), and then explores the book chapter by chapter with great attention to detail (see 76 ff.). For a critical perspective, see Westman (1994).

[111] See Nash to Kuhn, 10 October 1956, TSKP 25.46–47.

[112] [Kuhn], copy of the Author's Statement Questionnaire, 7 November 1956, TSKP 25.46–47, 1. Although the original is not preserved here and questions are not included, the answers are clearly by Kuhn. The questionnaire was submitted by Harvard University Press some months before; see A. B. Hastings' letter to Kuhn, 25 July 1956, TSKP 25.46–47, which had the questionnaire attached.

Physics, E. C. Kemble, who congratulated him for having overcome that problem—Kuhn seems to have had a hard time in his dealings with the press.[113] Less than one year later, however, in 1957, Kuhn received his first copy of the book and prepared a list for distributing complimentary copies. He asked the press to receive Nash's and Conant's copies in order for him to write a dedication before sending each book to them.[114] Kuhn had received Conant's foreword in December 1956, and had made a few minor changes in a paragraph—which he communicated to Conant by letter—and then submitted it to the press. He nevertheless did not ask for Conant's approval prior to submitting it and told him about these changes later. Conant did not complain; on the contrary, he was happy with Kuhn's initiative.[115] This familiarity with Conant is a sign of mutual respect and reliance, but it contrasts with Kuhn's still very formal way of addressing the Harvard ex-president in his letters as "Dear Mr. Conant" (while the latter used to write simply "Dear Tom").[116] As is evident from his correspondence, Kuhn was on a first-name basis with his close colleagues, especially with those who had played a key role in his life, but that was not the case with Conant. He continued to address his mentor with this formal style, even many years later.[117] The contrast is remarkable.

In the first two chapters of his book, Kuhn introduces the fundamentals that help us to understand Copernicus's innovation in subsequent chapters. Kuhn's introduction plays the role of an ancient astronomy "textbook," if you will. It serves as a resourceful summary of background knowledge that sets the reader in context and, at the same time, provides that reader with some necessary technical and conceptual tools. He often said that his book had the kind of technical information that some good historical approaches lacked, and had the kind of historical views that some readers could miss in some more technical approaches. Instead of choosing between a cultural-history book and a strictly technical description of the essential elements in ancient and modern astronomy, he had tried a combination of both. Besides that, he had emphasized for the press the contribution of an underlying structure of a rather *philosophical* nature, which made it a singular contribution to the literature.[118] It was not a popular-science book, of course, but he intended to teach some facts and some new ideas—philosophical, in his case, and related to his overall view, more or less as expounded in *The Quest for Physical Theory*—in order to transform the layperson image of Copernicus and the revolution in astronomy. This goal is approximately related to Conant's pedagogic plan, but Kuhn was never a loyal Conantian. His way of

---

[113] Kemble to Kuhn, 25 July 1956, TSKP 25.49.

[114] See W. J. Schwarz to Kuhn, 15 May 1957, and Kuhn's reply to Schwarz (together with the "distribution list"), 20 May 1957, both in TSKP 25.46–47.

[115] See Conant to Kuhn, 7 December 1956 and 7 January 1957, and Kuhn to Conant, 15 December 1956, all of them in TSKP 25.46–47.

[116] See some examples in the previously mentioned letters Kuhn–Conant, 1956–1957. Reisch (2019, 271) notices this difference of treatment in a 1961 Kuhn–Conant exchange of letters—the one on the first full draft of *Structure* that we will examine at the end of Chap. 5.

[117] See, for example, Kuhn to Conant, 29 June 1961, TSKP 25.53.

[118] See [Kuhn], copy of the Author's Statement Questionnaire, 7 November 1956, TSKP 25.46–47, 3.

understanding science was unconventional for somebody like Conant, and the years from 1947 to 1962—especially those that followed *The Copernican Revolution*—showed to what extent Kuhn's thinking was indeed unconventional in that precise sense.[119]

The underlying philosophical theme helps to conduct the historical description in the book, but it does not come to the forefront as in *Structure*—which, accordingly, needs more than one historical case to present its main theses. The history of science weighs more here, though at some points it shifts, as noted, to a sort of ancient astronomy textbook, which almost simulates a hands-on training within the reach of a wide readership. At that point, historical information becomes practical, and historical description exhibits the working of geometrical resources. At some moments, the reader must apply the knowledge acquired in the book from both sources, historical and technical, such as when Kuhn explains the reasons for the acceptance of the Copernican system on the part of Copernicus's successors such as Johannes Kepler. From the "economy" of the Copernican system in qualitative terms, Kuhn launches into a precise description of the aesthetical reasons for which a selected group of astronomers of a further generation accepted its innovation. The preceding historical description in the book is crucial at that point, but understanding those specific reasons—that is, those of the historical actors involved—requires the previously acquired practical understanding of the geometrical resources in astronomy.[120] A key virtue of *The Copernican Revolution* is its smooth transition from one aspect to the other.

The underlying philosophical theme is of help here, too. Kuhn is careful to explain how the observations that were acquired over more than one millennium before a certain arrangement made them easy to handle, and liable to a unified treatment, were accumulated on the basis of different worldviews.[121] At some points, even in *Structure* five years later, Kuhn would show that astronomy shows a case in point of observations that seem common to a number of different cosmological interpretations; they may be considered almost naked-eye observations.[122] However, he argues that these observations, which a further model would *unify*, did not condition the cosmologies, which were to some extent independent of them, and they did not by themselves convey "structural information" to their interpreter.[123] The unifying model, which he terms "the two-sphere universe," receives considerable attention from the first chapter, in which the model is introduced after talking about

[119] See the last sections of our Chap. 5.

[120] See *Copernican Revolution*, 168–181.

[121] See, for instance, *Copernican Revolution*, 8, 47.

[122] See *Copernican Revolution*, 25; *Structure*, 117.

[123] *Copernican Revolution*, 25–26.

the already available observations.[124] He shows that this is a general unifying framework that permitted a huge number of variations from the ancient world to Copernicus' time.[125] A similar strategy is followed in Chap. 2, when the two anomalies of planetary motion are introduced and the geometrical resources provided by geometricians and astronomers from Eudoxus to Ptolemy are described.[126] Chapter 3 returns to that model from the Aristotelian point of view that had already meant so much to Kuhn ten years before.

William Stahlman, a friend from Kuhn's meetings with other young historians while at Harvard, and also a student of Otto Neugebauer (and so, well trained in ancient astronomy), saw some of Kuhn's decisions as scarcely accurate in historical terms, while some others appeared to him to be mere mistakes. Stahlman let him know about these by way of a detailed letter and a number of conversations. They engaged in a cordial debate on what to do with some concepts that Kuhn had used in the book.[127] An initial cause for concern for Stahlman was the idea of a "two-sphere universe," which he did not consider as historically plausible. Kuhn argued for his pedagogical utility, which Stahlman accepted. In contrast, Kuhn told Stahlman, historian of science C. Doris Hellman had also criticized Kuhn's use of the expression, but in that case she "castigated my use of the phrase," he added.[128] "The universe of antiquity was *not* a two-sphere universe," she said, "but a multi-sphere universe, and by his very exposition Professor Kuhn contradicts his use of the term. Let us hope that this phrase is not taken up and perpetuated!"[129] Hellman's criticism had been sharper, indeed.

The pedagogical intention on Kuhn's part included other questionable descriptions. Stahlman, for example, claimed that there were some inaccuracies in Kuhn's description of the geometrical devices employed in Ptolemaic astronomy. Stahlman argued that epicycles were not a resource employed in Ptolemy's solar theory, and that there was no equant in it, for that matter. Neither did he approve of the expression "Ptolemaic astronomy" without further qualification. Otherwise, some solutions could be attributed to Ptolemy that were not correct.[130] Kuhn's reply by letter was that,

---

[124] See *Copernican Revolution*, 45.

[125] See *Copernican Revolution*, 28; to see the difference between such parts, compare 8–25 to 25–44.

[126] See *Copernican Revolution*, Chap. 2, and compare 45–50 to 50–77.

[127] It was cordial, actually, but Kuhn told him that, after a previous conversation in Madison, Wisconsin, he had grasped a worrisome criticism of "historical irresponsibility," and that, at that moment, he was almost at his "nadir of confidence in the book." That was the reason why he had just finished that long (5 pp.) letter. (Kuhn to Stahlman, 16 February 1959, TSKP 25.46–27, 5.) Stahlman was at pains to reassure Kuhn about his convictions on Kuhn's effectiveness as a historian, and to apologize for any misunderstanding he could have caused in sending him the corrections. In parallel, Kuhn was not less worried about his own response. Fortunately, everything ended well, and the tone never became less than friendly. See Stahlman to Kuhn, 20 February 1959, and Kuhn to Stahlman, 25 February 1959, both in TSKP 25–46–47.

[128] Kuhn to Stahlman, 16 February 1959, TSKP 25.46–47, 1.

[129] Hellman (1957, 220).

[130] Stahlman to Kuhn, 5 February 1959, TSKP 25.46–47. See also *Copernican Revolution*, 64–73, for all these cases.

indeed, the problem lies in how the expression "Ptolemaic astronomy" is understood. For him, that phrase was different from "Ptolemy's astronomy," and the difference is explained by the same pedagogical strategy that, in parallel, made "the two-sphere universe" appropriate for him. In the first case, though, it is possible to talk about a *tradition* of Ptolemaic astronomy; yet, for Stahlman it is a phrase to be particularized in Ptolemy's work or in a further development, while for Kuhn it is a reference to "a tradition of astronomical practice."[131] In that part, Kuhn was trying to show how to solve the problems of planetary motion with the resources that he had just introduced: a sort of replacement for hands-on training. As he told Stahlman, "What I was trying to do in the section was to illustrate the sorts of devices developed from the second century B.C. to the time of Copernicus in an effort to make the basic one-epicycle one-deferent technique work."[132] Kuhn's replies to more particular problems were based on the kind of pedagogical strategy that he had assumed throughout the book.

On this and other points, Kuhn took account of Stahlman's comments and those of his friend and historian of science Harry Woolf, as well as of those from reviews, and made changes in the Random House edition of the book.[133] Long paragraphs were introduced: one of them on the "two-sphere universe," explaining its artificiality and, at the same time, Kuhn's reasons for using it; and another one on "Ptolemaic astronomy," also describing its nature and scope in the book; and there were other additions besides.[134] Indeed, it seems to have been a matter of concern for Kuhn to be taken seriously by his colleagues, the historians of science.[135]

Despite that, Kuhn's plan in *Copernican Revolution* was bold, because he combined the three levels of detail that I have already noted: historical and technical precision mixed up with a pedagogical purpose, aided, in this latter case, by a certain philosophical intention that helped to give shape to the whole narrative. As Hellman noticed, the mention of conceptual schemes was recurrent, and seemed to act as more than a catchphrase, giving shape to the meta-historical interpretation of the Copernican case. For Hellman, the catchphrase in question was a trace of Conant's views, which were present in this book. She added that "No one that had not grown up in the atmosphere of Harvard University would have written the book in the terminology employed."[136] The term "conceptual scheme" was a signal of that influence, and the problem was that, "when overused as here, can be annoying."[137] I am not sure that Kuhn truly "overused" this term, but it was indeed derived from the Harvard context, as we have seen.

---

[131] Kuhn to Stahlman, 16 February 1959, TSKP 25.46–47, 3.

[132] See Footnote 131.

[133] See Woolf to Kuhn, 5 February 1959, and Kuhn's reply 25 February 1959, both in TSKP 25.46–47. Woolf's commentaries were not as extensive and substantial as Stahlman's.

[134] See Kuhn's "Note for the Random House Edition," February 1959, together with the 9 pages of corrections; see 2–3 for the ones mentioned above; all of them in TSKP 25.46–47. See *Copernican Revolution*, 27–28, 66 for the relevant passages as they can be found today.

[135] See again the closing passage of Kuhn's letter to Stahlman, 16 February 1959, TSKP 25.46–47, 5.

[136] Hellman (1957, 220).

[137] See Footnote 136.

Nevertheless, Kuhn employed the phrase "conceptual scheme" in a different way than Conant did. For Conant, experimentation led to a conceptual scheme and this latter, in turn, led to a new set of experiments. Concepts, on the other hand, were understood in a pretty strict operational way, and their arranged schemes were different from mere explanations.[138] In Kuhn's hands, the conceptual scheme is more than a good strategy for producing further empirical exploration. It is different from a mythical worldview in that it has explanatory properties, which for Kuhn is one of the key functions of a conceptual scheme; in that case, a *logical* function. Any worldview whatsoever might satisfy the human need for a place in a larger scenery, a *psychological* function that, for him, is nonetheless important, but it must also fulfil the *explanatory* function in order to become a conceptual scheme. This notion, in Kuhn's hands, plays a key role in cognitive and in epistemic terms insofar as it helps to explain as much as it provides understanding.[139] In addition, it is a guide for future behavior and expectations about the natural environment. Prediction is based on it. It provides new insight into nature that, as he says, is not simply gained from inspecting observations. And, of course, it shows its failures and the gaps to be filled, and therefore where new investigations must be conducted. "Typically," Kuhn says, "a conceptual scheme provides hints for the organization of research rather than explicit directives, and the pursuit of these hints usually requires extension or modification of the conceptual scheme that provided them."[140] Behind all this, as Kuhn also suggests, there is a "commitment to the scheme," and this commitment is to its truth, despite its recurrent falseness.[141] Despite that, insofar as it is productive of new knowledge, and insofar as it leads, as in the Copernican case, to a full new conceptual scheme, the blind trust in the promises of the conceptual scheme is basic for the advancement of science.[142]

In Kuhn's hands, therefore, the idea of conceptual scheme holds a key role in the unification and arrangement of scientific activity and of scientific knowledge in human societies, whether ancient or modern, that his later philosophical thinking would find in other concepts such as the paradigm itself. In that sense, Kuhn's approach to the notion transcends the one provided by Conant in his more modest usage of the term. Yet, it would be an anachronism to find in this notion a precedent of the notion of the paradigm. Even though it plays a similar role in the description of a given example of scientific research, it still lacks the kind of vision of scientific learning and of the differential advancement of the scientific disciplines according to their internal arrangement of training and research that ultimately led to something

---

[138] See our Chap. 2.

[139] See Kuhn, *Copernican Revolution*, 6–7, 36–40.

[140] *Copernican Revolution*, 40.

[141] *Copernican Revolution*, 39. Here, Kuhn seems to be aware that talking about truth in this case is problematic: "Such commitment or belief is always rash, because economy and cosmological satisfaction cannot guarantee truth, whatever 'truth' may mean" (39). However, I apply the true/false distinction here, though only as a simplification.

[142] *Copernican Revolution*, 41.

like the concept of paradigm.[143] At that time, Kuhn's reference to conceptual schemes was based on the kind of functions that orientations played in his vision of science in the Lowell Lectures: as underlying aids for scientific world-views and research. "Paradigm" was later more than a useful term. It involved an understanding of the nature of scientific communities that Kuhn did not yet have when *The Copernican Revolution* was published.

In Chap. 3, Kuhn shows how Aristotelian physics and cosmology provided a conceptual scheme for these disciplines, and how they coexisted with the Ptolemaic tradition in astronomy. Then, in Chap. 4, he shows that coexistence was troubled by the "inconsistencies" between both perspectives, and the search for a solution for them paved the way for Copernicus's work.[144] In Chap. 5 Kuhn shows Copernicus's work, and how it emerged from the problems of the theories that has been examined in previous chapters of Kuhn's book. In the following chapter, Kuhn shows the acceptance of Copernicus's work by the "converts" to the heliocentric system—who obtained new evidence on behalf of that new theory—as well as the arguments of some of his critics.[145] Chapters 6 and 7 show the rise of a whole new view about nature. The description of the Copernican innovation that Kuhn provides here is dynamical, and not only because it describes how one certain theory was overcome by another; it is dynamical, too, because of the underlying philosophy of science that catalyzes this pedagogic and historical description. As he would often say in his later years, the philosophy of science that would be the target of his criticism only appeals to the rationality of empirical belief on the basis of its justification by the facts. Kuhn's vision in this book already displays a dynamic picture in which the acceptance of a new hypothesis, the process by which it is established, the role that its predecessor plays in the process, and the social and cultural milieu in which all of that takes place are intermingled. Scientific rationality can only be accounted for *relationally* and by resource to the *social* milieu in which such processes take place. This is a message that, though implicitly, *The Copernican Revolution* provides. In the years in which Kuhn wrote *The Copernican Revolution*, his exploration of the social dimension had already started, and at a higher level than in the early 1950s (more on this below). But his vision of the relational nature of the validation and acceptance of hypotheses had already started to be an integral part of a systematic theory at the beginning of that decade, in the Lowell Lectures. Besides being the kind of book that he could now write on the basis of his own lectures, *The Copernican Revolution* was a first primer to a new view of scientific development, though it was one that was not highly thought-provoking. *Structure* would be a different story.

---

[143] We return to this in the next chapter.

[144] See *Copernican Revolution*, 104–105. Barker (2021) talks about the Islamic work on such inconsistencies and much more, as Kuhn does not.

[145] See *Copernican Revolution*, 188.

4.7 The Lowell Lectures and the Social Milieu of Science

The example from the history of thermodynamics in Sect. 4.5 allows us to illustrate a component of Kuhn's historical vision that, although present in the Lowell Lectures, was not as evident as in this more recent historical case. The component is the social environment in which that vision is framed. It is a component that helps to describe the causes and the process of change. Yet, we saw that in the Lowell Lectures Kuhn offered some details related to the social arrangement of scientists within their groups when he talked about the social distribution of revolutionary change among age groups, for instance. The correlation between age and attitude towards change could be explained by virtue of the different access the younger and the older members of a group have to a well-established theory, or to its successor. The role of outsiders, as against the insiders, is another well-known example of this kind. These were social phenomena that his dynamics of scientific revolutions seemed suitable to explain. The same happened with the idea of simultaneous discovery, which Kuhn also mentioned in the Lowell Lectures, and of which the emergence of the principle of conservation of energy is a splendid example; Kuhn's theory explains it without difficulty.[146]

Yet, in the Lowell Lectures he does not pay attention to the complexity involved in the social context of change. He shows his limited attention to the dynamics of change based on the role of orientations, ignoring aspects suitable for a sociological examination of science.[147] Correspondingly, the diversity of experimental results that serve as an observational basis and of the involved conceptual schemes and vocabularies that are characteristic of a revolutionary process are not as evident in the 1951 lectures as they are in his 1959 paper on the history of the energy-conservation principle. In the Lowell Lectures, the social milieu—educational, linguistic, conceptual, practical, and institutional—that provides the internal causes of scientific change and the emergence of a new unified vision of nature is not portrayed in the same way. In the example of the appearance of the vision of energy in nineteenth-century physics we observe communities that are conceptually very distant from each other, with different educational and professional traditions, whose combination gives rise to a complex structure of research work. The state of knowledge of physical and chemical processes favored contact with a somewhat dispersed fact, the conversion of forces. In order to find a unified method of research, the viewpoint based on the notion of energy required a good quantitative adjustment and adequate conceptual vocabulary. That viewpoint was born from work conducted on various fronts. Gradually, the first theoretical expressions for conversion and the first equivalences among those expressions emerged from them. All this involves a process in which the barriers that come from a diversity of practices, languages, theories, viewpoints, and methods must be overcome. With the benefit of hindsight, once the connection is captured from a unified point of view, the link of energy may appear as an inescapable connection, Kuhn indicated. This may be true, whether it is the vocabulary of energy or another that establishes this link. But the establishment of such a link follows a social milieu

[146] See Lowell 5, 42–43; GR 93.

[147] See Lowell 5, 45; GR 93–94.

that has its own tempo. Science only develops in that kind of complex and dynamic setting. The social milieu, thus, not only offers resistance, but it is also the vehicle itself of communication. The physics of energy could only be born through theoretical unification, and this does not occur through other channels than the communication routes that are socially established.[148] The Lowell Lectures roughly illustrated some main features of the process, but mainly focused on the epistemological foundations of revolutionary change. Meanwhile, the nature and influence of the social milieu in which an orientation was developed and in which it weakened and struggled with others for a central place in scientific research were left out of those lectures.

By the mid-1950s, Kuhn was encountering new ideas related to scientific development that he did not count on in 1951. Although his vision of the contingency of belief, and therefore of its social root, were already present in an essay like "The War and My Crisis," written roughly a decade before, it does not imply that he showed signs of wanting to go deeper into such roots beyond the interests he showed in 1951. However, he soon found a goal and a communication framework to do just this. Owing to his philosophical inclinations, both were somewhat unexpected. The framework was that of the social studies of science promoted by the very logical empiricism he was trying to fight. The goal was none other than *The Structure of Scientific Revolutions.* In this new context, the plan of the Lowell Lectures could be developed, but he would have to do this in a monograph on the sociology (or social history) of science, as this was the plan of its promoters. Kuhn was invited to join this plan and, as we will see next, he did so willingly. In this project, his search for an alternative in the interpretation of science was combined with the exploration of a view on the socio-historical nature of science that was gradually becoming more present in his historiographical and philosophical perspectives. Next, we look at the invitation that came from logical empiricism, mainly from Philipp Frank, to join a project in sociology of science, and also what Kuhn intended to contribute once he accepted. Around that time, he also accepted to write *Structure.* In the next section, we will examine what we know about such an invitation. Then, we will briefly examine his consolidation as a supporter of the importance of the social milieu. All this activity is important for us to understand the origins of the argument in *Structure*, as this book bases its arguments on a conception of science with a social dimension. We will then see the relationship between his first plans for *Structure* and Kuhn's pro-sociological activity, as he communicated it to the editors of the publishing project in which his book was to be framed: the *International Encyclopedia of Unified Science*. Let us see, first of all, how this productive relationship with the members of logical empiricism began.

[148] A further example is visible in Kuhn's exploration of the emergence of the quantum view of energy and the social channels that it required in the early years of Planck's radiation theory (roughly 1900–1906 and later). See Kuhn's *Black-Body Theory* and our Chap. 7 for a summary.

4.8 Kuhn, the Sociology of Science, and the Institute for the Unity of Science

On December 2, 1952, Philipp G. Frank, former member and founder of the Vienna Circle and at that time professor of physics and philosophy at Harvard, sent a letter to Kuhn inviting him to collaborate in the organization of a sociology of science research project for the Institute for the Unity of Science. Earlier that year, Frank had already talked with Warren Weaver, from the Rockefeller Foundation, about the plans for such a research project.[149] The committee was initially made up of two people—Frank himself and Ernst Nagel—and would be three if Kuhn accepted. Merton's signature was also present on the project's memorandum that Frank sent Kuhn along with the covering letter "Research Project in the Sociology of Science." In the letter, Frank presented the project's goal as social influence on broad scientific issues, that is, those that do not follow directly (that is, inductively) from the evidence. Kuhn was enthusiastic about the invitation. In a draft of his response to Frank, he said: "Thank you very much for your invitation to join the organizing committee of the Institute [for the Unity of Science]'s project on 'Sociology of Science.' There is no group in which I should rather participate [...]." He joined the committee and began to contribute his own ideas to the initial plans.[150] In July 1953, Frank reported to Weaver that the activities to promote and pursue research in the sociology of science, helping scholars and students to deal with different suggested topics, had taken place during the previous year. The committee in charge of selecting themes and researchers included Nagel, Frank, Merton, Barber, and of course Kuhn. A year later, Kuhn reappeared in a new report with the same function.[151]

The perspective that Kuhn began to bring to that group expanded the one that Frank and his collaborators had presented in their initial project. They distinguished two types of factors involved in the acceptance of a scientific theory: those of an empirical nature and those considered to be of an existential nature. The correspondence of the theoretical statements and the observation statements was, for them, a factor of an empirical nature, while the technological applicability of a theory or its compatibility with some social doctrine were part of the second type of factors, the existential ones. We can say that the former factors were internal and the latter external to science (the distinction is mine). Both, the institute members said, were clearly distinguishable in certain episodes in the history of science. In some of these episodes the choice

[149] See Frank to Weaver, 30 January 1952, in Rockefeller Foundation Records, Projects SG 1.1, Series 100 International, Rockefeller Archive Center, Box 35, Folder 285 ("Unity of Science"). See also Reisch (2005, 294 ff., and Chap. 15) for more details about the institute and Frank. See Reisch and Tuboly (2025) for a detailed study of such a project and its surrounding context—including Kuhn's role in this project. I thank George Reisch for providing me with a copy of this interesting paper.

[150] Kuhn to Frank, draft, n. d., 1; Frank to Kuhn, 2 December 1952; Frank, Merton and Nagel, "Research Project in the Sociology of Science," n. d.; all of them in TSKP 25.53.

[151] See Frank to Weaver, 30 July 1953 (including the attached report for the Rockefeller Foundation) and 29 July 1954, both in Rockefeller Foundation Records, Projects SG 1.1, Series 100 International, Rockefeller Archive Center, Box 35, Folder 285 ("Unity of Science").

of a theory had been made using the first (the empirical) as a criterion, but there were others in which the choice had been made based on existential factors without taking into account the empirical correspondence of the consequences of a theory or the validity of the logical derivation of the same. Frank's plan was to deal with this second series of historical cases through a theory that would explain the external influence on the development of science.[152]

The conception that empirical correspondence is the *only* inherent factor in the development of science is the foundation of the search for an empiricist criterion of meaning. As we know, Kuhn had doubts about the various forms of the principle of verifiability precisely because they turned certain crucial factors in the development of a theory with good empirical correspondence into factors alien to science. His idea was that empirical correspondence is part of the nature of traditional sciences like physics, but not as an independent factor. This correspondence is established on the basis of other factors of very diverse origin and nature, which come from points of view that allow the organization of human thought for purposes of pragmatic efficiency. In other words, there is no logical distinction of an internal/external type that allows us to affirm that sometimes a theory is chosen only on an empirical basis. There is no physical doctrine that does not exhibit correspondence of some kind, nor any that does not require other factors to determine the kind of correspondence.

For this reason, Kuhn outlined a correction to Frank's point of view that, although we are not completely sure he sent, surely must have appeared in at least some of his communications with the group. This correction, which in the aforementioned draft has three different versions, can be presented as follows (this is the first of these versions):

> Let me raise one question about the draft prospectus which you enclosed. Would it not be appropriate to include in the committee's terms of reference an examination of those sociological factors which impinge upon an individual scientist not by virtue of his membership in a national community (say the United States), but by virtue of his membership in a narrower professional group (say the American Physical Society)? I suspect that in twentieth-century western science socially conditioned, implicit, professional "faiths" have assumed many of the roles in the guidance of research and in the determination of the acceptability of scientific theories which religions and metaphysical systems played in the physical sciences of the seventeenth century.
>
> I believe this professional consensus has an important bearing upon the problems which a scientist considers worth attacking, the experiments which he employs to resolve his problems, the abstract aspects of his experiments which he considers relevant and the logical and experimental criteria which he demands of "valid" argument. It is my own guess that an examination of such factors in historical context would cast real doubt upon the possibility of distinguishing "between the acceptance of a theory because its consequences are in agreement with observed facts, and the acceptance of a theory because of other factors..."[153]

[152] Frank, Merton and Nagel, "Research Project in the Sociology of Science," n. d., TSKP 25.53, 1–2.

[153] Kuhn to Frank, draft n. d., TSKP 25.53, 1; the mentioned alternatives appear on 3–4. The quote from Kuhn reproduces a phrase from Frank, Merton and Nagel, "Research Project in the Sociology of Science," n. d., TSKP 25.53, 1.

As we see, Kuhn is subtle when showing his correction of the empiricist point of view that Frank and his collaborators present. (And the fact that he drafted three alternative versions of the same argument indicates that this subtlety was premeditated and very calculated.) Kuhn indicates that the first thing that influences theory choice is the criteria of the community of specialists dedicated to the discipline. As we have said, it is not possible to set an empirical correspondence criterion without that influence, so it is also not possible to differentiate between the types of factors that determine the choice except in their historical and social origin, because, for example, on some occasions religious beliefs have been part of the internal criteria of the group and more recently it has not been so. For example, such beliefs have been part of crucial elements in the development of Copernicanism, but they have not been so in the development of the notion of energy, although in this case the analytical tools and the available experimental work largely depended on research traditions that were different, parallel, and only moderately connected (sometimes scarcely related), that is, on criteria of a communitarian type. For this reason, for Kuhn, the examination of factors that originate in the consensus of the professional group, of those beliefs shared by its members, is more appropriate than the unqualified study of the impact of religion or politics on scientific development.[154]

In this draft, Kuhn mentions for the first time a term that will become crucial on the road to the writing of *Structure*: "consensus."[155] In the Lowell Lectures, the idea that the practice of science requires a consensus about current problems and modes of solution is almost a natural idea. Scientific knowledge and language are a natural but highly specialized language and knowledge. Their primary objective is the elimination of ambiguities in describing phenomena, which in turn allows for a theoretical structure that establishes a clear relationship between terms, expressions, and empirical referents. It is a structure with little flexibility, which must be torn down to allow the inclusion of new pieces, especially at its base, but which has a great capacity for prediction and explanation. As we know, for Kuhn, scientific knowledge does not increase from a sum of qualities, their interpretation as entities, and the application of appropriate names given to them by each individual, but from the fixation of objects and from a common collective use of vocabulary that communicates, with some vagueness involved, a core of more or less clear meaning. In scientific language, the tendency will be to ascertain the use of terms, to assign them a clearer, stable meaning, and to expand the resulting core of meaning by specifying features and constructing a network of relationships which is as simple as possible. For this result to be possible, scientists must set certain objectives and resort to orthodox methods to reach them. Therefore, it seems that there is a form of consensus in which young researchers are immersed from their integration into a scientific community. This is a consensus that they must master and that they often help to consolidate.

To speak of such consensus is to refer to "social factors"; even to "existential factors" we can say, because values govern the search for consensus in a scientific

[154] Frank, Merton and Nagel, "Research Project in the Sociology of Science," n. d. TSKP 25.53, 4.
[155] See *Structure*, for example, 11, 15, 172.

community. For example, the truth of statements, their simplicity and their logical neatness, as well as other factors that vary from one historical time to another, from the more positivist viewpoints to those that privilege overarching cosmological points of view, are values (or are factors that work like them). But from Kuhn's point of view, this consensus is not a bridge to elements external to science, since both "empirical" and "existential" factors participate in it. The study of religion or politics and their relationship with science belongs to a sociological study of consensus in scientific communities only in those cases where such forces with a socio-cultural nature are involved in scientific language and work. This study of consensus is of the same type as that carried out in cases such as that of energy conservation, in which what is studied is, for example, the roles of technological development and of its conceptual and material resources in the fixation of certain new concepts, new technical resources, both theoretical and instrumental, and new educational and research paths that are crucial in the subsequent development of science.

The institute was prepared to receive these ideas. A frequent member of the institute, present on the board of trustees and a participant in its activities, was Quine, who showed perspectives on logic and language that were inspirational for (and influential on) those that Kuhn would soon start to contribute.[156] As is well-known, Quine often discussed his ideas with other main members of the institute such as Carnap, an acquaintance from the old times of the Vienna Circle, owing to their common philosophical interests. Likewise, another main member, Charles W. Morris, was an American philosopher who was similarly interested in pragmatism and in logical positivism. The institute was also visited by Conant. However, the person responsible for opening the institute to ideas like Kuhn's seems to have been Frank himself, with interests very close to Kuhn's, as we have seen.[157]

Frank's institute kept trying to publish a collection of monographs that formed the *International Encyclopedia of Unified Science*. The *Encyclopedia* was an old logico-positivist project that tried to satisfy this current's ideology, that is, to free science from any pseudo-scientific burden and to contribute to establishing a link between its very diverse branches. Mainly a project of Otto Neurath, the *Encyclopedia* had been published in Chicago since 1938 with the collaboration of Morris and Carnap and under the supervision of the institute. As George Reisch says, under Frank's direction, the institute also cared about historical and sociological research on science.[158] The monograph that they offered Kuhn to write, and for which Kuhn chose the title *The Structure of Scientific Revolutions* (not to his liking at the beginning, as we shall see later), fitted squarely into that extended plan. We explore the making of *Structure* in the next chapter, but, as Kuhn's collaboration with the institute led to the origin of that monograph, let us explore its beginnings a little.

---

[156] On Quine and such roles and activities at the Institute, see the documents at the Rockefeller Foundation Records, Projects SG 1.1, Series 100 International, Rockefeller Archive Center, Box 35, Folder 285 ("Unity of Science"). See especially some calls for papers for essay contests, or the lists of seminar meetings around the period 1950–1956, all of them in such folder.

[157] For the data in this context, see Reisch (2005, 39, 294 ff.); Holton (1999, S102). See also Carnap (1963, 34 ff.).

[158] See Reisch (2005, 9–11, 229–230, 295); Stadler (2007, 18–21).

The plan for a monograph on the history of science for the *Encyclopedia* was an old plan that had suffered from certain problems. These were not infrequent in the already troubled history of the *Encyclopedia*, as five of the ten monographs for the first volume underwent changes, either of author or of thematic character. The same happened with six of the ten monographs for the second volume. The monograph devoted to the history of science always appeared as the fourth in Volume II of the *Encyclopedia.* In February 1937, the historian Federigo Enriques was invited to write it, but he did not accept. Louis Wirth did it in October 1937, but he wished to devote it to the sociology of science. Wirth did not carry out his commission and in June 1952 Frank tried his luck with historian of science I. B. Cohen. The subject of the monograph should have shifted back again to the history of science, as is evident. Cohen, however, did not accept the commission, either, but recommended a young historian of science from Harvard, the young Tom Kuhn of course. We know very little about the interactions that occurred during those months from June to December 1952, and if the invitation for Kuhn to become a member of the institute's project in the sociology of science involved the writing of the monograph for the encyclopedia, which had been so often rejected. What we do know is that, in February 1953, Kuhn took it on.[159]

As we already said, in its definitive version, *Structure* introduced a clear variation with respect to the Lowell Lectures: it spoke more clearly of the social nature of science. The practice of science depended, more than on a method, on a consensus like the one I have just mentioned. The new social dimension that Kuhn brought to the philosophical study of the scientific method lay in said form of consensus. Along with Frank, Nagel and, later, Merton, Kuhn promoted the sociology of science, and that work would be the roots of the future arguments of *Structure.* This monograph upheld the usefulness of sociological studies of science without pretending to be in itself a sociological study or argument. By virtue of historical examples, *Structure* showed a social dimension of science that was responsible for the type of discontinuous development of science that Kuhn also described there.

The usefulness of the sociology of science for Kuhn is evident in his activities from the period 1953–1959, the years before *Structure*, which we will see in the following pages. We will see how that attitude is reflected in *Structure*, too. But let us start first with his plans with the members of the institute. Kuhn, Frank, Nagel, Merton and Barber made plans for the kind of sociological studies that could be significant for a good understanding of science, and that could have priority over others. The scheme I am about to describe, titled "Possible Research Topics: Sociology of Science," shows what they had in mind.[160] The scheme has a clear Kuhnian bias that will be evident

[159] See Reisch (1995, 52–53; 2019, 109, 389); Interview 1995, 291–292, where Kuhn says that he "was approached by Charles Morris" and that Morris "asked [him] whether [he] would take over a volume in the encyclopedia"; Kuhn, "Plans for Research," 22 October 1953, UCBTK, edited by Hufbauer (2012, Appendix, esp. 458). See also Galison (1995, 29–32).

[160] I shall comment on said document as it appears in TSKP 25.53; however, there is a second, slightly different version—but almost identical except for the absence of point III.7 and for the also different arrangement of items per page—in the folder for the Unity of Science at the Rockefeller Foundation Records, Projects SG 1.1, Series 100 International, Rockefeller Archive Center, Box 35,

as we proceed. That Kuhnian bias allows us to assign him a partial authorship; an influence that seems likely, at least. The scheme, however, is unsigned. Yet, some of the terms used, such as the idea of a conceptual scheme (in any case often used by other authors, as already noted), some of the suggested cases, and, undoubtedly, the approach given to the socio-historical research, constitutes evidence in favor of a partial authorship by Kuhn. If we had to risk a conjecture, the most plausible hypothesis would be that he contributed this scheme to a first meeting with the previously mentioned members of the group.[161]

Let us start with the examples. The scheme lists them at the end, but they are significant and general enough to start with. In the scheme, there are a couple of examples that we may consider typical outside of Kuhn's views, as they highlight the existing tension between the more general ethical and philosophical perspectives and scientific development: they are those of the theory of relativity and quantum theory. This would fall within the study of science in relation to the broader social framework (Section III of the scheme)—that is, not limited to, as Kuhn would say, the scientific community, which in this scheme is reflected in a previous section (Section II). In that group of problems, one can see science within the social forces in which it is generated and that affect it externally.[162] In the scheme, a section is devoted to this.

However, there is another group of examples whose relationship with the type of sociological research proposed in the first place is much closer. They are also more typical of Kuhn's views. In this group, the key figures of Carnot and Joule are mentioned, who we know are part of Kuhn's immediate interests. Through their research, and also, in a second example, with the École Polytechnique as a primary historical target, the scheme suggests the study of how two different traditions, that of engineers and that of physicists, came into contact.[163] This double plan of a social history of science (or even of a history of scientific traditions) has a clear Kuhnian

---

Folder 285 ("Unity of Science"). Galison (1997, 788, fn. 16) already mentions this second version. Reisch and Tuboly (2025, Sect. 6.3.3) also examine this document and Kuhn's likely participation in it as I do.

[161] See two pages of handwritten notes by Kuhn that begin with "Meeting Hotel Prince George," New York (n. d. TSKP 25.53). This brief and schematic group of notes shows ideas by Kuhn that later appeared in the scheme. So, "The age problem," as he says there—that is, the relationship between age and innovation, already discussed in the Lowell Lectures (as we saw)—later appeared in the scheme with other words as point I.6. His suggestion to study how innovative work is communicated first by means of journal reports and then by textbooks reappears somewhat transformed in the scheme in points II.5–6. Even an idea that somebody asked him to "reformulate" on the "role of verbal difference in directing and stimulating research" reappeared then as the very first point in the scheme (I.1). For further details, see [Anonymous], "Possible Research Topics: Sociology of Science," n. d., TSKP 25.53, and compare it with Kuhn's handwritten notes. Frank reported to Weaver about these meetings in his letter from 30 July 1953, Rockefeller Foundation Records, Projects SG 1.1, Series 100 International, Rockefeller Archive Center, Box 35, Folder 285 ("Unity of Science").

[162] [Anonymous], "Possible Research Topics: Sociology of Science," n. d., TSKP 25.53, 2–3, III.1–8.

[163] [Anonymous], "Possible Research Topics: Sociology of Science," n. d., TSKP 25.53, 3, IV.1–2.

inspiration. Likewise, the plan to inquire about the relationships between a science like physics or mathematics and some branch of philosophy like epistemology, or between the sciences themselves, in various cases, might have been influenced by Kuhn. The study of conceptions of space is one of the suggested examples of the first class, for instance. The studies that are proposed always have a clear historical dimension, which corresponds, once again, to Kuhn's point of view.[164]

In this regard, it is worth noting that the only specific examples of the whole set belong to the historical research developed by Kuhn as a result of his seminar on thermodynamics. If we examine the other more general sociological problems listed in the first two sections of the scheme (Sections I–II), we see that those specific historical examples referred to earlier help to illustrate the more general sociological research as particular historical cases from a typically Kuhnian point of view, because these sections refer to what we have seen in the Lowell Lectures on the influence of orientations and languages in the selection and treatment of problems. I shall illustrate this with a brief examination of Section II.

This section presents the perspective that Kuhn suggested to Frank for a work on the social factors influencing scientific development. In this section, various objectives for a potential sociological research study are pointed out, which include, on the one hand, the purely institutional aspects and, on the other, those that we could call "theoretical."[165] In this regard, this section of the outline proposes to study how theoretical knowledge is transmitted, what kind of support is used for this purpose, and who or what institutions have control over this transmission, without forgetting how control is exercised. All of this, just like the attention to the function and role of scientific institutions in their respective historical and political contexts, is part of those institutional aspects mentioned in the first place.[166] As part of the second kind of aspects, which I have classified as "theoretical," we find the contents of science, the attention to its mode of change (both the theoretical contents themselves and the norms that govern them, that is, the prescriptions of a methodological nature), and the relationship between such contents and the distribution of groups and subgroups. It seems that, around these aspects, there is a great interest in the relationship between the organization of knowledge and research and these forms of social distribution. It is easy to see that the investigation of such theoretical aspects is directly related to the investigation of the institutional aspects, which is confirmed by the fact that this Section II does not use the division that we have made here. Everything in it is part of a set of points to study.[167]

This whole plan fits well with Kuhn's perspective, and that seems to be what he is trying to convey in this text if (as I think is likely) he contributed to its writing:

---

[164] [Anonymous], "Possible Research Topics: Sociology of Science," n. d., TSKP 25.53, 3, IV.3–5.

[165] The classification is ours, including the denomination of "theoretical aspects" used below. In the scheme, the aspects that we will discuss below form a list of nine points, which is the content of Section II.

[166] [Anonymous], "Possible Research Topics: Sociology of Science," n. d., TSKP 25.53, 2, II.4–8.

[167] [Anonymous], "Possible Research Topics: Sociology of Science," n. d., TSKP 25.53, 1–2, II.1–3, 9.

how the study of science and its social dimension involves paying attention to the epistemic aspects of science in relation to the social organization of knowledge and research. In addition, there is notable attention given to change, especially when the scheme mentions the study of the interaction between groups and previously created theoretical, conceptual, and normative contents.[168] To put it briefly, the author or authors of this scheme are very interested in the form of internalism in the study of social history that is characteristic of Kuhn. It is possible to affirm that however those ideas eventually appeared in the scheme, they were what Kuhn tried to contribute to this group on the sociology of science formed by himself, Frank, Nagel, Barber, and Merton.

4.9 Lecture at Berkeley, 1956

These approaches to sociology were reflected in a lecture Kuhn gave at the University of California, Berkeley, in the fall of 1956 (so, shortly after having been hired by the university) for the Social Science Colloquium, in which he showed an example of scientific practice that qualifies for this type of sociological study: measurement in science. The revision of this lecture a few years later, in 1959, and then published in *Isis* as part of a special issue with Harry Woolf as guest editor, turned it into one of Kuhn's best-known articles: "The Function of Measurement in Modern Physical Science."[169] The lecture he gave in 1956 had a similar title: "Role of Measurement in Development of Science."[170] In this lecture, he defends the idea that the unfamiliarity with the measurement processes that really takes place in a natural science like physics sets up a bad starting criterion for a subsequent comparison—that of the social sciences with physics—in which the social sciences often fare worse than physics. Against this, Kuhn believes that a good knowledge of the nature of measurement in physics would show a closer proximity than is apparent.[171]

There are two points about measurement in this text that will sound familiar to those who have read the later article published by Kuhn, but which are worth briefly discussing. The first one is the Kuhnian explanation of the notion of "reasonable agreement." This expression is also addressed extensively in his 1959 article.[172] He explains to the social scientists something that many of them must have been aware of: reasonable agreement is variable, and what may be insufficient for one scientist from a given area of science, may for another, from a different area, be satisfactory. Experimental resources, their scope, and observation technology define these limits.

[168] That attention to change is the subject of the aforementioned Subsections 1–3 and 9.

[169] Its origins are described in the Preface to *Essential Tension*, xvii, and in Interview 1995, 295. See Kuhn (1961b). See also Bokulich and Bocchi (2024), a recent interesting study on the scientific context around the writing of this article.

[170] Kuhn, "Role of Measurement in Development of Science," 23 October 1956, TSKP 23.6.

[171] Kuhn, "Role of Measurement in Development of Science," 23 October 1956, TSKP 23.6, 1.

[172] Kuhn (1961b, 184–185).

Without reasonable agreement, therefore, there is no way to achieve confirmation of the theory by experimental observation. Moreover, exact checks are not only infrequent, says Kuhn, they are suspicious of fraud.[173]

This first point tells us about the nature of confirmation in a well-established science (in its classical period, we might say), in which measurement plays a central role. Reasonable agreement is set in science textbooks through comparisons that show proximity, not distance, between the theory and measurement results, which, as we know, has an educational function. For this reason, Kuhn vindicates the study of research results not yet converted into textbook material as a source for the study of measurement. This is the second point worth commenting on. Textbooks, in Kuhn's view, offer a distorted view of this scientific activity, however positive, educationally speaking, it may appear.[174] Once the ongoing research reports have been examined, he says, it is not apparent that observation and measurement precede theory; rather, they go hand in hand with it. As he said in the 1959 article, this leads to a related idea that, although widespread in the methodology that he criticizes, is misguided: that the route to generalization begins with the data, and that confirmation is just the reverse activity of the discovery process. In criticizing this idea, he once again attacks the Pearsonian, positivist prescription that this is the only rational pathway to a science devoid of arbitrariness.[175] For him, in contrast, science advances through a continuous interaction of theory with observation, and the science textbook helps to disguise the research activity where these roles are difficult to distinguish.

In these two points, one of Kuhn's key terms in this phase, "consensus," is again evident. When he turns to the consolidation of the scheme that emerges from the textbooks, or to the agreement on the confirmation of a scientific hypothesis, he gives evidence of the kind of consensus that directly affects the practice of science from within the community itself.[176] In his view, the sociology of science exclusively addresses the social factors that arise directly from scientific activity. He does not attempt to look beyond these limits to provide his theory with a social foundation. This is what he tries to make clear to Frank, and what ultimately influences *Structure*. When in 1983 he received the John Desmond Bernal Award from the Society for Social Studies of Science, he could not avoid mentioning that he saw his book as "exclusively internalist," and then acknowledged that, although there were sociological elements in it, these were only taken from scientific practice, not from its surrounding social environment.[177]

---

[173] Kuhn, "Role of Measurement in Development of Science," 23 October 1956, TSKP 23.6, 2–3; Kuhn (1961b, 184–185).

[174] Kuhn (1961b, 179, 183–186); Kuhn, "Role of Measurement in Development of Science," 23 October 1956, TSKP 23.6, 3.

[175] See Kuhn (1961b, 183, 197).

[176] See Kuhn, "Role of Measurement in Development of Science," 23 October 1956, TSKP 23.6, 3.

[177] Kuhn (1983b, 26 [quoted phrase], 28).

Kuhn's lecture for the Social Science Colloquium at Berkeley made it clear that the difference between physical sciences and social sciences was mainly due to an education process based on a form of *communication* of the working foundations, all of which (foundations and form of communication) is *agreed upon* within the community. The communication vehicle of ongoing research and the pedagogical vehicle for agreed-upon knowledge are, respectively, two different material resources: the specialized journal for research and the science textbook for education. The numbers in them and the agreement they often represent are indicative of a specific form of research organization. All of this is subject matter susceptible to sociological study, as Kuhn presents it. The reason why research *can be* organized in this way is a matter to study separately, but it does not seem easy to conclude that measurement itself, the accuracy of the numerical agreement it exhibits, marks the difference between social sciences and natural sciences. As we will see, he appeals to his own type of contrast of social nature between the natural sciences and the social sciences as a key in his search for a theoretical perspective on the kind of consensus in mature science that he explores in the drafts of *Structure*.[178] This fact is significant for us: it indicates the social nature of a central piece of Kuhn's argument (a piece and a nature that, as we can see, were relatively new to Kuhn's thinking). I am referring, of course, to the paradigm, whose relationship to the search for that type of social consensus inside a community is clear.

4.10 Kuhn's Collaboration with Sociologists of Science Before *Structure*

As a result of activities (and ideas) like those just examined, by the end of the 1950s Kuhn was a recognized member of a community of researchers involved in the development of this type of sociological study of the natural and the social sciences. He contributed again with a new version of his paper to a conference on the "History of Quantification" in New York, in the Social Science Research Council, 20–21 November 1959. A result of that meeting was Woolf's already mentioned special issue of *Isis* that included Kuhn's paper, and that was also published as a volume by Bobbs–Merrill.[179] When the following year, in 1962, A. M. Macbeath reviewed that volume for *Nature*, Kuhn's contribution was highlighted on two occasions, only followed by Paul Lazarsfeld's text, which was mentioned on a single occasion. Macbeath said:

> Measurement for its own sake is useless without a theory to give it direction, and all the writers insist that qualitative work must precede quantification in any science. T. S. Kuhn, rejecting the over-simplified text-book picture of a physicist measuring and then finding a theory to fit his measurements, goes so far as to write (p. 43): "Knowing what results they

---

[178] See our Chap. 5.

[179] See Woolf (1961a, 133; [ed.] 1961b); Herring to Kuhn, 7 December 1959, TSKP 23.6.

should expect from chemical analyses, chemists were able to devise techniques that got them."

Finally, two paragraphs below he repeated his reference to Kuhn: "Among individual essays, I was most interested by T. S. Kuhn with his provocative view of the role of measurement and 'crisis states' in the development of physics […]."[180] Merton and Pendleton Herring, President of the Social Science Research Council (New York), the body that organized the congress, communicated to Kuhn their optimism and the latter asked him for new ideas to promote a social history of science from that organization.[181] Kuhn responded to Herring with no less optimism about how to exploit the potential of the field:

> Perhaps I need not to elaborate again how very fruitful a research field I think this one could become in a very short time, given a few people working in it. I can think of no other field that so immediately offers so very much unexamined data. Nor can I think of a field in which examination of the data is so likely to promote a new understanding of issues vital to current national policy. You have heard me talk of misleading stereotypes about the nature of science before. Sociology of science seems to me particularly likely to do away with such stereotypes and to replace them with more viable concepts.
>
> […] What can be done to get the field moving? Because I do not altogether understand why there has not already been far more activity in it, I am far from sure that I know the answer. During the last decade a number of people with the requisite scientific background have gone into sociology. These are just the people one would like to have working on the problems in the sociology of science, but unfortunately they seem always to turn to problems in statistical analysis or to discussions of the nature of the methods capable of making sociology more "scientific." Our problem would seem to be that of persuading people like this that they can do more applying sociology to science than science to sociology. But how is that point to be made effectively?[182]

Kuhn continued to tell Herring in the same letter that this research field already had enormous possibilities at that time, and that in a not-too-distant future it would produce the kind of research that he already saw as necessary. Kuhn proposed to discuss this issue with some of his remaining colleagues in this field, starting with Merton, of course.[183]

This letter to Herring shows us to what extent Kuhn felt integrated among sociologists and optimistic (with his usual caution) about the future of social and historical studies of science. Kuhn considered himself part of a pioneering group that included Merton. Along with his primary goal for social studies of science, which was to understand science itself, it was possible to find another one: to understand the "issues vital to current national policy," as he says in the first of the paragraphs quoted from that letter. As on so many other occasions, the echo of Conant's work is felt here.

---

[180] Macbeath (1962, 506).

[181] See Merton to Kuhn, 23 November 1959, and Herring to Kuhn, 7 December 1959, both in TSKP 23.6.

[182] Kuhn to Herring, 21 December 1959, TSKP 23.6, 1.

[183] Kuhn to Herring, 21 December 1959, TSKP 23.6, 1–2. Reisch and Tuboly (2025, Sect. 6.5.1) also examine this correspondence between Kuhn and Herring.

4.11 The Social Dimension in Plans for *Structure*

After reading these plans and ideas, and observing Kuhn's impressions about the fertility of the sociological study of science, it is possible to better understand his plans for the monograph that has brought us to this point, *The Structure of Scientific Revolutions*. We have seen that, by that time, Kuhn already had a clear idea of the type of approach required by the sociology of science, as he told Frank and collaborators. When in 1953 he started to talk with Charles Morris about the details of his monograph on the history and sociology of science, his description agreed with his original point of view, as described to Frank, and had elements that were already visible in the Lowell Lectures. Yet, his journey through the history and the sociology of science that we have seen in this chapter reinforced the few elements from such points of view that were going to appear in that monograph, to the extent that they would become essential components in it, as we shall see in the next chapter. The "seven years" in the history of science, which he talks about in the preface to *The Essential Tension*, and his excursions into the sociology of science with Merton and other colleagues enriched that vision of science.[184] The beginnings of that process are already visible in the plans that he conveyed to Morris.

The plan that Kuhn developed in July 1953 takes us back four years, to 1949, when he criticized Ayer in his private diary while in the Society of Fellows. In that plan, which we will examine shortly, we see a thinker who perceives and considers the limitations of the empiricist criteria of meaning. We find a thinker who emphasizes what is ignored by verification and seeks to replace the positivistic vision underlying it. However, all this includes a perspective of science influenced by its marked social nature. He speaks of scientific theory as playing the role of an "ideology" that sets up reasonable agreement in measurement, that directs us towards a specific type of problems, and that closes the research field around some specific methods, standards, and goals. A theory has a clear *social dimension* on that account, which is added to the purely logical dimension that marks the scientific method.[185] He misses this social aspect in the empiricist account of theories, and it is what he tries to embed into the study of method. In this way, he reconciles all his previous intellectual trajectory with his increasingly deeper knowledge of the fertility of the sociology of science and also with Morris's goals. Let us see that plan for *Structure* from 1953.

In his letter to Morris, Kuhn briefly displayed the vision of science that we have seen so far. Theories are not reducible to their logical reconstructions without losing a large amount of information about them and their functions in the process. In this letter, he shows that in his vision of science, theories include the points of view or orientations that he talks about in the Lowell Lectures. In explicit reference to the experiment, he expounds the ideas that he will later develop in his talks on the function of measurement that we have just seen. Experimentation and observation depend, in practical and normative terms, on this broader vision of theories. Theories fulfill for

---

[184] See *Essential Tension*, xvi.

[185] Kuhn to Morris, 31 July 1953, TSKP 25.53, 1–2. See Reisch (2019, 206 ff.) for an interpretation in political key of Kuhn's appeal here to the word "ideology.".

him, indeed, an ideological function. They set up behavior patterns and goals for a profession. If we leave this out, he says to Morris, we ignore an essential component of scientific research. Ingredients of a psychological nature are part of that theory, Kuhn adds, and are not easily subjected to a treatment typical of rational reconstructions. However, there is no scientific research without them.[186] Kuhn says in this letter that, "as psychologically inevitable consequences of a formally verifiable theory, these unformalizable components of a theory have an essential role in the process by which new knowledge is acquired."[187] Kuhn intends to redirect the methodology of science to the study of such elements:

> One of the most striking results of the "ideological" portions of a professionally institutionalized theory is the relatively firm closure which it gives the field of scientific problems. Theories preserve themselves by restricting the attention of the profession to problems which can in principle be solved within the theory and by inhibiting the recognition of important incongruities in the application of the theory to nature. In some sense every theoretical orientation excludes the existence of totally unsolved problems. A significant theoretical reorientation replaces a complete, but retrospectively exclusive, schema of knowledge with a more inclusive one. An innovator does not add new knowledge to old; he rather imposes a new set of categories on nature, destroying and replacing an older set, which, embedded in the profession by training and practice, die hard. So intellectual discovery is necessarily intellectual revolution.[188]

Kuhn's letter to Morris, in summary, begins with a plea against the verificationism that characterizes the logico-positivistic tradition in which the project of the *Encyclopedia* finds its roots, and from which *Structure* has its framework. Kuhn does not hesitate when, boldly and bluntly, he introduces his anti-empiricist and anti-formalist credentials to the representative of that project. His determined character is similar to that offered in the draft of the previous letter to Frank. It is not unusual in Kuhn's behavior, though.

Kuhn had tentatively titled his monograph *The Structure of Scientific Revolutions*, but was far from confident about the phrasing. Carnap did not like the references to "structure" in the title, as Morris told Kuhn; the ideas of revolution and science, meanwhile, were much more acceptable for Carnap, who suggested a title based only on the latter. Morris, conversely, defended Kuhn's title.[189] It seems that Carnap's considerations were not much reflected upon in this case. The plan for *Structure* included examining the ideological function of theories with a historical approach and with cases obtained from his own field of study and work. "The viewpoint raises its own problems," Kuhn added, "and discussion of them will occupy the balance of the essay."[190] This approach required, therefore, a good dose of history of science and a study of science as a phenomenon that is, at least partially, social. This last

---

[186] See Kuhn to Morris, 31 July 1953, TSKP 25.53, 1–2, for what has been indicated in this paragraph.

[187] Kuhn to Morris, 31 July 1953, TSKP 25.53, 2.

[188] See Footnote 187.

[189] See Morris to Kuhn, 6 October 1953, TSKP 25.53.

[190] See Footnote 187.

requirement proved that the monograph thus planned was fit for the gap that Morris and Carnap had invited Kuhn to fill. Kuhn argued it in the following way.

To conclude his letter to Morris, Kuhn apologized for the modification of the initial plans. "I know that this is not quite what you were after when you approached me about the volume, but I hope that after our conversation it will come as no surprise."[191] He then noted that the methods of argumentation and the perspective that he adopted did coincide with Morris's initial plans: "The point of view of the monograph is drawn from the history of science and the examples will be historical, so I am not quite so far off as the above discussion may indicate. Also, my basic problem is sociological, since it arises because any theory which lasts must be embedded in the professional group by which it will be overthrown."[192] Kuhn knew that he was not doing exactly what Morris and the *Encyclopedia* needed. But he did know that he wanted his ideas in that place and trusted that the editors, Morris in particular, would see to what extent his plan fit well there.

Kuhn would no longer dissociate *Structure* from his current attention to the social dimension of science. Next, we are going to see his difficulties in presenting this perspective of the study of scientific method in a coherent way. First, we will see how he solves the problem of combining the discontinuity of scientific development with the fact that certain scientific resources (as the calculation techniques, for example) persist beyond a revolution. Then we will examine the way in which Kuhn presents the nature of scientific consensus, which was so important in order to talk about the social nature of this perspective of science and its method. We will then see how the well-known paradigm will make its appearance. Let us turn to the making of *The Structure of Scientific Revolutions*.

---

[191] Kuhn to Morris, 31 July 1953, TSKP 25.53, 3.

[192] See Footnote 191.

Chapter 5
The Construction of *Structure*

5.1 The Kuhnian Reconstruction

There is a constant in Kuhn's thinking that marks the whole of his theoretical, philosophical evolution. We saw it anticipated, faintly outlined, when he established the nature of the causal connection against Bertrand Russell's stance, back in our second chapter. It is what we could describe as a perspective of a modal nature, which considers not only real events, but also the possible, the necessary ones, and also those that, although potentially real, are not real—not to forget those that are not even possible, of course. To do this, we need a prior representation of nature, not just a good collection of real cases. To establish a fixed behavior for phenomena through a law, we establish an ontological classification of objects and essential features of them that, as we know, *excludes* the existence of entities that do not follow this pattern. This is, as we know, the way he contemplates physical law in its *ideal* state (and it is not a perspective exclusive to Kuhn).[1] Moreover, this is the way he conceives the nature of science itself. As Kuhn told Morris in the paragraph of his letter we looked at previously, "[i]n some sense every theoretical orientation excludes the existence of totally unsolved problems."[2] This analogy is important to understand what Kuhn had in mind. Let us see what it consists of.

The ideal state of a scientific law is the eventual fruit of a long process, as Kuhn says in *Structure*.[3] The exclusion of the thorniest, ambiguous aspects of an ontology is carried out gradually at the same time laws are applied and developed and a varied experimental work is carried out. That exclusion and the development of laws are part of one and the same process of joint development of a whole perspective of nature. In this respect, the consolidation of Aristotle's and Newton's physical theories is an extremely long process that does not necessarily start with a perfect understanding

[1] See, for instance, Hanson (1958, 99–112).

[2] Kuhn to Morris, 31 July 1953, TSKP 25.53, 2.

[3] *Structure*, 78–79.

J. V. Mayoral, *Thomas Kuhn*, Springer Biographies,
https://doi.org/10.1007/978-3-031-86512-1_5

of natural objects. Normally, such research delves into natural behavior at some intermediate point of what we could call the normal logical process of establishing a law—an only partial view of that law and of the phenomena that it governs—which develops through the investigations of more than one individual in a number of directions not previously examined. These directions seek to exclude features of objects, or relationships between them, that hinder the logical relationship envisaged between them and the law. As Kuhn said in the eighth Lowell lecture, the problematic features to be suppressed are those that involve the object in situations that cannot be simultaneous in a consistent view of nature. That is, such obstacles are semantic ambiguities ascribed to that object and usually due to their origin in natural language. The scientific research that he has in mind consists (although with greater complexity) of their elimination.

Kuhn understands the development of science in terms of an identical vision. The science of each period develops ontological, instrumental, and communication patterns and excludes others. The researcher—whether a historian, a philosopher, a sociologist of science, or from another field—is a theorist who, like the natural scientist, has to outline the details of key scientific objects, whose main features must be isolated. That theorist must establish a pattern of scientific behavior. To do this, the theorist, the interpreter, will have to exclude certain forms of science practice that do not correspond to the visible pattern under scrutiny. That interpreter will have to thoroughly understand the pattern of internal theoretical exclusion in that science, and the particularities of the social environment in which that theory develops. *Structure* was the first theoretical exposition of this perspective, which is one that Kuhn had already applied in his historiographical research. In Alexander Bird's words, *Structure* is "theoretical history"; there is an empirical part to it, which supports its claims, and a theoretical one, too—"a certain pattern," Bird says (now, of course, more generally speaking). Interpreted in Bird's terms, in this book I have been looking so far for the elements of that theoretical construction—of such a "pattern," in short—and now, in this new chapter, I will try to find the last of these elements and observe how Kuhn brings them together in *Structure*.[4]

However, the idea Kuhn has in mind is more complex than it seems. Science develops in social environments, with a multitude of factors related to the organization of scientific education and research and to the relationship between the individual and the collective in that environment and that development that need to be considered. All of this would be part of "the institutional structure of science," to use Bird's apt words again.[5] When talking about a complex framework of *social environments* (in plural), we are talking about various languages and ways of organizing research that are simultaneous in time, that partially coincide in objectives and methods (and that also sometimes diverge in this respect), and that constitute variables to consider in our attempt to understand the science of another era. Such social environments also have an important peculiarity: despite their frequent and sometimes increasing divergence, there is usually a filtration of points of view between them due to the

[4] Bird (2000, viii). Expressions into quotation marks are obtained from that page.

[5] Bird (2000, viii).

search for analogies capable of filling certain gaps and solving remaining problems in the (so to speak) receiving environment. Thus, the social environments diverge and converge throughout the history of science; traditions merge and then divide in another way, without this contraction and expansion being completely uniform. The complexity of science, therefore, does not acquire the image of a simple tree branching, but that of a non-uniform network that Kuhn himself, in *Structure*, called a "ramshackle structure." In his words: "Often, viewing all fields together, it seems instead a rather ramshackle structure with little coherence among its various parts."[6]

The study of scientific revolutions helps us to make sense of this process, as these are changes in the foundation and widespread use of laws and other aspects of scientific theories, which were motivated by the inability to resolve certain ambiguities without starting to see nature (and to behave in it) in a different way. Such research allows us to explain these macro-historical phenomena because revolutions are the turning points of such expansions and contractions; they are focal points of attraction and dissociation of traditions. *Structure* aimed to explain these episodes. To do so, Kuhn had to question the epistemological and methodological bases of the traditional view of science—which, in any case, as we know, he intended to do from the outset. Having the intention to do it and doing it were not the same, of course. He faced important challenges in the task of theoretically fixing this modal conception, this perspective about the behaviors that are regular, almost mandatory, and those others that are excludable by the historian because of their incongruity with the ordinary ones.

In this chapter we will study the challenges that he encountered in his preparation of *Structure*, as well as the last pieces that were central to his perspective.[7] In the next pages, we will examine some unfinished drafts of *Structure* and the main differences between the published version and the first unpublished draft—which some of his closest friends and colleagues were able to discuss with him.

5.2 Time Delays

Throughout 1953, and even in 1954, Kuhn did not work on any new research articles, and only published five reviews, all on books of methodology and history of natural philosophy from the seventeenth century.[8] From the end of 1950 (the months of preparation for and presentation of the Lowell Lectures) until 1953, Kuhn devoted most of his time to carefully preparing his classes. As part of this preparation, the possibility of composing *The Copernican Revolution* emerged.[9] He never believed that the task would take him as long as it finally required. His application for a Guggenheim Fellowship indicated that his sabbatical year 1954–1955 would be key

[6] *Structure*, 49; see also *The Road*, 117.

[7] All this development is also brightly summarized in Nye (2011, 242–244).

[8] See Kuhn (1953a; 1953b; 1953c; 1954a; 1954b).

[9] See our Chap. 4 for details about *The Copernican Revolution*.

to bringing this small project to a successful conclusion. During the summer of 1954 he would not have to prepare classes for the following year, so he could devote himself entirely to *The Copernican Revolution*—a draft of which, at the beginning, in 1953, there were versions of the most technical chapters, and of some historical chapters, circulating among Kuhn's students. During the summer of 1953, he had already prepared drafts of the first two chapters, and a half of the third; by July 31, 1953, he had completed the first two. According to the original scheme, *The Copernican Revolution* consisted, as in the version that we know, of seven chapters plus an introduction and appendices. Kuhn planned to write the fourth chapter around June 1954 and to be able to deliver the manuscript on the first of September of that year, so that the rest of his period as a Guggenheim fellow could be dedicated to the monograph for the *Encyclopedia*, which he planned to finish in July of 1955.[10]

These plans, however, went awry. In September, 1955, a visibly distressed Kuhn responded to a friendly reminder by the always polite Morris with pessimistic news. The period of the Guggenheim Fellowship, which was granted to him, had passed without him being able to dedicate more than brief moments and notes to the monograph for the *Encyclopedia. The Copernican Revolution* had claimed all his attention, which included rewriting a part of the book. He had sent the text of that book to the publisher only three days before. He did not trust in being able to do anything about *Structure* due to his classes, proofreading the new book, the preparation of additional material for it, and so on. The worst part was that his job as an assistant professor at Harvard would end that year, and it was possible that the following year he would already be elsewhere, as eventually happened. Moving to a new university, he estimated, plus the work that was left with *The Copernican Revolution* would delay *Structure* by a couple more years. This, he knew well, could be unacceptable to the editors of the *Encyclopedia.* He shrewdly asked Morris not to replace him to write the text, because he wanted to do it himself; he really *wanted* to see his ideas embodied in that text, initially a long article, for the *Encyclopedia.*[11]

Kuhn successfully forecasted his delay in the preparation of *Structure.* There were voices that warned him of the reduced significance of the book on Copernicus as compared to *Structure*; for example, his colleagues from the debate group at MIT: Boas, Stahlman, Lurie, Dupree, etc.[12] As already reflected in the previous letter, in 1956 he moved to the University of California, Berkeley, which indeed delayed the project again. During the 1955–1956 academic year, he was quite busy with the revisions of *The Copernican Revolution*, in which Harvard University Press had introduced numerous corrections. However, luckily, Kuhn had an important ally: the great patience and experience of Morris, that had been honed by the long preparation of the *Encyclopedia*, with its associated problems.[13] Kuhn was certainly not the only

[10] Kuhn to Morris, 31 July 1953, 3; Kuhn, "Plans for Research," 22 October 1953; both documents in TSKP 25.53. For the second, see the edition by Hufbauer (2012, 458–459).

[11] Kuhn to Morris, 15 September 1955, TSKP 25.53, 1–2.

[12] Boas Hall (1999, S78). Boas recalls that in that seminar Kuhn's theses were discussed over and over again and that only he persistently pursued *The Copernican Revolution.*

[13] On Morris and the *Encyclopedia*, see Reisch (2005, 10 ff.).

one who was delayed with a monograph. Morris preferred to wait a little longer, so he granted Kuhn a period until October 1957 to finish his text.[14] This gave Kuhn a full academic year from the time he moved to California in June 1956. And although Kuhn's intention was to exceed the common average of monographs, which was about fifty pages, up to eighty (a calculation that we now know to be somewhat naïve), it was certainly enough time.[15] Yet, he spent the summer of 1957 preparing classes and a talk. In the 1957–1958 academic year he was going to start a new course, and we already know the importance that Kuhn gave to that fact. He devoted the rest of the academic year mainly to his lectures on the history of science at Berkeley.

On July 11, 1958, Kuhn wrote to Morris again. First, he told him the reason for his new delay throughout the last academic year. However, this time he told Morris that this chain of continued delays might come to a halt, for he was writing the monograph. He had little written yet, but for the new academic year he might be able to have a complete draft. He assured Morris, however, that nothing would stop him finishing. The following year he would be in Palo Alto, at the Center for Advanced Study in the Behavioral Sciences, he said, and had high hopes for that period being almost completely dedicated to *Structure*.[16]

These hopes were in vain, however. Kuhn finished his year at the Center with a couple of longish drafts of the first chapter—which was actually part of a long book that our author had started to plan—plus two articles that contained part of what was going to appear in the rest of *Structure*: "The Function of Measurement in Modern Physical Science" and "The Essential Tension: Tradition and Innovation in Scientific Research."[17] If we leave aside his partial drafts for Chapter II, highly significant for the monograph though they were (more on this below), he had nothing else to show in this regard. On December 17, 1959, he wrote to Morris again to communicate his, again, somewhat gloomy prospects—especially for Morris's plans, not so much for Kuhn himself. In Palo Alto he had worked hard on *Structure*, indeed, but more on its content than on the writing itself. Kuhn sent Morris the aforementioned material (Chapter I and the two papers mentioned) in order to try to convince him that it was worth the wait. In this case, however, the prospects were truly excessive, insofar as Kuhn asked Morris for five years to finish both his newly planned book and *Structure* itself.[18]

Setting the submission date as 1964, around eleven years after accepting the assignment, must have been too much even for Morris, who had his limits; and, of course, so did Carnap. Neither of them was willing to postpone the completion of the *Encyclopedia* five more years for Kuhn's monograph. Morris proposed an honorable way out: to transform the already existent Chapter I into the long-expected monograph, and to leave the subsequent, broader and deeper book for later. This was

[14] Morris to Kuhn, 26 September 1955, TSKP 25.53.

[15] Morris to Kuhn, 10 August 1953, TSKP 25.53.

[16] Kuhn to Morris, 11 July 1958, TSKP, 25.53.

[17] On the contents of the former, see our Chap. 4; on the latter, in which Kuhn talked for the first time about paradigms in the *Structure* sense, see below, in this chapter.

[18] Kuhn to Morris, 17 December 1959, TSKP 25.53, 1–3.

partly Kuhn's intention, but in the monograph of the *Encyclopedia* he intended to summarize the most important aspects of that book and, to do that, he first had to fix issues that he could not address in such a short time. Morris invited him to consider December 1960 as the final submission date.

This time Morris would get lucky, although the solution was not the one he had proposed. Between the last letter and February 1960, Kuhn said that he had completely rethought *Structure* and that he now saw himself capable of writing it at the end of the academic year, during the summer period. And so he did. Kuhn finished the first definitive manuscript in September 1960, and the typewritten copy was sent to Morris, Carnap, and others in April 1961. That new version was 178 pages, much too long for the *Encyclopedia.* In spite of that problem, there was now a result to negotiate with, and Kuhn wanted to do so because he considered that his text was extremely valuable the way it was. In fact, for the moment, he was willing to give up the plan for a longer book, since in the course of its preparation he had solved the pending problems and *Structure* was, in that state, the explanation of his philosophical ideas that he wanted to see in print from the beginning.[19]

5.3 The Early Plans for *Structure*

In 1962, *Structure* was a completely different text from the one Kuhn began to plan shortly after speaking with Morris for the first time. In 1962, the text was much more extensive, a book in every respect, as Kuhn would ultimately present it to the publisher. But it was also a text with a more accurate, solid, and explanatory argument. This 1962 version demonstrated more ideas than those with which Kuhn had made his early plans.

To begin with, we can make a rough distinction between versions. The first outlines and notes, which I am going to comment on next, refer us to the problem of explaining why a scientific revolution has something to say against the usual idea of cumulative progress, and also against the empirical validation of theories as a major criterion for said progress. The final version of *Structure*, meanwhile—that is, as we can read it today—is a book about the psychological link between those who work on the same series of problems in a science in a state of paradigm-based consensus, and about its sociological phenomenology. Scientific revolutions and the view of science as a set of traditions with partially independent histories are studied as consequences of that perspective; actually, this is indeed the way they are presented.

An important, distinguishing detail is the different emphasis on the notion of consensus—more precisely, on a specific, paradigm-based kind of consensus. In a first phase, around 1953–1958, the study of such a type of consensus is auxiliary to the discussion of scientific revolutions and the profile of science. In 1962 that aid becomes

[19] For what has been said so far in this and the previous paragraph, see Kuhn to Morris, 17 December 1959; Morris to Kuhn, 4 January 1960; Morris to Kuhn, 4 February 1960; Kuhn to Morris, 13 February 1960 and 13 October 1960; all in TSKP 25.53. See also SSRD, and our Sect. 5.7.

the central motive. We must also bear in mind that in the first phase Kuhn is still developing his perspective on the usefulness of sociology that we have seen before, and that for a long while his main concern was *The Copernican Revolution* and the many elaborations required during its late stages. Only after that early phase would that perspective on the kind of consensus needed for mature science be completely available to be incorporated into a theoretical vision like that of *Structure.*

A series of undated schemes show Kuhn's journey towards the monograph. The usual list includes a first section devoted to "The Problem of Scientific Revolutions," or to formulate and then answer the question "Why 'Revolutions'?", or simply "What Is a Scientific Revolution?"[20] The usual second section is already devoted to the kind of stable stage that then undergoes a crisis, which is usually covered in a third section. The reference is to a "commitment", or "consensus," but also to the vision of "[t]heory as [an i]deology," as it appears in at least two schemes.[21] The appeal to the term "ideology," which he already used in his exchange with Morris in July 1953, is clear.[22] Kuhn also uses that term later, in the early drafts of Chapter I of the monograph.[23] The rest of the monograph is usually planned to deal with the revolutionary results of the crisis state and the new kind of stable state.[24] These might be from that previously mentioned early phase, though the lack of dates make them pretty difficult to place in their proper contexts.

It is much clearer that, in March 1959, Kuhn had a specific profile of *Structure* with five chapters. The first, "What Are Scientific Revolutions[?]" was dedicated to the idea of scientific revolution and would deal with the description of scientific revolutions, which included a plea for the perspective obtained thanks to the history of science and some features of revolutions that we examine in the next section, which includes the combination of accumulation and revolution. The second chapter, on the "Normal Practice of Science," would deal with normal science, with the distinction between textbook science or theory and creative or research science, as well as with the emergence and nature of consensus. The third chapter, "The Crisis State," would talk about this kind of episode, its origins and its conclusion. The fourth, "The Confirmation Debate," would deal with the real role of confirmation in science

[20] See the outlines in TSKP 25.53. In particular, the first title appears in "Notes Towards Unity of Science Monograph," which, though undated, shows "The Structure of Scientific Revolutions" as the title for the third section, which suggests that this scheme is pretty early; the second title, the first question, appears in "Unity of Science" and in another one titled "Structure of Scientific Revolutions," this time for the whole of the monograph (both undated); the third one, finally, can be seen in another outline titled "The Structure of Scientific Revolutions," undated as with the others.

[21] See, respectively, "Notes Towards Unity of Science Monograph," 1, on "commitment"; "Structure of Scientific Revolutions," 1, for "consensus," and that scheme as well as "The Structure of Scientific Revolutions," 2, for the last phrase. All of them, again, undated and in TSKP 25.53.

[22] See the last section of our previous chapter.

[23] I examine the drafts of Chapter I in the next section. On the use of "ideology" in the drafts of Chapter I, see Pinto de Oliveira (2017, 750, fn. 9). On the term and *Structure*, see Reisch (2019, 25, 56, 79, 109, 130, 212) among other passages. See also the references in our previous chapter, Sect. 4.11.

[24] This is pretty clear in the outline "Structure of Scientific Revolutions," n. d., TSKP 25.53, 1. See also the scheme titled "New Org[anization]," also undated (TSKP 25.53).

according to Kuhn and with the relationships between theories and of these with evidence. Finally, the fifth, "Revolutions and Scientific Progress," would talk about a new concept of progress of an evolutionary nature, which includes a partial continuity at the data level; the idea of accumulation would be dealt with here as well.[25] All this included a mix of philosophical argumentation with historical examples. It is striking that Kuhn considers that some credit can still be given to positivism, which he adds after mentioning the idea of partial accumulation.[26]

As is clear in this summary, the current version of *Structure* is quite similar to this scheme; much more so than the Lowell Lectures. However, there are some details to comment on here. In 1962 there are thirteen sections that expand the contents of the five that we have just seen. Sections XI–XII of 1962 mainly incorporated the material planned for the first chapter of the 1959 scheme, although there is material scattered through other sections, such as IX. Part of the argument on the revisionary role of the history of science, which corresponds to the first chapter of 1959, would also appear in Section I ("Introduction: A Role for History"). The second chapter of 1959 also has a first point about textbook science versus creative science that in the 1962 version appears mainly in Section XI and would also be related, after the changes that we are going to study next, with Sections II–V of 1962, where normal science and paradigms are studied. Additionally, the third and the fourth chapters of 1959 would be approximately Sections VI–XII in 1962, which are dedicated to the processes of crisis and scientific revolution, although Sections XI–XII also include material from the first two 1959 chapters. Finally, the fifth chapter of 1959 and Section XIII of 1962 would deal with the same issue: Kuhn's model of scientific progress. All this redistribution is not completely precise, naturally, but it can help us to check how the topics were redistributed in the version given to the press for publication.

This refers to the *schematic* plans that Kuhn had in March 1959. In fact, there are only two draft versions of the first chapter, which we examine in the next section. Along with this first chapter we have very brief and initial parts of the second chapter. The rest of the material was only written in its penultimate version, the one Hoyningen-Huene dubbed *Proto-Structure*,[27] which precedes and is very similar to the one that reached the press and that we all know, only with one less section, and some passages as yet unwritten. I will talk about that later. That next version is also quite different from what we have just seen because it is built around the concept of paradigm. The usefulness of paradigms for the writing of the penultimate version would become evident to Kuhn just before coming up with it. We will soon be able to see where the paradigm comes from and what place it occupies in *Structure*.

[25] Kuhn, "SSR—New Outline," 12 March 1959, TSKP 4.4, 1–7.

[26] Kuhn, "SSR—New Outline," 12 March 1959, TSKP 4.4, 7.

[27] See Hoyningen-Huene (2006a).

5.4 The Old Chapter I: Revolutions and Incommensurability

The first draft chapter of *Structure* was titled "Discoveries as Revolutionary" and was dedicated to explaining how to make the ideas of accumulation and of revolutionary breakup in the historical succession of theories compatible and to describing the characteristics of the second kind of change and its consequences for our vision of science.[28] A second version of this first chapter had the title "What Are Scientific Revolutions?" and has some similarities with the first one.[29] More precisely, in the first two sections both deal with the evidence on behalf of revolutions in science as a kind of change that is more commonly visible in other disciplines such as the arts, and argue about the relationship between that idea and the one just noted: the frequent evidence of accumulation of knowledge in the sciences.[30] Further sections of both drafts are diversely arranged and sometimes include different details, but they deal with similar points, from the problem about the nature of change itself and qualifications about the role of evidence in it, to the rewriting of the immediate past and how it affects our view of the history of science, including the processes of revolutionary change and their results.[31] The attention to the textbook vision of science and the differences with the idea of creative science, which we comment on below, appears in both cases, but it is clearer in the first draft, while the second emphasizes the pedagogic but disguising role of textbook science, without appealing, as in the first one, to the true face of theories as guidelines for creative research.[32] In that sense, the first draft still echoes the point of view in the Lowell Lectures, while the second version seems more concerned with the dividing line that a scientific revolution means for the history of science—a dividing line that the specification of consensus periods and the roles of textbooks help to draw in a distinct way in the following chapter.[33]

Old terms like the "re-orientation" of a discipline re-emerge in both of these drafts.[34] And, as we shall see below, this is not the only remarkable piece of vocabulary that appears here. So too, are "paradigm" and "incommensurability" further additions that start to adopt the sense that they acquire in the published version of three years later (more clearly so in the case of the latter term than in that of the former). Many aspects of the influence of revolutions on standards and accepted

---

[28] See SSR1A.

[29] For a thorough study on the argument concerning the comparison of science and art that is visible in these drafts, which I also examine below, see Pinto de Oliveira (2017).

[30] Compare SSR1A, 1–17 and SSR1B 1–17. The partition into sections and the complete extension of both drafts are strikingly similar.

[31] See SSR1A, SSR1B, 17–37 in both cases.

[32] Compare, for instance, SSR1A, 17–26; SSR1B, 25–37.

[33] On the plan for subsequent chapters in the second draft, see SSR1B, 37–41.

[34] See SSR1A, 36; SSR1B, 17, 21, 33. They will do so in *Structure* as well; see *Structure*, 138. It is a passage, after all, almost reproduced verbatim from the second draft, as is clear below.

problem-solutions in every stage of science are discussed.[35] On the whole, many of the phenomena involved in Kuhn's views on revolutionary change are already present here, more or less developed. Moreover, some passages would be similarly reconstructed, and at some points plainly reproduced, in the final draft of *Structure*. This is the case, for example, with the well-known passage on the relationships of meaning between Newtonian and relativistic physics, now visible in Section IX, which was already written here in two different versions, one in each draft (the second one very similar to the published version).[36] Similarly, many passages of Section XI, "The Invisibility of Revolutions," of the published version were already part of the second of these drafts.[37]

We should now turn to the argument that Kuhn pursued in these drafts concerning the nature of change, especially revolutionary change, in science. One way to understand the modes of change in the natural sciences was to compare them with those taking place in other areas of culture and society. We can think of politics, for example, or other areas of the sciences, such as the social sciences, to name but a few. But he first tried to establish a comparison with what is observable in the fine arts. Art history was particularly useful for Kuhn (especially in the first draft), because in art history the disruptive change was linked to creation, to invention. This allowed him to talk about a feature that he wanted to highlight about discovery in the physical sciences, and to compare it with an area where it had always seemed indisputable. Kuhn discusses this in the first section of this chapter, dedicated to accumulation and revolution:

> [F]rom a debate between, say, Rembrandt and Picasso, we anticipate no conclusion except, perhaps, *de gustibus non est disputandum*. In the arts—prototypical for the non-cumulative disciplines—one set of standards, techniques, images, and tastes can and does displace another. There is development. But, except perhaps in the realm of technique, the new does not embrace the old as part. Successive stages in the development of art seem incommensurable and, as creative idioms, incompatible in a way that successive stages of science do not. Only in the sciences does change seem unequivocally to occur by accretion.[38]

This is from the first draft; the second says something similar, in particular: "The creative idiom of a Rembrandt, Bach, or Shakespeare resolves all its aesthetic problems and prohibits the consideration of others."[39] This is a message that Kuhn aims to apply to the sciences as well.

It is interesting that he uses the term "incommensurable" here to refer to the relationship between phases in the historical development of the art of painting. With this, he seems to try to create a vision of the irreconcilable differences between artistic styles that is similar to the one he would later try to apply to science itself.[40] This latter message already exists in these pages, although his main challenge is

[35] See, in particular, SSR1A 26–37; SSR1B, 17–25.

[36] Compare SSR1A, 12–16, and SSR1B, 13–17, with *Structure*, 99–102.

[37] See SSR1B, 31–34, 35–36; *Structure*, 137–139, 141–142.

[38] SSR1A, 3.

[39] SSR1B, 3. See also Pinto de Oliveira (2007, 749) for a comment on these two extracts.

[40] See also SSR1B, 3.

to make it compatible with the residue of accumulation and commensurability that is also visible in science. In subsequent chapters, we can see to what extent that challenge would remain a problem for his view on science. This was a real challenge for him. It often reappeared, thereby confirming his worst fears that his discontinuous model was in dispute with a well-founded common-sense view about science. He was already aware of that problem in the early stages of his preparation of *Structure*, and his response in this first version consists of revealing, to those not familiar with the history of the sciences, that some form of accretion of knowledge, of accumulation, is forced to coexist with incommensurability and theoretical discontinuity, whose roles should not be minimized as they had been in the methodology of empiricism.

To show how discontinuity may be grasped, in the first draft he again establishes his distinction between textbook science and creative science and argues that the breadth of the latter disappears owing to the view that emerges from the textbook version. That distinction between forms of science is less visible—it rather disappears from sight—in the second draft, where the disguising power of textbook science, which conceals the true exercise of science, particularly in its historical past, are mostly emphasized, as already noted.[41]

What does creative science include? For him, it contains many of the ingredients that are missing in the textbook, without which science would not have the dynamic, lively, creative character it presents. He shows that the style of scientific communication, especially in recent times, involves a disappearance of the agent, the subject, and his or her research as a project, as a motivated endeavor, and as an effort through a multitude of obstacles. There are certainly echoes of Conant in all this, incidentally. As Conant said:

> The stumbling way in which even the ablest of the early scientists had to fight through thickets of erroneous observations, misleading generalizations, inadequate formulations, and unconscious prejudice is the story which it seems to me needs telling. It is not told in courses in physics or chemistry or biology or any other of the natural sciences as far as I am aware. Take up a textbook of any of these subjects and see how very simple it all seems as far as method is concerned, and how very complicated the body of facts and principles soon becomes.[42]

Conant aims to reveal the mind behind the textbook: *who* investigates, pursues a result and writes research articles; *who* fights with problems to gradually obtain solutions. Kuhn recalls, in a similar sense, that we no longer have access to this information, because the textbook hides it, although not maliciously. There is a contrast with the fine arts and literature in this respect. He mentions renowned artists and writers in the history of the fine arts and of literature, from Michelangelo Buonarroti to his beloved William Shakespeare, and refers to the availability of the works that constitute a milestone for the aesthetic progress they brought about. Our knowledge of art and literature puts us in direct contact with past works like *Hamlet* or the Sistine Chapel. Real science, on the other hand, is not like real works of art. Its functional aspects—what its creators observed, handled, or acted on—disappear by virtue of

---

[41] See SSR1B, Sect. 4.

[42] Conant (1947a, 15–16).

the textbook view and its design. "Science," he says, "offers no equivalent for the way in which the works of Shakespeare, Bach, or Michael Angelo live in the present consciousness of the artist and his audience."[43] In the sciences, access to the past is "through textbooks," which continuously report about it but, in turn, "render it functionless, and [...] destroy it as the arts do not and cannot." [44] If, as indicated in the Lowell Lectures and also after *Structure*, our education has a component of direct contact with certain objects or situations mixed with a simultaneous use of language, descriptions, and the application of terms and names, our education in the arts, according to his view, enjoys a contact with the results of the past of which scientists are not aware during their formative stage.[45] Everything is presented, he indicates, as if it had just been discovered. His repeated criticism of Whig historiography is again clear here.[46]

History reveals a less-ordered appearance than that provided by the textbook. The structure of a textbook is functional. It exhibits unity and shows accumulation of knowledge. The experience of traveling back beyond scientific revolutions, he asserts, shows us that this appearance is the most genuine. When, for example, we think about the Copernican Revolution, he says, or the one caused by the theory of relativity, we tend to think that there are no losses of any kind that may seem central and contrary to the vision of accumulation that we usually hold, with the instrumental character for theories that is often emphasized for this form of progress. However, there are such losses and they are mainly conceptual. In Sect. 2 of the first draft, in his presentation of his main arguments in favor of incommensurability (a term he applies later in this chapter), he asserts that although the Ptolemaic and Copernican perspectives speak equally of the Earth, the characteristics of the object designated by that name are not the same in both cases. The same happens with some concepts in physics, for example in the transition to the theory of relativity. In many of the cases where we can observe instrumental continuity, we can also see that there is a conceptual discontinuity.[47]

In the latter pages of that section, he also discusses the idea that a theory that succeeds another reduces the latter to a particular case of it—the thesis of reductionism. He repeats this in Sect. 2 of the second draft. He does so with an argument that is very similar to the one he uses in *Structure* when dealing with the rupture between Newton's and Einstein's theories and which reaches the same previous conclusion.[48] As we already know, Kuhn argues in *Structure* that Newton's dynamics

[43] SSR1B, 28.

[44] See footnote 43.

[45] SSR1A, 3–4.

[46] See SSR1A, 33–35, where Kuhn discusses this redefinition. Interestingly, Pinto de Oliveira (751–754) explores a contrast art/science like this in other figures, not only Conant, but also George Sarton.

[47] See SSR1A, 8 ff. (esp. 14–16); see also 27 for the use of "incommensurability"; I shall return to this passage below.

[48] See SSR1A, 12–16; SSR1B, 13–17. As already noted, compare those pages with their equivalent in *Structure*, 101–102. The basic argument in the former ones are preserved in the latter with more or less significant variations.

are not subsumed under relativistic dynamics, but rather that there is a conceptual divide between them.[49] This example reveals that, for Kuhn, revolutionary change involves meaning variation—a conceptual change. In the same place he also refers (though only in the first draft) to the contextual theory of meaning, distinctive of logical empiricism, according to which, the change of theory affects the meaning of theoretical terms, which they obtain from the former that constitutes their context.[50] For him, the idea of conceptual change accepted in the philosophical view he attacks helps to show that, behind instrumental growth, we can observe a discontinuity between theories, and that perhaps for this reason we find the kind of opposition that sometimes occurs against revolutionary change.[51]

Kuhn's way to demonstrate the source of discontinuity is to offer a description of science that goes beyond the textbooks, that is, the scientific enterprise that emerges from the historical study of scientific practices and theories in which not only language, but also the mind, the individual, the group, the established theory, learning, and practice are shown in continuous interaction. That description is sharply argued in the first draft. The second draft opts for showing the revisionary nature of textbook science, though its similar description of true past science shows this creative science and its true face. That negative message in the second draft is surely more representative of the overall purpose of *Structure*, which could be understood as an emphasis on the sources of the "normalization" of scientific behavior that leads to the kind of progress that is visible in well-established disciplines (mature disciplines, as he says in the book). In the first draft, however, there is still room for some interesting details, besides the historical examples, about the kind of science that is grasped when delving deeper, historically speaking, beyond the limit of the textbook version.[52]

In this description, we re-encounter the Kuhn that we have been examining in the previous pages. The real theory includes the normalized version of the textbook as a tool for communicating knowledge mainly in the early stages of education.[53] But, for him, this real theory surpasses the realm of learning and, within its new, broader limits, a dialogue with the immediate past is much more likely. This may happen because, in this way, we find the management of open problems, those still to be solved, and even some uncertainty about how to solve them. All this involves

---

[49] See *Structure*, 98–103.

[50] See SSR1A, 15. Kuhn cites Carl Hempel as a source, although the title (with some minor variation) of the text he mentions is that of J. H. Woodger, *The Technique of Theory Construction*, not the *Fundamentals of Concept Formation in Empirical Science*, both in the *International Encyclopedia of Unified Science*, vol. II (Woodger 1939; Hempel 1952). Kuhn had read both, as we know, which must have caused the confusion. Regarding his reading of Woodger's text, see our Chap. 2; about his contact with Hempel, see our next chapters, as well as *The Road*, 225. See Newton-Smith (1981, 155) and Bird (2002, 445, 452), for a couple of mentions of this connection between the theory of meaning visible in Kuhn and that of the logical empiricists.

[51] See SSR1A, 16.

[52] On the kind of science hidden behind the textbook version, see SSR1B, 34.

[53] For an informative description of textbooks in this drafts, see especially the second one, SSR1B, 26.

dealing with sources that are no longer mediated by the formalized structure of the textbook. Although the sources do not go beyond the last process of scientific change (the textbook marks the limits of the origins by fixing them in the most recent revolution, from which it itself is born), there is contact with them that reminds us of the contact that is typical of other areas of culture, such as the fine arts and literature, for instance.[54]

In the pages of the first draft of Chapter I, more clearly than in the second one, there is a bridge between the treatment of creative science that is visible in the Lowell Lectures and the way Kuhn would later talk about it in *Structure*, which is closer to the view of the second draft. Between both, there seems to be a slight but significant transition to a clearer emphasis on the concealing nature of what he would soon call "normal science." The basic concept for the study of science in these drafts remains the classic notion of *scientific theory*. Textbooks are here the object that mark a minimum reference work for learning and for conveying new results. It is something that must not be contradicted by new research. The textbook, on that account, contains the semantic cores, the central, essential beliefs of each theory, which leaves the margins of said theory free for the scientist to intervene in them and to expand the first through them. Not only that, the textbook contains key examples of problems and solutions. Kuhn already uses the term "paradigm" here to refer to them.[55] However, even though the term is already present in this draft, it does not seem to fulfill the same central role that it would play in the published version of *Structure*. Kuhn uses it now to point out representative cases of a type of thesis or resource of widespread use, whether in the sciences or in philosophy. In the second draft, the reference to "paradigms of achievement in a modern text" is a similar usage.[56] Meanwhile, the textbook and the theories it contains assume a much more central role for the moment. It is more central in the second draft, and the notion of paradigm seems to take its place in the definitive, published version of *Structure* as the key resource for education and research. Clearly, in these drafts Kuhn's is diffidently introducing some of the characteristic vocabulary terms of *Structure*, from paradigms to incommensurability, but with a less specialized and idiosyncratic use adapted for the philosophy of science. In the last stages at the Center, while he was struggling with the nature of normal science, as we shall see below in this chapter, a new *sense* for an old term in his vocabulary, as "paradigm" was, would emerge.[57]

Nevertheless, by virtue of this mode of expression, in the first draft Kuhn starts to introduce two central ideas. The first idea is that there are revolutionary processes whose significance is more worthy of a careful examination than had been considered in the traditional empiricist philosophy of science. Accumulation must be reexamined, reinterpreted, he seems to tell us. The second idea is that these revolutionary processes are better appreciated from a correct explanation of the distinction between

[54] See SSR1A, 17–20.

[55] See SSR1A, 34–36.

[56] SSR1B, 37.

[57] Pinto de Oliveira (2007, 749–750, 758, 761) also talks about the different uses of such terms in these drafts and in the published version of *Structure*.

textbook science and creative science, that is, from that old distinction that was essential in the Lowell Lectures, and increasingly significant (and useful) for Kuhn. He speaks of his new conception of theory and of the research practice that develops with it in the following way:

> To apply textbook laws in research the scientist may introduce additional or limiting assumptions; or he may partially restate some of the textbook laws; or he may extend the laws by metaphor or alter them in some other way besides. But whatever its nature in a particular case, some such imaginative *ad hoc* modification must be introduced when the scientist uses his school science for research. In professional practice, textbook theories and concepts are no longer full and rigorous prescriptions for prediction and problem solving. Instead they become more or less flexible guides in the search for novel predictions and solutions. As guides, they limit the range of solutions that the scientist will conceive or accept. Furthermore, within the limits that they themselves impose, research theories are a constant source of metaphors, maxims, and other suggestions for solutions. But they do not determine those solutions themselves. They do leave room for the creative scientists. When transferred to research, textbook theories function as styles of thought. Only in texts and to students (perhaps also to engineers) do they supply the thought itself.[58]

The last lines of this paragraph deserve comment, as they allow us to explain how he would combine the idea of accumulation and revolution in the same theory. Seemingly following a perspective from Fleck, and using Fleck's vocabulary,[59] Kuhn considers that "textbook theories function as styles of thought," as he says in the paragraph that we have just seen. We can also understand that a scientific revolution can be explained as a substitution of one style of thinking for another. This is an interesting usage on his part, though it did not endure in this place beyond this draft. In the second one, there are no longer any references to "styles of thought." Kuhn, however, seems to have thought that this part belonged in the second chapter—although he never wrote the corresponding sections. In a scheme for Chapter II that already includes the notion of paradigm as a central concept, the reference to the "style of thought stuff" is still included and related to the "flexibility of paradigms." In further notes, he wrote that one of the main points in that chapter should be related to "[p]aradigms in texts. [They c]an't be the [the] same in research," after which he adds "[s]tyles of thought. Copernicus and Copernicanism, etc."[60] This point of view is key to his approach, but it seems that he sees it as more appropriately expounded in the following chapter.

Let us try to understand what he is saying here. The fact that the same theory appears in a textbook with the standardized format that he ascribes to it, while, at the same time, constitutes something as loose and fruitful as a style of thinking is explained by a distinction of a functional nature between two approaches to the concept of theory. For him, theories may appear as specific solutions with a clearly differentiated nomological and conceptual content, empirically well supported by means that would satisfy a supporter of operational analysis. But they can also appear, beyond that restricted sense, as a key that sets the pattern of research without

---

[58] SSR1A, 20.

[59] See Fleck (1979, 2, 6, 9, 36, 39, 41, 64, 99–100, 125 ff.) among other significant passages.

[60] [Kuhn], "II.2. Normal Science—Adapting Paradigms," n. d., TSKP 4.4, 2, 3.

completely restricting it; on the contrary, in this second sense, theories are as inspiring of creativity as artistic currents are for the imaginative work of the artist. Whether theories appear in one way or another depends on who studies them and in what context they are studied. If it is a student, their main function is to replicate the solution and all its virtues to learn how to manage it. For this student, the theory is a more rigid medium than it is for the researcher, who must go beyond what is established, but for whom the theory of the textbook acts as an anchor with a common goal and as a source of inspiration. It is the same theory, but in the first case we are talking about a theory that is limited to its presentation in the textbook and in the second we are talking about a "research theory," as Kuhn said, with broader and blurry limits. In this second way of seeing them, theories do indeed mark styles of thinking. As an example, Kuhn shows that Copernicus's work is a theoretical resource in the first sense, while Copernican astronomy tells us about a theory in the second sense and would be a guide for research that would bring together solutions as disparate as those of Kepler or Newton.[61]

The distinction between theories thus introduced helps Kuhn, in the later pages of this first draft, to distinguish the areas of accumulation and revolution that he had pursued from the beginning. The design of the textbook, with its somewhat forced reconstruction of a coherent and finished vision of the theoretical resources it offers, allows us to conclude that there is clear confirmation of the progressive and cumulative nature of science. He shows, however, that the theories whose comparison allows us to infer this kind of progress cannot be understood in historical terms, but, we should rather say, in methodological terms. That is, the theories that are confronted are always expounded from the point of view of a privileged observer capable of comparing their finished versions and their confrontation with the facts, always more effective for the successor in a revolutionary process. We know that in the classic constructions of the scientific method, the ability of a theory to predict and explain the same facts as its predecessor and, above all, its virtue of dealing with other facts that the previous theory did not consider, or could not describe theoretically and predict, always speak in favor of the successor; it is rational to choose the latter instead of the previous and better established one. Kuhn illustrates this methodological thesis with the example of Copernicus versus Ptolemy, but, once this is done, he shows its irrelevance in a historical sense and, therefore, in understanding the true nature of the scientific revolution. Neither the theories nor the evidence in their favor are at the disposal of scientists at all times in a phase of revolution—rather, they scarcely are. The comparison is subsequent to this phase and always offers a comfortable comparative position at the mercy of a privileged observer, who, evidently, is not the historical actor.[62]

---

[61] See SSR1A, 17–21, for the distinction outlined by Kuhn in this paragraph. In *The Copernican Revolution*, "Copernican astronomy," we must remember, leads to a sort of general view, not ascribed to a specific solution, but to a sort of root for a family of them. In that book, Kuhn aimed to introduce a wide audience, beyond pure experts, to such a root of heliocentric approaches to the problem of the planets and other heavenly bodies. See our Chap. 4 for further details.

[62] See SSR1A, 23–25.

The process of revolution does not usually involve the cases of better confirmation of the successor. What comes into play in the rare moments of revolutionary science is rather a confrontation between the theories as styles of thinking. The theories, now seen as these larger and looser research areas and resources, mark the development of scientific research. What scientific research consists of, what it is like to be an active researcher, how it is (and how it *should* be) well-done science, are essential characteristics of scientific practice in each historical era that rely on the style of thought marked by an existing theory. This latter is the foundation for such characteristics, not the other way around, as this theory is the only foundation on which this practical and, above all, normative message is based.[63]

Certainly, that is an explanation of the way science evolves according to the view Kuhn had been cultivating since the Lowell Lectures, more sympathetic to the real nature of science, and more akin to the notion of style as applied to science. In fact, he returns to the use of "ideology" in both drafts of Chapter I in order to show what the dynamics of science is—he was going to write about such a dynamics in subsequent chapters (which he didn't in this version). The reference to crises and revolutions in the established ideology of a given historical age remains a good way to list the contents of further chapters in the book—a kind of reference that he would not use later in the published text.[64] In the second draft, "ideology" remains a term in use, while the notion of style vanishes from its pages—seemingly in order to return in the following chapter—and he accentuates the significance of textbook science. Therefore, the dynamics of change is introduced as a fracture and replacement of a well-established version of that kind of science.

In this sense of breaking the normative unity of all science, later in the first draft he introduces the term "incommensurability" again, after it first appeared a few pages back. Incommensurability occurs here between styles of thought.[65] In the second draft it appears a bit earlier but in a similar context, and the association with styles is no longer present.[66] With that notion, he refers to the reconstruction of the entire group of problems and methods available in a past science—the one that is abandoned after a revolution. If we understand theories in the creative sense of the first draft, the rupture between styles is the visible outcome of revolutions. If we turn to the idea of theory, then scientific theories are incommensurable. As he would go on to say in *Structure*, and similarly in these drafts a few years before, "The normal-scientific tradition that emerges from a scientific revolution is not only incompatible but often actually incommensurable with that which has gone before."[67]

Kuhn's drafts of Chapter I thus synthesize much of what he already wanted to say about scientific revolutions, and about his wishes to show the true face of science. Some elements of the picture that *Structure* was going to show are already there, and I am not only (not even primarily) referring to paradigms and incommensurability.

---

[63] See SSR1A, 22–24.

[64] See SSR1A, SSR1B, 37–41 in both cases.

[65] See SSR1A, 27.

[66] See SSR1B, 18.

[67] *Structure*, 103; SSR1A, 27; SSR1B, 18.

The closed nature of normal science, based on the kind of training device—also a selective repository of information, not only about the methodological arrangement of science but also about its history—that the textbook is, is an emerging message that is more present in the second than in the first draft. The trace of Fleck's notion of "style of thought" was still intended to play a role in relation to the function of paradigms in creative science, though he did not write that part of the book in this way. In these pages, the conceptual framework for the concept of paradigm seems to be taking shape. Its place at this point is, however, mainly covered by the notion of the textbook. In any case, Kuhn's main message in this chapter is to cast doubt on the unchallenged assumption that the natural sciences are a kind of cognitive enterprise that follows a linear development towards truth. Further chapters were supposed to offer stronger proof for that thesis. However, some previous steps of a conceptual nature still awaited to be taken before doing that, and they related to the kind of problem that ultimately led him to emphasize the notion of a paradigm.

5.5 Consensus: A Comparative Study

During his stay at the Center for Advanced Study in the Behavioral Sciences, Kuhn dedicated himself to working on *Structure* with an intensity and continuity that he had not been able to achieve until then. At the end of the period, however, he wasn't satisfied with the results. On the last day of July 1959, Kuhn summarized his results to the director of the Center, R. W. Tyler, and lamented that *Structure* was not yet fully realized, not even in a percentage that was sufficiently reassuring for him. He had written several texts: a couple of reviews, the article on measurement and two of Kuhn's now better-known texts, "The Essential Tension: Tradition and Innovation in Scientific Research" and "A Function for Thought Experiments."[68] The importance of the latter for his thinking later in his career, especially after 1975, cannot be exaggerated, as we shall see in Chap. 8. As for *Structure*, however, Kuhn was still on the first chapter. He even had two drafts of it. This delay, as we know, was what really bothered him. He wanted to have finished *Structure* at that point, and, in its place, he only had one of its five parts, which, on top of that, was also excessively long. The result was therefore understandably frustrating. We discussed earlier the very gloomy prospects that Kuhn communicated to Morris shortly after; however, this was just one more step in the long journey that led Kuhn to *Structure*.[69]

The result from an intellectual point of view was not so negative. Kuhn believed that the year at the Center had been intellectually beneficial. This second and more positive assessment was due to a process of going deeper into the structure of scientific development from a more appropriate point of view for the philosophy and the sociology of science than he had achieved until that moment, and with greater coherence. The previously cited articles written at the Center were a testimony to

[68] Kuhn (1959a; 1961b; 1964).

[69] Kuhn to Tyler, 31 July 1959, TSKP 14.15, 1.

this result and, from that point of view, he was much more satisfied. The notion of "paradigm," for instance, as used in *Structure*, started to become evident in "The Essential Tension: Tradition and Innovation in Scientific Research."[70] In that sense, the year at the Center might well be, he believed, a key period in his intellectual development. With hindsight, I would venture to say that he was not mistaken. Kuhn had managed to gather very scattered pieces of his thought that had been accumulated over the years and was very optimistic about what would emerge once the complicated process of composing his detailed vision of scientific development came to an end.[71] In the process, the term "paradigm" had found a key place that soon became crucial for his argument.[72] However, for (some) friends and (many) foes, he had given far too much attention to this "paradigm." We shall see, for example, Conant's reaction to it near the end of this chapter.

In any case, Kuhn affirmed that a pattern had emerged and that he had pursued it to try to specify it as much as possible. This pattern had appeared in direct relation to his stay at the Center, mainly due to the kinds of scientists that he had encountered there. These were, according to Kuhn, highly professionalized classes of researchers—psychologists, sociologists, linguists, anthropologists, economists, and some historians—many of which had a strong disciplinary identity.[73] His attendance at the workshops with his colleagues had not been very profitable for him and he made it clear that he had soon stopped attending them. Kuhn had established good relationships with only four members at the Center, but he was nevertheless satisfied. In this way, he continued to tell Tyler, he was able to dedicate himself to his own interests and obtain the positive results for his thinking that he had expressed before.[74] If, as historian Jamie Cohen-Cole claims, a well-defined objective for the Center was "exposing its members to disciplinary cross-fertilization that was supposedly impossible at their home institutions,"[75] Kuhn did not satisfy this objective and, as he explains, seems to have tended towards a certain isolation.

An exception to this seems to be Michael Polanyi's lecture as a visitor at the Center. Both in 1967 and later in 1995, Kuhn doubted whether that lecture was an influence on his thinking in the days he was working on the idea of normal science

[70] See *Essential Tension*, xviii.

[71] Kuhn to Tyler, 31 July 1959, TSKP 14.15, 2.

[72] See *Essential Tension*, xviii.

[73] During that year, there were illustrious collaborators at the Center. For example, among the social scientists were D. Bell (sociology), C. DuBois (anthropology), R. Firth (anthropology), M. Fortes (anthropology), E. H. Galanter (psychology), C. Geertz (anthropology), G. C. Homans (sociology), R. Jakobson (linguistics), J. R. Levenson (history), G. A. Miller (psychology), K. Pribram (psychology) and J. J. Schwab (biology and education). Among the philosophers were Quine and Davidson. Kuhn and Quine seem to have had some friends in common, such as D. Aaron (linguistics) and L. Benson (history) along with Homans and Levenson. Kuhn and Quine interacted at the Center. Quine arrived at the Center prepared to correct *Word and Object*, still unpublished, something on which Kuhn collaborated, as he made comments on its Chapter II. See "1958–59 FELLOWS," n. d., TSKP, 15.16, 1–3; Quine (1985, 281–283; 1960, xi); Cedarbaum (1983, 203).

[74] See Kuhn to Tyler, 31 July 1959, TSKP 14.15, 3–4.

[75] Cohen-Cole (2014, 88).

and consensus.[76] He admitted to W. H. Poteat in February 1967, in a nowadays well-known exchange, that he had read *Science, Faith, and Society* and *The Logic of Liberty* at the beginning of his career as a historian of science, and that he had mixed feelings about those readings.[77] On the one hand, he was glad to know of somebody with a similar concern with science; on the other hand, he did not think that "a close scrutiny of scientific behavior is an appropriate base for a defense of democratic values," and was of course troubled (as he still was in 1995) by "the extent to which Michael [Polanyi] resorts to something very like ESP when trying to explain how it is that the dedicated scientist gains his insight into nature."[78] In the same vein, after listening to Polanyi's lecture at the Center, "though impressed by Michael's emphasis on the importance of apprenticeship," he noted, he "did not make much of the idea or see why he pushed it so hard."[79] For him, it was "another manifestation of the occasional quasi-mystical elements in Michael's thought."[80] In September of that year, 1967, Kuhn was asked to comment on a paper by Polanyi in a symposium of the American Psychological Association; I examine this commentary in Chap. 7. At that point, in the fall of 1967, Kuhn tried to find the common points between them just as some differences, and he seems to be trying to go deeper into the notion of intuition—which he thought Polanyi did not do.[81] In a letter to Poteat in the previous February, however, he did not doubt his decision to exclude it from the influences on his thinking that he reflected in the preface to *Structure*, but he did think that he had in common with Polanyi a group of "concerns," to use his term; those that had led him to paradigms and those that had led Polanyi to tacit knowledge were similar.[82] That may have been what he obtained from that lecture at the Center, and which he thought, in 1995, might have contributed to his arrival at the notion of a paradigm.[83]

Concerning *Personal Knowledge*, Kuhn did not read it when it was first published, as he told Poteat and then said again in 1995.[84] Kuhn argued that he could imagine that a book like that would make him reflect on foundational issues, and it was published right at the moment he was fighting with the central issues of *Structure* and was under the pressure of time—a battle that he felt he was losing, as we can see in this chapter. I[85] can imagine that he expected to read it when that work was finished. He must have

---

[76] See Kuhn to Poteat, 28 February 1967, TSKP 4.13, 1; Interview 1995, 296. Kuhn's exchange with Poteat in February and March 1967 is available in the appendix at the end of Jacobs (2009, 99–101; see esp. 100).

[77] On this exchange, see Molenski (2006–2007); Jacobs (2006–2007; 2009); Nye (2011, 242); Timmins (2013, 311–312). All these references deal with the relationship between Kuhn and Polanyi, and discuss the likely influence of the latter on the former.

[78] Kuhn to Poteat, 28 February 1967, TSKP 4.13, 1; Jacobs (2009, 100). See also Interview 1995, 296.

[79] Kuhn to Poteat, 28 February 1967, TSKP 4.13, 1; Jacobs (2009, 100).

[80] Kuhn to Poteat, 28 February 1967, TSKP 4.13, 1; Jacobs (2009, 100).

[81] See our Chap. 7 for details; see also Postscript—1969, 190–191.

[82] Kuhn to Poteat, 28 February 1967, TSKP 4.13, 1–2; Jacobs (2009, 100–101).

[83] Interview 1995, 296.

[84] Kuhn to Poteat, 28 February 1967, TSKP 4.13, 1; Jacobs (2009, 100); Interview 1995, 296.

[85] Interview 1995, 296–297.

done so, because he mentioned the book in his 1961 replies to comments on "The Function of Dogma in Scientific Research"—Polanyi acted then as a commentator—and then in *Structure*, Section V.[86] Yet, his mention is coherent with his resistance to allude to it as a source—which it wasn't—in the preface. Likewise, shortly after publishing the book, he told the mathematician Alexandre I. Wittenberg, from the Université Laval, Canada, who had asked about the connection, that his "draft was largely complete when Polanyi's book came out," which was not accurate, as the book had been published in 1958 and, as we see in this chapter, the draft was far from being finished at that time.[87] His statements about the lack of influence on him and its causes seem more plausible, however:

> I find almost all of his argument extraordinarily difficult to follow in any detail. I repeatedly find myself applauding his stance, and then losing him very quickly when he starts to develop it. And at a number of points, where I do follow his argument, it seems to me either very weak or quite wrong. His book seems to me extraordinarily important, but I find wrestling with it a very frustrating experience, indeed. Have you had any similar experience with it?[88]

Indeed, as a conclusion, it seems likely that Kuhn's attitude towards Polanyi and his books was that of a younger colleague with a similar point of view, who therefore agreed with the ultimate intention of his work and could admire the effort, but who also thought that he could not agree with the specific key points and explanations in his discourse. That seems to be the case in 1967, too, when he commented on a paper by Polanyi, as already noted.[89] Strategically, he simply ignored *Personal Knowledge* in 1958, and when he did look at it in detail, he did not really like it.

An important aspect of Kuhn's theoretical achievement at the Center was his work on the concept of consensus in science, insofar as the kind of agreement that emerged in a period of normal science required a focal point, which the notion of paradigm fulfilled. That process is clear by exploring the partial drafts of the (never finished) Chapter II—from the original, five-chapter plan for *Structure*—as I shall do in the next section.

Kuhn did not start to write Chapter II as it appeared in the scheme of March 1959. In fact, all the versions that remain, incomplete and with interspersed comments (more on them below), start from the same point of view: trying to isolate the kind of scientific consensus typical in the natural sciences through a comparison with the social sciences. Thus, it seems that the seed of the stay at the Center had germinated in these partial versions. As Kuhn confessed to Tyler,

> Being among social scientists forced me to think harder than I had before about the nature of and reasons for the apparent differences between the natural and the social sciences. Furthermore, the environment inevitably provided a convenient source of data. As a result, my thoughts on stages in the maturation of a science or of a scientific specialty have developed

---

[86] See *Structure*, 44–45 (n. 1); Kuhn (1963a; 1963b, 392–394).

[87] See Kuhn to Wittenberg, 24 April 1963, TSKP 4.16, 1.

[88] Kuhn to Wittenberg, 24 April 1963, TSKP 4.16, 1.

[89] See our Chap. 7.

> more and assumed a more central position than they would have if my free year had been spent under other auspices.[90]

If we remember his reading of Max Weber in 1949, Kuhn criticized him for an excess of confidence in the quantitative methods as used in the natural sciences. Kuhn also disagreed with Weber on the insurmountable barrier the latter saw between the natural and social sciences due to the possibility in the natural sciences of applying a formal, mathematical approach to a subject matter that, like the physical, natural world, was not subject to unpredictable individual attitudes and behaviors. Kuhn never agreed with Weber on this matter, though he found Weber's view on the method of the social sciences as accurate as, and closer to, that of the natural sciences than Weber himself could appreciate. This idea remained in Kuhn's mind for a decade, and his stay at the Center confirmed his impressions.

His perspective on the methodology of the social sciences erased boundaries between them and the natural sciences. His rather different views on scientific research, as compared to well-established approaches, explained this. The more traditional sciences were manifestations and products of a behavior that he would consider *mature* and that could be achieved over time.[91] It took many centuries for some sciences like chemistry, electricity, geology, or mathematical economics to achieve that type of social behavior. We should speak of "social behavior" because the maturity of a science is achieved through the establishment of modes of education and communication that are only understandable at a social level. As he said in a talk given at Berkeley in November 1959, titled "Is Social Science Science?", what differentiates a mature science from one that is not are not the individual research methods but the modes of communication within the group. That is something that sociological research can trace, and which he drew attention to in his previous plans for a sociology of science, as we have seen.[92] When talking about modes of communication, he mainly refers to the transmission of results and the education of future researchers. The observation of the daily discussion of the natural-sciences researchers could also be illustrative, according to him.[93] The distinction had to do with the depth reached in the exchange of information. A researcher of a mature science, he showed, would not discuss the foundations of the discipline or debate its main objectives. By contrast, in the social sciences, any element belonging to the foundations and the discipline itself could be reconsidered. The kinds of text involved in education and the transmission of results were also very often different. In the mature sciences there was a gradual shift from textbooks to research papers that any student had to traverse carefully; in the social sciences this gradual transition from the core of established knowledge to the frontiers of research was not so clear and could be, and often was, missed out.[94]

[90] Kuhn to Tyler, 31 July 1959, TSKP 14.15, 2–3.

[91] See, for example, *Structure*, 10, 24.

[92] See our Chap. 4 for details.

[93] See Kuhn, "Is Social Science Science?", 3 November 1959, TSKP 3.12, 2–5, 7.

[94] See Kuhn, "Is Social Science Science?", 3 November 1959, TSKP 3.12, 1–5.

This perspective on the two major groups of disciplines is surely arguable, but there is no doubt that it underscores something important for the vision of the scientific method and something central in his upcoming vision in *Structure*: the social structure, characteristic of each discipline or interconnected group of them, defines ways of relating itself to its foundations, and hence, modes of normalized behavior for their members. Against the profiles of a science characterized by common traits, which is visible in the empiricist perspective of the method and also in the sociological vision of the scientific enterprise as a whole,[95] Kuhn supports a *plural* vision of the sciences that, though it surely contributes less to the demarcation of science from non-science, provides a down-to-earth conception, more confirmable by historical evidence, of the scientific enterprise.

In 1969, in his postscript for the second edition of *Structure*, Kuhn acknowledged that: "If this book were being rewritten, it would [...] open with a discussion of the community structure of science."[96] The importance, even the priority, of the scientific community is present in *Structure*, of course, but in an excessively diffuse way and relegated to the background by other terminological elements—especially by the omnipresent paradigm, about which I will soon speak. Thus, in Section XIII, in 1962, Kuhn points out that despite the "losses" often involved in scientific advancement,

> the nature of such communities provides a virtual guarantee that both the list of problems solved by science and the precision of individual problem-solutions will grow and grow. At least, the nature of the community provides such a guarantee if there is any way at all in which it can be provided. What better criterion than the decision of the scientific group could there be?[97]

Kuhn felt that pages 35–40 of *Structure* provided a more or less sociological view of the scientific community.[98] However, the first sections of the final version, and especially the most appropriate one, "The Route to Normal Science," Section II, do not dedicate the necessary space to this issue. Instead, they show how work is done in a mature science, not what the new subject is—the scientific community, of course—and how it explains that way of working. And as he noted in 1969: "Scientific communities can and should be isolated without prior recourse to paradigms; the latter can then be discovered by scrutinizing the behavior of a given community's members."[99]

The community reaches a state in which the type of education becomes gradual and in which the student learns to solve problems that progressively become key to the subsistence and understanding of the theory. In the student's eyes, the logical and empirical support of the theory, that is, the logical and empirical coherence of the scientific world in which that student works, becomes more fragile, as the latter goes deeper into research work. There are no greater problems for that student than

---

95 See, for example, Merton (1942).

96 Postscript—1969, 175.

97 *Structure*, 169. See also *Structure*, 163–169.

98 Kuhn to J. S. Carlson, 19 February 1968, TSKP 6.1.

99 Kuhn, Postscript—1969, 175.

those in which he or she works to suppress such deficiencies. But there are a couple of questions that are still beyond the reach of this image[100]:

1. What are the elements of the education and work of a science in its mature phase? We know they are the foci of the community's consensus, but, what *are* they in particular? Are they methodological rules? Or, are they perhaps psychological worlds?
2. Where does the pattern of mature science come from? What process is followed to reach the stage that we have just described?

Structure was written in full when Kuhn found a response to both problems, in particular when he took the answer to the first question (the paradigm and its nature) as a starting point for the entire work. The idea of scientific revolutions, of the kind of incompatibility that he was going to term "incommensurability," and even of crises and anomalies, seem to be more or less understood by him at this point. The main open issue was to demonstrate the missing piece that answers questions such as those I have formulated in the first place, which are related to the kind of consensus that we find in the mature sciences. Let us approach the emergence of this piece first, before turning to the question about maturity. We shall approach both in the next section. It is clear that Kuhn found answers to such questions comparatively late in the day when compared with other elements in his perspective. Indeed, such questions were not easy for him to formulate and answer. Let us see just how that happened.

5.6 The Rise of Paradigms

During his period at the Center, Kuhn emphasized the point that, except in a period of scientific revolution, science works to solve problems, which from his perspective is equivalent to saying that scientists try to establish statements in the realm of vagueness of the research theory. In the reworking of the lecture on the function of measurement that he was going to give in November 1959 for the meeting on quantification that I have already mentioned, he referred to this way of working as *normal science* or, rather, as the "normal practice of science." Normal science showed a way of working that is based on a kind of consensus that is characteristic of some sciences in a mature stage of development. In his opinion, the step taken in that lecture was central in the development of this typically Kuhnian concept.[101]

Normal science would henceforth be the common term of mature science. It was not a leap that was not prepared in advance. The version of Copernicus's work that we have already seen shows us that research within the same tradition uses the resources of the established theory, and mainly attends to its needs for clarity and

[100] These two questions are part of my own explanation; they are not a quotation from Kuhn.

[101] See *Essential Tension*, xvii–xviii, 182, 188; Interview 1995, 295; also see the draft of the paper in question, "The Role of Measurement in the Development of Natural Science," n. d. (ca. 1959), TSKP 23.6, 5 (where he adds by hand "normal" right before "scientific practice"), 12.

application to nature which, in Kuhn, as we know, involves a shared psychological world. This is the characteristic kind of scientific work in its classical phase, as explained in the Lowell Lectures. Normal science is, consequently, that classical period. Between one and the other designation there seem to be nothing more than nominal differences. However, according to Kuhn there was a significant step in applying that denomination: "somehow or other I saw normal science as puzzle-solving, although it wasn't already all there; it was something that came out at that point, and which helped me be ready, I thought, to write *Structure*."[102]

"Puzzle solving" was a useful metaphor. In Kuhn's vision of science, terms and concepts are already present for the members of the entire community. In fact, they are so present in the minds of researchers that they are rarely mentioned. Scientific research aims to reduce the margin of vagueness around these concepts. Normal science is a puzzle-solving activity because it tries to fit together all beliefs and explanations of concepts, along with available observations, without leaving gaps between them. Normal science tries to make the theory an uninterrupted description of nature. These pieces are already available, and it is also conceivable that they can fit together. If we look at the organization of measurement in a science like physics according to Kuhn, in its normal practice that activity requires a standard of measure, a quantitative margin in which numbers indicate significant differences or coincidences. This fits well with the idea of the puzzle: only one type of piece serves to assemble the image. In examples like the Ptolemaic astronomy with which Copernicus's predecessors—and even Copernicus himself—worked, the research process could be understood in these terms. Yet, what were such pieces? What sanctioned this kind of *convergent* work towards the same end and the same language? In short, Kuhn had to find the crux of a kind of scientific work that, at the same time, acted as a source of research values in science. That central piece not only restricted the vocabulary, but also the research goals, the search itself, of a whole group of scientists.

This problem in need of a solution makes us return again to the scheme that Kuhn had in March 1959. There the pressing problem was precisely this, and some changes in the second chapter of this scheme reveal to what extent he did not stop working on it until the last moment. The emerging notion of paradigm, already present in the pages of former partial drafts, and even in his own usual vocabulary, provided him with the clarity he had been seeking, when he gradually (but quite rapidly) observed the useful explanatory role it could play if it occupied a central place in his argument. That was the motive behind giving it so much prominence in *Structure*. The concept of consensus is explained in the 1959 scheme in Sect. 4 of Chapter II, after describing the type of work in science according to a style of thought, and by contrast with the sciences that do not follow this pattern.[103] In a later scheme of Chapter II, meanwhile, the notions of consensus and mature state of a science are introduced in Sect. 1, and other aspects of consensus are dealt with in Sects. 2 and 5. There, Kuhn propounds

---

[102] Interview 1995, 295.

[103] [Kuhn], "SSR—New Outline," 12 March 1959, TSKP 4.4, 3–4.

to explore these notions in depth. To do this, he must relate the consensus to the style of thought, and the previous step is to specify what the *object* of consensus *is.*[104]

The concept of paradigm, once adapted to its new functions in *Structure*, managed to fill that explanatory gap. We have already observed its appearances between 1951 and 1959, especially in the draft of Chapter I that we saw previously, where it is used to talk about a representative case in certain areas, from philosophy and psychology to the natural sciences, although without occupying the foreground of his description as it will do later.[105] The reference to paradigms, therefore, has precedents in very general and habitual uses in his way of speaking. It is evident that, throughout most of the partial draft of Chapter II, the notion does not change much; it is only in the second half of the longest draft that we find something more specialized, and which in his hands would become the notion of paradigm as it later appears in *Structure*. I shall show below how this changed while Kuhn was writing Chapter II. The paradigm in that book is an example, case, or model that is, or can be, integrated into a theory as representative of a certain part of it plus its already well-known applications, usually successfully confirmed. In terms of his normal science, the paradigm is part of a theory composed of an orientation (or style of thought, as he said for a while), a solid theoretical core (including its law statements), and an application of that theory. Nevertheless, the *function* it plays in creating (or promoting) a kind of *practice* of science within the community is surely more important for him than the pieces out of which it is formed. The way this appears in Kuhn's manuscript for Chapter II, where it started to acquire the features it shows in *Structure*, is worth a little description. Here, I will give a physical description of that draft (or rather drafts) and then a description of its contents.

Among the partial drafts of that chapter that have been preserved in Kuhn's repository—three drafts plus some spare notes and second versions of certain pages—there is one of greater length than the rest.[106] Let us explore these other drafts before examining the longer one. Two of those shorter drafts are titled "The Nature of Consensus in Science." One of them starts on p. 42, that is, right after the end of Chapter I. The other one starts on p. 38, as if he had decided to suppress a part of Chapter I (all of Sect. 5, probably, the one devoted to introducing the contents of the monograph). While the former draft extends through p. 50, the latter one only includes a couple of full pages and is followed by some handwritten and typewritten lines, notes, and comments. These two drafts look like failed attempts to deal with the main line of his argument, still mainly based on consensus without further qualifications.[107] The third draft (the longest one) is different, and not only because its title is different, too. That partial draft, "The Normal Practice of Mature Science," albeit incomplete,

---

[104] [Kuhn], "New Outline II," n. d., TSKP 4.4, 1 (esp. Sect. 2, a–b). Although undated, the document mentions the outline of March 12, 1959, showing at the same time that it will include the same discussions of the nature of consensus, the form of scientific practice in its absence, and the idea of essential tension also developed at this time (remember his article with the same title, Kuhn 1959a). See [Kuhn], "SSR—New Outline," 12 March 1959, TSKP 4.4, 3–4.

[105] See also Lowell 5, 45; Lowell 6, 1–3; GR 93–99.

[106] See them in TSKP 4.4.

[107] Reisch (2019, 236 ff.) has carefully dealt with such drafts.

is 20 pages long. It is followed by two sets of items: (1) three re-starts of the same text from p. 43, which, even in the longest case, only extend to p. 49; (2) three pages of notes, the two last of which are a replacement for pp. 45–46 and are accompanied by a page of comments and instructions. Though ordering them is mostly arbitrary, I will refer to the three first rewritings (set 1) as "versions 2–4," and to the latter pages of notes (set 2) as "version 5." It is possible to distinguish the versions within set 1 thanks to their length: version 2 reaches p. 49, version 3 only p. 46, and version 4 ends on p. 47. Version 1 would thus be the original partial draft.[108] I do not mean to say that these notes were composed in that order (1–5); I number them that way only for convenience. Leaving aside those additional notes and rewritings, this third draft (i.e. the version 1) as a whole covers Sect. 1, "Consensus and the Maturity of Science," and only two pages of Sect. 2 (pp. 61–62, the second page of which, 62, is rewritten), whose title is "Normal Practice, I—Working on Theories."

This will do as a physical description of the material for Chapter II. Regarding the contents, some evolution from the original draft (version 1) can be observed. Kuhn starts to note that the consensus in the natural sciences involves an agreement among the group members about the problems to work on and about the criticism that is relevant. That kind of consensus produces good results, he says, like a greater "specificity and scope of the new scientific theories and techniques" and an isolation of problems, "usually esoteric," to deal with. These will later produce revolutionary change—as he adds, "[w]ithout consensus there would be no revolutions and vice versa."[109] In version 2, he starts to analyze what kind of consensus he is talking about, and his earliest attempt is to show that such "consensus," he says, "means something very like 'agreement sufficient for effective professional discourse.'"[110] He adds that this agreement affects the "group's problems, methods, and standards of solution."[111] He relates consensus with *behavior*. In version 3, for example, he takes it to be a property which is observable in the community members' research practice. So, he writes: "the diagnosis of consensus (or its absence) must normally be made not by examining the opinions of group members but by observing how, in the absence of special questions, they go about their own work and the evaluation of that done by others."[112] Their research practice is conditioned by that kind of consensus, which is thus observable in such practice. In version 4, he avoids the term "agreement" when he refers to the emergence of an "effective professional discourse," an expression he repeats here, but he only talks about its "existence," a feature whose recognition by the interpreter leads to the conclusion that there is consensus within such a community.[113] In short, in versions 1–4, he tries to emphasize that there is some agreement reflected

---

[108] Just as the rest of documents, they can be found in TSKP 4.4.

[109] See Kuhn, "Chapter II—The Normal Practice of Mature Science," n. d. (ca. 1959), TSKP 4.4, 42–43.

[110] See version 2 (n. d.) in TSKP 4.4, 43. Significantly, Kuhn substituted the last three words for the original "normal professional intercourse".

[111] Version 2 (n. d.), TSKP 4.4, 43.

[112] Version 3 (n. d.), TSKP 4.4, 43.

[113] Version 4 (n. d.), TSKP 4.4, 43.

in practice, in behavior, more than in explicit statements, and that it is not dependent on specific pieces of belief, method, or theory, although it affects the way problems and solutions are accepted.[114] "If few geologists [...]," he writes in version 1, "ever accuse their colleagues of publishing professional statements that are in principle vacuous, that need not mean that most geologists agree on what a vacuous statement is."[115]

Consensus is thus a property that is evident to an external observer (say, a historian of science) more by virtue of the scientists' activities than by explicit comments on them; and it affects the methodological behavior of the scientists. Kuhn illustrates this point on the basis of a behavioral difference between the natural and the social sciences, as previously mentioned. Yet, in version 5, in the sheet of notes that accompanies it, it looks as if he did not want consensus to depend on the absence of controversy on specific normative aspects of the discourse. Showing that social scientists often discuss philosophical matters regarding meanings, foundation, and methods of their science, does not mean that we can find the key to agreement in the absence of *such* a specific kind of discussion. He prefers to call our attention to something more apparent, but at the same time more difficult to isolate in terms of a traditional philosophy of science. He talks about a characteristic of consensus that seems to be key on that account: the "felt relevance" by a scientists of the "work [made] by another specialist working on [the] same aspect of nature."[116] He says that this *felt relevance* must be added to two key ideas, to wit, that community members, once they are such members, see how the problems they choose are, as a rule, accepted as legitimate by the rest, and that the outcomes of their research are only submitted by others in the group to the kind of doubts that they themselves would consider legitimate as well.[117] An idea like "felt relevance" has the virtue of showing that, in some forms of professional behavior, scientists work on a set of problems that they have in common, that their behavior exhibits reactions that make them converge to a point, while they have sufficient leeway to avoid saying that they explicitly agreed on the philosophical details of their discipline. It is not that he pursued *that* precise idea as a term of the art in his philosophy, but such an idea, passingly annotated while working on Chapter II, is a good summary of the kind of search that led him to the next step: the quest for the focus of agreement.

In his instructions to himself about how to proceed with version 1 (such instructions are part of version 5), he suggests to himself to withdraw the parts of the text (pp. 44–49) that are already devoted to the comparison between the natural and the social sciences and to the ordinary understanding of consensus. He seems to prefer to start talking about the nature of a mature science, which he deals with from p. 49. A piece that initially looks doomed to disappear from the earliest section of the chapter

[114] See such versions in TSKP 4.4, especially versions 2 and 4, 43–44 in both.

[115] Kuhn, "Chapter II—The Normal Practice of Mature Science" (version 1), n. d. (ca. 1959), TSKP 4.4, 46.

[116] Version 5, n. d., TSKP 4.4, unnumbered page.

[117] See version 5, n. d., TSKP 4.4, unnumbered page; "Chapter II—The Normal Practice of Mature Science" (version 1), n. d. (ca. 1959), TSKP 4.4, 43.

is, as he says, "my version [of consensus] based on paradigms." However, an added, handwritten note at the bottom of the page reads: "I don't think I can really get on without the paradigm stuff in [Section] I.1."[118] As he also says in these notes, the idea of a paradigm would appear later in the chapter.[119] With hindsight, we know that he did not follow through with that plan, but that in the final design of the monograph, the paradigm would indeed appear as a main notion from the outset.

Indeed, the paradigm changed it all. Yet, it does not seem to be a key piece of version 1 until Sect. 1 is fairly advanced. Initially, the notion appears adjectivally, and usually in plural, as paradigm "sets," which are contained in textbooks as guidelines for future practice.[120] However, in its first appearance a few pages before, it is applied to social science, where the kind of consensus he is looking for is not present. He had written there: "In several characteristic social sciences a piece of research that one sub-group considers a paradigm of successful scientific achievement may be received as 'pure speculation' by a second sub-group in the same field."[121] Not that Kuhn feels later in his career that the sciences that have not encumbered a solution to the height of an hegemonic, practice-guiding exemplar for the whole profession cannot have paradigms for each school; on the contrary.[122] However, at this moment he is looking for a key piece that distinguishes the kind of practice that, on the basis of a generalized consensus, can eventually lead to revolutions. To do so, he starts to focus on those paradigmatic items (problems and solutions) that appear in the sciences that he considers mature. This point of view starts to appear in reference to paradigms from p. 48 onwards. Consensus is a matter, now, of "paradigm problems and paradigm solutions upon which the individual scientist can model his own work."[123] They are located in textbooks and transform the profession making further solutions—which sometimes turn out to be paradigms of "succeeding generations," he says—attainable from them.[124] As he says here, in version 1 (other versions do not transform this part), a science becomes mature as there is such a piece for the whole practice, and as far as scientists conform their work to that achievement, they need not worry about the measure of conformity with the paradigm.[125] It seems that the intuition of a felt relevance, as noted above, is already present in Kuhn's mind here.

---

[118] Version 5, n. d., TSKP 4.4, unnumbered page.

[119] Version 5, n. d., TSKP 4.4, unnumbered page.

[120] Kuhn, "Chapter II—The Normal Practice of Mature Science" (version 1), n. d. (ca. 1959), TSKP 4.4, 48. As Hacking (2012, xxiv) recalls, Kuhn acknowledges such sets for the vision of *Structure* in "Second Thoughts of Paradigms" (Kuhn 1974, 297).

[121] Kuhn, "Chapter II—The Normal Practice of Mature Science" (version 1), n. d. (ca. 1959), TSKP 4.4, 44.

[122] See our Chap. 7, Sect. 7.4; see also Hacking (2012, xxv), who refers to Kuhn's emendation in "Second Thoughts on Paradigms" (Kuhn 1974, 295, fn. 4).

[123] Kuhn, "Chapter II—The Normal Practice of Mature Science," n. d. (ca. 1959), TSKP 4.4, 47.

[124] Kuhn, "Chapter II—The Normal Practice of Mature Science," n. d. (ca. 1959), TSKP 4.4, 48–49 (the expression in quotation marks comes from this latter page).

[125] See Kuhn, "Chapter II—The Normal Practice of Mature Science," n. d. (ca. 1959), TSKP 4.4, 47.

It seems that the notion of paradigm was intended to reform the whole of Chapter II. In a scheme that I have passingly commented on in Sect. 5.4, Kuhn considers starting the chapter by showing simply how the "[n]ormal practice" of science is "work guided by paradigms." Kuhn aims to show "[w]hat a paradigm is" on the basis of examples from the history of science, and then, in further sections, plans to show how the period of consensus is a matter of working under the guidance of a paradigm.[126] Paradigms seem to be "flexible" guidelines for research—a role in which he planned to return to the relationship with the idea of "style of thought," as already noted—but they also present some "rigidity" when they are playing the role of finished solutions for students in textbooks.[127] It seems that explaining contents like these together with examples, he was going to transform the whole of Chapter II, with paradigms as its main characters. As we know, the whole book would eventually be transformed, but that paradigms were going to be the leading idea did not change.

In later pages of Chapter II (version 1), he illustrates his views on the paradigm-based maturity of some sciences—chemistry and physics above all, with special attention to physics.[128] He describes the relationship of paradigms to textbooks and to classic books of some disciplines, and then to specialized journals. For him, such classic books do not resemble classic literature, for instance, insofar as they are liable to substitution once a better formulation is available, so they are sources of paradigms rather than the objects of paradigmatic attention themselves.[129] By that point in the chapter, the relationship between paradigms—or paradigm-solutions and paradigm-problems—and textbooks is well established and the connection to the notion of maturity is already clear. This seems to be, however, part of what he aimed to deal with later in the chapter. We will never know what this would have looked like. It is relevant to note that he ends the draft on p. 62 by calling attention to the degree of "talent" involved in normal science—as he writes:

> In the normal practice of science [...] every successful undertaking is "routine" in the special sense that all but the most concrete details of its outcome are known in advance. [...] This characterization, I discover, leads most non-scientists and too many students of science to conceive normal practice as a kind of scientific makework—an activity ideally suited to second-rate talents and possibly of some practical utility, but one which cannot, by its very nature, contribute anything really basic to scientific advance. This impression is, however, largely mistaken. [... M]any of the problems encountered in normal practice call for genius of as high an order as any encountered in extraordinary practice. More important, it is only

126 [Kuhn], "II.2. Normal Science—Adapting Paradigms," n. d., TSKP 4.4, 1.

127 [Kuhn], "II.2. Normal Science—Adapting Paradigms," n. d., TSKP 4.4, 1–3. The relationship of research theories and the notion of styles was already clear in the March-1959 outline for the chapter. That is probably the reason why he started exploring creative science in Chapter I, first version: he did so in order to prepare the reader for a thorough exploration of consensus and normal science. Paradigms were not present at that point yet, of course, but when they were, they were intended to play that double role of "rigid" guidance and "flexible" inspiration (or focus of analogy). See [Kuhn], "SSR—New Outline," 12 March 1959, TSKP 4.4, 3; "New Outline II," n. d., TSKP 4.4.

128 Kuhn, "Chapter II—The Normal Practice of Mature Science," n. d. (ca. 1959), TSKP 4.4, 49–55.

129 See Kuhn, "Chapter II—The Normal Practice of Mature Science," n. d. (ca. 1959), TSKP 4.4, 55–59.

> as such talent is applied to the problems of normal practice that occasion for extraordinary practice arises.[130]

That is a feature of normal science that he had many difficulties conveying to the readership of *Structure*. His emphasis on the nature of normal science, on the key educational role of textbook science, and on the way such texts help to train would-be researchers at the same time as they conceal the genuine past of a mature science, made the creative nature of science, and the former role of the idea of creative science very difficult to grasp. Even informed interpreters turned into critics of that vision of normal science as leading to an unreflective, irrationalist, dogmatic, and dull exercise. Kuhn's efforts with the true nature of science were not made in pursuit of that vision, however. The paradigm was intended to show what the practice of a mature science, one that attempts to understand nature better, truly is.

We still have to refer to the second question of the two mentioned at the end of our Sect. 5.5, and whose relevance in the completion of Kuhn's theory will seem more obvious now. The question was, where does the pattern of mature science come from? Translated into the vocabulary of the paradigm, the question is, how does a paradigm appear, or arise? This question occupied Kuhn's thoughts for a while, and it is even difficult to say that he completely resolved it. In fact, the issues about the difference between mature and pre-mature science reappeared in his afterthoughts, once *Structure* had already been published.[131] We have a written testimony of his initial struggle to achieve a satisfactory version for the book.

Among his notes prior to the writing of the final draft of *Structure* (1960–1961),[132] Kuhn believed that, in order to talk about consensus—one of his key pieces in the previous scheme, especially in Chapter II of his initial unfinished version—he needed a clearer understanding of how a consensus arose.[133] As we already know, there were sciences such as physics, characterized by a certain manifestation of consensus among the members of the community, as opposed to the social sciences, which did not enjoy the type of education and research that, according to Kuhn, was based on paradigms. At some point in their history, the former had made a transition to this form of consensus. From then on, their progress was cumulative except in periods of revolution, in which the transformation of research problems (the paradigm shift) made scientific traditions before and after a scientific revolution conceptually, ontologically, and methodologically incommensurable. In short, once this form of consensus was reached, scientific advancement was discontinuous in Kuhn's sense,

---

130 Kuhn, "Chapter II—The Normal Practice of Mature Science," n. d. (ca. 1959), TSKP 4.4, 61–62.

131 See our Chap. 7, Sect. 7.4, for more details. In undated notes located in the same folder as those of March 1959, probably of a later date, he speaks about "pre-professional" versus "prof[essional] stage[s]" of a science. See the notes that start with "1. Repriese." (sic), n. d. TSKP 4.4, 1.

132 See our next section for details.

133 [Kuhn], [Untitled Notes], n. d. (ca. spring–summer of 1959), TSKP 4.4, 1. These notes start "To handle II, 1 I must know more […]." Hereafter, I shall refer to them as [Untitled Notes]; unfortunately, they are undated.

and the reference to paradigms was also relevant. *Structure* was exclusively dedicated to describing the development of the sciences that had reached this form of development.

This transition does not become a distinctive feature that is intended to help us to draw a demarcation line between pseudo-science and science. The transition itself has another character. The acquisition of this form of consensus allows for cumulative development and collaborative work through the distributed solution of problems. The main phenomenon of the process is the isolation of the problems to work on and the exclusion of aspects that become irrelevant in that work. After a revolution, these could be central to the new science, however. A good example to which Kuhn could refer in order to illustrate this acquisition process is the well-known history of the law of conservation of energy, which started from the divergent study of various forms of conversion until a few pioneers found a more or less common vocabulary whose development seemed to open new doors and which led to theoretical and instrumental development, that is, to the exploration, elaboration, and clarification of this vocabulary and its consequences.

In the aforementioned notes, Kuhn briefly alluded to one of the keys to this process. In the sciences that acquire this form of consensus, we move from a more or less intuitive contemplation of phenomena, not very different from everyday life, alien to the practice of science, to a more specific conceptualization in which there is no room for a lack of consensus. He asserts here that this may explain an indication from Cavell about the distinction between scientific theories, properly speaking, and usual religious beliefs.[134] For Kuhn, therefore, in order to talk about that kind of consensus, it is crucial that the community focuses on particular kinds of uniformity, of scientific objects, of specific instrumentation resources and observation means to deal with such objects, and of a language able to express beliefs and central statements about all this. In other words, for him, it is key that the process of acquiring the kind of theory that he himself had been observing and conceptualizing since at least 1945 culminates in these forms of intra-community social consensus. This latter focuses on a representation of nature formed from some specific components and excluding others, and allows for specific practical resources and forms of description and explanation, also excluding others in the process.[135]

The state of mature science, the emergence of this form of consensus in the investigation of phenomena and, with it, of certain ways of observing, studying, exposing, and solving the problems they pose, comes, therefore, from a gradual process of unification. In this state, the process of revolutionary transition between states of consensus begins. The above-mentioned notes reveal the moment, unfortunately undated, when Kuhn found this important relationship between maturity and consensus, which is key to the present reconstruction of his dynamics of scientific ideas. After a dotted line that seems to close the reflections of this matter during the previous session, he wrote:

---

[134] [Kuhn], [Untitled Notes], n. d. (ca. spring–summer of 1959), TSKP 4.4, 1. Kuhn mentions only "Stanley" in these notes, who would hardly be other than Cavell. Thematically, it also makes sense.

[135] See our Chap. 1, Sect. 1.12, for an early perspective of this nature.

> I now suspect that yesterday's wrestling did the trick. My line seems to be as follows: Pattern of development from consensus to consensus is something sciences achieve with maturity. Earlier have far more divergence. In this early stage problems are presented by everyday life and are problems for laity and craftsman too. Lots of competing solutions. To make the transition some of these problems do get solved, but more get abandoned or by-passed. Simultaneously get a professional group that becomes arbiter of right and wrong; that's no longer left to all educated men, etc. Science wouldn't work that way. Then problems get esoteric, etc.[136]

The appearance of the concept of paradigm as a key piece of his full view will complement this central idea. In the definitive version of *Structure*, he describes this process early in Section II, "The Route to Normal Science," already based on the notion of paradigm, in order to later deal with the mechanism of paradigms and the revolutionary cycle. One more issue should be mentioned as it appears at the end of these notes. He knew very well that the process of acquisition of paradigm-based consensus as described in *Structure* was a simplification. His view of science is plural, not monolithic, so the theory thus outlined threatens to be overly simplistic. He returned to this problem after publishing the book, as we shall see in the next chapter. *Structure* could not deal with such complexity, though it was evidently an oversimplification. Kuhn seems to indicate that it simply could not deal with this in greater depth within the work as we know it.[137]

In the 1960–1961 draft, Kuhn poured all the contents we have seen so far—from the earlier vision of revolutions and continuity, including crises and incommensurability, to these latter crucial ingredients about consensus, rules, paradigms, and maturity—into a synthetic view. The sciences in *Structure* have a great diversity of forms. However, in Kuhn's theory, the idea of contemporary science had a recognizable prototype: theoretical and experimental physics. This representative of the sciences is identified in his work through a non-exclusively methodological but also *social* portrait that emphasized the role of the institutions internal to science itself. He also managed to show that, only under the conditions thus explored and thanks to a concept like the notion of paradigm, could he give a plausible explanation of the way a scientific discipline historically evolves, the way it changes, until it reaches a state of success in the form of theoretical and practical achievements and a specific form of practice.

The central piece of this explanation is the concentration of education and research of those individuals dedicated to trying to understand nature in a single group that focuses on an example of solved-problem that serves as the canon for such learning and subsequent work (which gives normative unity and creative flexibility to the social bond).[138] This basic piece has an empirical and sociological basis, and can be

---

[136] [Kuhn], [Untitled Notes], n. d. (ca. spring–summer of 1959), TSKP 4.4, 2. In reading this paragraph, we must take into account that it was not intended to be read by anyone other than Kuhn himself. I have corrected some mistakes, however.

[137] [Kuhn], [Untitled Notes], n. d. (ca. spring–summer from 1959), TSKP 4.4, 2. See also Kuhn (1959c) for such a pluralistic view.

[138] On the problems of circularity in the definition of community and paradigm that *Structure* creates, see Kuhn's awareness and how he dealt with this in our Chap. 7, especially Sect. 7.3.

summarized in the adoption of a few theoretical successes as a reference for future research behavior for the whole group. Once this piece is acquired, the mode of scientific development, however complex and somewhat vague it may seem, explains the current state of science without too many conceptual commitments (like a realistic notion of truth) and with a certain evolutionary tone, as he expounds at the end of *Structure.* And, by using the evolutionary vocabulary somewhat freely, Kuhn exposes all this through his most recognizable ancestors (examples from the history of science), in the way that Conant taught him that science could be explained to outsiders.

In its working scheme of March 1959, *Structure* had other parts that we have already looked at in Sect. 5.3, and that would end up becoming part of later sections of the book after the initial five that focus on the ideas we have already seen. To the best of my knowledge, no further notes in this collection show that Kuhn had the same kind of trouble with his description of crises, anomalies, and revolutions as he had with the paradigms, maybe because they were older in his own thinking. As we shall see in further chapters, many—or rather most—of his revisions deal with the role of paradigms and then with the underlying vision of which they became a part. I am not referring to the cycle of revolutions, but rather to the extent to which our understanding of nature depends on a net of concepts and terms (and thereby on a language and a scheme of things) that we acquire by membership in a group, and through a process that depends more on paradigms than on definitions and rules. Let us now take a look at the subsequent history of the first full draft of the book.

5.7 *Structure*: The "Penultimate Draft"

With the new piece, the paradigm, Kuhn reordered the scheme of *Structure* completely. This happened just after sending the letter to Morris in December of 1959, in which Kuhn suggested unrealistic deadlines.[139] In February of 1960, Kuhn wrote again to Morris and told him:

> One by-product of that difficult letter [of December 17th, 1959] was a minor morning brainstorm which left me with a new and reduced outline for the monograph. I think that it's sufficiently closely modelled on the pieces that I've already done so that I might well be able to push a draft through during the summer. There are, to be sure, a couple of significant sections that I've not yet thought my way through, and I'm not yet clear that I could keep the whole thing short enough, but I think there's a real chance.[140]

In this case, Kuhn's expectations were fulfilled. In a letter to Morris on October 13, 1960, Kuhn told him he had finished his first full version of *Structure*, without footnotes and at about 160 pages long, more than a month before, on September 11, at around two in the morning. Kuhn was enthusiastic about the draft, which he said he had written in a continuous manner, without stopping too much to revise

---

[139] Kuhn to Morris, 17 December 1959, TSKP 25.53, 3.

[140] Kuhn to Morris, 13 February 1960, TSKP 25.53, 1.

the text, after the first 40 pages. His plans at that time included sending a copy of this first version, in December of 1960, to various colleagues and friends in order to correct it during the summer of 1961.[141] He actually submitted the draft to his friends in April, 1961, and between May and June of that same year he received responses, some of which we will see in the last section of this chapter. Kuhn sent copies of this version to Barber, to Merton, and to Shils on 14 April, 1961; to Nagel on 15 April; to Morris on 16 April, 1961; and, finally, to Conant and to Kubie on 22 April, 1961; Pierre Noyes must have received his copy at an earlier time, as he responded on 16 April with some comments.[142] Kuhn also probably distributed it among other colleagues. We know that he provided one to Paul K. Feyerabend, too.[143] Feyerabend, as it is well known, was a philosopher who defended the importance of the history of science for philosophy of science as forcefully as Kuhn did, and who also propounded and explored in depth the notion of incommensurability. He had become associate professor in Berkeley's Department of Philosophy in 1959. They became friends from 1960 onwards, although they strongly disagreed in many philosophical matters. In April 1961, Feyerabend was in close contact with Kuhn, so he received a copy of the draft.[144]

The draft that he sent them, with twelve sections, not thirteen as in the one he published, and 178 pages, without footnotes, is the one available in his MIT archives. This version, in those archives, is preceded by a note that reads "Penultimate Draft of Structure before June 1960." There is no reason to think that it is not the *penultimate* draft before the final one, but there *is* reason to believe that this date is not correct. A small dating error would not be surprising if the note was not written by Kuhn, something that seems plausible since it is handwritten and it does not appear to be his handwriting. However, what he said in his letter to Morris in October 1960 is more decisive: if he finished his first version on 11 September, 1960, there is no reason to date it as prior to June of the same year. The copy archived at MIT is also the same one that Paul Hoyningen-Huene renamed as *Proto-Structure* and that, according to his estimates, Feyerabend also received and commented on.[145] In this way, to avoid confusion, I will refer to this text by its usual title, *The Structure of Scientific Revolutions* (or simply *Structure* when abbreviated), with the caution of

---

[141] Kuhn to Morris, 13 October 1960, TSKP 25.53.

[142] See TSKP 25.53 for all these letters (the same one for all the addressees, in fact, with occasional minor differences in the contents). At the suggestion of Noyes, he submitted a copy to the physicist Sidney Bludman, who made some interesting comments, too; see Noyes to Kuhn, 21 April 1961; Bludman to Kuhn, 23 May 1961; Kuhn to Bludman, 20 June 1961; all of them in TSKP 25.53.

[143] See Hoyningen-Huene (2006a).

[144] These lines on Feyerabend are based on Matteo Collodel and Eric Oberheim's recent investigation into Feyerabend's correspondence and manuscripts. See Collodel and Oberheim (2024, 36–53, esp. 38, 48 ff.). On the letters he sent to Kuhn to summarize his (mainly critical) attitude towards the draft of *Structure*, see Hoyningen-Huene (1995) and Collodel and Oberheim (eds.) (2024, Chap. 9, esp. 276–293, 300–313).

[145] Hoyningen-Huene (2006a, 611). See also Hoyningen-Huene (1995); Collodel and Oberheim (eds.) (2024, 276–293, 300–313); the letters in these editions show that this was the version that Feyerabend commented on. See also Hoyningen-Huene (2015, 187–188); Collodel (2014, 7–8).

pointing out that it is a draft from the archive when relevant, for which we will assign it a date from September 1960 to April 1961.[146]

There are variations in this version, as I said. The preface of the published version is not present and the first section is titled, simply, "Introduction," without the subtitle "A Role for History." This section does not start with the same famous phrase of the published version ("History, if viewed as a repository [...]"),[147] although a slightly different version is the second sentence of the introduction. Additionally, this first section does not yet contain the criticisms of the philosophy of science born from the historiographical approach; specifically, those that questioned the generality of a distinction like the one traditionally exposed between the context of discovery and the context of justification, and that irritated some of Kuhn's colleagues in philosophy.[148] According to Hoyningen-Huene, Kuhn introduced these paragraphs at the suggestion of Stanley Cavell, but he seems not to have been very satisfied with having done so in the long run.[149] In the present draft there is no sign of these statements. In their place, Kuhn writes that, at that point, he will include elements more typical of a preface.[150] There are other differences, but the most significant pertains to Sections IV and V of the published version. The sections of the 1962 version of *Structure* that today have the titles "Normal Science as Puzzle-solving" (Section IV) and "The Priority of Paradigms" (Section V) were a later expansion of a single section in the 1960–1961 draft, whose title was "Normal Science as Rule-Determined" (Section IV).[151] The contents of this are a little different from those later sent to the press. I will examine this in some detail below.[152]

5.8 Paradigms Versus Rules: The Old Section IV

The main difference between the primitive Section IV of the draft of *Structure* from 1960–1961 and its final version, published in 1962, is one of emphasis and development of a main idea: traditions of normal science do not need to be based on sets of rules, especially if these constitute the basis of education and research. We can explain this idea of Kuhn's in the following way. Education requires a process of greater

[146] See Kuhn, *The Structure of Scientific Revolutions*, draft, ca. September 1960–April 1961, TSKP 4.5. Hereafter, in these notes, we will refer to it as SSRD (see abbreviations). I thank Pierre Noyes, Paul Hoyningen-Huene, George Reisch, and Matteo Collodel for fruitful conversations about these drafts a few years ago. My current version is indebted to them.

[147] *Structure*, 1.

[148] See *Structure*, 8–9, especially the two last paragraphs of the section.

[149] See Hoyningen-Huene (2006b, 124–125). Hoyningen-Huene (2006b, 130, n. 1) recalls that Wesley Salmon did not like the comments finally included by Kuhn in those pages. The reference cited by Hoyningen-Huene is Salmon (1990, 325).

[150] SSRD, 10.

[151] See SSRD, 36–44.

[152] Previous studies of this section include Isaac (2012a, 224–226; 2012b), Hoyningen-Huene (2015, 187–188), and Reisch (2016, 15).

convergence, we can say, towards a group of theoretical and material resources, while research demands some divergence; and note that I say "some," that is, not a complete departure from the theory, but a means of clarifying it without surpassing it. This, incidentally, requires from the individual a certain ongoing process of convergence towards some theory (be it the established one or a revolutionary alternative), motivated by the stimulus to generate a psychological, phenomenal, or behavioral world (whatever you wish to call it to follow Kuhn's views), a conceptual scheme and a language that are increasingly better articulated. Even the mature processes of problem-solving, or puzzle-solving for Kuhn, can be understood as a means of improving this cognitive and linguistic scenario in the best possible way. Although coherence in one's own world does not always involve convergence towards an established theory, this individual thrust of research works well in periods of dominance of a certain theory. Rules, if they are understood as the traditional method says they should be, are too rigid to promote the corresponding freedom of thinking, decision and action and, at the same time, the kind of ontological, semantic, and methodological bond that an orderly and progressive practice of science requires. For this reason, paradigmatic examples and solutions can be more useful than the classic, more rigid rules that scientific methodology sought.

In the primitive Section IV, this idea, which is central to Kuhn's argument, is exposed in a way that is perhaps too fast and disorganized. Let us follow Kuhn's argument a little. As he will also say in the published version of *Structure*, rules are not usually linguistically visible. There is a general level that presents rules, commitments, "without which no man is a scientist." The draft reads:

> The scientist must, for example, be concerned to understand the world and to demonstrate previously unnoticed aspects of its order. That commitment must, in turn, lead him to scrutinize some aspect of nature in empirical detail. And, if that scrutiny displays pockets of apparent disorder, then these must challenge him to a new refinement of his observational techniques or to further articulations of his theories.[153]

The one mentioned here by Kuhn is an unwritten rule but basic enough so that it is not subject to a revolution. There are other rules that do not have that luck. In them we discover the way in which we must study some aspect of nature. On certain occasions we find them, for example, in laws. He talks about Newton's laws as a good example of this series of rules.[154] Newton's celestial mechanics was developed in the eighteenth and nineteenth centuries by mathematicians such as Leonhard Euler, Carl Friedrich Gauss, and many others, all of whom dedicated their efforts and ingenuity to the development of mathematical techniques that would better adjust the differences between observations and theoretical predictions, in addition to solving theoretical puzzles like the stability of the Solar System.[155]

In the draft of *Structure*, Kuhn here ends his description of the new understanding of the rules. He has followed the order we just looked at and does not continue beyond

[153] SSRD, 40–41. See also *Structure*, 42. The variations between the two versions of that passage are minimal.

[154] SSRD, 41.

[155] See SSRD, 33, 41.

that point.[156] In the final, published version, meanwhile, he devotes a good part of the new Section IV, "Normal Science as Puzzle-solving," to showing to what extent this new version of the rules is present in scientific practice. To begin with, he shows there a clear parallelism with the idea of "established viewpoint" or "preconception" that constitutes a bridge with his notions of orientations and of styles of thought that he adopts as central in his perspectives from 1951 to 1959.[157] After showing that scientific research needs that kind of firm but flexible tie, he makes a more exhaustive review of the new idea of rule. We find rules, in that new sense, in the central statements of a theory that serve as a guide in reconstructing its ontological, conceptual, and normative implications.[158] These rules have little relationship to the typical ones sought by scientific methodology, since they go beyond the strict methodological field. These rules help to reconstruct the community's cognitive and linguistic scenario; pure methodological norms are a subgroup of it. He adds these latter at the end of his description, not at the beginning as he did initially. His perspective in the published version is more exhaustive, better ordered, and places his idea of rules in the place that corresponds to his broader theory—that is, the one we have been exploring in our previous chapters.

Kuhn does not wish to belittle the idea of rules. That idea, he indicates, in fact marked his first conceptions about the link that brings together the members of a scientific community under a tradition of normal science; paradigms, as we know, appear quite late as a central concept in his theory. In the draft, he says that in the first section of *Structure*, one can see how he still takes advantage of an approach based on rules rather than on paradigms.[159] "That is the position," he says in the old Section IV, "implied, for the sake of preliminary simplification, in the introduction to this monograph, and I still find it extraordinarily tempting."[160] So, he says in the introduction, for example, that "the new theory implies a change in the rules governing the prior practice of normal science."[161] He has, however, already discarded the idea, and only uses it for a better understanding of his new viewpoint by virtue of a comparison with the older one. His paradigms explain this path of normal scientific development in a natural way. Normal science is puzzle-solving, as it explores the world implied by a paradigm, where the latter is also the vehicle for education, that is, for the introduction of the aspiring scientist to a theory. Therefore, for him there is no insurmountable barrier between education and research. As he says in the old Section IV of the draft of *Structure* ("Normal Science as Rule-Determined"):

> But if sample problems and applications are part of what makes possible a tradition of normal research, then that tradition need not be entirely determined by discoverable rules.

---

[156] See SSRD, 41.

[157] *Structure*, 39.

[158] *Structure*, 39–42.

[159] SSRD, 39. See Isaac (2012a, 224; 2012b, 105–106), where the author comments on this same passage from the draft of *Structure*.

[160] SSRD, 39.

[161] SSRD, 8; identically in *Structure*, 7. See the following paragraph for a continuation of the same idea.

> Applications and problems need not imply rules in order to determine normal science. Rather than learn rules the scientist can, and in some part clearly does, learn by practicing on paradigm problems. In contemporary scientific education this process of education through problem-solving begins in high school or in the first year of college and continues steadily to, or more often through, the doctoral dissertation. Subsequent professional research is no sharp and sudden departure. Its problems are more elaborate, and they are ones that have not quite been done before. Usually, unlike his dissertation subject, they are invented by the scientist for himself. But they continue to be closely modeled on the existing applications of the paradigm, a paradigm which is thus both more and less than a set of rules for the conduct of the scientific life. It is because they learn in this way that scientists can so regularly agree in their evaluations of particular problems and particular solutions without manifesting any similar agreement about the full set of rules that appear to underlie their judgments. One can model work upon a paradigm or recognize work modeled on one without being entirely able to say what it is that gives the model its status.[162]

The paradigm, in short, serves as a link with a theory in development. Learning and research must show coherence with such a theory by means of such a link. At the same time, that link is loose enough to allow the pursuit of research with one's own resources, with the behavioral world, and with the theoretical language thus created through education and subsequent research. This kind of link, which helps us to see that science does not need to explicitly commit itself to any decalogue of rules, is briefly explored in the old Section IV of the first full draft of *Structure*, but is developed in greater depth in the published version through Section V. There, Wittgenstein's work supports Kuhn's views. It is present there as a source of inspiration that lends philosophical soundness to a perspective on learning without rules that is already, as we know, well established in Kuhn's intellectual development.[163] It would not be the last time in which Wittgenstein was a central influence in Kuhn's views, and in which the former lends support to the latter's views. In our subsequent chapters, we shall see how there would be other occasions on which the Viennese thinker and his newly published texts would be of great help to Kuhn.[164]

A second source of support often goes unnoticed: Kuhn cites his conversation with, and a paper of, James K. (Kuhn) Senior at the end of Section V of the published version of *Structure*.[165] Senior speaks mainly from his own experience, as he was a chemist and worked for the Kent Chemical Laboratory at the University of Chicago. Senior's text is an exposition and defense of the internal knowledge of the kind of language that is specific to a highly specialized scientific environment such as that of a chemistry laboratory, which is marked by semantic vagueness and, in his words, by a frequent "change in meaning" that occurs in the experts' exchange of scientific knowledge. None of this affects those who share such specialized languages, whose exchanges are based on a common, though rarely explicit, theoretical knowledge,

---

162 SSRD, 42–43. For a similar perspective, although shorter, see *Structure*, 47.

163 See Isaac (2012a, 225, 229–230; 2012b, 95–96, 103–107), who discusses this connection with Wittgenstein on the basis of the new Section V of *Structure*, and tries to place it in its correct context.

164 See, in particular, our Chap. 8.

165 *Structure*, 50–51.

which becomes broader as the distance between the speakers' sub-specialties gets smaller. Kuhn mainly refers to this at the end of Section V.[166]

There are several points of commonality between Senior's stance and Kuhn's. To begin with, they have in common the rejection of the rational reconstruction of scientific language without the appropriate cautions, and without some skepticism as to the final results and their usefulness. In addition, Senior suggests that this reconstruction is typical of certain initial sections of textbooks, where the appearance of formal order shows a language that has little relation to the true vehicle of specialized communication.[167] Senior's vision is so similar to that defended by Kuhn since 1951 that it is not strange that Kuhn appreciated his comments. But there is more in Senior's view of interest for us. Senior talks about how a "native talent for bilingualism" would be useful if one wished to take advantage of rational reconstruction.[168] I have doubts that Kuhn would have expressed it in a very different way, particularly in terms of his more mature theory, where the concept of *bilingualism* is the resource he offers to overcome the barrier between incommensurable languages after considering them partially untranslatable (a barrier that is very similar to the one Senior establishes between specialized scientific languages and their formalist reconstructions).[169]

5.9 Unexpected Reactions

With respect to the *Encyclopedia* project, Kuhn soon realized that the old logical positivist project was dying. He had no qualms, moreover, in making this clear to Morris. Despite the latter's infinite patience, waiting for Kuhn to clarify his thoughts and allowing a notable extension of the text, Kuhn told Morris that, for some time now, the *Encyclopedia* had not been reaching everyone and, especially, the philosophical profession—Kuhn's main target.[170] A few months later, Kuhn said essentially the same to Carroll Bowen, from The University of Chicago Press, in an attempt to separate the publication of the book from the positivist project.[171] In his defense, he offered the opinion of those to whom he had sent the manuscript.[172] Through Bowen, Kuhn achieved the agreement of his publisher. *Structure* would be published in hardcover separately from the *Encyclopedia*, although it would still appear as part of it, and in paperback with the format of this collection, regardless of its publication

[166] See Senior (1958, 166) for the expression "drift in the meaning".

[167] See Senior (1958, 163, 167–168) for these two mentioned points.

[168] Senior (1958, 167).

[169] See *The Road*, 238.

[170] Kuhn to Morris, 13 October 1960, TSKP 25.53. Galison (1995, 31) also mentions this letter and Kuhn's argument. See also Reisch (2019, 248).

[171] Kuhn to Bowen, 18 June 1961, TSKP 25.53.

[172] See our Sect. 5.7 in this chapter.

in its second volume.[173] This second version, however, followed Morris's insistence, who from the beginning had defended the validity of the *Encyclopedia* project and who subsequently tried to ensure that *Structure* did not break its link with it.[174] From then on, understandably, Morris was much less generous with Kuhn in his correspondence. However, both Morris and Carnap offered their support to Kuhn's text between March and April 1962.[175]

In practical, not only theoretical terms, therefore, Kuhn voluntarily separated himself from the logico-positivist project, with which he shared (at least in his version of it) very few points and in whose possibilities as an accomplished tool for interpreting science he no longer believed at all. In his opinion, circumstances had changed a lot since the invitation to write *Structure* back in 1952. It is worth us looking, now, at the last steps of *Structure* on its already long journey to the University of Chicago Press.

On July 24, 1961, Kuhn told Bowen that it would take him about two months to finish his final version. He was in a bit of a hurry to do so, as in 1962 he would start chairing a group that was going to search for (and would try to preserve) sources of the recent history of quantum physics, and he anticipated that it would monopolize all his attention for a few years.[176] But habits do not change easily, and his constantly overly optimistic expectations are a good example of this. In February 1962, the book had not yet been sent to the University of Chicago Press.[177] Finally, on February 25, 1962, Kuhn told Roger W. Shugg, also from the press, that the book would be in their hands by the first day of the following month, which must have happened, because, as we have already said, Morris had already read the text by the end of March. Kuhn told Shugg in that letter that the pressure of work meant that the latest version needed a good editor. *The Structure of Scientific Revolutions* appeared in the fall of 1962.[178] By the end of September, he had seen the book, though he had not received a copy as yet (he was in Europe at that time).[179]

Kuhn strived to get the recognition for *Structure* that he believed it deserved. He complained to the publisher about insufficient promotion of the book.[180] He even convinced A. C. Crombie to change the title for the proceedings of the symposium that this latter had organized at the University of Oxford in July 1961, in which Kuhn had participated with a text later little appreciated by him, "The Function of Dogma in

---

[173] See Bowen to Kuhn, 26 June 1961, Morris to Bowen (cc. Kuhn), 29 June 1961, Bowen to Kuhn, 5 July 1961, Bowen to Morris, 5 July 1961, Morris to Bowen, 19 July 1961, Bowen to Morris, 24 July 1961, all of them in TSKP 25.53.

[174] See Morris to Kuhn, 23 October 1960, TSKP 25.53.

[175] See Morris to Kuhn, 26 March 1962, and Carnap to Kuhn, 28 April 1962, both in TSKP 25.53. On Carnap's letter, see also Reisch (1991, 266–267), and Galison (1995, 32).

[176] Kuhn to Bowen, 24 July 1961, TSKP 25.53. See our Chaps. 6 and 7 for more details about this project.

[177] See Kuhn to Bowen, 31 August 1961, and Kuhn to Shugg, 19 January and 25 February 1962, all in TSKP 25.53.

[178] See Kuhn to Shugg, 25 February 1962, TSKP 25.53.

[179] See Kuhn to Nancy S. Romoser, 25 September 1962, TSKP 25.55.

[180] See Kuhn to Bowen, 17 December 1962, and Kuhn to J. H. Storm, 25 March 1963, TSKP 25.55.

Scientific Research."[181] Crombie had thought that *The Structure of Scientific Change* would be a good title for his collective book, but it was also a title that Kuhn thought he had obtained from him. Kuhn managed to change Crombie's mind in any case, and in the end the book was titled, as is well known, *Scientific Change*.[182] Kuhn's high opinion of his own book was beyond doubt, and he also considered it an important milestone in his career.[183] He hoped that others would see it that way, especially his colleagues and closest friends. In particular, he expected a warm reception on the part of his mentor, Conant. However, that was not exactly what he got from the former Harvard president.

To see the impact of Conant's letter, dated June 5, 1961, on Kuhn, we can contrast it with the reaction to the letter Nagel sent him just one day earlier, on June 4.[184] Nagel was critical of four points, mainly: first, of Kuhn's lack of precision about the object of his criticism, both authors and currents; second, of the vagueness of the notion of paradigm, a recurrent criticism in the reception of *Structure*; third, of the current Section X on "Revolutions as Changes of World View" and its psychological interpretation, which, according to Nagel, did not represent the implications of a scientific revolution well enough; and, finally, of the uncertainty about the cognitive status of the Kuhnian idea of scientific theory. Nagel also defends the reduction of a theory like Newton's mechanics to a more general one like Einstein's relativity, which, as we know, Kuhn attacks in Section IX of the book.[185] Kuhn nevertheless agreed with these criticisms, especially with the first two, while the last two, along with the question about reduction, simply reveal, for him, opposing positions between them. He had no objection to Nagel's position, therefore. Kuhn seems to agree with Nagel because their disagreements represent the kind of division between two ways of practicing the philosophy of science that he thinks are difficult to reconcile.[186] They are natural differences, in short, and this did not seem to hinder their relationship at all.

---

[181] Kuhn crossed paths with Polanyi once again at this point, as the proceedings show; see Crombie (ed.) (1963, Chapter 11 [Kuhn 1963a] and subsequent commentaries). Kuhn had dinner with him and his wife at their Oxford home. At that point, Kuhn recalled, they did not have the amount of "time for talk about our mutual problems" that he had wished. Kuhn sent him a copy of his replies after the conference, as he had not done with other commentators, because he wished, in his case, "to be sure that I am not misinterpreting your intent or your position any more than the inevitable condensation makes necessary." His exchange with Polanyi had made him re-examine some of his points more deeply. Kuhn to Polanyi, 1 September 1961, TSKP 25.13. By September 21, 1961, Polanyi had replied twice to Kuhn, very kindly and, in the second letter (the first one, handwritten, is undated), with many commentaries on Kuhn's reply for the volume; see Polanyi to Kuhn, undated and 21 September 1961, both in TSKP 25.13.

[182] See Kuhn to Crombie, 14 May 1962, and Crombie to Kuhn, 24 May 1962, both in TSKP 25.53. See Crombie (ed.) (1963) and Kuhn (1963a; 1963b).

[183] See Kuhn to Bowen, 18 June 1961, TSKP 25.53.

[184] On the Kuhn–Conant letters on the draft of *Structure*, see Hershberg (1993, 410, 860, n. 84); Driver-Linn (2003, 273–274); Reisch (2012, 338–340; 2019, Chap. 14); Hamlin (2016, 303); Wray (2016, 13, 17–18; 2021, Chap. 2); Mayoral (2005, 388, 405–406; 2008, 54).

[185] See Nagel to Kuhn, 4 June 1961, TSKP 25.53.

[186] See Kuhn to Nagel, 22 June 1961, TSKP 25.53.

Kuhn took a little more time to respond to Conant—about a week more, specifically. And his response reflected some surprise at Conant's reaction and also some sadness.[187] Kuhn thanked his former mentor for the criticisms, although in this case it seems to have been pure politeness.[188] In reality, Kuhn rejected Conant's criticisms—in contrast to his response to Nagel's. And Kuhn was probably right. In the remainder of this section, I will examine in depth a key point of disagreement in which Conant and Kuhn show incompatibility and irreconcilability between their respective approaches. In fact, it seems to be that a difference that was already present, but that Kuhn's high respect for Conant helped to remain hidden, came to light, and not in a completely affable manner.

The object of the controversy were paradigms, of course: Kuhn's seemingly arbitrary use of the term in the book. This main criticism is very similar to Nagel's on the same point, albeit with some difference in presentation. Nagel tells him that on p. 23 of the draft of *Structure*, whose contents no longer appear in the published version, there is an identification between the term in question and the idea of a conceptual scheme, which is true, while on p. 39, in a passage that we no longer have in the book either, although we discussed it a few pages back, Kuhn assigns it a less common additional role. In that passage, let us remember, Kuhn argues about the normative importance of the paradigm. This point is not only present in *Structure* currently, but better developed, as he expanded the old Section IV into two sections with more content: the current Sections IV and V. The paradigm, he said on page 39, does not represent, let us say, a mere possible *explanation* of the theory—which would be governed in that case by its own normative structure (a permanent, or constant, vision of scientific method, for example). On that page, the theory establishes indissoluble links with whatever the paradigm is. Kuhn simply tells Nagel that he has not yet fully explored paradigms, though he will. And the expansion of the old Section IV along with the changes on p. 23 could well have been due, at least in part, to Nagel's stimulus on precisely this point.[189]

Conant's tone is much less analytical than Nagel's. To begin with, he treats the idea of a paradigm as a convenient resource and clearly shows that he does not believe that Kuhn knows exactly what he is using. He also believes that the term is used excessively in the social sciences and in pedagogy (and mentions a colleague from Columbia University, who must have been Merton, to support this opinion).[190] Conant also compares the aforementioned p. 23 of the old draft with p. 138, to show that, while in the first place there is no strong clarification of the term (unbeknownst to him, he agrees with Nagel on this point), in the second place there is a sample of what Kuhn tries to communicate in the book, but without the term in question. On that page 138, there was previously a long paragraph that introduced Section X (currently Section XI), in which Kuhn displayed the break that occurs between two scientific communities separated by a revolution, and how the old resources of the empiricist

[187] See Reisch (2019, Chapter 14) for an interesting reconstruction of that reaction.

[188] Kuhn to Conant, 29 June 1961, TSKP 25.53, 1.

[189] See Nagel to Kuhn, 4 June 1962, and Kuhn to Nagel, 22 June 1961, both in TSKP 25.53.

[190] On Merton's use of the term in question in previous years, see Wray (2011, 52–55).

method—like the different confirmation of two theories, which allows for rational argumentation in favor of only one of them—no longer work in a Kuhnian scenario.[191] For Conant, there is no trace of paradigms in that paragraph, which is, for him, *a good thing*. In sum, for Conant, the idea of a paradigm could be used simultaneously with other more precise concepts, specifically with the classic reference to scientific theories, but not the way Kuhn used it, as a standalone resource.[192]

In his response to Conant, Kuhn presents several lines of defense against the criticism. The first has to do with the *unity* of reference through a concept that serves as a guide. Kuhn does not wish to diversify in this respect, because if he used the old idea of theory, he would still have to preserve the new one of paradigm. This unity, on the other hand, is due to the fact that the term in question is used in two senses that do not seem to be very clear to Conant. Perhaps this was what Kuhn deduced from his exchange with him, because the changes later introduced in the aforesaid pages had less to do with clarifying a definition of paradigm than with highlighting these two senses, which are the *historical* and the *sociological* aspects of the paradigm.[193] Kuhn observes, correctly, that for Conant the paradigm is a timeless scientific achievement. For Kuhn, however, it has a more defined historical validity. Kuhn's criticism makes sense, because Conant's pedagogical program assumes that the examples from the past serve as models for the present. Kuhn adds to this that, for Conant, these examples still retain their relevance for the scientist of today. The historical barrier that Kuhn raises is not so present in Conant, who does not fully take into account long-past periods of history. We can recall the historical periodization of science characteristic of the operational vision of Bridgman, to which Conant subscribed, and compare it with the historical relativity characteristic of Kuhn.[194]

That would be the *historical* aspect of Kuhn's use of the word "paradigm," which he does not want to lose and which he does not see so clearly in Conant. Additionally, for Kuhn, the paradigm allows us to unify several aspects of science that in practice are not as differentiated as the defenders of the traditional method try to show. Conant avoided talking about theories, a very hackneyed use, but he replaced that term with his conceptual schemes, a term that was not very far from that, after all.[195] For Conant, the difference between theory and observation was reduced to a distinction between experimentation and conceptual schemes, which was, again, not so very different in the first place. Kuhn's paradigm, despite its identification (on the previously mentioned page 23) with conceptual schemes, is more than that due to its *functions* of a *social* nature, which mainly include, as he emphasizes in

[191] See SSRD, 138.

[192] See Conant to Kuhn, 5 June 1961, TSKP 25.53.

[193] As we shall see in Chap. 6, Masterman also helped Kuhn with new dimensions for the concept. I shall return to these two senses in Sect. 6.3, though I deal with Masterman's points later in that chapter; see Sect. 6.8. One of them has to do with the sociological nature of a paradigm as well. I shall return to that at that point. See also Postscript—1969, 174 ff., on the association of "sociological" to "paradigm" from his point of view. There are differences between Kuhn's association in that text and the one I suggest in this section.

[194] See the letter from Kuhn to Conant, 29 June 1961, TSKP 25.53.

[195] See Conant (1947a, 47).

his letter to Conant, the education of new scientists and the direction of research (the reasons why the term was finally adopted by him, as we have seen). To talk about this, Kuhn does not need, nor does he want, the strange and extemporaneous nuances of philosophers; he does want, however, that resource to fulfil a function in real science—a function that is not only theoretical, but also social. For Kuhn, therefore, the paradigm had to acquire the specific central role that it had acquired for him.[196]

The modifications introduced by Kuhn in the draft, if we pay attention to the pages cited by Conant and Nagel, are consistent with these responses. Kuhn had partly defined the paradigm on p. 23; admittedly insufficiently, but at least endowing it with certain characteristics. For Kuhn, the paradigm was a solution to a specific problem, which included a particular representation of the natural aspects being dealt with and the basic beliefs that govern it, even if these did not have to be restated in practice explicitly. The representation also provided uniformity to the field of application of the theory. Kuhn emphasizes again the significance of the paradigm providing the meaning of terms, but not through specific and explicit contents for them. Scientific concepts are learned in practice. He removed this part and transferred all this second aspect (paradigm-based learning of meaning) to the new Section V of the published version. In fact, the following passage from the published version of *Structure*, Section V, had previously appeared—more briefly and with variations in expression, none of them substantial—at the end of the (eliminated) p. 23 of the previous full draft:

> If, for example the student of Newtonian dynamics ever discovers the meaning of terms like 'force,' 'mass,' 'space,' and 'time,' he does so less from the incomplete though sometimes helpful definitions in his text than by observing and participating in the application of these concepts to problem-solution.[197]

Kuhn finally suppressed the previous definition of paradigms and offered a framework in the new Section V to explain a paragraph like the one I just quoted. Instead, on the current p. 23, we find the kind of profile for the paradigm that Kuhn specified for Conant, which shows not so much what it is, or what it contains, as *what it is for*. It is not difficult to notice that he mentions there that his use of the term is, in point of fact, an "appropriation" coming from another field, such as linguistics, only he adds new meanings; specifically, not because it is "an object for replication," but because "it is an object for further articulation and specification under new or more stringent conditions."[198] We see there an implicit response to the presence of the term in other scientific fields. Finally, regarding p. 138, which was praised by Conant, the text of the primitive version is no longer in that place; the current paragraph on p. 136 is the second in the 1960–1961 version, with slight initial modifications.

---

[196] Kuhn to Conant, June 29, 1961, TSKP 25.53.

[197] *Structure*, 47; see SSRD, 23, where he writes: "Students of Newton's *Principia*, like those of any other scientific classic or text, learned the meanings of concepts like force, mass, space, and time less from the paradigm's explicit definitions than from the way in which these concepts were applied in problem–solution.".

[198] *Structure*, 23.

Alongside his critique of the idea of paradigm, Conant presented others. Almost all of them had to do with the, for Conant, excessive simplification of some of the ideas that shaped *Structure*. Conant told Kuhn, for example, that scientific revolutions do not all have the same magnitude and that not all create completely new cognitive environments. This seems like a critique of the famous idea, in Section X of the published version of *Structure*, that after a scientific revolution scientists seem to be inserted into a completely altered world, not only conceptually, but also perceptively and practically. Similarly, Conant warns that the community is a more complex group than it appears in this draft. In addition, for Conant, *Structure* is a vision of the theoretical aspects of science, but its relationship with techniques, which particularly worried him and that he explored in his writings on pedagogy, such as in *On Understanding Science*, was barely dealt with.[199] Finally, in addition to these and some other criticisms, the former president of Harvard showed Kuhn the pragmatic aspects of observation, emphasizing its relationship with action, thus pointing out something that, for him, was missing in Kuhn's book.[200]

In answering these criticisms about simplification, Kuhn tried to avoid them by showing the limits of the text's extension, or by suggesting that a reading with the help of pure common sense would help to see the compatibility between what was said and the variations themselves. For example, he accepts the variety of human groups, orientations, and modes of revolutionary change. Yet, he also tries to indicate to Conant through his individualized responses something that we can summarize by saying that the simplified approach does not prevent the reader from understanding that *Structure* is not an obstacle to observing the relative dimensions of each of these aspects of the scientific community. And Kuhn is right on this point, because, as he himself emphasizes in his response letter, the use of paradigms and the exclusion of rules is an invitation to leave a door open to the variety of viewpoints within groups, even though practice contributes to the stabilization of a certain theoretical and practical resource. This same variety is also not excluded in terms of the cognitive impact of scientific change. In the next chapter, we shall see, however, that oversimplification would become a major area of concern for Kuhn after the publication of the book. So, to some extent, Conant's criticism in this sense would, in the end, be vindicated. Finally, Kuhn does not quite understand the last criticism mentioned, insofar as his Section IX (currently Section X) *defends precisely* what Conant accused him of failing to notice. Maybe the difference between them is that Conant does not consider perception significant by itself (the practical context is more important for him), while it remains central for Kuhn, although, in his opinion, it should be understood in a less traditional manner. Kuhn takes the opportunity to show himself reluctant to adopt, as Conant suggests, a pragmatist approach. He rejects this last suggestion and uses Dewey as an example, who is not, for him, reliable in these matters, because, he adds, he still depends a lot on a too traditional perspective on perception.[201]

---

[199] See Conant (1947a, 22–23, 73–74, 103).

[200] Conant to Kuhn, 5 June 1961, TSKP 25.53, 2–3.

[201] See Kuhn to Conant, 29 June 1961, TSKP 25.53, 3–6; Conant to Kuhn, 5 June 1961, 3. In Kuhn's letter to Conant (29 June 1961, TSKP 25.53, 6), Kuhn says that, although the pragmatists'

In summary, this exchange marks a deep disagreement and is a signal of a certain independence of Kuhn's views with respect to Conant's. In almost all his criticisms, Conant clings to the constancy of scientific thought from its origins, which are much closer than Kuhn would accept. He also clings to a traditional distribution of the roles of theory and experiment and sees accumulation as the form of progress.[202] Kuhn, on the other hand, has already broken with all of this. We already noted this in our Chap. 2, where we advanced a paragraph from the letter to Conant that we have explored here (from June 29, 1961).[203] In it, he said that after examining the history of science he had found discontinuity where Conant only saw a less developed and simpler monolithic enterprise in its own past. Conant's pedagogical plan is built on this continuist premise, while Kuhn's is based on a vision of the periodicity and discontinuity of the scientific enterprise that presents a clear analogy with Koyré's vision of scientific change, with the stages of individual formation in Piaget, and with his own views, cultivated during his education back in the 1930s and 1940s, on the autonomy of the stages of personal development that every individual experiences.[204] For Kuhn, as Piaget would have accepted, each stage in the development of individuals, or of science, holds an autonomy and independence that should not be erased by an interpretation based on their later manifestations. This idea marks a huge difference with Conant and with most of what his work represents, which is, in some key aspects, a view of science that is not completely detached from the empiricists' perspective. Thus, even though *Structure* exhibits a Conantian vocabulary and phase scheme, and although some of the focuses of interest had already been evident in *On Understanding Science*, the way Kuhn exposes all this marks a frontier with Conant's thought. In 1961, with *Structure* essentially constructed, Kuhn's emancipation from Conant took place. There are, therefore, no traces of empiricism or operationalism in Kuhn's essential, professional, or personal commitments. From this point on, the revolutionary we all know is born. Despite that, this would not be the end of his attention to the themes in *Structure*, and of his attempt to build a solid explanatory structure for scientific development. At that point, Kuhn would start to fight a different battle: one in defense of *The Structure of Scientific Revolutions*. We will see how he did this in the next chapter.

---

ideas "have been immensely fruitful in a number of applications, their failure to make a perceptible dent in traditional epistemology seems to me symptomatic of their failure [to] carry the attack far enough." For this reason, he also says, "I always find Dewey's treatment of science so very very far from the mark, though in talking about mores and morals and social theory he says many things that he ought to continue to say when he talks of science" (6).

[202] However, Conant agreed with Kuhn (by letter, in December 1962, i.e. after reading the published version of *Structure*) that scientific progress does not need to be connected to the belief on the successful scientific theories' gradual approximation to truth—a critical position that is, of course, also visible in *Structure*, Section XIII. See Conant to Kuhn, 19 December 1962, TSKP 4.8, 2.

[203] See our Chap. 2, Sect. 2.7.

[204] Recall his essay "The War and My Crisis," as examined in our Chap. 1.

Chapter 6
In Defense of *Structure*

6.1 Kuhn and Berkeley's Senior Philosophers

By the time Kuhn saw *The Structure of Scientific Revolutions* published, his status at UC Berkeley had started to change significantly. In July 1961, he had been promoted to full Professor of History and he was in charge of the Program in History of Science, originally ascribed—just like his own appointment—to the Departments of History and of Philosophy on the basis of a part-time status. Yet, in the process, he had been transferred *completely*, together with the program itself, to the Department of History, leaving behind his part-time appointment in the Department of Philosophy.[1]

The Department of Philosophy had supported Kuhn's promotion at all times, but they preferred to have him promoted in History, only. The philosopher Karl Aschenbrenner, the chairman of that department, had written a long letter in November, and then another shorter one in December, 1960, to Lincoln Constance, Dean of the College of Letters and Science, advocating for a full transfer of Kuhn's duties and funds to the Department of History.[2] According to Aschenbrenner—who wrote a report in his defense after this affair was all over—it was Kuhn himself who had started that debate. Kuhn had been offered a job at Johns Hopkins University and his promotion to full Professor at Berkeley was on the table. The Department of Philosophy supported such a promotion. At the same time, they had to debate a parallel issue. According to Aschenbrenner's memorandum, Kuhn wanted to know about the department's plans for the Program in History of Science that he chaired. Aschenbrenner told him that, whereas his promotion was not in question, the affiliation of that program to their department was quite a different question, and that this one *had*

[1] See Clark Kerr's letter to Kuhn, 27 June 1961, UCBTK. Concerning this episode in Kuhn's relationship with the Departments of History and of Philosophy at Berkeley, see Hollinger (1998, 48–49), and Andresen (1999, S64). On Kuhn's period at Berkeley, see Andresen (1999, S63 ff.); Hufbauer (2012, 439 ff.); Collodel (2014); Collodel and Oberheim (2024, Sect. 1.3).

[2] See Aschenbrenner's letters to Constance, 15 November 1960 and 1 December 1960. Both in UCBTK.

J. V. Mayoral, *Thomas Kuhn*, Springer Biographies,
https://doi.org/10.1007/978-3-031-86512-1_6

to be discussed. The Department of Philosophy (or rather, a part of it—the senior members alone, as we shall see below) agreed that the program *should* be transferred to the Department of History in full. Aschenbrenner presented Constance with both plans—support of promotion and transfer of the program—in the above-mentioned letters.[3]

In the former letter (November 15), Aschenbrenner did not spare any effort to make it clear that Kuhn was not a philosopher, and that he did not seem to have the intention of becoming one.[4] He also affirmed that, at the same time, Kuhn's plans for combining the history of science with philosophy in their program were also ill received by the senior philosophers (i.e. the full professors).[5] In his opinion, his departmental colleagues had not shown any interest in Kuhn's major area of research. In fact, Aschenbrenner himself had frowned at the possibility of a Program in History and Philosophy of Science, which Kuhn had suggested—with the programs at Indiana and Princeton Universities as his model.[6] Additionally, as a historian, Kuhn would have few acquaintances in the American academic world from the field of philosophy, so he was not ready to take on an important role, for instance, in selecting new talents for the department, a role that the other members did play.[7] Aschenbrenner combined this argument with his impression that a double appointment involved doubling Kuhn's work load, which was a particularly heavy (and unnecessary) load for Kuhn.[8] However, it is difficult to avoid the impression that this latter point was rather a pretext.

Kuhn later said that all this was the attitude and decision of senior members of the department, and that they did not ask the opinion of those colleagues who were not full professors as yet—at that time, that included John Searle, Stanley Cavell, and Paul Feyerabend—who, of course, did not share that view or support that decision.[9] Aschenbrenner confirmed this point, but defended his position as the only admissible one, because full professors were the only authoritative group to make such decisions. Less convincingly, he added that he would not have been able to guide the debate over the program transfer alone with a group of fifteen colleagues, perhaps more.[10]

According to David Hollinger, Constance did not inform Kuhn about this position when he had the opportunity (in particular, in a conversation on 5 December, 1960).[11] Six days later, Delmer Brown, the Chair of the Department of History,

---

[3] See Aschenbrenner, "Memorandum on the advancement of Professor Thomas S. Kuhn to the professorship and his transfer to full status in the Department of History," 8 June 1961, TSKP 14.11, 1–2.

[4] As Hollinger rightly says, this "assertion [...] would shock anyone familiar with Kuhn's career" (Hollinger 1998, 48).

[5] Aschenbrenner to Constance, 15 November 1960, UCBTK.

[6] See Aschenbrenner to Constance, 15 November 1960, UCBTK, 2. See also Hollinger (1998, 48).

[7] Aschenbrenner to Constance, 15 November 1960, UCBTK, 3.

[8] Aschenbrenner to Constance, 15 November 1960, UCBTK.

[9] Hollinger (1998, 49); Kuhn's letter to Strong, 5 May 1961, TSKP 14.11, 2.

[10] See Aschenbrenner, "Memorandum...," 8 June 1961, TSKP 14.11, 1–3.

[11] Hollinger (1998, 48).

expressed to Constance his concern about Kuhn's reaction upon noticing the delay. The Department of History's attitude was radically different from that of the senior philosophy professors. The historians hoped that Kuhn would remain not only at Berkeley but also in their department. Indeed, Kuhn was on very good terms with many members of that department, and—as Aschenbrenner said—most of his courses were offered in History, not in Philosophy.[12] For Brown, despite his willingness to remain in the Department of Philosophy as a part-time member, Kuhn might accept to remain full-time in the History Department if the transfer were presented *soon* as a good opportunity for him (and separately from his promotion). Brown was afraid of the contrary effect if it were not: Kuhn might react by accepting the job at Johns Hopkins.[13] Despite Brown's suggestion, Constance did not follow his advice, and the result was not good. A letter from the Department of Physics on 18 April, 1961, to Vice-Chancellor E. W. Strong shows that, at that time, nobody had yet informed Kuhn about the Department of Philosophy's plans.[14] Finally, Strong informed Kuhn about the proposed plans—both the support of the promotion *and* the petition of transfer—on May 3.[15]

On May 6, three days later, Kuhn wrote an interesting report titled "Reconstruction of Events," which was based on his own research about this issue. The mere existence of this report attests to the importance that Kuhn attached to this chain of events. According to this report, upon finding out about the Department of Philosophy's plans, after hearing these from Strong, he started to gather information later in the day—in conversation with Aschenbrenner, and the philosopher Benson Mates, and with Brown, Constance, and the historian Kenneth M. Stampp the following day (May 4). Kuhn's description, he annotated, was based on his interaction with Brown.[16] This report shows that, in his view, Constance was wrong to withhold the information for that long, though there was no sign of bad faith. Constance had concluded that the transfer was a "separate matter" altogether, and that it would be discussed later on.[17] As Kuhn wrote:

> This is the unfortunate beginning of the apparent "conspiracy of silence". Obviously it was a mistake, but equally obviously it was a considered decision made in good faith. [...] In any case, they proved to be quite mistaken. It was not, however, until late in January that any evidence that they were mistaken began to accumulate in the Dean's office. By that time I had long since turned down the Hopkins offer.[18]

In a letter to Strong on May 5, Kuhn rhetorically asked, "was it either wise or fitting that I be kept in ignorance of the full professors' recommendation for over five months after it was made?"; and he also recalled something still worse: "how was I

---

[12] Aschenbrenner to Constance, 15 November 1960, UCBTK, 3; Aschenbrenner, "Memorandum...," 8 June 1961, TSKP 14.11, 1.

[13] Letter from Delmer Brown to Constance, 11 January 1961.

[14] See the letter from the Department of Physics to E. W. Strong, 18 April, 1961, UCBTK.

[15] See the letter from Kuhn to Strong, May 5th, 1961, TSKP 14.11, 1.

[16] See Kuhn, "Reconstruction of Events," 6 May 1961, TSKP 14.11, 1.

[17] Kuhn, "Reconstruction of Events," 6 May 1961, TSKP 14.11, 2.

[18] See Footnote 17.

expected to feel when asked for a response to the unexpected news within forty-eight hours so that a recommendation for my promotion could go forward?".[19]

According to Kuhn's report, it seems that, in his opinion, the Department of Philosophy's full professors had acted unilaterally. "Paul [Feyerabend] did not know of it until I told him last night [May 5th, 1961]. "Yet," Kuhn wrote, "he is the person in the Department [of Philosophy] (Associate Professor) most involved professionally with history of science; he and I have repeatedly discussed joint programs."[20] In his letter to Strong, Kuhn said: "The impending transfer was rendered a *fait accompli* by a group which neither included nor, so far as I can now determine, consulted those members of the faculty who know most about the present program on this campus or about relevant precedents elsewhere."[21] Kuhn again mentioned Feyerabend as well as A. Hunter Dupree, and, of course, himself, in that regard. He also cited the "longstanding tradition [... that] all policy decisions in the Department of Philosophy are the business of the entire department"—thus contradicting Aschenbrenner on that point; despite that, he went on, "neither the Associate Professors nor the Assistant Professors were consulted as they have always been in the past and as they had recently been assured they would be in the future."[22] Additionally, the Department of History made the decision to promote Kuhn, although some of its members remained unaware of the Department of Philosophy's plans.[23] In short, a group of philosophers decided, one-sidedly, that Kuhn and the program in the history of science would not be ascribed to the Department of Philosophy in a near future, and that Kuhn's promotion was a good opportunity to get rid of them both.

Kuhn doubted that Aschenbrenner and his colleagues had acted according to the university rules, and even suggested that their behavior might be subject to a claim.[24] In the end he did nothing, though. "After considering all of the above as best I could [...]," he wrote, "I decided that it would only make sense to insist on staying in the Philosophy Department if I meant to fight. But that would hurt the program [in History of Science ...]." Kuhn wished to preserve his good relationship with students from the Department of Philosophy, so "the good of the program seemed to dictate a peaceful withdrawal."[25] As he said to Strong on May 5, "since that program [in History of Science], which is my primary University responsibility, would clearly suffer if I did not accept their recommendation, I reluctantly concur in the program's transfer, together with that of my entire appointment."[26] As Hollinger says, however, Kuhn was deeply distressed by the Department of Philosophy's decision.[27] As Kuhn wrote, "the information [that the report] contains was

[19] Kuhn's letter to Strong, 5 May 1961, TSKP 14.11, 2.

[20] See Footnote 17.

[21] See Footnote 19.

[22] See Footnote 19.

[23] Kuhn, "Reconstruction of Events," 6 May 1961, TSKP 14.11, 3.

[24] See Footnote 17.

[25] Kuhn, "Reconstruction of Events," 6 May 1961, TSKP 14.11, 4.

[26] Kuhn's letter to Strong, 5 May 1961, TSKP 14.11, 1.

[27] Hollinger, (1998, 49, fn. 10).

hastily gathered [...] under considerable emotional stress."[28] In two years' time, when considering a new offer from Princeton, Kuhn annotated "Some Indictments of Berkeley," and, under the heading "Philosophy Department," "Its treatment of me" comes in the third place, and, among the "Reasons for leaving," the "'Distance' from philosophers—my research" is in second place.[29] Indeed, Berkeley's Department of Philosophy had hurt him deeply. Indeed, this rupture between them was a reason for Kuhn to accept the job at Princeton—though there were others, as we will see later.[30]

6.2 An Offer from Princeton

In July 1961, another significant change took place for Kuhn. On July 1, the project for gathering original sources of the history of quantum physics began, and it would last for three years, ending on June 30, 1964.[31] Also in July 1961, it seems he started to correct the first full draft of *Structure*—the one he had sent to some of his friends and colleagues in April—and these corrections took him until early March 1962 (see Chap. 5). In short, a series of significant events began when he was promoted to full Professor in the History of Science at Berkeley and was transferred from the Department of Philosophy to that of History.

The publication of *Structure* arrived in the fall of 1962, the year in which he moved to Copenhagen. As the chair of the project on quantum physics, he, together with his team, settled into the Danish city in order to gather new sources for his project and to conduct more interviews. Previous interviews had been conducted in different places of the United States—mainly California and New York (see Chap. 7, below, for details). He stayed in Denmark until September 1963, and then returned to Berkeley in order to resume his duties and to finish the project itself in June 1964.[32] In the meantime, however, an offer from Princeton changed his plans and moved him to the East Coast again.

In April 1963, at his office in Copenhagen, Kuhn received a letter from his old fellow student at Harvard, Charles Gillispie.[33] Gillispie had just reviewed *Structure* in *Science*, but Kuhn had mixed feelings about the result.[34] On the one hand, he

[28] Kuhn, "Reconstruction of Events," 6 May 1961, TSKP 14.11, 4.

[29] [Kuhn], "Some Indictments of Berkeley," 7 May 1963, TSKP 14.11, 1–2.

[30] See Gillispie (1999, S90).

[31] See Wheeler, "Final Report of the Joint Committee of the American Physical Society and the American Philosophical Society on the History of Theoretical Physics in the Twentieth Century," 29 December 1972, TSKP 6.8, 2.

[32] For these dates, see Corner et al., "Request for $202,996 for Period 1 July 1961–1 July 1964 to Record Interviews with Men Central in the 1913–1938 Quantum Revolution and to Retrieve Personal Documentary Material on Quantum Theory and Related Scientific Developments, 1898–1938," 7 June 1961, TSKP 6.12, 5, 8; Kuhn, "Sources for the History of Quantum Physics," Memorandum, 8 August 1962, TSKP 6.12, 1.

[33] See our Chap. 4 for other details on Gillispie and Kuhn as still young historians of science.

[34] See Gillispie (1962).

confessed to Noyes by letter that he "was delighted and pleased by Gillispie's review of the book." He felt its highly detailed account was good promotion for *Structure*. On the other hand, Kuhn complained: "I am sorry that his final attitude is reserved skepticism."[35] In any case, the former impression outweighed the latter, so he was happy with the result.[36]

In his April letter, Gillispie informally invited to Kuhn to become Professor of the History of Science at Princeton. The letter was not yet a formal proposal, but Gillispie had the intention to turn it into such—a process that he started in early May.[37] He did his best in order to seduce Kuhn with an attractive environment. Stimulated by his presence, the historians of science would turn to the history of quantum mechanics. So, it would not only be *Structure* that was the main attraction of Kuhn's presence there—his expertise in the history of recent physics was most valuable for them. In addition, students would not only be in history, but also in philosophy (half and half, Gillispie said), and Kuhn could count on a close relationship with the Departments of Philosophy, of Physics, and of Mathematics. If something like that were possible, it would be critical for Kuhn to leave Berkeley behind. John Wheeler heartily supported Kuhn's appointment at Princeton, and J. Robert Oppenheimer approved it upon hearing from Gillispie himself. These two leading figures in physics exemplified the fact that Princeton scientists were probably going to receive Kuhn with warmth.[38]

On May 6, 1963, Gillispie gave Kuhn the context for this offer. Harvard had offered him the role of Professor of the History of Science, and it had also offered the historian John Murdoch a position as associate professor there—Murdoch eventually accepted Harvard's offer. Gillispie, however, decided to remain at Princeton. After all, he had spent more than a decade and a half there. Moreover, shortly before, in 1959–1960, he had founded the Program in History and Philosophy of Science at that university. So, Harvard was interested in attracting two renowned scholars to its own program, but Gillispie did his best to preserve the Princeton program. Kuhn was a sure option to reinforce the presence of the history of science—even more so when the collaboration with the Department of Philosophy was strong. Gillispie expected to rival Harvard in that common field.[39]

As a result of the first letter (April 18), Kuhn started to consider the offer seriously. From Copenhagen, he answered Gillispie on April 30 and told him that he had "spent all recent spare moments trying to weigh incommensurable strengths and weaknesses, and the result has been neither comfortable nor conclusive."[40] Actually, Kuhn's handwritten text, "Some Indictments of Berkeley," is dated May 7, 1963,

[35] Kuhn's letter to Noyes, 21 January 1963, TSKP 4.13, 1.

[36] See Kuhn to Noyes, 21 January 1963, TSKP 4.13, 1.

[37] Gillispie's letter to Kuhn, 6 May 1963, TSKP 14.11, 1.

[38] See the letter from Gillispie to Kuhn, 18 April 1963, TSKP 14.11, on which this paragraph is based.

[39] Gillispie's letter to Kuhn, 6 May 1963, TSKP 14.11, 1–2; Gillispie (1999, S89).

[40] Kuhn's letter to Gillispie, 30 April 1963, TSKP 14.11.

that is, only seven days later and still shows the conflicting views at play.[41] In his answer, Kuhn says that "a week of wrestling with the issues convinces me that I shall have to get back to Berkeley before I can seriously consider whether to leave," and that, despite his "discontent" with that university, he still "owe[d] both my friends at Berkeley and the institution itself a chance to talk back."[42] He and Gillispie agreed that the answer should wait until the following fall.[43] Meanwhile, on June 12, Gillispie extended the formal invitation for Kuhn to become professor at Princeton. Kuhn returned this more formal call (June 15) as they had previously agreed—they would wait until Kuhn's return to California the following September.[44]

In the meantime, Kuhn had started to try his luck at Harvard, because as far as he knew they were looking for historians of science. He wrote to Crane Brinton in order to find out if Harvard might be interested in him. He cited Gillispie's recent informal offer, and he also mentioned the recent publication of *Structure* on his behalf:

> I suppose that Harvard now also has appointments to make, and I imagine I may be among those considered. My interests no longer overlap [I.] Bernard [Cohen]'s as much as they did. In any case, if reactions continue as they have started, my recent book should begin to make me an ornament to, as well as a working member of any faculty. (Is this inexcusable immodesty? It is relevant, and I cannot be sure you yet know. My first claim to fame has been a time in coming.)
>
> There is, of course, an obvious reason why my name may never go on any list Harvard compiles in history of science. The same reason—my relation with Bernard—makes me quite unsure that I would accept even a Harvard offer.[45]

In the subsequent correspondence during June and July, Kuhn received confirmation that Cohen's attitude toward him was a difficulty for him returning to Harvard, and not only on the part of Cohen himself.[46] In fact, though his name had emerged in a committee in which new candidates were considered, some votes against his nomination were a result of Kuhn's likely attitude against Cohen.[47] That difficult relationship was indeed a problem for Kuhn, as was the fact that the Harvard context was not exactly as promising as Princeton's. Brinton and Kuhn exchanged letters on the same days that Gillispie and Kuhn did (June 12 and 15), and Kuhn was, at that point, clearly reluctant to cross the door he had himself opened shortly before.[48] In the end, the problem was that, at the request of Gillispie, Kuhn wanted to respond to Princeton by mid-October. Kuhn's presence at the university might attract postgraduates, but only if it was publicized early enough.[49] Yet, at that point, Harvard

---

[41] See [Kuhn], "Some Indictments of Berkeley," 7 May 1963, TSKP 14.11.

[42] Kuhn's letter to Gillispie, 30 April 1963, TSKP 14.11, 1–2.

[43] See Gillispie's letter to Kuhn, 6 May 1963, TSKP 14.11, 1.

[44] See Gillispie's letter to Kuhn, 12 June 1963, and Kuhn's answer, 15 June 1963, both in TSKP 14.11.

[45] See the letter from Kuhn to Brinton, 1 June 1963, TSKP 14.23.

[46] See Brinton's letter to Kuhn, 12 June 1963, TSKP 14.23.

[47] See Brinton's letter to Kuhn, 27 June 1963, TSKP 14.23.

[48] See Brinton's letter to Kuhn, 12 June 1963, and Kuhn's replies, 15 June and 8 July 1963, TSKP 14.23.

[49] See Gillispie to Kuhn 26 June–1 July 1963, TSKP 14.11, 5–6.

might still be in the process of deciding whether to make him an offer or not. Brinton still expected it to happen eventually.[50] On September 13, Kuhn replied to Brinton saying that he would be at Harvard by the end of that month, adding that he would "take a Harvard offer damned seriously"; Cohen's "presence in the program," in addition, "would not be more than a factor in my decision."[51]

Kuhn was concerned with the structure of Princeton's program from the start. In his "first round of the promised questions," on June 6, 1963, he had asked Gillispie about students at every level: "Has increased demand pushed the I.Q. up as much as it has at Harvard?" What's more, "how is the motivation?", he asked. He argued that "From my distance, [...] Princeton preserves its reputation as the country club of the big three. How much truth is there in it?"[52] He wished to know where history of science students came from—from the natural sciences alone, or also from the social sciences? And what did these students' original disciplines used to be? He also wished to know about the faculty, especially about the forthcoming members. Would there be room for historians of biology—in particular, for "pre-Darwinian technical biology," which he found "vastly more interesting than you do"?, or, for that matter, for the history of medicine, whose "role [...] (institutional) is terribly important and under-rated."[53] Kuhn's interest ranged from facts like these to those concerning the resources available to the students at every level, and, of course, to the current state of the blend of history *and* philosophy of science as a whole program at Princeton.[54] At Berkeley, he had praised the virtues of that program before a skeptical audience (as Aschenbrenner was). Now, Kuhn was eager to fulfil his high expectations, and to find room for his own improvement of the program. As we shall see later, Kuhn was involved in the improvement of the program from the very beginning.

All that time, Gillispie reassured him. In answering that letter in particular, he talked about the high quality of Princeton's students, whether undergraduates or graduates. Compared to Harvard, Princeton was, for Gillispie, a smaller community, but it enjoyed all sorts of activities and amenities, and the size did not affect the quality of their students and faculty. It was second to none, Harvard included. G. E. L. Owen, who was being considered for a post at that time—he ended up accepting a job at Harvard—told Gillispie (who repeated the former's words for Kuhn) that he admired how Princeton attracted prospective students by emphasizing the importance of the subject-matter of their interest, not only by granting them studentships.[55] It was a

[50] See Brinton to Kuhn, 27 June and 12 July, and Kuhn to Brinton, 8 July, 1963. TSKP 14.23.

[51] Kuhn to Brinton, 13 September 1963, TSKP 14.23.

[52] Most probably, Kuhn indirectly refers to F. Scott Fitzgerald's famous description of Princeton as "the pleasantest country club in America" (as quoted from Axtell 2006, 117), which appears in *This Side of the Paradise*. In his reply to Kuhn (June 26th, 1963, TSKP 14.11, 2), Gillispie critically referred, this time explicitly, to that description.

[53] All quotations and facts mentioned in this paragraph come from Kuhn's letter to Gillispie, 6 June 1963, TSKP 14.11, 2.

[54] See Kuhn's letter to Gillispie, 6 June 1963, TSKP 14.11, 2–3.

[55] It should be noted, parenthetically, that Kuhn expressed some reservations about Owen as a future member of the program. He told Gillispie: "it's absurd to pretend—as too many people have—that one can do ancient science without paying real attention to the Presocratics, Plato, Aristotle, and

distinguishing feature. In general, Gillispie tried hard to show Kuhn how his future audiences would have the quality that Kuhn was looking for. Gillispie's primary audience usually came from the natural sciences and engineering, and to a lesser degree from mathematics and philosophy. Gillispie also referred to the possibility of hiring new faculty members that Kuhn recommended. As regards the history of biology, Gillispie did not have a candidate, but he was open to that idea (together with the idea of hiring someone else for the history of ancient and medieval science).[56] Kuhn suggested Frederick L. Holmes, but they kept discussing other names as well.[57]

Kuhn also suggested the possibility of bringing Merton to Princeton. In Chap. 4, I showed Kuhn's interest in the sociology of science from his days in the sociology-of-science project with Philipp Frank for the Rockefeller Foundation. Now, a decade later, he was still interested in promoting the field—and he would continue to do so while at Princeton. Hiring Merton for Princeton was a good possibility, though junior faculty was probably the most feasible option. He nevertheless took advantage of this exchange and asked Gillispie about it, who agreed with that possibility, though he preferred to discuss it later.[58] Kuhn wrote:

> I think that the most neglected field of all at the moment is sociology of science, and I'm much interested in seeing the right sort of thing done about it. Could the Princeton Department be persuaded if a likely young candidate could be found? [...] Alternatively, is there by any chance the sort of enthusiasm and money around that might attract Bob Merton? (He's the only senior figure who seems to me on the right track.)[59]

Kuhn's interests in the sociology of science were strong while he was at Princeton. We shall see in Chap. 8 that he participated in seminars with social scientists during his period at the Institute of Advanced Study. He was also acquainted with a part of the recent literature—which included disagreements with developments in some cases. The social nature of his work in *Structure* and his continuing good relationship with Merton had helped him to maintain those ties all that time.[60]

A further question that Kuhn asked Gillispie is of a more mundane nature, but tells us something about Kuhn and his family's preferred way of life. At an early

---

the Stoics. But it's at least so far off to suppose one can neglect Euclid, Archimedes, Ptolemy, and Galen. (Actually, the latter shortcoming has even more serious effects when the purpose is to prepare students to study modern science.) Is Owen interested in the ancient technical sciences or is he willing to work them up?" (Kuhn to Gillispie, 8 July 1963, TSKP 14.11, 2).

[56] Facts described in this paragraph are based on Gillispie's long letter to Kuhn 26 June–1 July 1963, TSKP 14.11. Gillispie started to write this letter on June 26 while in Avalon (New Jersey) and, after a lapse in between, he finished it a week later at Princeton. It is full of details for Kuhn. Gillispie's mention of Owen can be found on pp. 4–5, and questions concerning the hiring of a historian of biology are on p. 3.

[57] See Kuhn to Gillispie, 20 July 1963, and Gillispie to Kuhn, 25 July 1963, both in TSKP 14.11.

[58] Gillispie to Kuhn, 25 July 1963, TSKP 14.11, 1.

[59] Kuhn to Gillispie, 20 July 1963, TSKP 14.11. Kuhn admitted in that letter that the chances of moving Merton away from Columbia were rather slim, though Princeton was not far away from Manhattan.

[60] We shall see more about this in Chaps. 7 and 8.

point in this correspondence, he asked Gillispie: "what is life in Princeton like?"[61] Though Cavell had reported a little about it, Kuhn wished to know more. After all, he and his wife, he wrote,

> tend to be urban types, exploiters of foreign restaurants, the theaters that run foreign movies, and the corner supermarket and delicatessen. Also, we value the absence of semi-compulsory faculty socializing that tends to go with, to take the extremes, small country colleges. Partly, then, the question is, how accessible is New York? (I know the time required to get there, but do people go?) [...] Why did Madame Koyré dislike Princeton so, or is that just a rumor?[62]

Clearly, Gillispie reassured him again, Kuhn would not have the same problems that Madame Koyré had had. She had been far from Paris, her homeland, and her family. She had usually been alone, because her husband used to spend the whole day at the Institute of Advanced Study (in Paris, he had worked at home), and she had gradually been losing her sight. Kuhn, by contrast, would find a friendly place for his whole family, Gillispie told him—a quiet place in which "faculty socializing" was just an option, and very close to New York. It was a convivial atmosphere in which highly technical natural sciences and mathematics nevertheless progressed. Kuhn would feel at home there.[63]

In late September 1963, Kuhn and his wife travelled from Berkeley to the East Coast; they had returned to California from Europe after a trip to France.[64] They went to New York first and visited the American Institute of Physics. Then they went to Princeton (September 27–30), where they would meet Robert Goheen, the President of Princeton, J. Douglas Brown, the Dean of the Faculty, and then some students and professors.[65] Kuhn wished "to chat briefly with Oppenheimer," because they had had a brief meeting in Denmark that he wished to extend.[66] Gillispie set the date for their meeting.[67] Finally, they travelled to Cambridge for a ten-day stay in which they visited Brinton. Kuhn asked him to stay at the Society of Fellows or, if unavailable, at the Faculty Club.[68]

Seven days after returning to Berkeley from that trip, around October 10, he sent a new letter to Gillispie (October 17) and then another one (on 22) asking for more details about academic and economic matters. The former letter, in particular, showed his emphasis on graduate training. For him, Princeton's program in history and philosophy of science was, first and foremost, a graduate program. "The

[61] Kuhn to Gillispie, 8 July 1963, TSKP 14.11, 2.

[62] See Footnote 61.

[63] Gillispie to Kuhn, 15 July 1963, TSKP 14.11.

[64] Kuhn explains the difficulties for not simply stopping on their return from Copenhagen in his letter to Gillispie, 8 July 1963, TSKP 14.11.

[65] Gillispie to Kuhn, 17 September 1963, TSKP 14.11, 1.

[66] Kuhn to Gillispie, 13 September 1963, TSKP 14.11.

[67] See Gillispie to Kuhn, 17 September 1963, TSKP 14.11, 1. In November, Kuhn returned to Princeton in order to talk with Oppenheimer again (and also with Eugene Wigner); see Kuhn to Gillispie, 29 October 1963, TSKP 14.11, 1.

[68] For details on this journey, see Kuhn's letter to Gillispie, 30 July 1963, TSKP 14.11, and Kuhn's letter to Brinton, 13 September 1963, TSKP 14.23.

program you and I have been discussing is, however, principally directed to graduate students. [… I]t is the thought of building a graduate program with you and the two other historians of science the University has promised to invite which attracts me so very strongly."[69] Yet, he saw Princeton as a "predominantly undergraduate university," a level in which he would spend most of his teaching and supervision time.[70] Though answering this question compellingly was not a necessary condition for him to accept, he seems seriously worried about it in that letter.[71] After all, his interests in education were not those of Conant's.[72] He was interested in training professional historians of science according to his own (pretty technically-inclined) vision. In building a program in the history of science he emphasized the training of new experts; particularly, those who could play a decisive role in philosophy—it would be like the realization of *Structure*'s main theses. This program was a good place for achieving such a practical goal. However, to do so, he had to emphasize graduate training. In fact, he would devote most of his attention while at Princeton to graduate training.

Gillispie answered both letters at length on October 21 and 26.[73] The former dealt with Kuhn's latter concern, in particular. In it, Gillispie confirmed that the program was designed for graduate school, though maybe in the future, the history of science could have more of a presence at the undergraduate level. However, his impression concerning the prospects for a program was different from Kuhn's. At the undergraduate level, interdisciplinary programs were not alien to the university's history. Yet, this program in particular, with a three-year history so far, had influenced others in plasma physics, biochemistry, linguistics and archaeology, for example, and that's because of their successful pioneering experience at the graduate level, something that was unusual at Princeton. Kuhn's professorship, together with the two future jobs for other historians of science, was intended to reinforce Gillispie's labors in that sense.[74]

At that point, however, Kuhn's decision was probably made, or soon to be made. Shortly after Gillispie wrote the second letter (of a more practical nature than the former one), he received a call from Kuhn.[75] It was October 27, and Kuhn wished to accept the job offer. Two letters from October 28 confirm this point—one addressed

[69] Kuhn to Gillispie, 17 October 1963, TSKP 14.11, 1.

[70] Kuhn to Gillispie, 17 October 1963, TSKP 14.11; quotation is from p. 3.

[71] Kuhn to Gillispie, 17 October 1963, TSKP 14.11, 3.

[72] Gillispie echoed Conant's emphasis on the education of future administrators and politicians in order to illustrate why undergraduate education was so important at Princeton, as indeed it was. See Gillispie to Kuhn, 21 October 1963, TSKP 14.11, 2. Clearly, Kuhn wished to distance himself from Conant in that sense too.

[73] See the letters from Gillispie to Kuhn, 21 and 26 October 1963, TSKP 14.11.

[74] See Gillispie to Kuhn, 21 October 1963, TSKP 14.11, esp. 1–2. Gillispie, however, also told Kuhn that some disciplines (such as Physics, Art, or Mathematics, for instance) placed strong emphasis on graduate education. See the same letter, 2–3.

[75] It seems that Kuhn received the letter from October 26th after his call to Gillispie; see Kuhn to Gillispie, 29 October 1963, TSKP 14.11, 1.

to Brown, the Dean of the Faculty, and the other to Gillispie.[76] At that point, he had agreed to start his new job at Princeton in September 1964.

He received welcome letters from Jerome Blum (Department of History) and from Gregory Vlastos (Department of Philosophy). Days later, Vlastos wrote another letter in order to find out whether he would like to be listed on the masthead of Philosophy as an external member. He, Vlastos said, would be like Alonzo Church—an honorary member (Kuhn from History, Church from Mathematics). Kuhn agreed, of course, and greatly appreciated the gesture.[77] "As you know," he wrote to Vlastos, "the opportunity to work with good students of philosophy of science is one of Princeton's great attractions for me. [...] I shall gain much personal satisfaction from having my name on the particular list."[78] It is pretty clear to what extent that point was true. He had checked with Cavell—who, at that point, was a member of the School of Historical Studies of the Institute of Advanced Study, Princeton—about the Princeton setting, and in his answer to Cavell's (very detailed) May report, Kuhn said that "though I love and admire my departmental colleagues, I get almost nothing intellectually important to me from them."[79] Hollinger says that Kuhn communicated his decision to his fellow historians in a letter in which the availability of students in philosophy in Princeton weighed favourably.[80] Carl Hempel's reaction was also unmistakably positive and promising. Hempel wrote a very kind letter to Kuhn just one day after the latter's acceptance. The historian William Bouwsma—incidentally, Kuhn's fellow student at Harvard in the early 1940s—had given him the good news, and he told him that he, as well as any other in Philosophy, were eager to start collaborating with Kuhn (as is well known, Hempel would end up teaching a course in collaboration with him).[81] The prospects for Kuhn to leave behind the bad taste that Berkeley's senior philosophers had left in his mouth looked good. Indeed, it seemed to dissipate when Kuhn returned to the East Coast.

[76] See the letters from Kuhn to J. Douglas Brown and to Gillispie, 28 October 1963, TSKP 14.11. See also Gillispie to Kuhn, 28 October 1963, TSKP 14.11.

[77] See the letters from Blum to Kuhn, 28 October 1963, from Vlastos to Kuhn, 29 October and 8 November 1963, and Kuhn's answer to Vlastos, 14 November 1963, TSKP 14.11.

[78] Kuhn to Vlastos, 14 November 1963, TSKP 14.11.

[79] Kuhn's letter to Cavell, 6 July 1963, TSKP 14.11, 1. Cavell was a member of the IAS during the academic year 1962–1963. Cavell says himself that he "spent the year between moving from Berkeley and installing myself back at Harvard" (Cavell 2010, 300). See also the IAS web page (https://www.ias.edu/scholars/stanley-louis-cavell) for further data.

[80] See Hollinger (1998, 49–50). See Kuhn's letter titled "To: My Friends and Colleagues in the Department of History," 27 October 1963, TSKP 14.11.

[81] See Carl Hempel's letter to Kuhn, 29 October 1963, TSKP 14.11.

6.3 Articulating Paradigms

In our previous chapter, we left Kuhn in 1961 right after discovering that "paradigm" was a term that helped to show to what extent his view of science was different from the view inherited from the empiricist tradition in methodology. For him, the notion of "theory" that this tradition assumed did not convey the kind of consensus that could be found in a scientific community in the mature sciences. "Paradigm," meanwhile, was more appropriate, albeit less explicit in content. Kuhn's use of that term is connected to various aspects of the nature of science. First of all, the notion plays a role in education and in investigation that "theory" cannot. Paradigms guide both education and research in absence of explicit or implicit rules, while a theory assumes the availability of rules. Either in the training context or in the research context, the learner or the researcher reason by analogy, and the focus of the analogy is the prevalent paradigm for their subdiscipline. Rules may be inferred, but their role is minimized once those agents learn the discipline they practice—once they get trained as scientists. Moreover, explicit reference to rules is, rather, a signal of dispute. It is often used when the agreement about what makes a problem or problem–solution acceptable is broken.

Secondly, the notion of paradigm plays three important roles as to the kind of field of philosophy that philosophy of science must be. First, it must be *historical*; second, it must be *social*; and third, it should handle *logic* with care. In the previous chapter, I specified the two former characteristics. Let me briefly emphasize them for the sake of this argument. The paradigm has a historical nature because of its limited validity—it is relative to a historical context. Conant does not readily admit this point, because for him, scientific achievements are timeless, this being a reason for using many of them in education (even though the practice to which they belong has long been outdated). Besides that, a paradigm shows that scientific practice is based on a particular kind of agreement or consensus that is conspicuous among mature sciences—to wit, practicing scientists learn to recognize a new instance of well-practiced science (e.g. a new solution or the discovery of a new problem) without having to agree explicitly on the reasons for declaring that instance an example of well-conducted practice. Finally, I said that logic must be, for Kuhn, handled with care. The seventh Lowell lecture is a case in point of that attitude. The plans for formalizing science are not altogether beneficial for a theory of science—particularly for the one that emphasizes its development. The notion of theory that Kuhn criticizes is still too attached to that program, while his own sense of "paradigm" is rather *detached* from that program. We shall see, however, that in later years (1963–1965) Kuhn attempts to model learning on a mathematical basis, and that a certain formal approach is present in his explorations in that sense (more on this below). Yet, these are attempts to simplify a complex situation to the point of making it manageable although eventually false or, at least, unlikely.

Kuhn was certain that the notion of paradigm, despite being a last-minute addition, had transformed *Structure* for the better. In the years with which this chapter has begun (roughly 1961–1964) he was already pursuing the clarification of that notion of whose importance he was convinced. In a reply to very early criticism by Dudley Shapere and N. R. Hanson from December 1963, he wrote that a great part of *Structure* did not really depend on paradigms, but it did not render the central concept of paradigm useless. "During that time before I spoke of paradigms," he said there, "I still spoke of scientific revolutions. [...] Most of the book could still be written in these terms."[82] Yet, after showing what was involved in the concept of paradigm, he said that all of it "seems to me terribly important and I'm not yet at all prepared to give it up." He acknowledged that the concept probably had problems of the sort Shapere suggested, but—he went on—the "existence of problems like these [does not] make me think that the heart of the notion of paradigm is wrong"; for him, it only required further articulation and research.[83] Days later (ten days into the new year) Kuhn still persisted in this position and in his firm defence of paradigms, although with the aim of understanding it better (more on this below).

His new reflections on paradigms started very early, in fact. He had already made some annotations in early May 1962—in March, he submitted the book to the Chicago press. At that time, his work for the Quantum Physics Project had already begun, but he had not yet moved the headquarters to Copenhagen. He took some notes on May 1, 1962, and considered Stanley Cavell's critical remarks on Section XII of *Structure*. Already in New York City, at the Rockefeller Institute, where he was interviewing some scientists like Richard Courant, Ludwig Edelstein, Fritz Reiche, George Uhlenbeck, or Peter Debye for the project, he took notes (not many) on his own book, related, this time, to some things Debye had said in the interview.[84] Although Kuhn was working on the quantum-history project, his attention to issues in *Structure* had not diminished in the least. A constant in Kuhn's activity over the time he spent on the project was that, even though the history of quantum physics required almost all of his attention, he never completely abandoned thinking about the themes of *Structure* (more on this below).

Paradigms remained his main concern. In his private research notes from 1962–1964, Kuhn seems worried about the extreme rigidity of the notion. In order to explain this a little, we can recall how he talks about his sense of "paradigm" at the beginning of Section III of *Structure*. There he says that he is adopting the word from fields like the teaching of Latin grammar, where it is a "model or pattern" that "functions by permitting the replication of examples any one of which could in

[82] [Kuhn], "Cleveland Paper," 27 December 1963, TSKP 3.13, 4–5. He also says there that "[m]ost of the main ideas in my book are ten to fifteen years old *with me*. Paradigms are about five years old and most of the book was developed without their aid" (4). Concerning this paper, note that this reply was not to the published version of Shapere's well-known review of *Structure* (Shapere 1964), which was published later, so this paper was probably a reply to an in-person meeting between them in Cleveland, though Kuhn does not mention any other detail about the occasion. Nevertheless, Shapere (1964) could give us an idea of the kind of problems that were debated.

[83] [Kuhn], "Cleveland Paper," 27 December 1963, TSKP 3.13, 9–11; see also Shapere (1964, esp. 30–33).

[84] See Kuhn, "Sources for the History of Quantum Physics," Memorandum, 8 August 1962, TSKP 6.12, 8–10 ("Magnetic Tapes Collected. July 1, 1961 to July 31, 1962"). The notes he took in May can be seen in T2E (see abbreviations) under the title "Towards 2nd Edition—Struc. Sci. Revs.," dates: 1, 4 May 1962.

principle serve to replace it."[85] His "appropriation" of paradigm, as he says, has to do with a sense related to the practice of science, in which "a paradigm is rarely an object for replication. [... I]t is an object for further articulation and specification under new or more stringent conditions."[86] The question is, how this process of articulation, which is the activity through which we observe that a given problem–solution is paradigmatic, is related to the existence of a recognizable example? If he cannot answer that question, his description of paradigms remains pretty incomplete and "too vague."[87] He notes this in May 1962, when he writes down:

> The treatment of paradigms is, of course, still too vague, and also it's wrong.
>
> Too much rigidity. [...]
>
> The line between what is part of paradigm and what's available for articulation can't be drawn. (I allow for this—by saying paradigm = achievement, which remains very good—but whole vocab[ulary] still implies the contrary.)[88]

One year later, in May 1963, he insists on the same point in his private notes. After considering that his defence of paradigms as "concrete scientific achievements" remains truly important, that there is not a ground of "stable facts"—which suggests that research on paradigms involves working on the theoretical and the observational basis of theories at the same time—and that a scientific revolution must be understood as paradigm-change and world-view change, he returns to the same point: how to differentiate "articulations" from revolutionary work.[89] Paradigms are reference achievements for the scientific group; most probably, they are what bind the community members together. Yet, the essence of their nature is that they are not simply "reissued" in every new problematic situation, but rather, so to speak, "reinvented" (my expressions) in the normal process of research. Their essence is change, whereas, at the same time, it is also (some kind of) stability. When, therefore, does change drift towards a revolution?

In 1964, Kuhn had still not managed to answer that kind of question. He took four pages of notes on 10 January, a few days after his reply to Shapere and Hanson in Cleveland (I previously mentioned them), and the former philosopher's criticism seems to stay in his mind, for he annotated: "Recent discussions—particularly those at Cleveland, led perhaps by Shapere[—]make me realize the need for the following sorts of revisions of *Structure*."[90] Kuhn then refers to paradigms once again, and states that "the biggest series of problems are here," after which he goes on:

> The term should be reserved for what led to it, the actual concrete model. Explicitly agreed upon generalizations—whether physical or metaphysical, whether rationalized from paradigm or imported from *outside* or other—should be treated separately. To make any and

---

85 *Structure*, 23.

86 See Footnote 85.

87 See Kuhn's notes in T2E, 1 May 1962, 1.

88 T2E, 1 May 1962, 1.

89 T2E, 30 May 1963, 1.

90 T2E, 10 January 1964, 1.

> every element of a tradition into a (part of a) paradigm makes things look too rigid, obscures my main point, and makes important distinctions too hard to draw.[91]

As far as it goes, it was pretty clear for him that "paradigm" was a "concrete achievement" and a "concrete model." However, there were problems when it came to distinguishing the model itself from its different manifestations—more specifically, from the different results of using it and articulating it in the context of other problems and further developments. He says that:

> The implication that there is *a* paradigm that underlies but is not equatable with all its divergent articulations smacks too much of idealism or hypostatization and must be removed. There's no problem about doing it except that of careful writing. We go from one specific model to another. Each is a paradigm.[92]

In other words, Kuhn does not wish to abstract the paradigm from a series of equivalent or quasi-equivalent manifestations of a given problem–solution. The paradigm is any of the valid articulations of that problem–solution and it changes as the way to read (i.e. to reinterpret) an achievement changes as well. In his example, successive articulations of a given theory, such as the Hamiltonian and the Lagrangian formulations of Newtonian dynamics, make for a new paradigm for the individual that, say, first learned that theory without them—at which time, in Kuhn's words, "elementary Newtonian formulation of dynamics" was his or her "paradigm."[93] This or that model, this or that achievement, plays the paradigmatic role for a given individual in his or her training or research practice. Successive transformations of a solution are entitled to that role. Note that I refer to "transformations": for Kuhn, the result of that process was "not a mere addition," he says; in the end, those formulations changed "the original Newtonian paradigm."[94]

We may think of a sort of genealogical relationship between these articulations of a given problem–solution that leads to the acceptance of them as successive paradigms for an individual or group. We may also think of the relevance they exhibit in a certain state of play—how useful they are at that moment. But Kuhn does not say anything about features such as these. Rather, he notes his difficulties in drawing the line between normal science and revolutionary science on the basis of the nature of paradigms. "I must have some way to distinguish articulation from revolution," he says in 1964. "This is [the] same problem," he goes on, "as defining what part of given work is paradigmatic, what part available for change and variety of opinion."[95]

The "concrete" aspect of the paradigm and the use of "model" on the part of Kuhn should not lead us to think that we are talking about some kind of fully explicit model as it may sometimes be understood. When, in correspondence, the psychologist Edwin G. Boring tried to describe what Kuhn's idea of paradigm was, the latter

---

[91] See Footnote 90.

[92] See Footnote 90. Once again, compare this paragraph with Shapere's (1964) criticism.

[93] T2E, 10 January 1964, 2.

[94] T2E, 10 January 1964, 2. That these formulations play such a role seems still clear in "Second Thoughts on Paradigms," written a few years later. See Kuhn (1974, 300, n. 10).

[95] T2E, 10 January 1964, 2.

qualified the extent to which a paradigm was not a model.[96] Kuhn told him that there were

> respects in which a paradigm is both more and less than a model. More, in that it includes a variety of methodological and instrumental commitments as well as commitments to theory or model; less, because so very much of it is implicit—the question is less whether the model actually exists than whether one could possibly describe what it would look like if it did exist.[97]

A paradigm is therefore a model (or the promising schema of one) that attracts the attention of scientists in a scientific group, and which may be transformed into new versions of itself, thus affecting successive stages of the same individual and sometimes successive stages of a scientific group. The concrete and model nature of paradigms must not conceal the fact that they go through historical development.

This historical character is difficult to match with the idea that change is sometimes destructive (i.e. revolutionary) or, more modestly, the cause of a crisis. Evidently, for Kuhn, this problem is important, but there are, at least, two more problems in his position. I will comment on the first one in this section, and will explore the second in Sect. 6.4. The first one is how such variations are related to the ultimate components of a scientific community, that is, the individual scientists themselves. On the basis of a narrow understanding of a paradigm, as an achievement that leads the whole community to fulfil its implicit expectations, the group may look like a strongly united group. However, we know that, for Kuhn, the composition of the community in a mature science pretty much depends on the idea of paradigm we have just seen—that is, a model (or something like it) that is articulated and undergoes historical development.[98] Individuals that work with a paradigm, who are involved in normal-science problem-solving, also evolve during the process and during their entire lifetime; they sometimes suffer crises themselves, he argues.[99] Even their individual perceptions of the changes that the achievement undergoes may be different within the community.[100] Kuhn notes in 1962: "Groups [are] not that unified; there aren't always texts in any literal sense. There's more freedom of approach."[101] So, not only do paradigms evolve during their lifespan, the community is itself a heterogeneous group with different and often changing attitudes toward the work they develop. Kuhn's notes show that he was aware that while the scientific community remains the unit of epistemological analysis in *Structure*, the individual might be an ultimate component of the social mechanism. In 1963, he says that, among the issues

---

[96] Kaiser (2016, 82–83, 85–86, 93–94) examines his correspondence with Boring.

[97] Kuhn to Boring, 29 November 1962, TSKP 4.7, 2. See Boring (1963) for the text on which Kuhn made this and other comments. Unfortunately, Boring altered the relevant passage in the published version according to Kuhn's suggestions, so it is no longer available. But see Boring (1963, 9), where Kuhn's indications are well represented.

[98] How Kuhn reconsiders and attempts to suppress that dependence of the idea of community on that of a paradigm is examined in Chap. 7.

[99] See T2E, 1 May 1962, 1 (point 2).

[100] See T2E, 1 May 1962, 1 (point 2); 30 May 1963, esp. points II.1, II.3.

[101] T2E, 1 May 1962, 1.

he is "probably in deep difficulty with," the following one features: "'Subsections' of the scientific community. Can I avoid being forced back to individuals as [the] only unit?".[102]

6.4 Scientific Realism

A second problem is more traditionally ascribed to Kuhn's position since *Structure*, and he was aware of it at this time—the problem of scientific realism. Does that position mean that we cannot count on asserting that our best theories are true? Should we avoid the idea that our best physical theories represent reality as it is? In fact, is the idea of a "reality in itself," or something similar, a meaningful expression? In the notes we are examining, Kuhn also seems worried about the kind of answer that, in full compatibility with *Structure*, he is willing to give to such questions.

Recall that this is an old problem for Kuhn, and that he did not only deal with it in *Structure*, Section XIII (and elsewhere), but even in his earliest speculations about scientific realism when he was still an undergraduate student at Harvard (see our Sect. 1.10). Although *Structure* closes by showing that the view that science progresses cumulatively towards truth is hopeless, Kuhn's account of his replacement position remained a topic for discussion. So, for example, a historian of art like Ernst Gombrich offered Kuhn some critical points by letter in August 1963, which Kuhn answered the following October; in 1964, Kuhn defended his position before the historian of science David Joravsky by letter; and, after a discussion during a Princeton summer seminar arranged by the Telluride Foundation in 1965, Kuhn discussed this with Robert Nozick—who had a really positive impression of *Structure* after having read it years before—and some students.[103] These are only a reduced sample of the set of replies to Kuhn's rejection of scientific realism that he received right after publishing *Structure*. Many others came eventually, of course. However, these three replies form a significant sample that Kuhn took seriously, especially the latter two, so I will now turn to them.

His interaction with Gombrich gave Kuhn a chance to show how his position is different from an opposition between an instrumentalist and a realist perspective. *Structure* shows how the ways in which a theory (or a paradigm, in this case) succeeds another involves a replacement of specific commitments about what the world is like. That replacement brings about the elimination of a whole worldview, as a result of which *only some* of the elements of the earlier theory are preserved—usually, the more instrumental ones (more on this below). Indeed, this position seems akin to instrumentalism, but it is *not*. At least, it is *not* similar to the kind of instrumentalism

---

[102] T2E, 30 May 1963, 1.

[103] See Gombrich to Kuhn, 11 August 1963, and Kuhn to Gombrich, 24 October 1963, both in TSKP 4.9; Kuhn to Joravsky, 9 January 1964, TSKP 4.10; T2E, 7 July 1965. On Nozick's favorable attitude towards *Structure*, see Cavell's letter to Kuhn, 28 May 1963, TSKP 14.11, 2. On the Gombrich–Kuhn exchange during that year, see also Pinto de Oliveira (2007, 757, 759).

that a logico-empiricist philosopher might defend (though not all of them assumed such a position, of course).[104] Gombrich noted this aspect of his position, but—precisely for that reason—he did not accept that Kuhn had set Popper on the same side of those philosophers from Logical Empiricism that Popper had criticized.[105] For Gombrich, Popper had attacked instrumentalism in a way that would have been useful to Kuhn.[106] However, that position involves there being assertions that may be considered true, while there are others—their competitors—that remain false at all events. Gombrich suggested two examples. First of all, Copernicus' heliocentric model states that the Earth moves around the sun, which *is* the case; opposite statements are simply *false*. Secondly, there are alternative representations of a certain object or person that could be considered more or less credible. Winston Churchill, Gombrich's example, may of course be portrayed as he actually was, but he may also be caricatured, and some of those depictions may be more or less close to the model. There is no doubt that caricaturing him as a giraffe, he says, would not only be inexact but also pointless. Closeness of fit to the model makes for good or bad depictions, even in the case of caricaturing. In short, when it comes to representing a certain object or phenomenon, there is always some progress towards truth involved—or, at least, a definitive elimination of falsehood.[107] Vacuous instrumentalism may be rejected on the basis of some commitment to realism, accordingly.

Kuhn, however, disagreed with Gombrich that progress towards truth is the only conclusion available, and that, even if it has been accepted, the only position that remains is pure instrumentalism. First of all, we can think of several representations of an object, event, substance, or process on the basis of the representation's closeness of fit to the appearances. For him, the "examples" discussed should belong to a "historically plausible situation," and the Churchill example clearly fails in that regard.[108] A caricature of Churchill and a depiction of the same man with the intention of being faithful to the original cannot be compared on the basis of closeness of fit. That comparison is pointless. Most probably, their designers do not intend to produce the same effects or reactions by means of them. Though Kuhn does not mention it in his letter, Gombrich's former example of heliocentrism is far more appropriate. We may agree that Ptolemy and Copernicus *had* similar intentions, and they did not prepare their models for the sake of finding alternatives and then deciding which one is closer to the appearances. In real cases of paradigm-comparison there seems to be no place for the idea of progress towards truth such as that Gombrich assumes as correct and undoubtable. Secondly, this alternative is not another example of instrumentalism, despite Gombrich's perspective. For Kuhn, it does not mean that Ptolemy's

---

[104] Concerning Herbert Feigl's and Hans Reichenbach's affinities to scientific realism, for instance, see Putnam (2001), Salmon (2005, Chaps. 1–2), Neuber (2011), Psillos (2011a; 2011b), and Mayoral (2017a, 120–121). See also van Fraassen's (2002, 220–225) contrary interpretation of Reichenbach.

[105] See *Structure*, Sect. XII, esp. 144–146.

[106] Gombrich mentions Popper (1956).

[107] See Gombrich to Kuhn, 11 August 1963, TSKP 4.9.

[108] Kuhn to Gombrich, 24 October 1963, TSKP 4.9, 2.

and Copernicus's attitudes are only a matter of offering the proper instrument of reckoning.[109]

We can read this interaction as an opportunity for Kuhn to show that his paradigm-based account of the comparison of scientific representations should be historically based. The analogy with fine arts must be handled with care precisely because of that reason. Let us recall that Kuhn employed that analogy in early versions of Chap. 1 of *Structure*.[110] But let us also recall that he emphasized *differences* as well as similarities, and that there were *key* differences between both activities. Fine arts and their history are a good source for understanding the idea of incommensurability. However, they are less useful if we turn to them to find out what a paradigm *is*, because—as we have just seen some pages before—Kuhn is clear in *Structure* that a paradigm is not just a model and it is not just an object or scheme used for simple replication, either. In articulating a paradigm, scientists commit themselves to a specific way of practicing science in a phenomenal world. Gombrich criticized the scientist's dependence on the paradigm in Kuhn's account. He did not understand the reasons for Kuhn to represent the scientist as a dogmatic individual, as such a narrow-minded expert, unable to imagine alternatives to the paradigm—different paradigms actually. Indeed, Gombrich wondered why the Kuhnian scientist cannot prepare different paradigms and then decide which one is correct—which one is closer to truth. He did not understand why the Kuhnian scientist is not much more open-minded—a free mind, in fact.[111]

We can understand the intellectual differences between Kuhn and Gombrich as those that exist between a historian of science and a historian of art. Gombrich was able to see an attitude in the stance of scientists towards their own discipline that for Kuhn would be unusual in the mature natural sciences of his interest. For example, Gombrich finds that, in the history of the fine arts (and it is probably something that he was willing to extrapolate to other disciplines as well), a historian may look for an alternative to a well-established paradigm in his or her own discipline—like Erwin Panosfky's iconological method, which Gombrich, stretching Kuhn's sense, also considers a paradigm—because he or she is bored of it.[112] Kuhn, meanwhile, does not see that situation as likely in science:

> Your remark about boredom as a source of crisis in the arts interests me particularly because I think nothing quite similar happens in the mature sciences. Good scientists do get bored, and they do change their careers as a result. But that change seldom or never takes the form of an attempt to upset the existing paradigm. Instead, the bored scientist is likely to change his field—say from nuclear physics to bio-physics, a transition that has been relatively popular since the war, or even from physics into administration or politics. In the absence of a lead from nature, there is no way to go about unseating a paradigm. And when there are such leads in the form of anomalies, science is seldom boring.[113]

[109] See Kuhn's letter to Gombrich, 24 October 1963, TSKP 4.9, 2. My interpretation goes well beyond Kuhn's explicit argument in that letter.

[110] See our Chap. 5. See also Pinto de Oliveira (2017).

[111] See Gombrich to Kuhn, 11 August 1963, TSKP 4.9, 2.

[112] See Gombrich to Kuhn, 11 August 1963, TSKP 4.9, 1.

[113] Kuhn to Gombrich, 24 October 1963, TSKP 4.9, 1.

Kuhn might easily have thought of himself when writing about what a scientist in a mature discipline does when he or she gets bored of his or her research field. It is probably true that what the scientist does not do in that case is to revolutionize the discipline. For Kuhn, looking for alternatives is something a scientist does when the world in which he or she practices science is itself too upset by difficulties. So, the scientist has to look for a solution. Clearly, Gombrich was right that the Kuhnian scientist is too committed to the world he or she lives in.

This difference, however, would not only be a matter of a disparity in their respective disciplines. Kuhn and Gombrich did not agree on the philosophical basis, either. This is clear from the problem that has led us to this point. Gombrich's vision of science is indebted to Popper's theory of science, whose free, rational thinker is closer to the ideal of the creative artist and the democratic individual that Gombrich defends than Kuhn's. Kuhn, however, always attempted to show that his scientist does not work under the oppression of a ruler. His or her commitment is cognitive and social in nature, and the theory he or she develops only makes sense in a certain world. Only if we consider the scientist's phenomenal world a source of oppression may this behavior be considered artificially constrained and narrow-minded. In Popper's (or Gombrich's) view, a quick calculation of the prospects for realism in a theory-choice situation is always possible; in Kuhn's view, there is no such possibility at all—or, at least, it would not be so "quick." But his position was often difficult to show. It was very difficult for Kuhn to argue for this to Popperian thinkers. Insofar as Gombrich was a supporter of Popper's views, this exchange with Kuhn was a sign of what was to come.

During the next few years, Kuhn became progressively aware of the difficulties that his perspective on science and revolutionary paradigm-change was going to create, even for the more basic, hidden, and naïve realistic commitments of many philosophers and other kinds of scholars. Shortly after writing his response to Gombrich, in January 1964, a response to the historian of Soviet science David Joravsky led Kuhn to think about the same problem again, though this time from another perspective. It would not be the last time he thought about this point.[114] It may have been significant, because he annotated more ideas for the second edition of *Structure* the day after. Among his reflections were those referring to realism and to his answer to Joravsky. This time, Kuhn was concerned with the idea of a pure idealism in his position—an alternative to instrumentalism in what, for many readers, started to be a fully-fledged anti-realistic position in *Structure*. So, he wrote to Joravsky that:

> I have never meant to suggest that there is not something (nature or the world) outside of the scientist's mind. Perhaps "outside of" is the wrong locution, but there is surely something besides the scientist's mind. Anything else would be absurd. Science is an empirical occupation, and nothing in the book is intended to be incompatible with insisting that scientists do their experiments on nature. Only, I am not clear that there is any single possible one-to-one correspondence between what is "outside" and "inside" the mind. There may not, that is, be a criterion which would select, even in the very long run, a single correct answer to the sorts of questions that scientists ask.[115]

[114] See, e.g., Kuhn (1979b, 206–207).

[115] Kuhn to Joravsky, 9 January 1964, TSKP 4.10, 1.

The last phrase almost seems addressed to a much younger Kuhn who, in the early 1940s, wondered about the possibilities for scientific realism. He might thereby confirm that his old suspicions that pluralism may be right were correct. Leaving that fictional connection aside, this part of the letter is intended to defend *Structure* against the charge of idealism, while he preserves the very idea of the natural sciences as typically and essentially empirical.

The day after writing this letter to Joravsky, Kuhn registered his doubts about that problem concerning the mind-independent world and its nature. He wrote: "I don't know whether I can get anywhere on sense in which nature is actually changeless though men live in different worlds. Must continue to agonize."[116] There were, however, some auxiliary ideas that might lead to a coherent position. Among them, the episode of the history of chemistry with which he closes Section X of *Structure* and his mathematical model for paradigm-based learning (more on this below) were promising.[117] But Kuhn also includes the point of view he expounded in the previous letter to Joravsky as a good point. "Perhaps 'besides,'" he wrote, "is a better way to think of it. [...] I need addition which does not make the whole equal to [the] sum of its parts."[118]

During the summer of 1965, Kuhn found another motive to return to reflections on the role (if there is any) that scientific realism may play in *Structure*. He returned to his idea of the world (or nature) and to its relationship to paradigm-change, but he also took up a problem that seemed to be solved in the book—the simultaneity of cumulation and rupture during scientific revolutions. I shall begin this latest part of the section by dealing with this latter point first.

In *Structure*, Section XIII, Kuhn shows that problem-solutions that are prospective candidates for new paradigms "usually preserve a great deal of the most concrete parts of past achievement," though, of course there are losses as well as new feats.[119] Since the beginnings of his preparation of *Structure*, Kuhn knew that granting that niche for simple accumulation of past achievements might be a problem for his perspective. In undated notes that come from the preparatory work for *Structure*, he shows that he was aware of this difficulty. On the one hand, "cumulativeness is real but partial," he writes. "Profession," he goes on, "won't let the core of what they've

[116] T2E, 10 January 1964, 4. For a later solution along Kantian lines, see Hoyningen-Huene (1993, Sect. 2.1).

[117] See *Structure*, 129–134, where the "Proust–Berthollet" case—as he mentions it (T2E, 10 January 1964, 4)—is fully developed as part of the Dalton case in the history of chemistry.

[118] T2E, 10 January 1964, 4. It is interesting to consider some of Kuhn's strategies when it comes to his reasoning on philosophical matters. For him, this very idea—the whole is more than the sum of its parts—is also applicable to the evolution (or articulation) of a paradigm as something more than adding up new pieces. The reform creates something new. Both his concept of paradigm and his concept of the world would be fruitfully approached on the basis of the same idea. Similarly, he sees the notion of chemical combination (probably through the lens of the Dalton episode in the history of chemistry of which he had just thought) as another promising idea. See these connections in the same place (T2E, 10 January 1964, 4, esp. III.2). Kuhn's uses of analogical reasoning are frequent.

[119] *Structure*, 168.

done before go, though they'll let much go."[120] He refers to "significant quantitative predictions," and to "techniques of numerical computation," as good examples of what the community usually preserves.[121] Kuhn shows that he is aware this is a problem for his theory. If true, then empirical and technical accumulation turn out to be criteria for theory-choice.[122] If, as an average, the scientific group opts for the theory that satisfies those criteria, the resulting perspective is close to a traditional view of scientific method. Even then, Kuhn showed that he was worried about this conclusion, because "there's still too large a residue of cumulativeness directly in it. As a result, I feel that the novel portions of my approach have too indirect a role in explaining why science works."[123] Kuhn solved such doubts by showing that the transition to a new paradigm is rarely as simple as that. Years later, in *Structure*, he showed that, in paradigm-choice situations, decisions can only be comparative, and, very often, a new aspirant can offer little evidence on its behalf. Verification (or confirmation) of a theory as the main and only argument supporting it is, Kuhn said, part of paradigm-based argument—that is, it is made once the problem–solution has been accepted as a paradigm. Paradigm-choice is, therefore, something made on the basis of a number of factors.[124]

In 1965, the problem of cumulativeness still remained. The day after talking with Nozick and some students at Princeton (July 6, 1965), he annotated on a typewritten sheet of new notes for *Structure* that accumulation in science was different according to a three-level approach:

> At the level of hard computations, science grows in a virtually if not entirely cumulative fashion. At the next higher level of explanations, the gains are probably net but there are losses. At the highest level, what is the world like, there seems no accumulation at all.[125]

In short, cumulativeness descends according to the following ranking:

Level 1. Computations—science *is* cumulative.
Level 2. Scientific explanation: science is only *partially* cumulative.
Level 3. Ontological commitments (or worldviews): science is *not* cumulative.

Although a careful reading of *Structure* leads to this ranking pretty easily, Kuhn obtained the impression—for example, from a conversation with Nozick—that it had to be emphasized more. In Kuhn's hands, this ranking involves a degree of incommensurability that makes old views supporting the unified and ahistorical nature

---

[120] [Kuhn], [Untitled notes], n. d., TSKP 25.53, 1. These notes start with the expression "Pattern of evolution".

[121] [Kuhn], [Untitled notes], n. d., TSKP 25.53, 1.

[122] Kuhn does not mention paradigms in these notes, so they must be prior to 1959, at least. This difference must be kept in mind when comparing statements from *Structure* and notes like these. The word "paradigm" has not yet acquired the priority it shall have after 1959.

[123] [Kuhn], [Untitled notes], n. d., TSKP 25.53, 1.

[124] See *Structure*, Sect. XII, esp. 147, 150 ss. See also [Kuhn], [Untitled notes], n. d., TSKP 25.53, 1, particularly the last (typewritten) paragraph. In those notes, in which the *Structure* version is still in progress, Kuhn emphasizes there that theory-choice always involves, in his view, theory-comparison ([Kuhn], [Untitled notes], n. d., TSKP 25.53, 1–2).

[125] [Kuhn], "Notes for book" (typewritten), 7 July 1965, T2E.

of scientific method senseless, and the vision of progress by accretion of evidence towards the same theoretical truth unnecessary and plainly wrong. However, Kuhn shows less surprise at finding contrary opinions and more certainty that *Structure* should argue better for his position. He cleared up his own vision on nature and reality that day as follows:

> I suppose [...] that nature is entirely stable. I suppose that man engages it (in a tug of war) and that reality is something that emerges from the engagement. Thus nature is not the sort of thing that is real or that consists of electrons or what you will. That sort of question cannot be asked about it. It's something else entirely, perhaps a *Ding an sich*. Reality is what emerges at the intersection or through the interaction of man and nature, and there are different realities depending on the mode of interaction.[126]

This perspective agrees with the one he sent to Joravsky more than a year and a half before. It shows some sign of stability as regards his attitude towards scientific realism. After all, he is saying something he would still be saying long after these years—that is, that some kinds of questions cannot be properly formulated. Kuhn is referring, in particular, to those questions related to the theory-independent "reality" of those theoretical concepts like "electron," which play a key role in our more cherished theories. Reality—and its electrons—is already granted in the kind of (ontological, semantical, cognitive) setting in which a scientific community performs its practice.

6.5 The Open Texture of Theoretical Terms

Making sense of paradigms and answering the questions related to the concept of nature and progress that Kuhn propounds in *Structure* are not the only targets of his new research on the themes of that book during these years (roughly, 1962–1965). From May 1963, at least, and for the rest of the decade, Kuhn returned to an old, pre-*Structure* theme, which had not a real presence in the book. Despite that, that theme traverses Kuhn's thinking from the late 1940s to his very last days. It has a significant presence in the last two Lowell lectures—it also affects the general perspective he provides in the rest of them—and underlies his vision of learning, language, and testing as it appears in several sections of *Structure*. Later on, Kuhn's key texts such as "Second Thoughts on Paradigms," *The Presence of Past Science*, and *The Plurality of Worlds* are based on it. During those years—more particularly, from 1963 to 1969, if we reduce the span to the written texts he has left—Kuhn took that theme up, improved the view he had developed for the Lowell Lectures, and paved the way for later thoughts on the vision. That theme helps us to understand the general viewpoint that underlies his view on science, of which *Structure* is, so to speak, an early sample, and that is progressively improved and detailed up to the mid-1990s.

---

[126] See Footnote 125.

The theme I am talking about is his view on language; both of the scientific language and of ordinary language. Probably impressed by the attention that philosophers at Berkeley paid to the ordinary-language school in philosophy—Cavell, his close friend, is probably the best example of this—Kuhn improved his criticism of the logical empiricism's views on meaning partly on its basis. Concerning his interest in ordinary-language philosophy, Kuhn talked about it in a long (and very interesting) self-biographical letter to the members of the Princeton Institute for Advanced Study (more on this letter below, Chap. 7). Kuhn says there that he developed some aspects of his paper, "A Function for Thought Experiments," thanks to the philosophers' discussion of the ordinary-language philosophy (and especially, of J. L. Austin's work) at Berkeley; Cavell's work is mentioned. In fact, an ingredient in Kuhn's perspective—the idea of an "empty space" between natural families—has a parallelism, he says, in that school, and in Austin's perspective in particular. I shall return to this. As far as that self-confessed influence is as old as the earliest draft of that paper ("late 1958," he says), we can situate it at least as early as 1958.[127]

His own views on language are much earlier, however—they come from the Lowell Lectures, as we saw in Chap. 3, especially from lectures VII–VIII. At that point in his development, the influence of the ordinary-language school of philosophy was almost nil. He had probably read Wittgenstein's drafts (see our Chap. 2), but that was all. However, there are points in common that he must have taken up again in order to make his position already well-grounded in the early 1960s, right after *Structure*.

Of course, his point of departure is the most finished version of scientific language as it appears in *Structure*. The main activity of the scientist is puzzle-solving, and this kind of practice does not need the involvement of rules. The scientist, that is, does not need to take rules on methods into account. The scientist's approach to unsolved problems are *solved* problems. They guide research in the absence of rules, which are often invoked only in the case that this focus on analogy is submitted to doubt.

How to learn the practice of science on that basis is expounded in Section V of the book, but he does not particularly refer to language-learning at that point. This latter subject, however, remains a key theme for Kuhn from late 1949, and it is the basis on which the role of paradigms (as exemplars) is properly explained. During the years after *Structure* on which I am focusing in this, the following sections, and the next

---

[127] See Kuhn's letter to the members of the IAS Anthropology Seminar, 5 May 1976, TSKP 5.16; expressions within quotation marks are located on pp. 1, 2, respectively, of that letter. Indeed, the influence of that school on the preparation of that paper can be found in its drafts. One of them, "The Status of Thought Experiments," is undated, but it seems to be the original one (so it is probably of 1958). It reflects that influence on pp. 28–29 (TSKP 25.14); similarly, another (handwritten and much shorter) version of that paper, titled "A Function for (Some) Thought experiments," dated 15 May 1963 (TSKP 3.13) says the same on p. 5, the last one (though it does not mention that school explicitly). Feyerabend vehemently criticized that connection with ordinary-language philosophy, as Hoyningen-Huene's edition of Feyerabend's letters to Kuhn show (see Hoyningen-Huene 2006a, 611–612 and Appendix 2, 629–631). As Hoyningen-Huene (2006a, 611) notes, Kuhn did not include that connection with the ordinary-language school in the final essay. I shall return to that connection later, in Chap. 7. A new edition of such letters is available in Collodel and Oberheim (eds.) (2024, Part III, see esp. 261–263).

chapter (1963–1970), Kuhn tries to show how we acquire a language that categorizes the world either from scratch or from the straight indication of an instructor. Whether on the basis of the linguist that approaches a different culture (an example that is key for Kuhn's own developmental and historical approach to philosophy of science), or on the basis of the child who is being instructed about ordinary-language terms, Kuhn tries to cover the cases in which a given individual acquires the main (kind) terms in use for a given community of speakers—scientific or not.

During these years, he pursues the key thesis that, in doing so, common-use terms in natural and scientific languages do not need definitions and that vagueness of meaning surrounding them—a property he assumes as central in the Lowell Lectures—is not an undesirable consequence of natural languages. Although Kuhn no longer supports terms such as "vagueness," he still fights for the old idea that some fringe of vague meaning is part and parcel of any functional vehicle of communication. His way of approaching this issue is slightly different, however, as we shall see. Besides that, it involves a mathematical model that, years later, would lead him to try to simulate in a computer how language-acquisition processes, and thereby world-acquisition, take place. However, as he said in 1967, in lectures delivered at the Rockefeller University, that model and computer simulation do not aim to truthfully represent real processes—they only aim to understand them a bit better.[128] Kuhn does not seem to have achieved ultimately publishable results concerning that simulation, as is well known.

Two sets of notes from this period show Kuhn's efforts for coming up with a fresh model of language. I shall arrange these two sets thematically. There is no temporal order for them because one of those sets shows no date of composition. In contrast, the other one is clearly dated, and the research notes it contains were composed from May 1963 to July 1965. I shall not include in them Kuhn's notes for his lectures at the Rockefeller University in 1966–1967.[129] Although these latter include ideas that are related to the latter set, they do so as part of the main goal of the lectures—namely, a full seminar on his main theses in *Structure.*

The former, undated set belongs to a folder in Kuhn's repository titled "Family Resemblance, 1965," which includes both groups of documents.[130] The dates of its documents span a period from 1963 to 1965, but it is difficult to say whether all of them belong to this period of time.[131] Because both sets are related on the basis of some crossed references, it would be reasonable to assume that they were composed

[128] See Rockefeller 4, 9; Rockefeller 5, 6.

[129] See esp. Rockefeller 4–6.

[130] See FR65. Not all the relevant documents can be found there. One of them, pretty significant for our purposes and dated, is located at T2E, 20 July 1965 (3 pp.).

[131] There is an important exception: a dated letter. A copy of Kuhn's letter to Kubie of 17 March 1955, which may also be found at TSKP 25.53, is included here (FR65). It is obviously almost a decade earlier than the period 1963–1965. This is evidently a source of insight for Kuhn on some themes related to the psychological nature of scientific reasoning (in particular, to creativity and intuition). He must have recovered it from his records in order to include it here because of its thematic importance. So, it is not out the question that Kuhn had composed some of the undated texts at a previous date, that is, before 1963. More on this letter in Chap. 7.

more or less at the same time. Yet, I shall be cautious and leave the issue about the date open. After all, it does not affect the content of my discussion or the document's relevance.

The undated set includes a significant manuscript that is, at the same time, a good example of the latter situation. It is a series of handwritten notes and an unfinished typescript titled "Cognitive Functions of 'Open Texture.'" This is a relatively long manuscript (19 pp.) in which Kuhn introduces his view on the "open texture" of the meaning of theoretical terms in scientific theories with R. B. Braithwaite's model—in his book, *Scientific Explanation*—as a contrasting view.[132] Kuhn's unfinished paper expounds Braithwaite's model (or "paradigm," as Kuhn calls it) for defining—in Braithwaite's words—"theoretical terms by means of the observable properties."[133] Kuhn shows that there is room for the open texture of theoretical terms in this model. As Braithwaite says, in order for theoretical hypotheses to *explain* rather than being mere "translations of the empirical generalizations," the theoretical terms within them must say more about the world than the content that the observational terms convey.[134] As that excess content is not determined, an explanatory theory in Braithwaite's model would involve an open texture of meaning for theoretical terms, Kuhn says.[135] Once he introduces Braithwaite's model in this way, he devotes the rest of the (unfinished) draft to showing "the indispensability of open texture to both science and language."[136] Because of its significance, I shall return to this text below. In the same set, there is a group of handwritten notes with the same title that complements this draft. These notes go on with the argument from the point Kuhn left it, though they are exploratory rather than conclusive.[137] He also devoted a few notes to Friedrich Waismann's "Verifiability," which is quite understandable insofar as the expression "open texture" first appeared in that paper. Finally, two more groups of notes are included in that set.[138]

The dated set is quite different. It includes two handwritten sets of notes of lesser extension, all of them composed in May 1963 (that is, fairly early) and with similar titles: "The Logic of 'Family Likeness'" and "Logic of Family Resemblance."[139] They were written during the same month (and, in one case, even on the same day) in which he also took some of the notes on *Structure* that we saw in our Sect. 6.3—30–31 May and 1 June, 1963. In what follows, I provide a reconstructive interpretation of these two sets of notes and drafts, dated and undated. In order to contextualize

---

[132] Braithwaite (1953).

[133] Braithwaite (1953, Chs. II–IV, esp. 62 ss.). See also Kuhn, "Cognitive Functions of 'Open Texture,'" n. d., FR65, 3, 5, 12–13, for the term "paradigm" as assigned to Braithwaite's model.

[134] Braithwaite (1953, 68).

[135] See [Kuhn], "Cognitive Functions of 'Open Texture,'" n. d., FR65, 5 ss.

[136] [Kuhn], "Cognitive Functions of 'Open Texture,'" n. d., FR65, 14.

[137] See [Kuhn], "Cog[nitive] Func[tions] of Open Texture," n. d., FR65.

[138] See the notes under the headings "Some Empirical Functions of Metaphysical Concepts," and "Open Texture in a Reconstructed Natural Language," both of which undated, FR65. The first of them is, in fact, a heterogeneous group of notes. See also Waismann (1945, 121 ss.).

[139] [Kuhn], "The Logic of 'Family Likeness,'" 16 May 1963; "Logic of Family Resemblance," 31 May–1 June 1963, both in FR65.

them, let us recall that they were written between Copenhagen and Berkeley, and began right after Gillispie's unofficial proposal for Kuhn to go to Princeton. At that point, Kuhn's mind was busy with a number of quite varied issues—from his work on the sources of the quantum theory to his potential departure from Berkeley.

In this section, I examine Kuhn's unfinished manuscript "Cognitive Functions of 'Open Texture'" a little more carefully. Before proceeding, however, I would like to say something about the difficulties in dating it with certainty. The unfinished draft (plus its companion notes) is devoted to the notion of open texture, its significance for the joint study of ordinary and natural languages, and its relationship to the associated notions of natural family and family resemblance.[140] The group of documents to which this one belongs spans from 1963 to 1965, so those dates would also show the period in which the draft was probably composed. In fact, in the "Logic of Family Resemblance" (May 30 to June 1, 1963) Kuhn returns to the idea of open texture as studied in that paper.[141] Yet, these latter notes do not refer to the draft itself. Besides that, the draft's vocabulary is not unequivocally based on *Structure*'s main concepts. Kuhn keeps referring to theories and problems, for example; although he applies the term "paradigm," he mainly uses that word to refer to Braithwaite's account of the meaning of theoretical terms, which is not Kuhn's classic usage in *Structure*—rather, it seems like the usage he exhibits in the 1959 drafts of the first chapter (see Chap. 5 for details).[142] Admittedly, Kuhn might be stretching the meaning of that term, but it is not out of the question that he was just mentioning in his notes from 1963 something he had drafted long before. Indeed, as we shall soon appreciate, his account of open texture is so close to the themes discussed (and the view he conveys) in the two last Lowell Lectures—especially the eighth one—that "Cognitive Functions of 'Open Texture'" might have been written only a few years later than those (though not before 1953 because of his discussion of Braithwaite's book). Let us not forget, in addition, that Braithwaite's *Scientific Explanation* is mentioned in a footnote in *Structure*, and that, at that point, he refers to the chapter he discusses in the draft.[143]

To sum up, Kuhn's draft on the "Cognitive Functions of 'Open Texture'" is hard to date exactly. All things considered, the evidence does not add up to anything

---

140 For all of that, see the documents in FR65.

141 See [Kuhn], "Logic of Family Resemblance," 31 May–1 June 1963, FR65, 8.

142 See also [Kuhn], "Cognitive Functions of 'Open Texture,'" n. d., FR65, 5, 12–13.

143 *Structure*, 100, n. 4. Kuhn also mentions Braithwaite's book in an endnote to the 1959 version of his paper on measurement, and then in the published version; see Kuhn, "The Role of Measurement in the Development of Natural Science," n. d. (ca. 1959), TSKP 23.6, 44, n. 2; Kuhn (1961b, 282, n. 4). In *Structure* (100), Kuhn refers specifically to Braithwaite (1953, Ch. III), esp. to p. 76, also mentioned in Hempel (1958, 204–205)—a paper highly relevant for the theme that Kuhn discusses in his draft, which also mentions Hempel himself, though not the paper itself. (See [Kuhn], "Cognitive Functions of 'Open Texture,'" n. d., FR65, 10, 12.) We know that, in October 31, 1965, Kuhn had started to read *Aspects of Scientific Explanation* (Hempel 1965), because he mentions its first chapter—as "the first essay in Peter's book"—in "Paradigm Model Notes" (31 October 1965, FR65, 1). However, though *Aspects* also contains Hempel's paper (1958), Kuhn might have read it before. In short, it seems that Kuhn had been working with Braithwaite's book (and perhaps with Hempel's paper) around the late 1959s, before writing *Structure*. Although these facts are worth to mention, they are not definitive.

definitive. I shall not speculate more, because I do not need to, either. Its relevance for the work of the period we are currently examining is clear and remains intact even though we do not know its exact composition date. Whether it was written in the period 1963–1965 or before, it bridges the gap between the two last Lowell Lectures and the ideas on natural families and family resemblance that he pursues at this later moment. It is, furthermore, sufficient evidence that some patterns and interests remained in Kuhn's mind from the outset until his later years, long after *Structure*, and that they delimit the kind of themes that accompany and, to some extent, underlie his discussion of paradigms and revolutions. I shall return to this later, both in this chapter and in the Epilogue of this book.

Let us return to the draft itself. Kuhn's plans for the draft were ambitious. He aimed to show that the notion of open texture is important in our account of scientific language. The way novelty is negotiated in that language requires making room in the definition of theoretical terms for a number of descriptions of perceptual situations that have not been experienced so far, but that might be part of our experience in the future if we wish to apply the term at that point. Closing—instead of opening—the texture of a term is something that the logical construction of the theoretical terms in a scientific language promotes, and that Kuhn suggests should be avoided.[144] His opinion in that sense was not, after all, very far from Braithwaite's, for whom previous attempts to define theoretical terms only by virtue of its observational consequences were unsuitable for a language intended to remain applicable on the basis of incoming evidence and novelty.[145] For Kuhn, Braithwaite's argument about the "arbitrariness" involved in the definition of theoretical terms in a language reconstructed on the basis of a calculus was an accurate representation of the open texture of concepts.[146] Once he introduced Braithwaite's reconstruction in some detail (Section I of his draft), Kuhn said that "Many different meanings" of the theoretical terms "are compatible with their role in the theory [...], and it is this flexibility or indefiniteness of meaning which constitutes the open texture of the corresponding concepts."[147]

In order to illustrate this point, it is useful to describe Kuhn's own analysis in the last pages of his draft.[148] This analysis also plays a role in further arguments in the next section, so this summary will be useful later. In his analysis, a phenomenological definition of a natural-kind term like "man" would be analyzable like this[149]:

---

[144] See [Kuhn], "Cognitive Functions of 'Open Texture,'" n. d., FR65, 10, 13, 15–18. See also pp. 12–13 for its application to ordinary language.

[145] See Braithwaite (1953, 68, 71, 75–76). See also Braithwaite (1953, 52–53) on this—and, allegedly, also on Ramsey's—position against Bertrand Russell's. For a critique of Braithwaite's interpretation of Ramsey's views in that regard, see Demopoulos (2013, 156–158).

[146] See Braithwaite (1953, 60–68; see p. 66 for the quoted word). Braithwaite does not mention open texture explicitly.

[147] See [Kuhn], "Cognitive Functions of 'Open Texture,'" n. d., FR65, 5.

[148] See [Kuhn], "Cognitive Functions of 'Open Texture,'" n. d., FR65, 14–18. My following reconstruction comes from those pages. Formulae (6.1)–(6.3) have been slightly modified here.

[149] I use Kuhn's notation in what follows. Formulae (6.1)–(6.3) can be found in [Kuhn], "Cognitive Functions of 'Open Texture,'" n. d., FR65, 15, 17–18 as formulae (6), (4), and (7), respectively. Here, I have reproduced them almost literally—they have been slightly modified.

$$\text{Man} =_{\text{df}} p_1 \cup p_2 \cup p_3 \cup \ldots \cup p_n \tag{6.1}$$

In this example, each p_i is a "particular percept," he says, to be found in the "perceptual world" of an observer. The perceptual world is likewise composed of a series of "perceptual elements," s_j, which are "phenomenologically defined"[150] and that in turn help to describe the p_i—that is, each "particular percept"—as follows[151]:

$$p_i = s_1 s_2 s_3 \ldots s_m \tag{6.2}$$

Each particular percept p_i is described by a *selection* (the conjunction (6.2)) of several s_j. Any particular s_j that does not appear in the selection is, on that account, a perceptual element that does *not* contribute to the particular percept in question.

So, (6.1) is the kind of definition that, in Kuhn's view, would be sought in order to achieve an "idealized natural language"—or, in other words, "a symbolic reconstruction of language by an idealization."[152] For Kuhn, such a project of reconstruction is "so unrealistic that [it is] probably useless for any general discussion of the properties of a natural language".[153] Yet, he uses it here in order to show how open texture—which in this kind of reconstruction would be unlikely to emerge, he says—"creeps in."[154] An open-textured definition of the same term with a similar structure would be rendered, according to Kuhn, as follows:

$$\text{Man} =_{\text{df}} p_1 \cup p_2 \cup p_3 \cup \ldots \cup p_n \cup \chi \tag{6.3}$$

Here, χ stands for the as yet unnoticed but nevertheless *possible* percepts, or consequences, and is also a conjunction of them. This way, a definition of "man" like (6.3) leaves some room (χ) for unnoticed percepts.[155] A definition is nonetheless available in (6.3), but it would be flexible enough to meet future novel experience.

In short, whether in Braithwaite's model or in Kuhn's analysis, we see that, once we reduce the definition of theoretical terms to a function of only observable properties, we obtain, in Kuhn's words, "complete and explicit" definitions for those terms that suppress that "flexibility or indefiniteness of meaning" involved in partial and implicit definitions.[156] Open texture would be, for him, the solution to that problem.

---

[150] According to Kuhn, the perceptual elements "are the various colors, tastes, shapes, smells, etc. which may concur in a particular percept" ([Kuhn], "Cognitive Functions of 'Open Texture,'" n. d., FR65, 14–15).

[151] See [Kuhn], "Cognitive Functions of 'Open Texture,'" n. d., FR65, 14–15, for the quoted expressions in this paragraph.

[152] [Kuhn], "Cognitive Functions of 'Open Texture,'" n. d., FR65, 14.

[153] [Kuhn], "Cognitive Functions of 'Open Texture,'" n. d., FR65, 16.

[154] See Footnote 153.

[155] See [Kuhn], "Cognitive Functions of 'Open Texture,'" n. d., FR65, 18.

[156] See [Kuhn], "Cognitive Functions of 'Open Texture,'" n. d., FR65, 4–5. See also Braithwaite (1953, 31, 54–55) for the distinction between observable and theoretical properties for which the corresponding types of terms stand.

Yet, he is clear that we must not associate open texture to a matter of vagueness or descriptive incompleteness.[157] Kuhn states that

> [t]he indefiniteness of meaning [...][158] is open texture. It has nothing to [do] with the vagueness or incompleteness of descriptions, phenomenological or otherwise, though vagueness, etc., may increase its extent. It is rather a linguistic expression of the freedom, permitted by the very lacunae of experience, to alter empirical definitions at will without affecting the pragmatic functions of the language in which the definitions are embodied.[159]

However, open texture is a concept that, in Kuhn's view, has some limits. In 1965, in a letter to philosopher Theodore Mischel (from Yale), he wrote:

> I do not doubt that many of our concepts have, in fact, a certain "openness" of texture, but I also suspect that the manner in which that openness is now used in philosophical discourse disguises problems and will prove pernicious. I should like to restrict the term "open texture" to those areas in which we are prepared to articulate our concepts in the light of further experiences. There are, however, many other areas in which we shall have to modify concepts if experience so demands, but in which our whole present method of employing our concepts as well as our success in so employing them commits us to the view that there will be no such experiences. Concepts are, I think, "open texture" with respect to only a small proportion of the experiences to which they have not previously been asked to apply. That does not, however, mean that we are free to handle the residual areas merely by extending rules when the time comes. The discussion of x[-]rays and of new planetary discoveries in my book is intended to suggest this point.[160]

So, for Kuhn, open texture is relative to a *limited* context of possible experiences that does not embrace *all* our possible experiences. It does not include, for example, the kind of experiences that we may have after a scientific revolution. The way terms are applied after a revolution is no longer related to their open texture.

Despite these considerations about open texture, a few years later, in his "Logic of Discovery or Psychology of Research?", Kuhn refrained from using the term "open texture" for representing his own thoughts on meaning and felt it was as "askew" as the term "vagueness of meaning" itself—which we may nevertheless remember as valuable for him in the years 1949–1951. He sets out the view that "incompleteness of definitions" does not involve there being anything "wrong with the meanings. That is the way meanings behave!", he says at that point.[161] Kuhn is not confused when he uses those terms in such an apparently incoherent manner. Rather, his thinking about the nature of the meaning of theoretical terms and its relationship with their definitions is evolving during those years. This is not a charitable interpretation on my part. Kuhn does not need any help here. His main thesis remains clear. Definitions

[157] See [Kuhn], "Cognitive Functions of 'Open Texture,'" n. d., FR65, 16.

[158] Kuhn associates this here with an example of a definition with open texture. I shall examine the case below.

[159] [Kuhn], "Cognitive Functions of 'Open Texture,'" n. d., FR65, 18.

[160] Kuhn to Mischel, 7 May 1965, TSKP 4.12, 1. His correspondence with Mischel followed Kuhn's visit to Yale in February 1965. In March, Mischel sent him a copy of a draft (together with some reprints of other papers), and Kuhn's remarks on it started an interesting exchange. I examine Kuhn's talk at Yale in the next chapter, Sect. 7.2.

[161] See our Chap. 3. See Kuhn (1970a, 19, fn. 1; also 1974, 316–318, fn. 21).

of theoretical terms must be open-textured; they should not be closed around a set of descriptions that are only based on our extant common experience. Yet, in later years he preferred not to associate terms such as "open texture" with "meaning" in order to avoid the view that there is an "imperfection" involved in the latter.[162] Kuhn is clear that the meaning of theoretical terms in science is not related to the existence of a certain list of properties, whether incomplete, imperfect, partial, open-textured, or not.

Insofar as Kuhn read Waismann's 1945 article—which he did—he must have known that vagueness and open texture are two different phenomena.[163] Vagueness is associated with the difficulties in applying a term such as "heap" or "bald," whereas "gold" or "man" are not vague in that sense, though they remain open-textured.[164] When it comes to those terms in whose definition empirical descriptions are involved—mathematical terms are quite another thing, because they are closed-textured—we can "foresee" neither *every* possible situation in which we can find a future instance of that kind, nor *every* possible but unnoticed feature or relationship.[165] That is why, for Waismann, verification of statements is too weak a method for granting meaning to a term. If meaning is reduced to the descriptions that condense our *extant* experience, its applicability (and so meaning itself) is utterly limited. Shall I then apply the term "man" to an individual that, despite fitting other descriptions for men, "is only one span tall"? That is one of Waismann's examples; another is: Shall I apply "gold" to a certain sample of a substance that "satisfied all the chemical tests for gold, whilst it emitted a new sort of radiation"? Even if we wish to answer "Yes!" to those questions, if we reduce meaning to a summary of descriptions only based on previous experience, such terms will not be applicable to the new, "unforeseeable" situations.[166]

For Kuhn, both Braithwaite and Waismann successfully show this property of terms in our languages—whether natural or scientific.[167] That is probably why he is careful to say that it is open texture, not vagueness, that he is supporting in that draft. In the draft, however, he also criticizes his allies. For him, neither of them "illustrate genuinely open texture at all"; he also includes Hempel in that group.[168] These philosophers criticize the empiricists' search for complete, explicit, and closed-texture definitions, and for Kuhn this is a project that is worth its salt. Yet, the exploration is incomplete, because, in a certain sense, even closed texture would be positive, he shows. Let us see how.

For Kuhn, Waismann, Braithwaite, and Hempel help us to discover the importance of *empirical* open texture alone, which is a good result even in metaphilosophical

[162] Kuhn (1974, 316, fn. 21).

[163] Kuhn's notes "Problems about Waismann's Paper" (n. d., FR65), referred to Waismann (1945), testify to that reading.

[164] Waismann (1945, 123).

[165] See Waismann (1945, 126).

[166] See Waismann (1945, 122–123).

[167] See [Kuhn], "Problems about Waismann's Paper", n. d., FR65, 1.

[168] [Kuhn], "Cognitive Functions of 'Open Texture,'" n. d., FR65, 12.

terms. For him, the presence of explicit definitions is just "excess baggage."[169] On the basis of Braithwaite's model, he says that the presence of such explicit definitions

> makes the theory conform with an empiricist's preconceived epistemology, *but that is all it does!* Open texture is, at the least, more economical than closed. Therefore, if Occam's razor is as applicable to formalisms as to forms, the empiricist's demands may be excessive.[170]

Yet, for Kuhn, *empirical* texture-openness may be simultaneous with a sort of *metaphysical texture-closure*, which, for him, is a good option. With metaphysical texture-openness/closure Kuhn refers to the range of *arbitrary* delimitations of the extension of a given theoretical term that we saw above. What is included in their extension (when they do not fully overlap with observational properties) and what is not help us to provide a delimitation of the concept, that is, a clear definition that, in that case, not only observational properties limit. The basis for that delimitation—that is, for drawing the semantic boundary around the theoretical terms—are different sources, from "taste," he says, to metaphysical commitments like those of the atomistic viewpoint about elementary matter.[171] Metaphysical commitments like this one, he says,

> impinge on experience and are therefore partially bounded by it, but [...] are also in part free constructions of the speculative imagination. Such choices are free with respect to the particular problem facing the scientist—they do not, that is, affect the applicability of his theory [...] to his data [...]. Nevertheless they may have an immense importance in his scientific work, for [...] it is precisely these free choices, governed primarily by taste and metaphysics, that relate one scientific theory to another and guide the scientist in his choice of problems and of experimental techniques. The scientist need not draw the boundaries, and he frequently does not draw them, at least in their entirety. But he has the option, and he often utilizes it fruitfully.[172]

Evidently, Kuhn is bringing the debate on open texture to his home ground, where he has some things to say. Many of these things resound in our ears as a melody that he has repeated since the early 1950s. Our commitment with the language that we have learned—whether natural or scientific—is also a commitment with a world in its entirety.[173] Kuhn had already talked about this in the Lowell Lectures and then, famously, in *Structure*.[174] Of course, that commitment is not with series of observations alone, but rather with the entities, substances and relationships that inhabit that world; with the behaviors in which we expect they are (and will be) involved; in short, with the ontological, not only the semantic aspects of our theories. It is, so to speak, a commitment with a world of scientific objects, their natural (i.e. physical) possibilities, and also with their "impossibilities"—that is, with those merely *possible* events that we expect not to happen (something that, in Kuhn's mind, is as old as his 1945 term paper on causality; see our Chap. 1).

---

[169] [Kuhn], "Cognitive Functions of 'Open Texture,'" n. d., FR65, 8.

[170] See Footnote 169.

[171] [Kuhn], "Cognitive Functions of 'Open Texture,'" n. d., FR65, 10–11.

[172] [Kuhn], "Cognitive Functions of 'Open Texture,'" n. d., FR65, 11.

[173] Kuhn mentions this thesis in "Cognitive Functions of 'Open Texture,'" n. d., FR65, 19.

[174] See Lowell 8, 35; *Structure*, Sect. X; and our Chaps. 3 and 5.

6.6 Family Resemblance: Simulating Learning

Kuhn's work on the open/closed texture of the meaning of theoretical terms is simultaneous with a bigger problem that frames the former: How is that world learned and transmitted without the aid of explicit and complete definitions? As noted in *Structure*, most of our terms in science are not only open-textured but, still worse, form "natural families"—they apply to classes of objects that remain together on the basis of "family resemblances."[175] If that is the case—that is, if definitions are, like generalizations, aids in the process of acquiring a language—if, in addition, they are often useless as compared to successful applications of an existing solved problem, and if they are only partial, implicit, and open-textured, then the process by which language is acquired—and the world (and its parts) is properly understood—must be something else entirely. It does not depend on rules and definitions. It must follow another pattern, so to speak.

In studying family resemblances, Kuhn attempted a different pathway for the first time. During his last days in Berkeley and his early years at Princeton, Kuhn tried to produce a mathematical model of (and later to simulate in a computer) the process by which natural-family terms (and thereby our knowledge of natural families themselves) are acquired and transmitted without resource to definitions and on the basis of inputs of series of qualities—which, in the end, are inputs of descriptions of properties of those objects that fall under a kind term.

At this point, the series of *dated* handwritten manuscripts that I referred to in the previous section are of particular importance. They help us to see *when* Kuhn got this project started. In the notes "The Logic of 'Family Likeness'" and "Logic of Family Resemblances" he repeatedly attempts to define a mathematical model that shows how natural families are created on the basis of a series of inputs of qualities; how, in addition, our vocabulary of kind terms is created on the basis of a model of the way groupings are made without the contribution of explicit definitions. In those notes, it is easy to see that Kuhn starts his model again and again, and that he progressively tries to clarify its basis.[176] Between 1963 and the early 1970s, Kuhn explored that model and its requirements. The two "Logic" manuscripts from May 1963, show his work in progress in that regard. A few new (and very brief) notes are from July and October 1965, and they show that Kuhn was devoting his thinking to that model.[177] In 1966–1967, he publicly presented its basis in the seminar he gave at the Rockefeller University.[178] Finally, he also mentioned it in the "Postscript—1969" to the second edition of *Structure*, and described it in a well-known paragraph of "Second Thoughts

[175] See *Structure*, 45; Wittgenstein (1953, §§66–77, esp. §67).

[176] See [Kuhn], "The Logic of 'Family Likeness,'" 16 May 1963; "Logic of Family Resemblance," 31 May–1 June 1963, FR65. I am using "qualities" as Kuhn does in such notes.

[177] See T2E, 20 July 1965; [Kuhn], "Paradigm Model Notes," 31 October 1965, FR65. A related document is also important, though it does not explicitly talk about this model: [Kuhn], "Natural Resemblance," 3 July 1965, FR65.

[178] See Rockefeller 4–6.

on Paradigms."[179] Although thinking about open texture remained important to him in this period, this model seems to have attracted most of his attention at this time.

In order to describe the model, let us start by considering an abstract presentation of a universe of qualities and objects. An initial, useful arrangement would be very similar to the basis of his own analysis of open texture above. Let us imagine that there is a number, N, of qualities in a universe of objects. Objects are the particular percepts we saw previously, and qualities are the perceptual features. Recall also that we may use a finite number of qualities in order to describe any particular percept. The percept, an object, is described by means of a simple perceptual language on the basis of those qualities that are its perceptual features. Generally speaking, each object in that universe is the result of the addition of qualities. Using the idea of a N-dimensional cube (a hypercube), he states that an object is a *vertex* on such a cube, "so there are 2^N vertices" in that universe, he says, and each of them "is a potential or possible object."[180] The location of such an object on the cube depends on the qualities that are associated with it.

Of course, in such a model, the role of complete (closed-textured) definitions is almost non-existent. Kuhn does not look for them. On the contrary, his main goal is to find out how such a universe of objects, qualities, and natural families can be learned and transmitted through language, ostension and examples, and without definitions.[181] In order to construct the model, we need inputs of strings of qualities. Each of these strings amounts to an object with certain qualities that is located in a vertex. On the basis of such series of strings (or objects, for short) we obtain "a given distrib[ution] of populated vertices."[182] We can think of a series of *potential* qualities some of which are *actual* qualities of a particular object, while some others are not. Kuhn expresses the string of qualities that characterize a given object as follows: "An object is, then, *completely* described by a conjunction like $1 \cdot \sim 2 \cdot \sim 3 \cdot 4 \cdot 5 \ldots \sim (N-1) \cdot N$ which says that it has quality 1, but not quality 2 and 3, etc., up to quality N. N is [the] maximum number of qualities."[183] Here, Kuhn symbolizes qualities with a number. From the extant possibilities, some are negated and others affirmed. These latter characterize a given object. His research pursues finding out how clusters (i.e. natural families) are formed in such a space (such a cube) only from the input of series of strings of qualities.

In order to illustrate how the basic N-cube is arranged, let us examine one of Kuhn's examples (a simple one) in Fig. 6.1. It represents a 3-dimensional cube in

179 See Kuhn, Postscript—1969, 191; Kuhn (1974, 310).

180 [Kuhn], "Logic of Family Resemblance," 30 May 1963, FR65, 1. See also Kuhn (1974, 310).

181 See Rockefeller 5, 2–6, 10; Rockefeller 6, 2.

182 [Kuhn], "Logic of Family Resemblance," 30 May 1963, FR65, 1.

183 [Kuhn], "The Logic of 'Family Likeness,'" 16 May 1963, FR65, 1. In further notes from the same day and the following, he opts for symbolizing any given vertex by "σ_i" ($i = 1, 2, \ldots N$). So, he writes that "$\langle\sigma_i\rangle = \sigma_1, \sigma_2, \sigma_3, \ldots \sigma_N$, where each σ can have only values 0 and 1" ("Logic of Family Resemblance," 30 May 1963, FR65, 1; even though this is the same day mentioned in previous references, this p. 1 comes from a second series of notes that reinitiates pagination; see also "The Logic of 'Family Likeness,'" 16 May 1963, FR65, 2). In these extracts, I have transformed some abbreviations into regular writing.

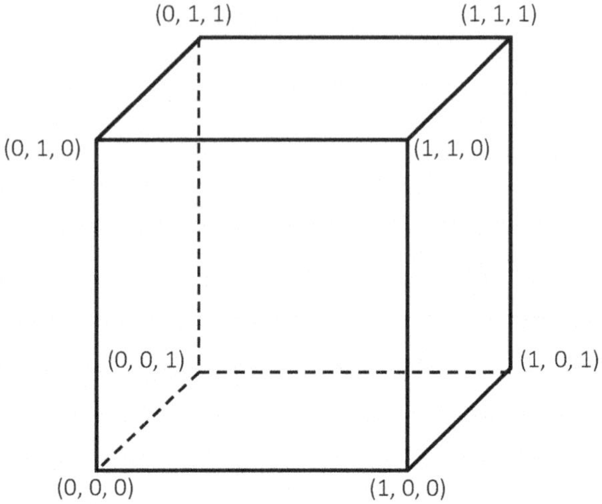

Fig. 6.1 An example of Kuhn's *N*-cube

which 8 possible objects might be located—a very limited universe.[184] Any vertex corresponds to a string of ones and zeros that represent true and false qualities, respectively. It is easy to see that strings like "0, 0, 1" or "1, 1, 1" are all that we need in order to see where to locate the corresponding object, and what qualities can and cannot be experienced of that object. Of course, "0" represents a false attribution of a quality to that object, and "1" would be a true one.

An important point for Kuhn is that there must always be "empty spaces" between natural families.[185] This is a feature that is always associated with his vision, and is present even when the model itself was obsolete in his own perspective.[186] So, he looks for a way to isolate groups of objects without the presence of definitions on the basis of some similarities between their members, and also granting a *limited* number of discrepancies. Once exceeded, these limits set the object within reach of some other group. So, objects are not only grouped on the basis of common features or qualities, but also by virtue of a maximal "distance" from the rest of their members. "Distance" is a term that, as we see below, Kuhn himself uses in order to refer to those qualities that are not shared between two objects.[187] These groupings are natural families, and their relationship between them is, of course, family resemblance (Wittgenstein's vocabulary is evident here). In Kuhn's example (Fig. 6.1), if membership to a group involves a variation of, at most, *one* quality

---

[184] For this example, see T2E, 20 July 1965.

[185] In the "Logic of Family Resemblance," 31 May 1963, FR65, 7, he refers to "empty spaces" as associated to the notion of "natural family.".

[186] See Kuhn (1974).

[187] See [Kuhn], "The Logic of 'Family Likeness,'" 16 May 1963, FR65, 1. "Distance" is later represented by symbols like "d(σ_i, σ_j)," and is of course the one between two vertices. See [Kuhn], "Logic of Family Resemblance," 31 May 1963, FR65, 5.

at a time, for example, it would be possible to form, with four objects, (a) a four-member group, (b) two two-member groups, (c) two groups with an unequal number of members (one and three members, respectively), or (d) four one-member groups; which option we select depends on the qualities of the corresponding objects.[188] In each case, at least a distance of one empty vertex would provide us with the required empty spaces.

In the notes he took during his investigations on May 30, 1963, Kuhn described his project as follows:

> Now imagine [that] we're given the cube and a given distrib[ution] of populated vertices. We want a vocabulary [...] for handling "natural clusters." [We w]ant to pick out groups of elements lying "next to" each other and distinguish it from other similar groups that are "separated" from the first by unoccupied vertices.[189]

Fourteen days before, in a similar set of notes, he had summarized this starting point more generally:

> If our qualities are conceived as forming an N dimensional space with all objects located at one of the vertices, then the maximum distance between two objects in the same family is $\sqrt{n}$.[190] Our technique is useful if in the natural world we find that objects fall into natural families such that most are near one or another cluster point and the environs are unoccupied.[191]

Kuhn also shows what the condition that produces the example of Fig. 6.1 is. In the case that—as in that example—the admitted variation is only of one quality at most, "Then we say, an object O belongs to the natural family F, if and only if, whenever O and O' are both in F, then O and O' differ in only one element of the conjunction."[192]

The problem is, of course, what that distance that separates two families *is* and how they (and that distance) would be *created*. "What proportion of vertices," he writes, "must be unoccupied if we are to get two, three, m, distinct natural families?" Similarly, he asks himself, "What is the probability that, distrib[uting] M objects at random over 2^N vertices we'll get any natural families at all?"[193] These are not the only primary questions that Kuhn knows must be answered—and that he tries to answer, or at least to formulate appropriately during those days. Questions about the qualities that must be preserved, or "essential," let us call them, as Kuhn often does, and which others are acceptable variations, or "accidental," are equally important, perhaps more. Finally, in Kuhn's viewpoint, these natural families must be able

---

[188] T2E, 20 July 1965, 1–2.

[189] [Kuhn], "Logic of Family Resemblance," 30 May 1963, FR65, 1.

[190] "For the moment," Kuhn says shortly before, "we shall say objects are alike if the conjunctions which represent them disagree at only n points" ("The Logic of 'Family Likeness,'" 16 May 1963, FR65, 1). Elsewhere, he shows that n is the "number of values of [... a vertex] that must be changed to go from" a vertex to another ("Logic of Family Resemblance," 31 May 1963, FR65, 5).

[191] [Kuhn], "The Logic of 'Family Likeness,'" 16 May 1963, FR65, 1.

[192] [Kuhn], "The Logic of 'Family Likeness,'" 16 May 1963, FR65, 1. I have slightly modified a couple of abbreviations.

[193] Both extracts come from the "Logic of Family Resemblance," 30 May 1963, FR65, 1.

to change. After all, scientific revolutions create such kinds of changes.[194] Let us examine these two latter problems in turn.

Concerning the distinction between "essential" and "accidental" qualities, Kuhn does not deny, as he writes in some notes on paradigms from 1965, that "we're programmed to look for essences, the only feasible way of handling the situation with a neural network of our limited size."[195] Despite that, he does not defend an essentialism like Aristotle's. His view favors the thesis of an average of qualities in common among the members of a given natural family.[196] However, it is the existence of a "fringe," of an "empty space" around each natural family, which really helps to divide up the collections of objects among the different natural families.[197] Kuhn writes that: "We can specify a natural family by specifying each member of it. This is, however, obviously an over-specification since it makes no use of surrounding empty space, i.e., it makes no use of the fact that what we're defining is a [natural family]."[198] In fact, for him, any member of the natural family will do in order to specify the group, though we should always bear in mind that there must be some room for variation—some qualities that will not be found in some objects.[199] Their internal relationship, after all, must be of family resemblance. However, it is also clear to him that there is something like a gradation of objects within a given natural family. There are, he says, "pure accidents" and "partial accidents," he annotates, depending on the null or slight role, respectively, that some qualities may play in specifying a natural family. "Common essences" and "discriminators" obviously play a more central role; the former help to specify qualities in common to two families, while the latter help to specify families by virtue of their differences.[200] In short, for Kuhn there is some room for the role qualities play in differentiating one natural family from another. Whether they are more or less common to a family and to the two neighboring ones is an important aspect of the question about the distance between them. It is also evident that Kuhn is confronting this problem with the possibility of comparing two groupings at the same time. In fact, the procedure of specification, the assumption of empty spaces between families included, requires such a comparative perspective.[201]

Concerning the problem of change, this second issue also involves a consideration of essentiality. This idea of "essential qualities," if acceptable at all, must be

---

[194] See these questions in "The Logic of 'Family Likeness,'" 16 May 1963, 1–2, and in "Logic of Family Resemblance," 30 May 1963, FR65, 1–2. The terms "essential" and "accidental" already appear there, though he often repeats them elsewhere, too.

[195] [Kuhn], "Paradigm Model Notes," 31 October 1965, FR65, 2.

[196] See [Kuhn], "Cog[nitive] Func[tions] of Open Texture," n. d., FR65, 3.

[197] See [Kuhn], "Logic of Family Resemblance," 31 May 1963, FR65, 8. Incidentally, referring to such space as a "fringe" (which he does on p. 8 of the previously mentioned document) reminds us of its similar usage in the Lowell Lectures. At the end of this section, I shall refer to them again.

[198] [Kuhn], "Logic of Family Resemblance," 31 May 1963, FR65, 7. In the original, the last expression within brackets ("natural family") is just "N. F.".

[199] [Kuhn], "Logic of Family Resemblance," 31 May 1963, FR65, 7–8.

[200] [Kuhn], "Logic of Family Resemblance," 31 May 1963, FR65, 10.

[201] See [Kuhn], "Logic of Family Resemblance," 31 May 1963, FR65, 8–9.

compatible with the possibility of transforming a set of natural families into another. According to this perspective, he says, "if O and O' were together in F before while O'' was in G, then afterwards I want O and O'' to fall together in F' [while] O' lies in G' or something of the sort."[202] For Kuhn, there are some interesting questions associated with some kinds of change. If the domain changes—if, for instance, the number of dimensions is greater—how does it affect the extant distribution of natural families? Besides that, will we have lost any kind of "information," he wonders, and would it be possible to recover it?[203] These are questions that are evidently related to his own work on scientific change and incommensurability in *Structure*, whose solution would depend on having a strong analytical basis for the problem of family resemblance, as he wanted.

However, to the best of my knowledge, there is no other finished work available that shows us further progress from all this groundwork. Kuhn seems to have used this work as the philosophical and mathematical background for simulating learning, and in later writings from 1967 to 1974 he told his audience that he was pursuing that kind of computer-based work. In fact, in December 1966, he told an audience in New York City that he had spent his time since the previous October learning FORTRAN. He really wanted to simulate the function and role of paradigms in learning on the basis of a computer program.[204] Meanwhile, his views on learning in "Second Thoughts on Paradigms," presented in terms of commonsense examples, was the bedrock on which he built his mature theory of scientific development. However, in 1992, he told philosopher of science Hanne Andersen by letter that he had eventually concluded that this kind of computer simulation would not help him to explore the problem about learning and teaching a certain worldview. Other available computer technology might have helped him, but he chose not to spend his time in learning it, and he preferred to continue his philosophical work. From that point on, he does not seem to have followed the progress of such a field with interest.[205]

And yet, it would be wrong to infer from all this that Kuhn's work on these themes during this period was a dead end, or even a failure. On the contrary, it was a stepping stone between the background that he offered in the years 1949–1951 and the reconstruction of *Structure* that he offered from 1975 onwards.[206] It was in fact a natural intermediate step. We saw that *Structure* had been enough for him when he published it (1961–1962) in order to explain his thoughts about scientific change. He had found out how normal science works by virtue of the presence of paradigms, and that was all he needed for that book. However, how paradigms really function was still something to explore more deeply. As we have

[202] [Kuhn], "The Logic of 'Family Likeness,'" 16 May 1963, FR65, 1. Again, I have slightly modified a couple of abbreviations. In the case of the word within the brackets ("while"), it seems to me a more likely term than the one Kuhn wrote: "which".

[203] [Kuhn], "Logic of Family Resemblance," 30 May 1963, FR65, 2.

[204] See [Kuhn], "Scientific Methods Colloquium," 4 December 1966, TSKP 3.14, 1–2.

[205] See Kuhn to Andersen, 5 March 1992, PHA. Andersen (2001, 44–46) gives an informative summary of Kuhn's efforts with this kind of computer simulation.

[206] See our Chap. 3 for the former period. I shall deal with the latter (i.e. 1975–1995) in the next two chapters and the Epilogue. Concerning the latter phase, see also Mayoral (2023b).

seen in this chapter, questions concerning how paradigms evolve without ending up founding a new view of nature, or concerning how they arrange the practice of science (from learning to research) around them, were still aspects that Kuhn had to understand better. Other aspects of paradigms required further thinking, such as the relationship between our psychological ways of thinking and the very idea of scientific rationality, something he had speculated about in a letter to psychiatrist and friend Lawrence Kubie in 1955.[207] At that point, Kuhn had been thinking about the true relationship between intuition and rationality. As we saw in his exchange with Boring, for Kuhn, the scientific mind works with aspects of problem-solutions that are only *implicit* in a paradigm. It is not evident how that work must be translated to standard reconstructions of rationality in the traditional philosophy of science. As he says in the "Postscript—1969," talking about intuitions does not involve depriving science of objectivity, or making intuitions themselves something alien to analysis.[208]

Since his days in the Society of Fellows, Kuhn had emphasized that scientific research had a distinctive historical evolution, and that the assumptions about perception and language on the part of methodology in the empiricist tradition would no longer do. So, he tried to show the role of psychology in scrutinizing the ways in which new theories emerge to provide solutions to hard problems in science. In his last year there, the Lowell Lectures allowed him to show how that process took place, and what changes the philosophy of science had to undertake in order to account for them—changes in the epistemology of science and in the philosophy of language associated to it, which involved rethinking the role psychology and linguistics play in explaining how the true language of science works and changes. But, of course, it was too soon for Kuhn to present his renewed views in public. In fact, he did not have a *finished* view in that regard. *Structure* was not definitive in that respect, either, but it showed the way to a new view of science more clearly, more specifically. So, the transitional step that we are exploring in this chapter is a natural evolution from those earlier assumptions and convictions. For him, cognition in science is, in his Lowell Lectures, different from the outlook that traditional epistemology had aimed to convey, and it paralleled the new views on logic and meaning that other perspectives in the philosophy of language had also provided. Piaget, Quine, Whorf and Wittgenstein provided him with inspiring works, and, around the time he published *Structure*, and later, those names returned to his notes.[209]

So, the work we have seen here returns to the main problems we saw in the Lowell Lectures, and to the views he had already attempted to communicate in them. For him, now and then, languages reflect a classification of the variety of objects, substances, and processes in the world in a way that fits in with our ways of experiencing it. Attached to those languages, and with only historical, contingent, validity, there are

---

[207] See Kuhn to Kubie, 17 March 1955, FR65. I return to this letter in Chap. 7.

[208] See Postscript—1969, 190–191.

[209] See [Kuhn], "Cognitive Functions of 'Open Texture,'" n. d., FR65, 6, 9, 13 for the case of Quine and Whorf. Wittgenstein is implicitly present in all his research on family resemblance, of course, and explicit in *Structure*, 45. Piaget is, of course, a recurring name in this and further work; see *Structure*, xl; *Essential Tension*, Essays 2, 10.

ontologies, metaphysical views about nature, that also condition not only our general explanations of scientific phenomena but also the scientists' ways of dealing with them—to solve the specialized problems that emerge from observing nature. Those classifications, with their metaphysical backdrop, change, and the world changes with them. Moreover, they are not introduced on the basis of definitions, and, if we talk about the meaning of their attached kind-terms, it must include not only those aspects of which every speaker is perfectly aware, but also those *vaguer* aspects of meaning that are only functional for a subgroup or a mere handful of speakers. If open texture is for Kuhn a valid phrase in the writings we have examined here, the phenomenon for which he considered it appropriate was also a phenomenon that was present in the Lowell Lectures, too, although he, of course, did not know about the term at that point. And the same can be said of questions about classifications, their historical contingency and modes of change, and about the restricted role of definitions. During these years, in the early to mid-1960s, Kuhn delved deeper into the role of, say, empty spaces between categories (or natural families) of objects and into how important the simultaneous learning of such classes is for acquiring them and the meaning of the terms that name its classes. Although his more mature vision on all that would have to wait for a few more years, the origins of that perspective—I am particularly referring to "Second Thoughts on Paradigms" as a first, most finished, presentation—trace a long way back to the Lowell Lectures and had an intermediate step in the notes we have examined here. In a way, despite the contribution of the paradigm-based view of science that Kuhn acquired between 1959 and 1961, he was working on a similar view on scientific cognition and language from the early 1950s to the late 1960s. His views after 1975 have a clear kinship with all that had been developed in those previous years, too. Kuhn, after all, never abandoned some of the basic ways of thinking about language and cognition that he had first explored in his youth.

6.7 From Princeton to London: Kuhn and Lakatos

In the summer of 1964, in August, Kuhn was ready to drive across the US from California to Princeton with his family. "[I]n lieu of vacation," he told Gillispie, "or as vacation, we mean to drive across the country with the kids."[210] They were going to live at 6 Queenston Place, in a house on which they had spent, in his words, "large sums of money," though it was not available until late November—a delay that required two temporary accommodations in between.[211] At Princeton University, he occupied an office at 70 Washington Road, close to Gillispie's.[212] Once there, he gave a seminar on the history of quantum mechanics, even though some students had asked for something arranged around the themes in *Structure*—the book that

[210] Kuhn to Gillispie, 4 August 1964, TSKP 14.11, 1.

[211] See Footnote 210.

[212] Gillispie to Kuhn, 2 July, 1964, TSKP 14.11, 1.

had made him a renowned scholar. Kuhn nevertheless opted for that historical case instead of less specific and more philosophical aspects of his work; at least, for the first semester at Princeton.[213] Concerning the project on the sources for the history of quantum physics, he and his co-workers had completed the key phase of interviews in the previous July, but he and other collaborators went on with the rest of the material under the auspices of another NSF grant.[214]

As we have seen in previous sections, Kuhn had spent the years before his return to the East Coast not only on the archival work for the quantum-physics project but also in pursuing further some aspects of *Structure*—particularly related to paradigms, learning, and progress. His communications with other philosophers such as Hanson, Hesse, Braithwaite or Cavell did not stop. Though he was willing to go on with the history of quantum physics (more on this below, Chap. 7), his concern with *Structure* and its themes had not diminished in the least. Accordingly, in November 1965, that aspect of his work figures prominently in official documents as research in progress.[215]

Perhaps more importantly, Kuhn at this point alludes to Imre Lakatos in his personal communications with Gillispie. As is well known, Lakatos is not only one of the three main figures associated with the historical school in philosophy of science, together with Kuhn and Feyerabend, he was also an important figure in Kuhn's career insofar as he brought him to his first public debate with Popper in London in 1965.[216] Let us look at this well-known episode in a little more detail. The previous year, Kuhn had already spoken about him to Gillispie as follows:

> Most important, but also most speculative is my desire to get you thinking about a man called Imre Lakatos, who has recently completed a four part series, "Proofs and Refutations," in the B.J.P.S. He's a very bright Hungarian with an appointment under Popper at L.S.E. I have met him and found him most stimulating. Also, I know he's got a first-rate reputation with Mary and the better young members of the British group. People have been preparing me for the B.J.P.S. piece for some time, and after ten pages of it (I just got it today) I think they're probably right. The opening pages, at least, are a remarkable combination of art, wit, and wisdom. However I may come to feel after finishing the piece and reflecting on it (remember, though, that I've spent some time with the author though not previously read his stuff), it's clear that he's someone to have around for a long weekend. [… H]e's very very bright. Also, […] he is to be visiting in the United States (somewhere on this coast)[217] for one semester next year, I think the fall. We really should get him around.[218]

---

[213] Gillispie to Kuhn, 2 July, 1964, TSKP 14.11, 2; Kuhn to Gillispie, 4 August 1964, TSKP 14.11, 2.

[214] See Wheeler, "Final Report of the Joint Committee of the American Physical Society and the American Philosophical Society on the History of Theoretical Physics in the Twentieth Century," 29 December 1972, TSKP 6.8, 2.

[215] See the recommendation form ("Form Z") for increasing salary, 24 November 1965, PUTK, 3–4.

[216] Kuhn himself calls it "confrontation" in a letter to Popper, though he was careful to put the word in quotation marks. See Kuhn's letter to Popper, 23 June 1965, TSKP 23.9.

[217] Kuhn was writing from Berkeley at that time.

[218] Kuhn to Gillispie, 29 May 1964, TSKP 14.11, 1–2.

Kuhn made it clear in that letter that Lakatos's training was as good as his mind, and that he really wanted to have him visit Princeton—whether officially or unofficially.[219] Indeed, at that point, the impression that Lakatos had made on him could not have been better.

That intellectual impression probably did not diminish during the following year, 1965, particularly preceding Lakatos's visit to Princeton in April 1965. Later on, however, the difficulties Kuhn found in Lakatos's arrangement of his encounter with Popper in London in July 1965 were disappointing for him, and Kuhn vented his anger on Lakatos. It was not the first time that Lakatos had suffered that kind of rage. Shortly before, he had felt the anger of two other key figures in his life: Karl Popper himself, and particularly his wife, Josephine A. Henninger, "Hennie." In that case, in April and May 1962, a misunderstanding with the Poppers during the year that they spent in California led to a hard, long letter written by Hennie Popper, full of harsh words against Lakatos. Hennie accused him of taking advantage of Popper's assistant in London, thereby damaging the progress of Popper's own research. In that letter, Hennie was indeed very harsh. She told Lakatos that he did not really care about Popper, and that he did not take responsibility for his faults.[220]

Kuhn's reaction three years later was similar to Hennie Popper's, although, evidently, for quite different reasons. Lakatos arrived in California for a stay of nine months at the University of California, San Diego, which must have begun in October, 1964. Already in La Jolla, he met Gillispie there. Around that time, Kuhn extended an invitation to Lakatos to present a paper at Princeton University (whose final date was April),[221] which the latter considered would be about his already famous four-part paper on the philosophy of mathematics, or on his more recent work on Popper versus Kuhn. Additionally, Lakatos had already told Kuhn, in May 1964, about the forthcoming Colloquium at Bedford College, London, the following summer. In October, Lakatos gave Kuhn further details and attached a formal letter inviting him to the conference.[222]

The conference initially had four sections, one of which was going to have Kuhn and Lakatos as the main speakers and Popper as the chairman.[223] Kuhn mentions

---

[219] See Footnote 218.

[220] See Hennie Popper to Lakatos, 9 May 1962, esp. 1, 3. See also Karl Popper to Lakatos, 17 April 1962; Lakatos to Karl Popper, 27 April 1962; Lakatos to Hennie Popper, 17 May 1962; and Hennie Popper to Lakatos, 22 May 1962; all of them in KPP 317.17. These letters show that Popper was suffering a heart condition while at Berkeley, and that he used to work hard in his research anyway. He needed help in the UK, and, seemingly, Lakatos was an obstacle in that sense—that was, at least, Hennie's version.

[221] See Kuhn to Lakatos, 23 June 1965, TSKP 23.9, 1. (Any of the two versions of this letter will do. This applies to my further references to the common parts of those letters.).

[222] See Lakatos' letter to Kuhn, 28 October 1964, TSKP 23.9. As noted, there are two letters—one formal and one less formal—with the same date. Facts included in this paragraph have been, all of them, obtained from those letters. See also Kuhn's letters to Lakatos, 6 November 1964 and 23 June 1965. In the latter, Kuhn says that Lakatos earliest invitation was in May 1964. Both letters can be found in TSKP 23.9.

[223] Lakatos to Kuhn (both versions), 28 October 1964, TSKP 23.9. See also Gattei (2008, 39–42), a good previous description of the changes in the conference.

that Feyerabend would be invited to the session.[224] Kuhn was going to receive a partial draft of Lakatos's paper around March 1965. It would summarize Popper's position and those of his critics—including Kuhn's views—so that he could prepare his own.[225] In November 1964, Kuhn agreed with that arrangement.[226] It is relevant that Lakatos's visit to Princeton in April 1965 was not as successful as one may have expected. In fact, in their later exchange, they both refer to some problem with a watch. I shall not speculate about it. Suffice it to say that it left a bad taste in the mouth for both, and that Kuhn brought it up later.[227]

The main problem arose with the change of plans that Lakatos communicated to Kuhn on June 18, 1965. Insofar as their joint session for the conference would take place on July 13, at that moment there were only a few weeks left (actually, only three)[228] for a change of that sort, which was unacceptable for Kuhn. Unfortunately, that was in fact what happened. Kuhn wrote a reply to Lakatos that was only slightly less harsh than Hennie Popper's letter. The main problem with the former is that Kuhn opted to make it a little more public. In fact, he wrote two long letters. One of them was submitted to Popper, William Kneale, Rupert and Marie (Boas) Hall, Mary Hesse, and Lakatos, of course. The first four copies were prefaced by a letter commenting the reasons for submitting the copy.[229] The other was different—an extended version of the former in which some more comments were included, and the watch incident re-emerged. This second version had a postscript addressed to Lakatos, which Kuhn ended by saying: "I have admired your paper and en[j]oyed our discourse too much to want to leave our relationship in this state. I very much hope […] that we shall succeed in reestablishing it. But the next steps will have to come from you."[230]

But what had gone so wrong, after all? Kuhn summarizes the steps taken so far in his reply to Lakatos. The previous April, Feyerabend, Lakatos and Popper had been confirmed—Popper being the chairman. Moreover, there would be two, and maybe three, sessions during the whole day: one in the morning, with the three main speakers, and one, perhaps two, more sessions devoted to discussion during the afternoon and the evening. Yet, in the interim, Popper, Feyerabend and Lakatos had dropped out of the program, and Kuhn's paper would be discussed in "a single three-hour session isolated from the few other subjects on the program that are even tangentially relevant to it," he said.[231] His paper would be discussed by John Watkins—as was finally the

---

[224] Kuhn to Lakatos, 23 June 1965, TSKP 23.9, 1.

[225] Lakatos to Kuhn (informal version), 28 October 1964, TSKP 23.9.

[226] Kuhn to Lakatos, 6 November 1965, TSKP 23.9.

[227] See Lakatos to Kuhn, 18 June 1965, and Kuhn to Lakatos, 23 June 1965, 2 (postscript), both in TSKP 23.9.

[228] Kuhn pointed this out to Lakatos in his long reply; see Kuhn to Lakatos, 23 June 1965, TSKP 23.9, 1.

[229] See Kuhn's letters to Popper, Kneale, the Halls, Hesse, and Lakatos, all of them from 23 June 1965 (and available in TSKP 23.9). The first four letters included very similar covering letters together with the main one.

[230] Kuhn to Lakatos, 23 June 1965, TSKP 23.9, 3 (postscript).

[231] Kuhn to Lakatos, 23 June 1965, TSKP 23.9, 1.

case—and the session would be chaired by, A. R. Hall, who, despite the fact that they were close friends, was, after all, a historian of science.[232]

As regards the speakers, Kuhn seemed willing to understand (at least to some extent) Popper's and Feyerabend's excuses. In the latter case—i.e. Feyerabend's—his understanding was rather a matter of "being used" to his behavior, so to speak. Feyerabend withdrew from the program and suggested that someone else read his paper. According to Lakatos, this suggestion was ill received by the organizers, who did not accept it.[233] Concerning this, Kuhn told Lakatos: "We have both known Paul for long enough to realize that he might decline at the last minute, for he has done it often before."[234] Kuhn's commentary was a rebuke to Lakatos, of course. Concerning Popper, the chairman had been invited to another session as a speaker, and he could not do both things, according to the organizers, Lakatos said. Kuhn accepted that, though reluctantly.[235] For him, Lakatos's withdrawal was the least easy to understand. Lakatos justified his abandonment on the basis of an alleged loss of confidence that the Princeton incident had caused, and also owing to his role as organizer—which was not new, anyway.[236] Understandably, Kuhn told him that he had "acquired an obligation," which was not suppressed by that double role. Besides that, his alleged loss of confidence could not justify his abandonment, either. Rather on the contrary, Kuhn said:

> If I can at all understand your withdrawal or your other failures of responsibility, it is not in the terms outlined in your letter but as a continuation of your performance at Princeton. You must by now realize that your malfunction here began hours before your paper was to be read and is not to be accounted for by the breakdown of anyone's watch, supposing that that occurred.[237]

As a result, Kuhn himself withdrew from the conference.[238] Or, at least, for a short while. Popper called him five days later and offered something different. The conversation convinced Kuhn that his participation in the session should go on, even though it would be much shorter—the afternoon and evening sessions were quashed—and Popper's participation was not granted. Kuhn wrote to him again two

---

[232] See Lakatos to Kuhn, 18 June 1965, 2; Kuhn to Lakatos, 23 June 1965, 1; both in TSKP 23.9.

[233] Lakatos to Kuhn, 18 June 1965, TSKP 23.9, 1.

[234] Kuhn to Lakatos, 23 June 1965, TSKP 23.9, 2 (postscript). Concerning the likely cause of Feyerabend's withdrawal, see Gattei (2008, 40–41).

[235] See Lakatos to Kuhn, 18 June 1965, 2; Kuhn to Lakatos, 23 June 1965, 2 (postscript); both in TSKP 23.9.

[236] He had told Kuhn that, during the time he had spent in the US, he had opted to give a lecture at Stanford University because György Pólya and Gábor Szegö were going to attend. Kuhn, of course, found this exception insulting: "your description of the one exception you did permit yourself (Stanford) is among the most devastatingly rude remarks ever addressed to me. (The fact that you do not mean it to be rude only makes the whole episode more depressing.)" (Kuhn to Lakatos, 23 June 1965, TSKP 23.9, 2 [postscript].) Lakatos' exception appears in his letter to Kuhn, 18 June 1965, TSKP 23.9, 1.

[237] Kuhn to Lakatos, 23 June 1965, TSKP 23.9, 2 (postscript).

[238] Kuhn to Lakatos, 23 June 1965, TSKP 23.9, 2.

days after that call, and repeated his preference for Popper's participation.[239] As is well known, Popper ultimately presided over the session and Watkins commented on Kuhn's paper.[240] So, at last, Kuhn got the attention that he clearly thought he deserved, especially from Popper himself. In his June letter, he summarized his own paper for Popper. Kuhn attached to that letter two sections of the paper in draft form—more specifically, the introduction and a further section. Kuhn hoped that they, the letter and the partial draft, together with *Structure*, would be enough for Popper to know what he was going to say and to prepare himself accordingly.[241] Their exchange was cordial, and Popper praised *Structure* at the end of his letter—though he did not avoid a slight criticism. Popper told him: "It seems to me that you overstress your central point; but one has to do this in order to get it across. And the point *is* important."[242]

Concerning Lakatos, after all these problems (and perhaps some more), Kuhn seems to have qualified his initial judgment about him in May 1964. Still in 1970, he told Roger Buck that "He [i.e. Lakatos] is bright and often extremely pleasant, but he is not one of humanity's more responsible members."[243] I guess Hennie Popper would have agreed.

6.8 From London to Princeton: Kuhn and Masterman

The Colloquium at Bedford College seems to have finished well.[244] A "lively discussion" took place—in Lakatos and Alan Musgrave's words in the editors' "Preface." The discussion involved Popper—as Kuhn had so forcefully insisted—as well as Stephen Toulmin, Leslie Pearce Williams, and, most important, the computer scientist Margaret Masterman.[245] Kuhn took notes about the debate. Concerning Pearce Williams' commentary, Kuhn registered some criticisms: that maybe the history of science would not "bear the weight" (which, it is easy to assume, refers to Kuhn's pattern of development); that Faraday had been a "revolutionary" despite his appearance of a normal scientist; or that Kuhn had not "proven [his] sociological thesis"—after which Kuhn simply wrote "Agreed."[246] Kuhn took and repeated some notes on Masterman, and their subsequent relationship shows evidence that he seriously agreed with some points in their conversation. Kuhn annotated some critical theses

---

239 Kuhn to Popper, 30 June 1965, KPP 317.17, 1.

240 See Gattei (2008, 42).

241 Kuhn to Popper, 30 June 1965, KPP 317.17, 2–3.

242 Popper to Kuhn, 7 July 1965, KPP 317.17, 3–4. Reproduced by kind permission of the Karl Popper Archives, University of Klagenfurt, Austria.

243 Kuhn to Buck, 29 April 1970, TSKP.

244 On Kuhn's, Popper's, Lakatos's, and Feyerabend's papers (as contributed to the 1970 volume), see Gattei (2008, 43 ff.). I return to Kuhn's papers for the volume in Chap. 7, Sect. 7.8.

245 Lakatos and Musgrave (eds.) (1970, vii).

246 See Kuhn, T2E, undated notes on "Pierce [sic] Williams." Below them, other notes on Margaret Masterman suggest that they were taken during or after the conference. See p. 1 for these notes and the quoted phrases in this paragraph.

that seemed important for him, and he repeatedly noted the first of them in particular: A "[p]aradigm can exist when a theory is not there."[247] The very same points, and in the same order (though at some points with different wording), are repeated in other notes.[248]

Masterman's words must have been particularly significant for Kuhn, who carefully annotated some of her points. One point emerges over the rest: that a number of different senses of the word "paradigm" appear in *Structure*—17 at first, of which, Kuhn noted, 5 were important. The question for them after that list—which changed later to 22 and then to 21—was "Have they anything in common?"[249] This question must have been added to the others we have seen before, and Kuhn surely thought (once again) that there were, as yet, open questions regarding paradigms. Recall, besides that, that this episode in 1965 took place while Kuhn was reflecting on some other open issues regarding paradigms and progress (see our previous sections).

In October 1965, Masterman finished a longish draft of a paper on Kuhn's concept of a paradigm and she sent him a copy from Cambridge. Its title was "The Nature of a Paradigm: A Study of T. S. Kuhn's 'Structure of Scientific Revolution[s].'"[250] At that point, the future of that draft was not clear. Her arguments extended through 57 typewritten pages plus heavily charged additional pages with endnotes for each section, and two long appendixes. Clearly, it was not intended for publication in that form. For Masterman herself, it looked like a part of a dissertation.[251] Kuhn thought that it would surely have to be cut down, and for their common friend Mary Hesse—the editor of *The British Journal for the Philosophy of Science*—it was five times larger than the papers they accepted.[252] Masterman had sent a copy of that draft to both and to Lakatos as well, and hoped to receive their reactions—and some help from Hesse to reduce it a little.[253] Besides that, she travelled to the United States in late November (i.e., a few weeks later), and took the opportunity to see Kuhn again on her way to Las Vegas, where a conference on computing awaited her.[254] At that meeting, they commented on the paper and, once she was back in Cambridge, she received a copy of Kuhn's notes on the draft. He thought about some improvements

---

[247] See [Kuhn], Notes on "Pierce [sic] Williams," T2E, n. d.

[248] See [Kuhn], "Results of Talk with Margaret Braithwaite," n. d., T2E, 1. (Margaret Braithwaite was, of course, Masterman's married name.).

[249] See [Kuhn], Notes on "Pierce [sic] Williams," T2E, n. d. That question is also Masterman's starting point in her chapter for the volume; see Masterman (1970, 65). See also her treatment in a previous draft: Masterman, "The Nature of a Paradigm: A Study of T. S. Kuhn's 'Structure of Scientific Revolution[s],'" October 1965, TSKP 14.11, 21–22. The list of 22 senses in the draft can be found in its Appendix B, while in the published version can be found on pp. 61–65.

[250] See Masterman, "The Nature of a Paradigm: A Study of T. S. Kuhn's 'Structure of Scientific Revolution[s],'" October 1965, TSKP 14.11.

[251] Masterman to Kuhn, 31 October 1965, TSKP 11.41, 1.

[252] See Kuhn to Masterman, 19 January 1966, and Hesse to Masterman, 3 January 1966, both in TSKP 11.41.

[253] Masterman to Kuhn, 31 October 1965, TSKP 11.41, 1.

[254] Masterman to Kuhn, 31 October 1965, 1, and Kuhn to Masterman, 23 December 1965, both in TSKP 11.41.

of the paper and included them in a covering letter with the copies.[255] In January, Lakatos invited her to publish the latter parts of the draft; meanwhile, Hesse would publish only the earlier part, devoted to the relationship between Kuhn's work and the notion of verification.[256] This latter publication eventually did not happen, but the parts on paradigms formed the paper that appeared at the proceedings of the conference, together with other contributions from Popper, Lakatos and Toulmin, to name just a few.

On May 20, 1966, Masterman sent Kuhn and Lakatos the new draft of her paper.[257] Lakatos was worried that some readers might find the new version somewhat lengthy—a worry that is all the more astounding, given the extension of Lakatos's own contribution to the volume.[258] Masterman replied that it was *Popper's* paper that should be cut down, not hers. Her paper, she told Lakatos, was the only chapter in that volume on Kuhn that truly dealt *with Kuhn*.[259] In the end, Lakatos himself did not consider it an obstacle, and Masterman and Kuhn agreed that there would be many references to Masterman's paper in Kuhn's replies, so no easy reduction would be possible.[260] In that regard, Masterman's new paper, with all her criticism of the notion of paradigms, was the most significant piece of them all for him. As he told her by letter a few days later, "I have not seen Toulmin's piece, but I do not expect to like it much better than Watkins', and I have not expected to spend much effort or space on that. So far, yours is the only paper I have wanted to talk about anyway."[261] In his notes on the first draft (those he seems to have sent to her in Cambridge after their meeting in the United States), Kuhn had written: "So far as I know, Mary is the only person who's read SSR as undercutting empiricism in an extended sense."[262]

Masterman complained in a letter to Kuhn that she did not understand the behavior of the group led by Popper.[263] Their attitude—and this was something that Kuhn himself had had the opportunity to discover—was hardly sympathetic. Indeed, Masterman's final version was the only chapter in the volume—perhaps followed by some parts of Lakatos's—that really aimed to clarify Kuhn's points. Her earlier draft from 1965 was not as explicitly "pro-Kuhn" as she said in the new version that she would be in the rest of the chapter, but the "pro-Popper" group, with Watkins as

[255] Kuhn to Masterman, 23 December 1965, TSKP 11.41.

[256] See Masterman's letters to Kuhn, 4 January and 24 January, 1966, TSKP 11.41. (The former letter is dated 1965, but this year is a typo, as the attached letter of Mary Hesse [dated 3 January 1966] that Masterman had just received attests.).

[257] Masterman to Kuhn, 29 May 1966, TSKP 11.41.

[258] Lakatos to Masterman, 6 and 9 June 1966, TSKP 11.41. (The former is Masterman's handwritten copy of a letter that is not in that folder.).

[259] Masterman to Kuhn, 8 June 1966, TSKP 11.41, 1.

[260] Lakatos to Masterman, 9 June 1966; Masterman to Lakatos, 13 June 1966; Masterman to Kuhn, 13 June 1966; Kuhn to Masterman, 16 June 1966; all of them in TSKP 11.41.

[261] Kuhn to Masterman, 16 June 1966, TSKP 11.41.

[262] [Kuhn], "Masterman Paper," n. d., TSKP, 11.41, 2.

[263] Masterman to Kuhn, 8 June 1966, TSKP 11.41, 1.

its main speaker, had already been pretty "anti-Kuhn."[264] This constructive function of Masterman's pieces—both the draft and the published version—is useful for me to finish this chapter. Some of the ideas contained there, and the effect they produced on Kuhn, help to summarize his research during the years that we have examined so far in this chapter. That research counts as an important part of the preparation for the postscript of *Structure* and the other new pieces that were published in the period 1970–1974, and with which I shall deal in the next chapter. Masterman acts as a catalyst for our summary.

Masterman's starting point was, as noted, the question concerning what all those senses of the word "paradigm" had in common. As she said, they were not necessarily incompatible.[265] If we pay attention to the list, we find that they are just qualifications of a previous application of the term. In other cases, they may be aspects of the functioning of a paradigm. And these are not the only examples of the variety of such senses. In that sense, however, Masterman's list is useful, because it is a sort of taxonomy of the many uses of the word paradigm in the somewhat *poetic* (rather than *philosophic*) style of Kuhn's *Structure*.[266] For her, those senses lead to three main groupings[267]:

1. "Metaphysical paradigms," or "metaparadigms."
2. "Construct" or "artefact paradigms."
3. "Sociological paradigms."

For Masterman, Kuhn's main contribution is the third sense, the *sociological* paradigm. The other senses of "paradigm" are philosophical. The two first senses have more often been the target of philosophers, while the third sense, in which Masterman is mostly interested, has hardly been discussed in comparison.

The *sociological* paradigm is, in her view, "a set of scientific habits." There are different kinds of habits—"intellectual, verbal, behavioural, mechanical, technological," she says—and, in her words, "[b]y following these, successful problem-solving

---

[264] See Masterman (1970, 61); the expression "pro-Kuhn" comes from that page. See also Masterman (1970, 61, n. 2) for an exposition and critique of Watkins's particularly aggressive attitude against *Structure*. The expression "anti-Kuhn" (specifically applied to Feyerabend's contribution to *Criticism and the Growth of Knowledge*) appears often in the Lakatos–Feyerabend correspondence from 1967–1968; see Motterlini's edition of some of Lakatos' and Feyerabend's writings and of their correspondence (Lakatos and Feyerabend 1999, 120–129).

[265] See Masterman (1970, 65),

[266] I am using here "poetic" in the sense that Masterman suggests as an option for Kuhn's style in *Structure*—as "a historian-poet," in her words, for instance; see Masterman (1970, 65, 76, 80, 83). In her draft, "The Nature of a Paradigm: A Study of T. S. Kuhn's 'Structure of Scientific Revolution[s]'" (October 1965, TSKP 14.11, 9), she opts for that qualification more clearly. In that draft (12–13), she relates Kuhn's particular style and its effects (e.g., that it may be obscure and confusing for many readers) to Wittgenstein's case by analogy; in particular, to the *Tractatus Logico-Philosophicus*.

[267] For the three next senses, see Masterman (1970, 65); "The Nature of a Paradigm: A Study of T. S. Kuhn's 'Structure of Scientific Revolution[s],'" October 1965, TSKP 14.11, 22–23. I have altered the order—she mentions sociological paradigms in the second place in both texts.

can go on."[268] These habits are based on research on a given achievement that is the second kind—this time philosophical—of paradigm in our list: a *construct* paradigm, which is, in turn, a "crude analogy."[269] She shows that there are two descriptions by analogy involved in the use of such achievement. In the first place, the concrete achievement is already "a way of seeing" things, and therefore a "gestalt-figure" of such things that populate our phenomenal world.[270]

A brief digression is in order at this point. It is in that sense—i.e., with regard to "a way of seeing" the world—that, for Masterman, Kuhn's concept of paradigm may have metaphysical implications that, in turn, may become the philosophers' business. Because the *concrete* paradigm involves a particular way of seeing things, it also conveys *metaphysical* consequences for our entire phenomenal world—i.e. it helps to *form* that world. So, the relationship between senses 1 and 2 in our previous list runs from 2 to 1 and not the other way around. That order—from 1 to 2—is, Masterman says, the usual way in which philosophers of science (like the Popperian audience in London) approach Kuhn's perspective. They ask about the way in which a problem–solution is obtained from a worldview. Masterman says that this is one of the main differences between Kuhn and Popper. The question for Kuhn is not how metaphysics is efficient in problem-solving, but rather the other way around—how puzzle-solving has the metaphysical consequences it has. That's why Kuhn's view is different from the previous "philosophy-of-science aetherialism."[271] In Masterman's words, "real science" is for Kuhn "a habit-governed, puzzle-solving activity, not a fundamentally upheaving or falsifying activity (not, in other words, a *philosophical* activity)," and that is the reason why "actual scientists are now, increasingly reading Kuhn instead of Popper."[272] Accordingly, it starts from the concrete achievement and the set of habits it produces, and explores the rest of the philosophical consequences that this down-to-earth perspective has. Kuhn's flight from "aetherialism" is based on that order of questions.

Let us return to Masterman's double description by analogy involved in Kuhn's view. If the concrete achievement—an "artefact," she says—is already "an organized puzzle-solving gestalt which is itself a 'picture' of something,"[273] it is then applied to something else (to another situation). So, in the second place, that achievement serves as a source of analogical thinking for a different situation, which it helps to describe.[274] That "way of seeing" the world on the basis of a two-step analogy helps to show that Kuhn's perspective is, as noted, tied to the idea that science works perfectly without the contribution of theories and rules—at least, when they are understood as logically arranged constructs in which meaning is (and must be) clear from the outset,

---

268 Masterman (1970, 66).

269 Masterman (1970, 76–79).

270 Masterman (1970, 76). I am using here "phenomenal world" somewhat freely. Masterman does not use that expression herself.

271 See Masterman (1970, 61, 72).

272 Masterman (1970, 60).

273 Masterman (1970, 77).

274 Masterman (1970, 76–78).

or whose legitimacy is gained by testing their deductive consequences.[275] In Kuhn's view, Masterman shows, science starts from examples of successful problem-solving that creates habits that turn problem-solving into *puzzle*-solving.[276] Of course, such examples are not easily "extensible"—that is they cannot be infinitely articulated—and they are incommensurable with one another. These are two central characteristics of Kuhn's bottom-up reconstruction of the scientists' activity that Masterman's picture makes understandable.[277]

Although Kuhn found some points in Masterman's interpretation arguable, her piece was the most interesting in the collection for him.[278] In the next chapter we see some of his critical points, but we will also see that some of his new moves toward the clarification of *Structure* involved answering to the challenge of making the meaning of "paradigm" sharper.[279] In fact, Masterman's paper and interpretation summarize Kuhn's concerns with *Structure* during the period after *Structure* very well, and also the main point of view from which he had started to think them through. I shall close this chapter with a reconstruction of that point of view.

For Kuhn, the idea that normal science was practiced by virtue of accepted achievements, paradigmatic for the entire community, and without the presence of a well-established theoretical structure in the sense logical positivism, was right. Paradigms, therefore, in the sense of concrete scientific achievements, were key for him; and they were sociological in nature, too.[280] So, in that respect, Masterman was right—although Kuhn did not think that those were separate aspects of his idea of paradigm.[281] Kuhn was also sure that the idea of a paradigm was different from the idea of a model, though it was intimately related—recall his letter to Boring. After all, there are aspects of the paradigm's representation of nature that are implicit and that call for subsequent research. The articulation of the paradigm during normal science is based on that characteristic of paradigms. So, Masterman's description of paradigms as involving a double analogy was not altogether right. The paradigm was something in between those analogies, he told her.[282] In short, his idea of science not only involves a concrete achievement, but also an unfinished one, and the notion of normal science essentially depends on that incomplete nature. As regards the concrete nature of paradigms as a central aspect of Kuhn's viewpoint, Masterman was altogether right.

After *Structure*, Kuhn assumed the challenge of distinguishing articulations from alterations of a paradigm—i.e., the kind of work that results in a scientific revolution. Whether changes in a given achievement give birth to a newer version of the same

---

[275] See Masterman (1970, 77) for the expression "way of seeing," also in quotation marks there.

[276] Masterman (1970, 70) also comments on that difference between puzzle-solving and problem-solving in Kuhn's *Structure*.

[277] See Masterman (1970, 79–84).

[278] See [Kuhn], "Masterman Paper," n. d., TSKP 11.41, for his initial critical points.

[279] See Postscript—1969, 174; Kuhn (1974, 294).

[280] See the beginning of this chapter and the closing parts of the previous one, too.

[281] See [Kuhn], "Masterman Paper," n. d., TSKP 11.41, 2–3.

[282] See [Kuhn], "Masterman Paper," n. d., TSKP 11.41, 4.

paradigm or to a brand new one was an interesting problem for Kuhn. However, even if he had not obtained a clear answer to that issue—and we saw that he had some quite clear ideas—the key point is that this issue helps us to understand the setting in which Kuhn's notion of a paradigm makes sense. Let me explain what that setting is. The paradigm is an example of good, true, and socially accepted, description of nature—of an articulated representation of a domain of phenomena according to some laws of the corresponding discipline. However, that example must be understood in *plural*, that is, as representing a class of concrete achievements, and not necessarily a unified theoretical structure.

Moreover, those achievements sometimes evolve themselves, so it is often a historically *dynamic* setting, too. There is a *family resemblance* between those examples—they do not respond to any kind of definition, or at least they do not need to. Making progress on a scientific view consists in developing new versions of the paradigm as applied to other domain of the phenomena. Similarly, getting trained in a scientific discipline involves having access to one or more—preferably more—examples of such kinds of example, or achievement (i.e. the paradigm). As he told Masterman in June 1966,

> the paradigm is a concrete example, [...] other concrete examples of the same thing are also paradigms, and [...] the relation between them is similarity, taken as a primitive. More concretely, the man who learns "swans" by exposure to six examples has been exposed to six paradigms each of them a paradigm for "swan." As a result, he has learned to see them all as similar and will apply this learned similarity relation whenever he meets other swans in the future. That is something I have learned in my seminar since you were here, and I find that I can make it even clearer against the version you provide. That, too, is extremely helpful.[283]

This perspective, as Masterman says, is already present in *Structure*, and his idea of science as based on such families of examples is part of the alternative he suggests for the philosophy of science.[284] Soon after that letter, he would develop a popular version for describing paradigm-based learning: the version available in "Second Thoughts on Paradigms," which, except for the absence here of the role that differences play in the learning process, is very similar to this one (more on that in Chap. 7).

Specifying that setting, making paradigm-learning and paradigm-articulation something more (something *clearer*) than an obscure—and probably subjective—intuitive process, was one of Kuhn's main objectives after *Structure*.[285] His work on a computer simulation that learns from nature on the basis of individuals that end up being instances of a class, and that teaches other simulations to do the same is intended to show how paradigm-based learning and research is possible, and is also explorable. That work on a computer simulation, which I have mentioned in a previous section, is related to the same investigation of paradigms. Masterman also showed him the problems involved in such simulations. Kuhn had asked about it in the past, and her paper

[283] Kuhn to Masterman, 1 June 1966, TSKP 11.41, 2.

[284] See Masterman (1970, 84–87); see also *Structure*, 45.

[285] Postscript—1969, 191. Masterman (1970, 84–85) also talks about that "intuitive inference".

finished with some interesting passages in that regard.[286] In December 1965, he had asked her: "I've recently described being trained in a paradigm as being programmed (for problem solving). Is there work on machines that program themselves by being led thru exemplary (paradigmatic) problem solution?".[287]

During the three or four years that followed the publication of *Structure*, Kuhn explored central aspects of the book—especially the concept of paradigm and the question concerning realism. We have seen here how Kuhn pursued that research, and also how computer scientist Margaret Masterman was a better, more sympathetic, and more productive conversation partner than many of those philosophers that were supposed to be his peers. Kuhn, we have also seen, always had difficulties in being taken seriously by the more senior, better-established philosophers—and not only in strictly philosophical themes. Masterman, meanwhile, a computer scientist, was readier to understand Kuhn's particular philosophical perspective than other philosophers. Their exchange was indeed productive. In the next chapter, we shall see how the 1960s ended and how the 1970s started for Kuhn, and how he published his replies to his many critics. After that, we shall see how his return to the history of science was, for many historians, just as polemic as his foray into the philosophy of science had been for philosophers.

[286] See Masterman (1970, 86–87).

[287] [Kuhn], "Masterman Paper," n. d., TSKP 11.41, 3.

Chapter 7
From Philosophy of Science to the Quantum Theory

7.1 Kuhn and Princeton's Program in History and Philosophy of Science

From 1964 to 1979, Kuhn was an active member of the Program in History and Philosophy of Science (HPS hereafter) and served as its director for three years from 1967.[1] In 1968, he was appointed M. Taylor Pyne Professor of the History of Science.[2] His involvement with the program was full, although he continually asked for leaves of absence until the early 1970s, when his part-time status at the Institute for Advanced Study (IAS) began, and his teaching at the HPS Program decreased considerably.[3] Between 1967 and 1968, he had conversations with Robert R. Palmer (who was Dean of the Faculty in 1967–1968) concerning the reduction of his duties. In October–November 1967, they discussed the possibility of reducing his teaching load, a proposal that, contrary to Kuhn, Palmer found excessive.[4] A few months later, in 1968, the University of Michigan offered him a job, and Palmer discussed with him and with President Robert Goheen the possibility of increasing the number of leaves of absence. They discussed reducing the terms between leaves from five to four. For Kuhn, that aspect of his future was more important than a pay raise. At that point the reduction of teaching hours arose again, and he had the support of the

[1] See Goheen's letter to Kuhn, 4 May 1967, and Kuhn's reply to Goheen, 10 May 1967, both in TSKP 14.11.

[2] See J. S. Finch to Kuhn, 10 June 1968, TSKP 14.11.

[3] Concerning leaves of absence, see Gillispie to Brown, 14 October 1965 (first leave for 1966–1967); the memoranda dated 21 January and 1 February 1966; Palmer's letter to Goheen, 21 March 1968, on future leave; Lester to Lawrence Stone, 11 February 1969 (leave for 1969–1970); and finally, his full Princeton record; all of them in PUTK.

[4] See Palmer to Kuhn, 28 November 1967, and Kuhn to Palmer, 6 December 1967; both in PUTK.

J. V. Mayoral, *Thomas Kuhn*, Springer Biographies,
https://doi.org/10.1007/978-3-031-86512-1_7

Department of History in that sense. Finally, Goheen, Palmer and Kuhn agreed the increase of leaves of absence as previously noted.[5]

Concerning his teaching, from 1968 to 1976, Kuhn seems to have taught very little to undergraduates. His main dedication was to graduate courses and seminars and PhD supervision, and he did quite well in that respect.[6] His courses and seminars were very demanding. In scientific terms, his research interests from the early 1960s as regards the history of physics were the history of quantum theory and its historical antecedents.[7] For Michael Mahoney, historian of science and, for some years, Director of the HPS Program, Kuhn's contact with students had reduced to graduates; he scarcely dealt with undergraduates. Moreover, in 1974, he reported to the Dean of the Faculty that, beyond his teaching, he was almost completely absent, which was a problem for the daily functioning of the program.[8] At that point, he was mostly devoted to his research, and to training new (and highly specialized) historians of science.

Kuhn received offers from different universities soon after having accepted the job at Princeton. The Massachusetts Institute of Technology, for instance, had already made him a first offer in 1967 in order to found a new Department of History of Science.[9] In 1968, as noted, the University of Michigan offered him a professorship devoted solely to graduate teaching—which would have been attractive to him, of course.[10] The University of Virginia did some research on him in 1968 as a likely candidate for Dean of the Faculty of the College of Arts and Sciences.[11] The University of Pennsylvania offered him the Benjamin Franklin Chair in the History of Science in 1971.[12] Even the University of Oxford offered him Chair in the History of Science in 1972. When he turned down the latter, he told the British historian Alan L. C. Bullock, who had invited him, that he had been there a few times since his first visit during World War II. Kuhn felt honoured by the invitation, but he told Bullock that he and his family were not ready for such a change. Besides that, he wanted to spend time in research, and this invitation meant to spend a part of it building an academic program in his discipline, the history of science—which he had already

---

[5] See Palmer to Goheen, 21 March 1968; Goheen to Palmer, 26 March 1968; Stone to Palmer, 1 April 1968; and Kuhn to Goheen, 8 April 1968, all of them in PUTK.

[6] See Gillispie's reports from 24 January 1974, and 17 February 1975, PUTK.

[7] See, e.g., Gillispie's report from 28 January 1977, PUTK, 1. Gillispie shows there that, in spring 1976, Kuhn had dealt with the history of physics from 1750 to 1930 in an especially demanding course.

[8] See Mahoney to Lemonick, 24 January 1974, PUTK, 1.

[9] See Gillispie to Brown, 30 May 1967, PUTK. I say a "first" offer because, as it is well known, Kuhn ended up at MIT, in that case in the Department of Linguistics and Philosophy, in 1979.

[10] See Palmer to Goheen, 21 March 1968, PUTK, 1.

[11] See Kolodziej to Murphy, and the latter's reply to the former, 11–17 December 1968, PUTK. E. A. Kolodziej chaired the Woodrow Wilson Department of Government and Foreign Affairs at the University of Virginia, and Walter Murphy was McCormick Professor of Jurisprudence at Princeton's Department of Politics. Murphy strongly supported Kuhn's candidature, although, it seems it didn't go anywhere.

[12] See Lester's letter to Goheen and Bowen, 3 March 1972, PUTK.

done and didn't want to repeat.[13] So, he turned down Oxford's invitation to occupy a chair in the history of science there, and he did the same with the rest.[14] His intention at that time was clear: simply to get more time for research. In fact, his last stint in Princeton, at the IAS, was devoted to his research. Though in many cases these offers were almost research professorships (Pennsylvania's chair, for instance), at that point in his career he simply used them to improve his position at Princeton.[15]

During those years, Kuhn did his best to serve Princeton. After all, it had improved his relationship with philosophers, which, in many cases, could not be better. He was well integrated into the university, and his work on behalf of HPS was doing well. His collaboration with Gillispie was fruitful during that decade and a half. In 1979, when Kuhn was already somewhat detached from HPS and spent most of his time at the IAS, two junior collaborators, G. L. Geison and Mahoney reported that, despite the problems that the program had been experiencing in the last few years (actually, it disappeared as a double program after Kuhn's departure), it had been one of the best in the country. In fact, the list of former PhD students during those years—for example, Larry Laudan, John Earman, Richard Grandy, Thomas Nickles, Philip Kitcher, Norton Wise and many others—testifies to its quality.[16] Kuhn's participation in the program was key for Gillispie, Geison and Mahoney.[17] However, as time passed, Kuhn's commitment decreased; he felt that he needed more time for his research. As Gillispie told Aaron Lemonick (Dean of the Faculty) in 1976, Kuhn assumed his duties with great dedication—another reason for his frequent petition for leaves of absence, and for long periods at the IAS.[18] In the early 1970s, as his replies to the philosophers' criticism of *Structure* started to surface, his commitment to the program gave way to his commitment to his own research—especially to *Structure* and, at least until 1975, to the history of quantum theory. So, from 1972 onwards, he agreed with Richard Lester (the Dean of the Faculty for some years after Palmer), that he would have a part-time status and would teach one semester only.[19]

---

[13] Kuhn to Bullock, 31 March 1972, PUTK.

[14] See Gillispie to Brown, 13 June 1967, as regards the offer from the MIT. Concerning Pennsylvania's offer, Kuhn obtained a part-time status with Princeton in order to spend more time at the IAS, which was key to remain at Princeton. See Lester to Goheen and Bowen, 3 March 1972, Concerning Oxford's offer, see Kuhn to Bullock, 31 March 1972. All of these letters are in PUTK. Concerning Michigan's offer, see above in this section.

[15] See Palmer to Goheen, 21 March 1968, PUTK, 1. Palmer, Dean of the Faculty, makes it clear there that, despite Kuhn's commitment with Princeton and the HPS Program, the attraction of such kind of post was hard to resist—he did it in the end.

[16] See Geison and Mahoney's report for Lemonick, 2 January 1979, TSKP 14.11; Gillispie (1999, S93–S94).

[17] See Mahoney to Lemonick, 24 January 1974, and Gillispie to Lemonick, 29 November 1976, both in PUTK. See also Geison and Mahoney to Lemonick, 2 January 1979, TSKP 14.11.

[18] See Gillispie to Lemonick, 29 November 1976, PUTK.

[19] See Lester's memorandum ("Proposal for Two-Thirds Time Status"), 8 March 1972, PUTK. Kuhn agreed with Lester by letter (Kuhn to Lester, 10 March 1972, PUTK).

Gillispie defended that agreement again in 1976 in order to prolong it (or something similar to it).[20]

7.2 Kuhn's Research Activity in the Late 1960s

In our previous chapter, we saw how Kuhn had exchanged ideas about paradigms with Margaret Masterman between 1965 and 1966. Upon returning from London, Kuhn resumed his usual research activity. For instance, he gave a talk for the Telluride Association Summer Program at Princeton, that the university supported economically, and that Robert Nozick chaired.[21] We must recall that some conversations with Nozick and with some students in that summer, just before Kuhn's departure for London in early July, had given him occasion for thinking about the problems for scientific realism involved in *Structure*.[22] Now, in August, he returned to that summer program, and gave a talk on the book. He repeated it in a colloquium at the Educational Testing Service, Princeton.[23] These are just a couple of examples of the kind of outside activities that arose from *Structure*. Kuhn was very active in that sense.

For Kuhn, those activities were good opportunities for revisiting the arguments in *Structure* from the points of view that he had been developing from 1961 to 1965, and that we examined in the last chapter. So, for example, in 1966–1967, and before a reduced audience at Rockefeller University, he returned to *Structure* almost section by section, step by step, in order to make some new points on the ideas in the book. For example, he introduced his audience to basic aspects of his new research on a computer simulation of paradigm-based learning. The points he made at these kinds of meetings helped him to prepare the groundwork for upcoming publications like the "Postscript–1969" to *Structure* and his "Second Thoughts on Paradigms." In the following sections, I would like to approach some of these in order to show what the origins of "Second Thoughts on Paradigms" were—it was a paper that ended up being increasingly significant for Kuhn as the years went on.

However, from 1965, and perhaps even before, Kuhn developed a line of research that had been highly important for him as a historian of science. As we have seen, Kuhn devoted plenty of time to developing programs in the history of science, either at Berkeley or later on at Princeton, and part of his research activity was connected to that discipline—after all, he was professor of the history of science. Before entering the Center for Advanced Study in the Behavioral Sciences, in 1958, he said that he would "spend some of the time [in the Center] extending my knowledge of problems

---

[20] See Footnote 18.

[21] See *Telluride Newsletter*, 51 (February 1965), 4.

[22] T2E, 7 July 1965. See also our Chap. 6, Sect. 6.4.

[23] See Kuhn's report on his "Publications and Professional Activities" for Princeton, 23 May 1966, TSKP 14.15, 3.

in the development of 19th century physical science and scientific institutions."[24] His work on the history of thermodynamics at the end of the 1950s and in the early 1960s was an adequate preparation not only for his archival and interviewing work from 1961 to 1964, but also for the subsequent work that he pursued in the history of quantum theory from its very beginnings with Planck.

In 1965, he presented some results of this line of research. In January 1965, for instance, he presented a paper at a meeting of the American Physical Society. It was titled "The Pursuit of the Quantum: A Progress Report."[25] More importantly, Kuhn often presented a paper with the title "The Crisis of the Old Quantum Theory." From at least March 1965, he did so very often.[26] It was, clearly, the core of his research into that period of the history of physics for a while. In fact, it was his tentative and surely preferred title for a book he later considered publishing. In a report on his own research for the Department of History in 1970, Kuhn said that his research on that subject was the basis for a future "small book."[27] I would like to emphasize the typically Kuhnian character of this initial plan, because his future long paper or small book would be about a period of crisis with "obviously intractable problems" that made the "profession" aware of "the status of their theoretical tools" and led the physicists to new concepts and, especially, new theories. "Matrix mechanics and electron spin," he says, emerged from that process as "two radically new physical theories" based on the new concepts.[28] It is worth noting that he does not employ the word "paradigm" in that short abstract of his research—actually, there seems to be no place for it in that summary. Despite that, the explanatory structure of scientific change that underlies his description—the meta-historical narrative involved, so to speak—is typical of Kuhn's accounts of some historical processes, and of *Structure*. Although the best-known result of this line of research, *Black-Body Theory*, differed from that "small book" in many aspects, it is clear that Kuhn approached that modern period in the history of physics from the point of view of his very recent account of scientific change. On the other hand, as we shall see below, that volume was only part of a more ambitious, broader plan that he eventually did not finish. It's likely that Kuhn's interests in renewing *Structure*, and perhaps some boredom with that historical research, led him to abandon the plan.

I shall return to this interesting line of research at the end of this chapter. I hope to summarize Kuhn's research into the history of quantum theory from 1964 to 1975 in Sects. 7.9–7.10. At this moment, however, I will focus on his philosophical work. The research initiated right after finishing the final draft of *Structure* produced a set of publishable results in the form of replies, a postscript, and a major paper that

[24] Kuhn, "Inventory of Professional Interests," 1958, TSKP 14.15, 2.

[25] See the report from the Department of Public Information, Princeton University, 2 March 1965, 2 (in PUTK).

[26] See again Kuhn's report on "Publications and Professional Activities," 23 May 1966, TSKP 14.15, 3, for that specific occasion in March, 1965. There will be many others.

[27] See Kuhn to R. D. Challener, 25 May 1970, TSKP 14.11. All previous quotations in the last few lines come from this letter. The letter itself was originally Challener's petition of information, and so sent by him (then, it was filled by Kuhn and sent back).

[28] Kuhn to Challener, 25 May 1970, TSKP 14.11.

were produced between 1970 and 1974, and we should turn to those before getting back to Kuhn's activity as a historian of science. We shall examine those well-known published results in Sect. 7.8.[29] Before doing so, in the next sections we must pay attention to the road that led to them in some detail.

The path that led Kuhn to those published results had already started years ago in his notes after *Structure*, as early as 1961. We examined those notes in Chap. 6 and reached the years 1965–1966. Some main themes emerged, mainly in the form of Kuhn's private reflections on open texture and vagueness, and on their relationship with a model for simulating paradigm-based learning.[30] His talks and lectures from 1965 to 1967 helped him to convey some of those ideas. Some progress in his views is evident, particularly in the way he wishes to communicate a main thesis: the construction of a model of classification of natural kinds of objects is based on family resemblances, and he rejects an overall, basic function for definitions and generalizations. As he said on one of those occasions, though explicit generalizations from cases may come to be useful, learning does not exclusively (nor mainly) depend on them; paradigm-based learning means that learning does not *essentially* depend on them. "Verbal generalizations at key points may help," he wrote in his notes for a talk at Washington in 1967, "but they're an inefficient way of doing the whole job."[31]

Some talks from this period are relevant in order for us to see that path. Let us look at some of them in chronological order. Invited by anthropologist Harold Conklin, Kuhn talked at Yale's Anthropology Colloquium about his "Thoughts on the Nature of Scientific Classification."[32] He introduced himself, rather diffidently, as somebody who was poorly versed in the field ("I know almost nothing at all about anthropology [and ...] its contemporary concerns"[33]), but he quickly turned to talking about how a natural classification of objects is formed, and how knowledge of nature is gained. It is, to the best of my knowledge, the first time he relates this general model, which was still in development, with the key role that thought experiments play in finding contradictions within a conceptual scheme and thus cause a transition to another without them. It would be in another anthropology gathering in the mid-1970s, this time at Princeton, when he would return to this interesting link, and though he would indicate at that point that he had just noticed that connection, its basis is already present here.[34]

Kuhn talked about all this at Yale in February 1965 and did something similar in December 1966, this time more focused on his plans for a computer simulation of the learning process, for a talk he gave in New York City, invited by Sidney Morgenbesser.

---

[29] I mean: Postscript—1969, and Kuhn (1970a, 1970b, 1974).

[30] We also paid attention to scientific realism, but it will not be the main target of our examination in the next few pages.

[31] [Kuhn], "American Psychological Association," 2 September 1967, TSKP 3.14, 8.

[32] See [Kuhn], "Thoughts on the Nature of Scientific Classification," Yale Anthropology Colloquium, 25 February 1965, TSKP 3.14, esp. 2.

[33] [Kuhn], "Thoughts on the Nature of Scientific Classification," 25 February 1965, TSKP 3.14, 1.

[34] More on this at the end of this chapter. See his "Thoughts on the Nature of Scientific Classification," 25 February 1965, TSKP 3.14, 11–13.

At that time (1966–1967), Kuhn was a visiting professor at Rockefeller University, and Morgenbesser might have taken the opportunity to invite him to talk, probably at Columbia.[35] Kuhn used the notes for his fourth session at Rockefeller for this talk.[36] Despite their brevity, the notes strictly written for that occasion are interesting because they show his plans for the computer simulation.[37] In February 1967, a few months later, he visited the Swarthmore College as a Distinguished Visiting Scholar and gave two lectures on different days: "Recurrent Patterns of Scientific Development" (12th) and "Paradigms and Theories in Scientific Research" (19th).[38] While the former lecture dealt with the usual Kuhnian cycle of normal science, crises and scientific revolutions, the second one is more significant for us. He presented there a revision of his account of paradigms in *Structure*, motivated by recent criticism (e.g., Masterman's), and advanced a well-known example he later introduced in published form in "Second Thoughts on Paradigms"—a child's ("Johnny") first acquaintance with different kinds of waterfowl. Finally, by the end of the following summer, in early September 1967, he took part in a Washington DC symposium organized by the American Psychological Association's Division 24, and in particular by psychologist Sigmund Koch, on Michael Polanyi, whose "Logic and Psychology" was discussed by Quine, Michael Scriven and, of course, Kuhn, who commented on Polanyi's notion of tacit knowledge.[39] Kuhn's handwritten notes for this symposium (14 pp.) are interesting, indeed, insofar as Polanyi's work is an excuse for revisiting Kuhn's new perspective on the acquisition of a natural classification once again and introducing it to the audience—which that day was composed of psychologists. His vocabulary is computer-oriented at some points, although he spends most of his time explaining the "Johnny" example again, as he had done on other occasions.

His seminar at Rockefeller University is particularly interesting for us. His audience was highly varied, and it included philosophers—and probably some distinguished philosophers such as Morgenbesser and Harry Frankfurt, and philosophers of science such as Nagel and Hempel.[40] Kuhn delivered his seminar in ten sessions,

---

[35] See [Kuhn], "Scientific Methods Colloquium," New York, 4 December 1966, TSKP 3.14, 1.

[36] See [Kuhn], "Scientific Methods Colloquium," 4 December 1966, TSKP 3.14, 5.

[37] See [Kuhn], "Scientific Methods Colloquium," 4 December 1966, TSKP 3.14, 1–2.

[38] See [Kuhn], "Recurrent Patterns of Scientific Development," 12 February 1967, and "Paradigms and Theories in Scientific Research," 19 February 1967, both at Swarthmore College, TSKP 3.14. The local journal, *The Swarthmorean*, published a notice of Kuhn's invited lectures; see "Scholar to Visit College Campus. Historian of Science to Speak Feb. 12, 19," *The Swarthmorean*, Vol. 39, No. 6, 5. Issue No. 7 also echoed the second lecture. Marcum (2015a, 85–90) is a study of "Paradigms and Theories in Scientific Research.".

[39] See Polanyi (1968), an extended version of his address there. Polanyi (1958) is, of course, the main source for that notion. Kuhn's handwritten notes for his contribution were simply titled "American Psychological Association," 2 September 1967, TSKP 3.14. See also Koch's letter to Kuhn, Quine and Scriven, 23 August 1967, TSKP 3.14. On Kuhn's contribution to the symposium, see also Kaiser (2016, 84–87, 94).

[40] This is, however, arguable insofar as there is no explicit mention of the names of some of these philosophers other than "Sid," "Peter," and "E. N."; Frankfurt is explicitly mentioned. Yet, Kuhn refers to the criticism of the present "prof[essional] philosophers": "I'd not counted on so large a representation of prof. philosophers" (Rockefeller 2, 1). I then assume that the presence of the

all of which were agreed at the beginning with the audience; they were usually on Wednesday and at 3:30 p.m., though there were some changes. The sessions followed the following schedule:

1. [November 2nd, 1966]?
2. November 16th, 1966.
3. November 30th, 1966.
4. December 14th, 1966.
5. [January 11th, 1967]?
6. January 25th, 1967.
7. February 8th, 1967.
8. March 1st, 1967.
9. March 15th, 1967.
10. March 29th, 1967.[41]

Dates for sessions 1 and 5 were not explicitly registered, but Kuhn confirmed them every two weeks, so we can assume that their dates may have been the days within square brackets.[42] It looks like he did not lead the sixth or seventh meeting and that Morgenbesser did it in his stead.[43] His visit to Swarthmore College was the reason for the lapse between the seventh and the eighth meeting. At Rockefeller University, the conversation was lively and the debate was very active. He was pleasantly surprised by the philosophers' contributions from the very beginning: "[I] hadn't supposed that the confrontation [with philosophers], which I welcome, would emerge so quickly."[44] His notes on the postmortem after the first session reflect key interventions from those philosophers.[45] As the meetings continued, however, he was not so pleased with the result. So, when the fourth session (which was the last one before Christmas vacation) finished, he "was fairly unhappy about the way things went," he annotated at the start of the following session.[46] And he went on: "Not so much that I failed to persuade anyone—my expectations in that respect are relatively modest. [...] But, as the discussion developed, I was uncertain that I'd even given anyone a glimmer of what I was after, what I meant."[47] If we pay attention to the content of such sessions (i.e. fourth and five), we can understand why Kuhn was so uneasy with the result.

The Rockefeller seminar, similarly to the other talks that I have previously mentioned, aimed to expound the main objective of Kuhn's new work since the

mentioned philosophers was real, though with that important proviso. See Kuhn, Rockefeller 1, "P-Mortem #1"; Rockefeller 1, 1; Rockefeller 3, 1.

[41] See Rockefeller 1, 1; Rockefeller 5, 1; Rockefeller 6, 1; Rockefeller 8, 1; Rockefeller 9, 1.

[42] See Rockefeller 1, 10; Rockefeller 6, 1.

[43] See Rockefeller 6, 1. Next to the title of that session, Kuhn annotated "Not read #7 was Sid." As there is no notes for session seventh, that was probably the session that Kuhn did not lead.

[44] Rockefeller 2, 1.

[45] See Rockefeller 1, 10 and "P-Mortem #1".

[46] Rockefeller 5, 1.

[47] Rockefeller 5, 2.

publication of *Structure*—the justification and reform of his appeal to paradigms. Our last chapter presented that plan in detail. Kuhn had tried to develop a model for paradigm-based training in science, had studied paradigms in more detail, and had discussed them with Masterman on the basis of her manuscript. He had even settled the basis for a computer simulation of that kind of learning. He was pretty sure that his model and the companion simulation would help to show the soundness involved in his depiction of normal science—the kind of growth-promoting research that was not necessarily practiced on the basis of explicit rules, and that he had displayed more clearly in the new Section V that he had added to *Structure* at the last minute. He would describe the model in his talks in those years, and more particularly in the fourth and fifth meetings at the Rockefeller University. Given Kuhn's character, it is not surprising that he felt a little worried and even depressed when he started to explain this idea, its basis, its consequences, and the relationship with the computer program, and he noted that the plan was not working as originally envisaged. The seminar and the contemporary talks are a testimony to Kuhn's efforts in making himself understood, and to the difficulties he would find in the process. The seminar in particular, with his repetitions of key points, shows that process clearly.[48]

Kuhn talked about many aspects of *Structure* in those days and discussed many of his apparent failures in the book with enthusiasm. There were five main themes: (a) the criticism of the logico-empiricist tradition and its defense of the idea of "rational reconstruction"; (b) his defense of a different kind of "rational reconstruction," this time based on historically reliable accounts of the science's past; (c) the independence of the scientific community with regard to the idea of paradigm; (d) a deeper analysis of the idea of a paradigm; (e) a model for paradigm-based learning and an account of its role in philosophy of science. Besides these, there were two more themes of interest for him—although they are less directly related to the previous five. First of all, Kuhn seems to have been dissatisfied with his perspective on revolutions as regards the importance of the scale of change and the point of view of the observer.[49] That counts as theme (f). Secondly, a significant last theme (g) would be the nature of historical knowledge, something he dealt with on the basis of his paradigm-based learning model. The core of his current research activity is, of course, point (e) and to a lesser extent (d). However, points (a)–(c) are the context in which that model makes full sense, so I shall begin with the three latter points in the next Sect. (7.3) and then turn to the former ones, (d)–(e), in Sects. 7.4–7.6. After that, in Sect. 7.7, I briefly examine points (f) and (g).

[48] In contrast, Kuhn did not emphasize other older themes in *Structure*, such as the theory-ladenness of observation, or the resolution and nature of scientific revolutions. Kuhn only repeated old arguments in the book on those points. See the notes for Rockefeller 8–10, which I shall not examine in what follows.

[49] See *Structure*, 6–7, 49–51, for some initial ideas about that issue. Kuhn's drafts for Chapter I (1959) of *Structure*—i.e., the one he did not publish—also mention it; see SSR1A and SSR1B, 39–41 (the same pages in both cases).

7.3 Rational Reconstruction and the Scientific Community

Ever since his days in the Society of Fellows, Kuhn was wary of—if not openly critical of—the phrase "rational reconstruction," which he readily ascribed to Carnap's and Reichenbach's works.[50] Sixteen years later, Kuhn was ready to say that "I am trying to do rational reconstruction too."[51] Of course, Kuhn was not at all a follower of either Carnap or Reichenbach at that point. He had simply found that those philosophers should not claim to be depicting the only possible vision of rationality at any period in the history of science. The crux of the matter was the difference between the concepts of *rationality* that their respective positions assumed—and which, of course, had nothing to do with an alleged defense of rationality on the part of the logical empiricists and an advocacy for irrationality and subjectivism on the part of Kuhn. As Kuhn himself wrote in 1966, "Where we differ—and I think the differences are real but only partial—is in the source of our notion of rationality and in our assurance that we know what it is."[52] Kuhn and the logico-empiricist philosophers were aware of the role that sometimes idiosyncratic and subjective factors played in the history of science. Reichenbach talked about them in *Experience and Prediction*, but he considered them material for a psychological scrutiny of scientific research. Here, Kuhn said that "The merely human and idiosyncratic bores me and is in any case irrelevant."[53] However, there were clear differences between the attitudes that they respectively considered "essential" and "idiosyncratic."[54] There lies the difference.

Kuhn considers Reichenbach as the opposite view's "clearest" spokesperson in his notes for the Rockefeller University seminar.[55] However, he also mentions Nagel and Hempel, who might have been listening to his arguments when he presented them that late-autumn afternoon at the Rockefeller University. His opposition to Reichenbach had been clear since his years in the Society, and had not diminished since then. After 1953—we do not know exactly when—he also read and strongly criticized *The Rise of Scientific Philosophy*, which had been published in 1951.[56] In his notes on the book, Kuhn disagreed with Reichenbach's basic premise that only if science is devoid of psychological bias and is based on the proper inference to valid conclusions shall we find the true method of science. The epistemology of science is his main goal, as he shows in *Experience and Prediction*, and as Kuhn notes, Reichenbach "says, 'If this isn't the way scientists behave it's the way they should

[50] See Carnap (1928a; 1928b; 2003); Reichenbach (1938). See also our Chap. 2 for further details.

[51] Rockefeller 1, 16.

[52] See Footnote 51.

[53] Rockefeller 1, 16. See also Reichenbach (1938, 3–5).

[54] See Footnote 51.

[55] See Rockefeller 1, 15.

[56] See Reichenbach (1951). Kuhn's notes on that book are in FR65. They are spare notes without title or date, though they show Kuhn's reactions to a book whose contents fit with Reichenbach's *Rise*. We know that he read it after 1953 because of his mention of Braithwaite's version of Ramsey (p. 12). Richardson (2007, 359–367) examines the negative reviews of Reichenbach's book in the early 1950s, and the context around them.

behave.'"[57] In contrast, for Kuhn, the psychological drive behind scientific ideas is not a source of misapprehension but rather something without which science barely works—recall his early notes in 1949 in that respect. Similarly, for Kuhn, analogical reasoning is not, as it is for Reichenbach, an outdated ingredient in an activity that we might call pre-science or pseudo-science; it is not restricted to the infancy of a science. Kuhn set himself against its alleged absence in contemporary science and also against the somewhat patronizing and definitely Whig position of locating it in ancient, superseded eras of the history of science.[58] Although critical, Kuhn's attitude towards Hempel and Nagel ("men I respect") seems a bit less radical, but they are for him, in any case, supporters of the "less strong form of the position" about rational reconstruction that Reichenbach represents.[59] They also share the basic premise, he says, though in an "implicit" form.[60]

Kuhn's position about Carnap in this criticism is different, too. For him, Carnap and Tarski represent a different approach to rational reconstruction that he considers "stronger" than Nagel's, Hempel's or Reichenbach's. For Kuhn, the three latter views still claim to have drawn some central, "essential" features from actual science. For them, reconstruction is based on that set of features. We should not forget, in any case, that, although the history of science may play a role for those thinkers, that role is incidental rather than essential. As Kuhn says, "If history has helped [them] in that [task], so much the better, but in so far as it's a record of error and idiosyncrasy, it can't be relevant."[61] In comparison, Carnap's and Tarski's views do not include any role whatsoever for the history of science, so, Kuhn goes on, "the problem of what's essential and non-essential vanishes," and their "[p]osition is logically impregnable."[62] An anecdote by the historian of science I. B. Cohen illustrates that stance.[63] Cohen remembers how he invited Carnap to give a talk to a group of graduate students in Cohen's field. Carnap politely declined after a long conversation with the Harvard historian. The reason for doing so was that Cohen, "had convinced him [...] that he really was as unhistorically minded a person as one could imagine. He [i.e. Carnap] had nothing whatever to say [...]," Cohen went on, "about the study of the *history* of scientific ideas that could possibly be of interest to historians!" As Cohen said, the history of science was not a component part of Carnap's analysis of scientific method. Neither "case histories of historians of science," nor "examples drawn from current scientific practice" counted among the empirical sources for his

[57] Rockefeller 1, 15; see Reichenbach (1938, 7 ff.). Kuhn's statement within quotation marks does not seem to be a literal quotation of Reichenbach's own words.

[58] See Kuhn's notes on Reichenbach in FR65, n. d. His notes begin with "I. My Problem (Confession?)" and run for 12 pages. Concerning Reichenbach's points, see Reichenbach (1951, 8–9, 13–14, 101, 117). These are passages to which Kuhn pays particular attention (see p. 5 of his notes).

[59] See Footnote 55.

[60] See Footnote 55.

[61] See Footnote 55.

[62] Rockefeller 1, 18. In the original, Kuhn writes "problems," in plural.

[63] The selection of this anecdote is mine, not Kuhn's.

philosophy.[64] And Kuhn would have agreed of course. For him, Carnap's "[o]nly concern is to produce a logical model which would produce results like those the scientist gets."[65] However, despite the "intrinsic interest" that he finds in such an approach, its results are for him "extremely powerful abstract mathematical and logical machinery whose realm of legitimate application is still in doubt."[66]

Indeed, Kuhn's problem with rational reconstruction as expounded by logico-empiricist thinkers was that either it could pick out features of scientific behavior that a historian might consider extemporaneous, and so irrelevant for a credible account of scientific rationality, or it could build an abstract model of rationality so detached from real practice that it could become ultimately irrelevant. In the best case—Nagel's, Hempel's and Reichenbach's weaker position—the result was far from being satisfactory. The reason for that discontent on the part of Kuhn was their reliance on a certain model of scientific rationality—the "hypothetico-deductive model for scientific theory," Kuhn recalls.[67] Despite its many virtues, he said, he was not "at all sure that it earned the right to be the exclusive arbiter of what's essential and what's idiosyncratic in the way scientists gain, check, and formulate scientific knowledge."[68] For him, the history and the sociology of science have some things to say about the way the philosopher of science marks the frontier between the inessential and the essential features of scientific rationality. A theory of scientific rationality requires empirical groundwork.[69]

A good example of that relevance is the role that the sociology of science may play in clarifying concepts that have usually been related to philosophy of science alone, and more particularly to key concepts in the recent empiricist tradition (broadly understood), as verification and falsification. For Kuhn, the sociology of science may contribute to our understanding of the scientific group with many interesting results. Although he is not completely clear at this point, he seems to be praising the sociologists for the intrinsic interest of their own field by itself, without regard to its application to other fields, when he says that the "main reason for working on those more sociological problems is indep[endent] of their relation to [the] phil[osophy] of science," a statement to which he adds: "Interesting." And concerning the kind of problems that would fall into that category of sociology-for-its-own-sake, so to speak, are "understanding of scientific life, perhaps science policy. [And s]cience as a social enterprise."[70] However, as noted, Kuhn does think that the sociology of science would be highly relevant for the philosophy of science when it comes to knowing more about verification processes. He often says that verification is a more complex process than is usually assumed. After all, science is a social activity. "As with language," he writes, "[you] don't have scientific knowledge unless you've got

---

64 Cohen (1974, 310, fn. 10). Cohen includes Quine in the same class.

65 See Footnote 62.

66 Rockefeller 1, 19.

67 See Footnote 51.

68 See Footnote 51.

69 Rockefeller 1, 17; see also 13.

70 Rockefeller 6, 13.

a group that shares it, applies it in similar ways, [and] recognizes its application."[71] This statement is not surprising, as it completely agrees with the reasons for him to introduce the very concept of paradigm in *Structure*. He goes on by saying that scientific knowledge, "in this sense, [is] intrinsically social," and that there is "[n]o private science as [there is] no private language. That's madness."[72] So, in the same sense, verification of theories, whenever that process takes place, is a complex course of events in which a group adopts a specific position and considers the significance of checking a theory with nature—and "with other theories," to use his words—within given limits. In that respect, sociology may be useful for a deeper understanding of the actual process—which is an intrinsically *social* process in spite of the philosophers' claims to have abstracted its essentials.[73]

In spite of his critical attitude towards rational reconstruction—which, we should not forget, he does not try to get rid of, but rather to reform—Kuhn continued his argument with its original logico-empiricist supporters insofar as he still found their associated models valuable, at least as starting points that would eventually be objectionable. His collaboration with Hempel followed this tacit rule. In examining Kuhn's account of historical knowledge in the Rockefeller seminar in Sect. 7.7, our point of departure is, once again, the logico-empiricist view of rational reconstruction. Yet, before turning to that point, we must examine how Kuhn recounts his new vision about paradigms and how they are a solution to some problems of the empiricist tradition. We look at this latter point in Sects. 7.4–7.6.

However, before that, there is a further point to examine. Our preliminary approach to Kuhn's attitude towards rational reconstruction in the current section has supplied some information about points (a) and (b) as introduced in our last section. Now we must inspect another recurring argument that he presented at Rockefeller University and elsewhere in order to grasp what was included in our point (c). That examination will help me to close this section.

One of Kuhn's concerns after *Structure* was the autonomy of his concept of scientific community with regard to that of paradigm. In other words, he wished to answer affirmatively to the question whether (and to what extent) the notion of scientific community is independent of a description like "those scientists that share a paradigm," avoiding the problem of circularity involved—the paradigm also seemed to need the notion of scientific community itself.[74] On top of that, the notion of paradigm also awaited further clarification, as we shall soon see. He was aware that he had assumed too hastily that the description of the scientific community that he had given in *Structure* was enough, so he started to revise that notion. He had understood in that revision that talking about the community did not involve talking about the *full* profession. That overlap might be helpful in order to see how *some* changes took place, but revolutionary change does not always involve a whole disciplinary field. Accordingly, it would perhaps be more appropriate to take into account the *size*

[71] Rockefeller 6, 15.

[72] See Footnote 71.

[73] See Rockefeller 6, 14–15; see 14 for the expression within quotation marks.

[74] See Postscript–1969, 175.

of the group, because the alleged community–profession identity should not always be assumed.[75] After all, alluding to the size of the event was also relevant in the case of scientific revolutions, too, as he had already mentioned in *Structure*, and as we shall see again below.[76]

An examination of his notes for the sixth meeting at Rockefeller University shows that he had started to become aware that the integrity of the community, so to speak, faded away at some levels of specialization. As a nuclear or solid-state physicist, an individual belongs to a greater community of natural scientists, but also to a lesser community of physicists (generally speaking). Its size is dramatically reduced beyond that point and it sometimes includes "half a dozen men who are the experts on [a] particular range of problems."[77] In addition, the members of such groups of limited size are simultaneously members of other groups of similar size in their vicinity. These groups may be "shortlived," he says, and they "[m]ay split, [they] may merge."[78] So, the underlying scientific community, as the new agent of scientific knowledge and development, is itself subject to historical change.[79] It may even be ephemeral at some points, and oddly enough, in some rare historical cases, they may be reduced to individuals.[80] So, if the scientific community is the group that experiences normal progress, a crisis, or is upset by a revolutionary change, and if the dangers of circularity prevents us from using paradigms to pick its instances out, it would be appropriate to identify it as a kind of historically changing agent itself. After all, identifying the individuals or group in charge of a scientific change, and also how a group experiences it (i.e., is affected by its consequences), are parts of the mechanism involved in a good explanation of theoretical change.

Kuhn started to talk about the sociologists' resources to identify scientific communities themselves. Recent research in the sociology of science became important for him, and not only Merton's pioneering work. His primary recommendation for the Rockefeller audience was Warren O. Hagstrom's *The Scientific Community*, a recent book (1965) whose preliminary version as a PhD dissertation he had read at Berkeley in November 1963.[81] According to Kuhn's summary, Hagstrom's 1963 dissertation, "Social Control in Modern Science," showed a deep empirical research into some "central norms in modern science" and the way they emerge and consolidate, and

---

[75] Rockefeller 6, 9–10; "Paradigms and Theories in Scientific Research," 19 February 1967, TSKP 3.14, 5, 9. See also his Postscript—1969, 175–180, where the results of this line of research on communities during these years were summarized.

[76] See *Structure*, 6–7, 49–51, and, as noted, SSR1A, SSR1B, 49–51.

[77] Rockefeller 6, 10.

[78] See Footnote 77.

[79] As he says in the second Swarthmore talk, "scientific knowledge is what they [i.e., the members of the scientific community], as a group, know." At the same time, he says, "If anything like a gestalt switch occurs, it is they who experience it, and it may be only they. By the same token, they're the bearers of norm[al] sci[ence]." [Kuhn], "Paradigms and Theories in Scientific Research," 19 February 1967, TSKP 3.14, 4.

[80] Rockefeller 6, 10–11.

[81] See Hagstrom (1965). Also, see Kuhn's letter to Marion J. Levy, Jr., 14 November 1963, TSKP 14.11, 1.

he had interviewed many scientists to do it.[82] Kuhn very much liked Hagstrom's research. His resources to talk about the scientific community were very useful in Kuhn's new case for the autonomy of the community with regard to paradigms.

Let us think for a moment about the groups that work on a certain problem in order to understand what the "fine structure"—to use Kuhn's expression—of the scientific community that we are looking for is.[83] A given problem may attract the attention of individuals from pre-existing groups. These groups may be unique and the individuals have similar training and vehicles for communicating their results, but the case may be different, too. Individuals that contribute with relevant solutions to the problem may come from groups that have no common media for their results and that may even have recourse to different methods and techniques.[84] If there is not a full communication breakdown between them, there may be some obstacles to it, and it is not possible to conclude that they form a unified community. Of course, the problem itself should not be the way to conclude that they originally formed a group, though they may end up forming one.

Kuhn's resources for isolating the relevant community—that is, the one from which the historian's research starts—involve some sociological techniques, such as Hagstrom's use of interviews. Kuhn had recently had experience with such a resource on the basis of his own work with the sources and pioneers of quantum physics. As he says, "you get answers":

> Research scientists do respond when asked[,] who works on the same sorts of problems? Whose papers they follow regularly? With whom they exchange pre-prints of their publications[?] With whom do they regularly discuss scientific questions by letter or in person[? ...] Many scientists do recognize themselves as members of special communities. In this area they do respond to sociologically oriented questions as they do not in others.[85]

Though Kuhn indirectly refers to Hagstrom's thesis here, his own experience with this activity also enriches this description and its success.[86] Another resource is, he says, the "examination of patterns of footnote citation in the published literature."[87] Here, Kuhn refers to the M. M. "Kessler group at MIT" and their systematic resources.[88] He is following that path, too. As he says, "I'm beginning to work towards a citation index of literature which I hope may be made to show groups and changing relations among them, checkable against patterns observed in other ways."[89] Although both techniques were only starting to produce results and there were open issues, Kuhn seemed confident that it would produce results in the long run. In fact, some results of that line of research appear in his *Black-Body Theory*, Chapter

---

[82] Kuhn to Levy, 14 November 1963, TSKP 14.11, 2.

[83] See [Kuhn], "Paradigms and Theories in Scientific Research," 19 February 1967, TSKP 3.14, 5.

[84] See [Kuhn], "Paradigms and Theories in Scientific Research," 19 February 1967, TSKP 3.14, 6.

[85] [Kuhn], "Paradigms and Theories in Scientific Research," 19 February 1967, TSKP 3.14, 7.

[86] See Rockefeller 6, 11. He mentions Hagstrom by name in the same place. At Swarthmore, he only alludes to the existence of such a work.

[87] [Kuhn], "Paradigms and Theories in Scientific Research," 19 February 1967, TSKP 3.14, 8.

[88] Rockefeller 6, 11.

[89] Rockefeller 6, 12.

IX, where he discussed the authors that, in the early twentieth century, devoted their efforts to the problem of specific heats as compared to those who pursued research on the black-body problem.[90]

This kind of study sets the stage for his new inquiry about paradigms. If points (a) and (b) showed that a historically and sociologically well-informed philosophy of science may be the proper way to understand scientific rationality and reconstruct its activities in scientific practice and development, point (c) shows how sociology helps us to find out the particularities of the new epistemic agent in science. Now, points (d) and (e) should show how to differentiate the pieces of scientific method, results, and applications that form scientific life within the proper group, and what kind of learning and research is involved in such practice. The key part of Kuhn's research on *Structure* after its publication seems to reach its peak during these years. The arguments he presented to small audiences are the basis for his well-known publications from 1970–1974, as we shall see below. Now, let us turn again to paradigms.

7.4 Paradigms: What They Are

In the talks and the seminar that we discuss in this chapter, Kuhn revisited the notion of a paradigm for a reduced audience. Its examination adopts the form of a series of reflections that he made out loud during those years (roughly, 1965–1967), especially in the case of the Rockefeller seminar, which was prepared as a discussion of key and controversial theses from *Structure* with the audience—the notion of a paradigm was one of them, without doubt. We should bear in mind, however, that the notion of a paradigm was also in transition for him at this point. Debates with Shapere and Hanson in late 1963 at Cleveland and, more importantly, with Masterman in 1965–1966 had brought some issues to light that he wished to fix.[91] So, when he gave a talk for anthropologists at Yale in February 1965, he was in the middle of a long reflection on paradigms that would intensify with his interaction with Masterman from the fall of 1965 to the spring of 1966. His seminar at Rockefeller in late 1966 was therefore a good opportunity to rehearse his new ideas—from roughly 1963 onwards—in public, and he would be doing the same for the rest of 1967. He would do so when he, for example, travelled to Washington, DC, in September 1967 in order to comment on Polanyi's philosophy of science. All the reflection and corrections pursued in those years were the basis for the new published revision of *Structure*'s themes.

Paradigms were clearly the focal point of his new reflections. Not only, nor primarily, because Masterman had discovered so many senses of the notion of paradigm—though that surely disturbed Kuhn—but rather because Masterman had showed him that she properly understood the importance of the notion for an alternative to extant philosophy of science in the empiricist tradition. The views he had been

[90] See *Black-Body Theory*, Chapter IX, esp. 207–210, 214–220.

[91] See our Chap. 6.

thinking about concerning the role of the Wittgensteinian views on family resemblance for the structure of paradigm-based learning made that piece of his theory truly important for his further development of the discontinuism involved in *Structure*. So, during those years—as noted in the previous chapter—his work on paradigm-based learning attracted most of his interest, and he left aside the other problems that we have been studying so far. His new viewpoint about that kind of learning as an alternative to logico-empiricist views of science included making explicit a better analysis of that kind of learning, of the notion of a paradigm and of its role in it. Most of his talk and seminar notes from these years (again, 1965–1967) include an account of parts, or of the whole, of that analysis as a core theme.

Of course, the starting point was to make it clear what a paradigm was. Kuhn had been worried about that point since his early confrontation with Shapere in 1963. In 1964, he had been thinking about the proper meaning of the term, which he assigned to the exemplary solved problem or model alone—leaving aside the qualifications that he made for Boring on the identity of paradigm and model, and the change in the assignment of the word to different solved problems or accounts of a similar solved problem as the community and its members evolved historically and theoretically.[92] After Masterman's challenge, in late 1966, and almost at the same time that he was displaying the complexity involved in the fine structure of the community, he conceded that the notion of a paradigm required revision. Admittedly, there are techniques to identify the community independently from the paradigm, but, on top of that, what exactly was the piece of theory on which that elusive semi-organized group (the scientific community) used to work? That is, what was the paradigm? "In [the] book," he wrote for the Rockefeller seminar, "all elements relevant to [the] practice of science are the group's paradigm, or are paradigmatic, or are parts of their paradigm or are themselves paradigms. [...] Now I want to separate out a special element for this phrase (though I doubt I can do it in political practice)."[93]

Kuhn had some doubts about the proper way to name the different aspects of the group of pieces of the scientific practice within a community that had fallen under the term "paradigm," but he was pretty sure of what a paradigm was not. For instance, the "metaphysics" underlying a given practice, "say atomism," he wrote, is not a paradigm—it is simply the *shared metaphysics*. He also wrote, we "[d]on't use [the term] paradigm just because [it] is shared."[94] "Shared laws," he added, were no such thing either—they are just the "shared laws."[95] As regards the group's metaphysics—which, he says, is "an ontology for the formalism, if you will"—he terms it "the community's metaphysics" as a denomination, or "the collective metaphysics of the group."[96] This part of the set that he formerly called "paradigm" includes "entities", and "forces"—elsewhere, he mentions "powers." It is the basis

[92] We examined his reflections on this point in our Chap. 6.

[93] Rockefeller 3, 14.

[94] See Footnote 93.

[95] See Footnote 93.

[96] See the latter in "Paradigms and Theories in Scientific Research," 19 February 1967, TSKP 3.14, 13. The rest of the quotations are from Rockefeller 6, 5.

for the proper functioning of laws and also limits "the regress of explanation," that is, it marks the ultimate pieces on which a scientific explanation is built—good old scientific objects in the early 1950s. Finally, there is also a normative role for them, as they are basic for choosing problems and for assessing problem-solutions. The group's metaphysics "[i]s part of the tool kit in research."[97] So, "men who thought that light was [composed of] particles looked for the pressure that it exerted when falling on dense bodies. Wave theorists had no room for such pressure until after Maxwell's theory made it part of physics again."[98] However, that role is not always "essential," he says: "19th century chemists work[ed] for some time without agreement on atoms."[99] Actually, the revolutionary or normal measure of a new problem–solution may depend on the extent to which the group has a real commitment to a particular metaphysics. Accordingly, this latter would be, in such circumstances, a necessary condition for revolutionary change.[100]

This vision is not far off the one that a younger Kuhn had developed concerning the role of orientations in science, with some of them intended for metaphysical and normative roles in science, back in the early 1950s, in the Lowell Lectures.[101] Moreover, the idea that physical laws necessitate an ontological background with causal powers and entities—some admissible and some others thinkable but physically inadmissible—as the domain of application traces back to his 1945 essay on causal connection in physics.[102] As in other cases, old ideas, properly adapted and revised, form the grounds on which Kuhn progressively built his perspective.

Concerning laws, in contrast with the previous component, they are shared by larger groups of scientists beyond the highly specialized ones around a given problem—recall our previous discussion at the end of the last section. Because of that extended acceptation and use, expressions like Schrödinger's equation or Newton's laws (which are among Kuhn's examples) are less likely to change; as Kuhn says, these are "changed only in major revolutions."[103] This sort of stability does not mean that the role they play in theory as definitions or as empirical generalizations is always the same. Here, Kuhn gives some glimpses of the (nevertheless brief) treatment of the changing semantic and epistemological status of laws that he would provide in the forthcoming "Postcript—1969" to *Structure*.[104] Laws are key in the "application of deductive logic and often of mathematics," he adds, and these latter—logic and mathematics—are also part of the group of elements within the

---

[97] All previous quotations are from Rockefeller 6, 5, except for "powers," which comes from "Paradigms and Theories in Scientific Research," 19 February 1967, TSKP 3.14, 13.

[98] [Kuhn], "Paradigms and Theories in Scientific Research," 19 February 1967, TSKP 3.14, 13.

[99] Rockefeller 6, 5. See also "Paradigms and Theories in Scientific Research," 19 February 1967, TSKP 3.14, 13–14.

[100] See [Kuhn], "Paradigms and Theories in Scientific Research," 19 February 1967, TSKP 3.14, 14.

[101] See our Chap. 3.

[102] See our Chap. 1.

[103] [Kuhn], "Paradigms and Theories in Scientific Research," 19 February 1967, TSKP 3.14, 11.

[104] See Postscript—1969, 182–183.

former concept of paradigm. In addition, the techniques of measurement and observation, which are often part of instrumentally complex lab work, share in these group commitments too.[105] Here we may notice the absence of another important ingredient in these commitments—the values that serve as basis for individual decision in theory choice.[106] Kuhn does not refer as yet to this normative element by itself, but it is involved, as noted, in his discussion about the metaphysical ingredient. He, however, would later pay attention to normativity in its own terms as he had done before—recall his normative orientations in 1951.[107]

Now, the question was how to name that set of commitments, insofar as Kuhn wished to show that the referent for "paradigm" should be a smaller key piece in his account of the kind of learning and research that takes place in normal science—a sample solved problem.[108] I shall return to this piece, in the context of its role in learning, in Sects. 7.5 and 7.6. Kuhn started his discussion at the sixth meeting of the Rockefeller seminar by showing his frustration at having applied "paradigm" indiscriminately for all these ingredients in *Structure* ("Wish I hadn't," he wrote).[109] His alternative could not be expressions such as "Professional ideology," or "Professional *Weltanschauung*, he said"; he even considered simply "Propaganda," but most probably for a very short while. An expression emerged: "professional matrix," and "matrix" is, in particular, a repeated term in all the relevant cases.[110] As we know, "disciplinary matrix" would eventually be the chosen phrase.[111]

Eventually, all these details concerning the meaning of "paradigm" formed the basis for Kuhn's Postscript to *Structure* and related essays (more on this below). One more detail helped to show what the real meaning of "paradigm" is—its ascription to mature disciplines alone. As he would show in the Postscript in a few years' time, it is not correct to say that difference in maturity is the emergence of a paradigm. Actually, there must be paradigms for groups of scientists even though these groups do not form a cohesive group all at once. After all, we have already seen how Kuhn talked about the complexity and the ephemeral character of small groups that do not overlap with an entire profession, and it would be too much to say that only sample solved problems that are shared by the members of the community as strictly overlapping with the profession are paradigms. Whether or not this latter is the explanation that Kuhn had in mind, the fact is that he wished to use "paradigm" for something more than the piece that emerges in that stage—the maturity of a discipline, on that account. Whether this piece exists or the profession is disaggregated into subgroups with an

---

[105] See [Kuhn], "Paradigms and Theories in Scientific Research," 19 February 1967, TSKP 3.14, 13–14; Rockefeller 6, 5.

[106] See Postscript—1969, 184–186.

[107] See our Chap. 3. See also Kuhn (1973) for a further separate treatment of values in theory choice.

[108] See Rockefeller 6, 4–5; [Kuhn], "Paradigms and Theories in Scientific Research," 19 February 1967, TSKP 3.14, 15 ff., esp. 17.

[109] Rockefeller 6, 4.

[110] See Rockefeller 6, 4, 7; [Kuhn], "Paradigms and Theories in Scientific Research," 19 February 1967, TSKP 3.14, 10, 15.

[111] Postscript—1969, 181. Kaiser (2016, 87–88) also talks about the last two phrases.

internal piece in common, Kuhn wished to be able to use "paradigm" in each case. So, the term "pre-paradigm" should rather be lost. "I want also to be able to say that the schools as well as the later community share paradigms among their members. [...] [']Paradigm['] is [a] wrong term to use in drawing [the] distinction between earlier and later developmental stage."[112] As Paul Hoyningen-Huene said, something like "pre-normal science" is probably better for such historical phases previous to the first emergence of a common paradigm for the whole profession.[113]

The proliferation of groups with a varied set of commitments and with different scales of shared examples (i.e. paradigms) makes for a more complicated scenario than the one depicted in *Structure*—and, undoubtedly, a promising and challenging one. It is, however, a scenario in which the classic depiction of a monolithic community that shares a common problem–solution that is replaced in a great scientific revolution becomes somewhat simplistic. Kuhn must have had in mind a picture like this one during that period. Yet, a more central piece of interest in this renewal of his former theory attracted most of his interest in these years: paradigm-based learning and the way it involved a full alternative to mainstream philosophy of science. This section has described theme (d) in our list, and we should now turn to theme (e).

7.5 Paradigms and Similarity-Based Learning

In his talk at Yale for the anthropology colloquium, Kuhn had the chance to offer some glimpses of his new, richer perspective on paradigm-based learning (as compared to his introduction in *Structure*) as something opposed to classification of objects, events and substances based on rules—and the kind of learning that follows from that point of view. In fact, at Yale, the key idea was that almost no classification is based on arbitrary decisions and that natural resemblance is the basis for those that work, or have worked in the past. Those, for instance, employed in natural history in the seventeenth century are "natural," he says; they are "not simply rand[om] facts arranged for convenience." Their bases are not correct for us, but it does not mean that such classifications—such as the Linnaean one, for example—are "arbitrary or artificial"; that "image of the natural historical or Linnaean period in the development of a science is largely a myth," he says. For him, all this criticism, of a rather historical nature, leads to the thought that "the classificatory problem is of a somewhat different sort," which he aims to demonstrate in this talk.[114]

This is a significant point at this stage in his intellectual development. I am not referring to this talk in particular, of course, but rather to the kind of hinge, or axis,

[112] Rockefeller 3, 14.

[113] See Hoyningen-Huene (1993, 188–189). Kuhn admitted that this issue was difficult: "[I n]eed to handle the entire maturation problem differently," he wrote. And added: "This goes deep" (Rockefeller 3, 14).

[114] All these quotations and ideas in this paragraph come from Kuhn's "Thoughts on the Nature of Scientific Classification," 25 February 1965, TSKP 3.14, 4–5.

that he finds during these years (roughly, 1963–1965) between two classical themes for him: the way we and scientists classify incoming information from the outside—something that traces a long way back to the years 1949–1951—and its relationship with the role of paradigms in learning, a point that can be located in the last stages of preparation of *Structure*.[115] When he demonstrates this axis connecting two aspects of his philosophy to an audience at Yale—as he would do soon at Swarthmore, at Rockefeller, and then before an audience of psychologists, Quine, Scriven, and Polanyi in Washington, DC, in 1967—he is showing a piece of genuinely new work. This common axis unifies these talks better than any other theme, so they are an informative signpost for Kuhn's direction for the next years. He returns again and again to that hinge as a basis for his own explanations, whatever the subject of a gathering to which he has been invited—whether anthropology, psychology, or philosophy of science. Classification, or taxonomy, would be his overall concern from this point on, and focusing on lexicons in the near future would be a way to deal with it. As noted, however, it was not a new theme at all. Yet, its relationship with paradigms *was* newer at this point, and that relationship was still awaiting to be better exploited. That was Kuhn's main task during these years.

If we return for a while to the research work in philosophy that he was pursuing during these years, we can recall that he not only aimed to achieve a proper analysis of paradigm-based learning, but also to obtain the helpful contribution of computer simulations. In fact, in his talk for the American Psychological Association in the very early days of September 1967, he used a different kind of vocabulary. He started to talk about "two [...] different modes of data processing" when he refers to the representation of scientific knowledge on the basis of definitions and explicit generalizations as something different from his own paradigm-based account and from Polanyi's "tacit knowledge," too.[116] As he said in Washington, Polanyi's view and his own "are not identical, at least not at present state of develop[ment]," but they "are seeing the same aspect of scientific education and procedure."[117] For Kuhn, this common, Kuhn–Polanyi view gives rise to one of those modes of data processing; the other mode was the kind of explicit representation that he criticized in authors like Reichenbach, Hempel or Nagel, and that we have seen in Sect. 7.3. For him, "both are to be seen as information processing programs, differentiated by the programming strategy, and both are equally capable [...] of being modelled on a computing machine."[118] In this talk, Kuhn goes on to say that it would be possible to change the strategy in data processing from one mode to the other; the tacit, or paradigm-based, knowledge may shift to a form of knowledge based on explicit generalizations and definitions. However, he shows that an adaptive property is lost in the process.[119] We will look at this later.

---

[115] See also our Chap. 6 for further details about Kuhn's private research on that common hinge.

[116] See [Kuhn], "American Psychological Association," 2 September 1967, TSKP 3.14, 3.

[117] [Kuhn], "American Psychological Association," 2 September 1967, TSKP 3.14, 1.

[118] [Kuhn], "American Psychological Association," 2 September 1967, TSKP 3.14, 3.

[119] See [Kuhn], "American Psychological Association," 2 September 1967, TSKP 3.14, 10–11.

An important detail that is worth emphasizing is that Kuhn does not give in to the temptation of polarizing the perspectives and identifying each pole with the "wrong" and the "correct" option, respectively—his and Polanyi's views being located at the latter pole, of course. At some points, making the tacit explicit is a way to apply "the tools of logic and mathematics," Kuhn says, which in the other sense "are scarcely applicable," and "you have little or no science, though perhaps much low-level knowledge."[120] As we shall see later, Kuhn would find arguments against accomplishing this latter option—at least, in full.[121] However, he is not a supporter of a contrary idea, either—namely, that scientific knowledge involves some hidden and inexplicable component that is not possible to analyze in some sense. Indeed, Kuhn is a skeptic—just like, in his opinion, Polanyi is as well—"as to whether this [i.e. Polanyi's tacit component and his own paradigm-based] component of knowledge can be reduced to formalisms, that is, reconstructed in a vocabulary of basic terms plus certain generalizations about them."[122] But it does not mean that he wishes to turn that sort of skeptical attitude into a different kind of faith in something even more mystifying: the activity of intuition, whose powers are quite mysterious when they are put into play. Instead, he wishes to

> Eliminate recurrent implication of what Dr. Polanyi and I have said in the past. [...] Tacit knowledge or paradigm-based knowledge has too often seemed to be explicitly formulated knowledge with some ingredient missing. The terms "indefinable" and "incomplete" recurred in Dr. Polanyi's talk last night. The gap they indicate is filled by a somewhat mystical power of trained and committed intuition. [...] I'd like to suggest that tacit knowledge may be just as concrete and analyzable mode of cognition as explicit knowledge. [...] Rather than tacit knowledge being explicit minus—the gap being filled by intuition—we may find that the two are simply different modes of data processing.[123]

Yet another way of expressing the idea of "essential tension" can be seen in this particular search for equilibrium. At the same time that the elements of a tradition in scientific knowledge start to be fixed, the systematic treatment of knowledge is not only convenient but also sometimes necessary. That is a pathway to the kind of establishment of scientific knowledge and tools that are part of normal science.[124] However, for Kuhn, that establishment should not suppress the adaptive properties of the kind of vehicle for approaching nature that scientific knowledge is, as we saw in his Lowell Lectures. Explicit knowledge should not obstruct the natural improvement of that vehicle. The road to divergence should thereby be left open.

In all this discussion, the word "intuition" appears more than once. For us, the term is full of meaning for Kuhn's intellectual development. In a 1955 letter to his friend, and psychoanalyst, Lawrence Kubie, he was very critical of the thesis that intuitive elements in the scientist's mental activity such as free associations, for example,

---

[120] [Kuhn], "American Psychological Association," 2 September 1967, TSKP 3.14, 10.

[121] See [Kuhn], "American Psychological Association," 2 September 1967, TSKP 3.14, 11–14.

[122] [Kuhn], "American Psychological Association," 2 September 1967, TSKP 3.14, 2.

[123] [Kuhn], "American Psychological Association," 2 September 1967, TSKP 3.14, 2–3.

[124] Hoyningen-Huene (2013) is a detailed study of that characteristic of scientific research that, for him, is a key feature of all science that is worth its salt.

should be suppressed, at least from the ultimate, explicit presentation of theories and their results. Reasoning must be logically, explicitly representable. For Kuhn, Kubie, who seemed to support such a thesis, was wrong. As usual, however, Kuhn showed that the roots of that problem could be found in hidden premises, or assumptions, that were customary when it came to thinking about scientific reasoning. For him, there were (ironically) free associations of the intuitive in common thinking with the preconscious, in the sense that this was a quick processing of information that would be slowly analyzed (e.g., in terms of a formalism) once it was explicitly expounded in the conscious and rational part of our thinking. This kind of work in parallel made the intuitive an automatized version of conscious thinking. It was the somewhat obscure, hard-to-explain, version of rational thinking. Neither were true for him. Intuitions, he said, were not mere extensions nor automatizations of the rational thinking. They were just another way of dealing with information. In 1967, more than ten years later, he seemed to have the proper tools for answering that early view on intuitions that he had conveyed to Kubie and for analyzing and throwing light on that concept. Intuitions were not a mysterious component of thinking, visible in tacit, or paradigm-based, knowledge. It was not just as if reason itself was playing fast and loose with information, something to be fixed in a further step. Rather, intuitive thinking was a basis for science, and it was not easily reducible to a logical scheme. As he used to say back in the late 1940s, when in the Society of Fellows, psychological plausibility was one of the key components of scientific thinking together with the logical analysis and the empirical testing of theories, and it was probably the *most important* aspect of it. Kuhn had stuck to that thesis and was now ready to express it more clearly. It is not surprising that his letter to Kubie appears among other drafts and notes on family resemblance in 1965.[125]

What was Kuhn's alternative vision? We have examined its early steps in the last chapter, as he was developing it from 1963. Now, we can make further points about the perspective that made it, in Kuhn's eyes, a valuable piece of theory for his own perspective. Besides that, one of his most famous examples, Johnny's learning of the common-sense classification of waterfowl, appears at this point as well, as a more understandable approach to his recent, more formal approaches to paradigm-based learning. Let us look at this in more detail.

We have examined what elements are shared by the community—whatever its size—when its members devote their time and effort to a given problem. Kuhn was at this time calling the group of shared elements its "professional matrix," and similar denominations, as we saw. There is a piece left; as Kuhn says, it is "an important residual component," and more importantly, it is an element "that's learned by doing standard problems, theoretical and experimental, a component that plays [a] fundamental role, but that's not articulable in these terms."[126] The "residual" element is, of course, the paradigm. As we know, he would later use another term to refer to it explicitly (more on this below), but at this moment it still has the old usage. Its relationship to learning is the key piece in all this. The formal model of learning

[125] See Kuhn's letter to Kubie, 17 March 1955, either in TSKP 15.53, or in FR65.

[126] Rockefeller 3, 15.

Fig. 7.1 Kuhn's scheme for processing stages

that he had been developing since 1963 served as a theoretical basis for a computer simulation that he had started developing in late 1966. There, he shows the basis of a kind of learning in which such examples, such residual pieces—in the end the most important ones—of scientific practice, are the pieces out of which a given picture of the whole environment of practice is built. And, of course, it is done without the contribution of explicit rules. This is a key piece of *Structure*'s old message, only that Kuhn's explanation is a bit more detailed.

As we saw in the previous chapter, Kuhn had the conviction that we can show that the percepts that form our whole perceptual world—recall his old idea of a behavioral world[127] as its forebear—can be arranged on the basis of incoming information, which can *later* be analyzed as a series of perceptual elements that may in turn be helpful to make a description of the percept. Recall, however, that he uses that perspective in order to argue against the idea of closed-texture. That is, the acquisition of perceptual knowledge about the world is of a different kind. It involves processing stimuli in order to produce a response—a process that produces *percepts* as the earlier basis for a verbal response. It is actually our primary source of experience of the world. In the talk to the American Psychological Association, he provided the scheme of Fig. 7.1.[128] In that talk, Kuhn pays attention first to the "Stage I," whose extreme to the left, the one that corresponds to incoming stimuli, is for him the least transparent of all the elements involved. Kuhn says:

> If the response is challenged—"Wasn't that a woodchuck, not a rabbit[?]"—it's not the stimulus but its intermediate trace, the percept, that's recalled and discussed. [...] The stimulus is not, even in the first instance accessible, except by a highly theoretical deduction to which the percept—which is given—is fundamental. [...] And one is proceeding from the percept, not from the stimulus, when one responds to a change by drawing on explicit knowledge: "No, it's a rabbit. It had long ears."[129]

For Kuhn, Stage I is particularly important, though its opacity makes it material for psychological research. Yet, the important point for him is that *percepts* are our *first* resource either to theorize about what truly happens in Stage I, or to correct the application of kind-terms in an ordinary situation.[130] So, the analysis of percepts in

---

[127] See our Chap. 3.

[128] See [Kuhn], "American Psychological Association," 2 September 1967, TSKP 3.14, 4.

[129] [Kuhn], "American Psychological Association," 2 September 1967, TSKP 3.14, 4–5.

[130] Kuhn notes that "Gestalt experiments suggest [that] the same stimulus can produce different percepts," while "[c]olor mixing experiments show that very different stimuli can produce the same percept. [...] So, there's much data processing, some innate but much of it learned, that goes on between stimulus and perception." As he says shortly before, "there's no one-to-one mapping" in the relationship between these latter. Extracts in this footnote are from Kuhn's "American Psychological Association" notes, 2 September 1967, TSKP 3.14, 5.

terms of perceptual elements, or qualities, need not be considered the basis of our structure of learning. Kuhn opts for considering the access to percepts as the earliest stage in our contact with the perceptual world, and to proceed from that point on. From that point of view, it is easy to understand that, even though his mathematical model and the computer simulation that he wished to build from it were intended to construct a universe of objects (i.e., a perceptual world) from *qualities*, these latter only represent incoming *information* for that kind of simulation. Despite their presence—which could have been replaced with series of stimuli—Kuhn's aim was, as he showed in New York, to argue for the possibility of a theory of perceptual learning not based on rules and definitions. Such rules and definitions, and having recourse to the quality-terms involved in them, could be helpful in the learning process, but he nevertheless tried to operate with something different, such as (to use his words at Rockefeller) the "psychologically and logically prior" step: "a learned similarity relationship" between those percepts.[131] This latter was the key step in his view. Percepts, and similarity between them, were more "primitive," as he also said.[132]

Kuhn's main thesis here is well expressed in this paragraph from his notes for the second session of the Rockefeller seminar:

> When I get to paradigms [I] will want to say: learning from paradigms is learning to recognize the similarity of the exemplars of the paradigm. To see concrete things as similar. [...] Furthermore, similarity is here a primitive relationship in that it's prior to any knowledge about properties with respect to which judgment of similarity is made. [...] And it stay[s] partially primitive in that it's never fully reduced to knowledge of properties. Or rather, that reduction changes its character, resulting ultimately in different future decisions.[133]

At Yale, in 1965, he talked about the formation of such natural families on the basis of "[n]atural resemblance," a relationship between members of a certain class that is based on family resemblance. "[T]hings," he says, "get put together, in a single classificatory box because they look alike." That relationship—the family resemblance between the members of an instance of such classes—is not "simply given," he also says. If we talk about children's perception, they only get at most "a bloomin' buzzin' confusion" (he borrows James's famous phrase once again). There is therefore a strong learning component behind our ability to perceive the world—to compose our perceptual world—the way we do. At the same time, though, it is possible to think that "others cut up the world differently, [and] see resemblance where we see difference and conversely." The gist of this thesis is that all this takes place in natural classifications, in the way we and "others cut up the world," as he says, without recourse to the kind of sometimes *arbitrary decisions* that are rendered in definitions of classes and attached empirical generalizations.[134] Although changing

---

131 See Rockefeller 4, 8.

132 See Rockefeller 2, 3.

133 Rockefeller 2, 3–4.

134 See [Kuhn], "Thoughts on the Nature of Scientific Classification," 25 February 1965, TSKP 3.14, 6–7. All quotations in this paragraph are from those pages.

and evolving, learned similarity between particular percepts holds priority over the latter component. At Swarthmore, he made this clear:

> [T]he learned perception of likeness or similarity is prior to and does not imply the existence of a set of criterion [sic] which would provide a basis for the judgment of likeness. [...] The question "Similar with respect to what?" is a question that need not be answered in full to give basis for the use of a paradigm as a model for a particular problem solution.[135]

A central point in this thesis is the idea, emphasized above, that the recourse to family resemblance helps to avoid the *arbitrariness* involved in making *decisions* about the proper contour of the classes we are alluding to. Kuhn introduced this point to a wide audience in these lectures and, on the basis of a simple example, later on in "Second Thoughts on Paradigms." In the end, this example represents the core of his project of a machine's creation of a universe of percepts and, then, its teaching of that universe to a second machine. It was clearly simpler to illustrate this point on the basis of a common-sense case—an individual teaching another to learn about the former's perceptual world; and it would be even easier if we thought about the way parents teach their children how their common world is arranged. Kuhn did it that way. At Yale, the case was based on the first kind of illustrative example—an individual teaching his or her very limited perceptual world to another; at Rockefeller and Swarthmore, it was Johnny (a child) and his father, who was teaching the former to classify different kinds of waterfowl. This latter, an example from 1967, would be the one that would end up filling the pages of "Second Thoughts on Paradigms," and of a key chapter from *The Plurality of Worlds*.[136] These examples are intended to illustrate the central point I have mentioned: making the distinctions between classes explicitly clear brings with it the danger of being unable to deal with future counterexamples, which will surely emerge in the future.[137]

This point is important for Kuhn. Recall that he had been arguing about the importance of including the feature of open texture for meaning whenever one wishes to make it clear through definitions. For him, Braithwaite's and Waismann's works help us see how the texture-closure of theoretical terms is counterproductive when facing unexpected counterexamples. In that case, previous arbitrary decisions concerning that closure—and Braithwaite's model was particularly helpful in that sense—prevent us from seeing how the application of such terms could be extended to those new cases. From his early work in the Lowell Lectures, that adaptive feature of natural languages should be preserved at any cost. I am referring, of course, to the *flexibility* (formerly *vagueness*) involved in natural languages. Preparing our linguistic vehicle for dealing with the perceptual world surrounding us on the basis of natural families created by virtue of family resemblances (the learned similarity relationship between percepts that Kuhn talks about) helps us to train that vehicle just

[135] [Kuhn], "Paradigms and Theories in Scientific Research," 19 February 1967, TSKP 3.14, 18.

[136] See [Kuhn], "Thoughts on the Nature of Scientific Classification," 25 February 1965, 8–10; "Paradigms and Theories in Scientific Research," 19 February 1967, 20–26; both in TSKP 3.14; Rockefeller 4, 5–9; Kuhn (1974); *Last Writings*, Chap. 5, esp. 240 ff.

[137] [Kuhn], "Thoughts on the Nature of Scientific Classification," 25 February 1965, 9; "American Psychological Association," 2 September 1967, 12–14. Both in TSKP 3.14.

as successfully with previous experience as it would on the basis of closed-texture definitions, only with the contribution of such a kind of flexibility.

There are, however, some interesting consequences of this strategy—this, as Kuhn said, data or information processing mode—when it is studied through the looking glass of a computer simulation such as the one Kuhn was trying to implement in those days. Let us turn to that simulation once again.

7.6 Johnny and the Computer Simulation

In Kuhn's example,[138] Johnny walks through a nearby park with his father and, when they bump into a population of diverse waterfowl—including swans, geese, ducks, etc.—the father shows his son how to differentiate, say, a goose from a swan on the basis of some exemplars of each species. Of course, the father is not using definitions at any moment. He only has recourse to his past experience with such fauna, and with some common-sense properties of the exemplars, such as the color of their plumage, their size, the length of their neck, and so on. The father may help Johnny to separate out the different instances in the only groups that are represented before their eyes in that situation.

Kuhn believes that there is no *essential* role that empirical generalizations or definitions may play in this kind of situation, which is a very ordinary one. Of course, the father—let us say that he is an ornithologist—may benefit from his previous specialized experience with such animal species and use some of those definitions and empirical generalizations, but in fact Johnny would most probably end up having obtained a proper classification of such waterfowl without them. Johnny will have shifted from his own original world in which geese and swans were part of the same undifferentiated class to a new world in which they form two separate species.[139] (Table 7.1 may help to illustrate the change.[140])

Each exemplar that Johnny and his father have perceived this afternoon are paradigms for learning that particular case of natural classification of a part of the local fauna.[141] Johnny has learned something about the use of terms involved, and also something about the percepts that emerge from the stimuli he receives, Kuhn

[138] In the present section, I have based my introduction to the Johnny example on Kuhn's manuscripts of his talks; in particular, "Paradigms and Theories in Scientific Research," 19 February 1967, TSKP 3.14, 20–24; Rockefeller 4, 5–9; "American Psychological Association," 2 September 1967, TSKP 3.14, 6–14.

[139] See [Kuhn], "Thoughts on the Nature of Scientific Classification," 25 February 1965, TSKP 3.14, 9–10; Rockefeller 4, 5–6; "American Psychological Association," 2 September 1967, TSKP 3.14, 8.

[140] The table is based on [Kuhn], "American Psychological Association," 2 September 1967, TSKP 3.14, 7. The expressions "Swan stimulus field" and "Goose stimulus field" follow from that manuscript.

[141] See [Kuhn], "Paradigms and Theories in Scientific Research," 19 February 1967, TSKP 3.14, 21; "American Psychological Association," 2 September 1967, TSKP 3.14, 8.

Table 7.1 Johnny's world change

Stimuli	Percepts	
	Phase I (pre-new world)	Phase II (post-new world)
Goose stimulus field	No difference	Goose
Swan stimulus field		Swan

says.[142] Moreover, he has not only learned something about geese alone, or about swans alone, but also about all of them at once. Furthermore, if he has also faced exemplars of other species, like ducks for instance, the examples of duck or of goose have also taught him some things about the meaning of "swan" as well.[143] All of it, Kuhn shows, is interrelated.

Not only are those species interrelated in such a way that watching them exhibits some similarity relationships that are enough for Johnny to learn how the fauna is distributed in boxes, as Kuhn said, the world also contributes in cases like this to make this possibility of learning feasible. When we are talking about such natural families, we, Kuhn says, can count on the exclusion of some unavailable cases, like the imaginary cases of a swan–goose, if something like that is possible at all. Previous, traditional experience with such waterfowl, that the father embodies, is a warrant that anomalous instances will not occur. Most probably, neither Johnny nor his father have even mentioned anything like them. So, for Kuhn, in that experience of learning, Johnny learns new language, new particular percepts, and new aspects of the world itself. In fact, Kuhn says, Johnny has gained new *knowledge*—about language and about the world, both at once; new knowledge, moreover, that is *inherited* from a member of the *past* experience *tradition*.[144]

Nothing in that process depends on explicit knowledge—on instructions that may be explicit, formalizable. It is, Kuhn says following Polanyi, *tacit* knowledge that has been acquired through paradigms, on the basis of a source of past experience—Johnny's father—and with the participation of a world in which no intermediate cases require the contribution of definitions and generalizations. It is *fallible* knowledge, Kuhn says, but it has features that other aspects of common-sense knowledge hold, such as the possibility of predictions. Once again, transforming this knowledge into a case of *explicit*, instead of tacit, knowledge would only result in a strategy of processing data, as Kuhn termed it, that would deal with past experience in a successful way, but that, unless it also makes room for open texture, might be uncertain when confronting some future cases. Johnny, meanwhile, will not have problems dealing with emerging anomalies by virtue of his currently trained abilities. There is no need for Johnny, for his father, or for their forerunners to *decide which* sufficient and necessary conditions an instance of local waterfowl must satisfy in order for

[142] See [Kuhn], "American Psychological Association," 2 September 1967, TSKP 3.14, 8.

[143] Rockefeller 4, 8–13.

[144] See [Kuhn], "Thoughts on the Nature of Scientific Classification," 25 February 1965, TSKP 3.14, 10; "Paradigms and Theories in Scientific Research," 19 February 1967, TSKP 3.14, 22–23; Rockefeller 4, 8; "American Psychological Association," 2 September 1967, TSKP 3.14, 8, 12.

terms like "swan," or "goose," or "duck" to be successfully applied in such a case, and there is no need for them to *alter* that initially *arbitrary* decision when facing an instance of waterfowl that has *some* characteristics of a goose, and some others of a swan, for example.[145]

This attitude of Kuhn against pointless arbitrary decisions goes back a long way, as we know. Recall our Chap. 3, when we examined the Lowell Lectures—especially towards the end of the eighth lecture. Kuhn was basically making the same point with which, years later, he finished both his commentary on Polanyi's talk at Washington and, a few years later, "Second Thoughts on Paradigms."[146] Our search for definitions as the only pathway to the meaning of terms leads us to decide where we should draw the boundaries of meaning surrounding the use of terms: what empirical generalizations should we preserve, and what properties must be present, and what others not, in order for us to apply the term correctly. Given that we often grasp similarity relationships between objects that may differ in some properties and that sometimes lead to empirical generalizations that are only true of a few examples of the class, the boundaries of these latter may be drawn in a number of ways—again, Braithwaite helped to illustrate that case. Why *should* we decide *where* those boundaries *must* be drawn?

This latter is Kuhn's main question. Indeed, given that we grasp such similarity relationships and that, owing to them, we obtain a functional categorical and linguistic vehicle to successfully deal with the perceptual world and its variety, who needs definitions? Drawing them only makes our future behavior harder. If we draw a boundary too close to the extant class members, we may not be competent to deal with difficult cases in the future—those that involve a slight variation in the defining properties a little beyond the limits that have already been established, but that, in other cases, would be included in the class. Or else, we may draw the boundaries so that they fill the entire universe of possibilities. As Kuhn said in 1951, this strategy deprives us of all the experience that we have accumulated about that class and its neighboring sets. As he used to say, even those features in the periphery of the meaning that are only present in some examples of the class are part of the meaning. They often play a role for some members to use the language successfully—to make true predictions about the behavior of some objects, for instance. This feature of our imperfect language is an adaptive virtue for us. Why should we deprive our conceptual and linguistic vehicle of a feature that behaves so well in the pragmatic, daily setting of language-use?[147]

In spite of these long-standing convictions, Kuhn's exploration of these issues on the basis of his computer simulation showed some challenges for his position.

---

[145] [Kuhn], "American Psychological Association," 2 September 1967, TSKP 3.14, 8–14.

[146] See our Chap. 3 for further details about the eighth Lowell lecture. See also [Kuhn], "American Psychological Association," 2 September 1967, TSKP 3.14, 13–14; Kuhn (1974, 313–318). These three texts are the basis for what I shall argue in this and the next paragraph.

[147] At Rockefeller, fifth session, once the computer simulation had been introduced, he also wondered about the possibility of shifting to a definitions-based mode of learning. At that time, however, he only made the questions; nonetheless, his position is clear, though implicit. See Rockefeller 5, 15.

Neither of them was a serious challenge to this thesis, but it raised some interesting questions, whose answers should be explored. Unfortunately, most of them remained unanswered; he did not fully address them, as far as I know. Nevertheless, they are worth our attention.

At Rockefeller, Kuhn introduced his audience to the "machine program," as he used to term it.[148] It would not be unfair to say that this machine turned Johnny's mind into a simulation with which Kuhn could explore some fundamentals of his arguments about paradigm-based learning. Of course, the computer simulation and its underlying mathematical model were older than the Johnny example—though not *much* older, to tell the truth.

The machine program was introduced in our last chapter, and as regards the details of its composition there is not much more to add to it. However, in his Rockefeller notes, especially for the fifth session, Kuhn explored the specifications of the model when it is put to work. In the Johnny example, both Johnny and his father satisfy some simple roles that do not seem to present many challenges. However, when Kuhn turned to the machine program, some simple but crucial issues arose that led to some interesting challenges. As mentioned, to the best of my knowledge, Kuhn did not solve these.

The process, as we know, was to have a universe of objects created by a machine and then to teach a second machine that universe on the basis of concrete objects. As noted above, Kuhn's main interest is to show how the process of learning can be based on paradigms alone, without the participation of rules of any sort during the process—of course, the machines need to be prepared internally—and by taking particular cases (the percepts, as seen above) as primitive. Such primitive elements are not labelled in any special way expect for the features, or qualities that characterize them. These qualities, or dimensions, help to locate the object in question in an n-dimensional cube, as we saw in our previous chapter. Owing to those qualities, those elements are located near others, or very far from them. It is, therefore, not a very different situation to that of two individuals which, being only slightly different in their size or in the white tone of their plumage, an observer can identify as close enough to form a simple group of two similar individuals. That two-member group could be considered a class. The more interesting example, however, is that of classes with more than two members. As Kuhn does not employ definitions, only family resemblance, transitivity is not warranted among their members; that is, member a may be similar to b, and b similar to a third member c, but of course a and c need not be similar. This would of course be arbitrary unless we had some measure of dissimilarity. In his drafts, he used to put that measure in terms of a maximal *distance* between two vertices within the n-cube, which, if empty—that is, if not covered by other intermediate object—would lead to a good indication of dissimilarity.[149] If the maximal distance is overcome, it is difficult to form a group with the vertices involved—that is, with the objects (as samples of ordered qualities) that the observer, maybe the machine, is considering. This measure must be part of the machine's original programming, as it

---

[148] Rockefeller 5, 6.

[149] See our Chap. 6 for details about that distance.

would be for us, human observers, without recourse to definitions. The psychology of Johnny's and his father's natural tendency to form natural families is thus being studied on the bases of a computer simulation.

The first set of problems that Kuhn finds here is related to the formation of the universe of classes for the first machine—or for Johnny's father, for that matter. In the machine program, as we feed it with random arranged sequences of ones and zeros that represent rows of active and inactive qualities, we see the different vertices of the *n*-cube populated.[150] Given a particular measure of dissimilarity, some initial groups are formed. Yet, the random process makes these groups unstable. As other rows of values for the sequence of dimensions are input, the extant groupings dissipate. That is definitely not what Kuhn was looking for. As he says, he is searching for a "universe of clusters," which he also calls a "well-behaved universe."[151] To do so, he knows that he must be more careful when preparing the machine, maybe with a probabilistic algorithm, so that stable clusters are warranted. It all depends on the stability of clusters. So, once some of them are formed, anomalous cases (i.e., distant from previous groupings) must either form their own class when they are far off the extant classes, or ignored if they are just in between two or more of them. New objects shall simply be included in the nearest family, thereby enlarging this latter's area.[152] As Kuhn says, in doing so, he "use[s] a probability criterion [.... He must a]dmit certain proportion of points depending on [the] ratio of distances to two families competing for it."[153] Once he does so, the well-behaved universe emerges, though the groups are "of rather odd shape," he says: they are "[n]ot particularly dense."[154]

We may ask why this is so, and why we need not expect something similar in the human case, except perhaps in the initial stages of learning. Kuhn does not seem to deal with this question himself—probably because the answer is pretty clear—though the answer remains in his hands. Of course, in Johnny's case we count on the idea that *natural* families emerge from *naturally* given stimuli. As Kuhn used to say, we must count on the world, on how it is, in order to accept that natural families are formed without the participation of definitions. A continuity of natural forms is out of the question in that case. This does not prevent us from assuming that a change in such natural distributions is possible, and it does not make Kuhn a realist about such natural families either.

Once the well-behaved universe is formed, the second step was for Kuhn to train another machine with only recourse to the former's machine universe, and without definitions—with only the elementary programming. The second machine has to learn without recourse to definitions, but now, at least, the problem mentioned in the last paragraph is solved from the outset, because the original universe is, so to speak, given. Johnny's stance is very similar now. We must train the second machine to acquire the original machine's universe. At this point, the second machine

[150] See Rockefeller 5, 10.

[151] Rockefeller 5, 11, 17, respectively.

[152] See Rockefeller 5, 12.

[153] See Footnote 152.

[154] See Footnote 152.

receives samples of each group within that already well-behaved universe. Kuhn thought that only one sample of each group was not enough to acquire the kind of family-resemblance relationship that would help the second machine to, for instance, ascribing a new unspecified member to its proper class, so initial information must include more members of each class to train the second machine. A couple of interesting questions emerge in this process: (1) Are the paradigms of each class *unique*? (2) Should we select a *particular* sample as the more convenient paradigm for the learning process? Kuhn expounds them as follows: "How nearly interchangeable are the elements we feed it first—the paradigms[?]" Besides that, "Could we as well have fed it others or are some paradigms more successful than others[?]"[155] It is difficult to admit that there is, for him, something like the "archetypal paradigm" (my expression) for each class, but he is aware that maybe some concrete examples could be more useful than others in the learning process, owing to their distance from neighboring groups; after all, he prefers "larger distances" when identifying "unnamed elements," for instance.[156]

To me, it seems difficult to find a better answer than this to those questions, given Kuhn's conception of the paradigm as something that may evolve over time and with the evolution of the community members themselves. Recall that, in our Chap. 6, we described his musings about the *changing* (one would also say the *plural*) character of the paradigm as the community itself and the resources its members employ to understand it also evolve. Accordingly, if P_2, a certain paradigm, is an improved version of another, P_1, we may wonder which the paradigm *is*. But his answer was, we should remember, that all of them are. The change affects the kind of world and practice that working scientists in each phase learn, but there is no incompatibility between their practice. This is surely true of this simpler case of learning to see the classes of objects that populate a perceptual world, too.

Kuhn also talks about the idea of an "aura" (his expression) around every natural family and its role in showing what the meaning of the kind-term is. He expounds this point as follows:

> Having given [the second] machine certain members of each family, I can ask it to scan the cube. [...] For each vertex tell me, which family would you have put an object here in, or would you say that this vertex ought not be populated at all? [...] This generates a set of vertices around each family in which other members of the family could turn up without making difficulty. [...] This is the family's aura, and it's in some sense the meaning of the family's name.[157]

There are similarities between this idea and his conception of meaning and its vagueness in the Lowell Lectures again. Recall that, at that point, he admitted that the meaning of a kind-term included not only a hard core, but also a periphery in which not so constant features of some samples of the same kind were also part of the experience of some groups. For him, those characteristics should be included in the meaning of such a term. In exploring the idea of the "aura," he wonders about the

---

[155] Rockefeller 5, 13.

[156] Rockefeller 5, 14; see also 13.

[157] Rockefeller 5, 14.

existence of such an area as well. Of course, a related problem is to what extent the meaning remains stable and to what extent it changes. Kuhn is, after all, worried about the possibility of an area of stability that does not need definitions to be obtained. As he says, "[t]his is the behavior in my old swan–geese diagram."[158] Recourse to definitions tends to suppress such apparently minor areas, but for Kuhn they are a source of adaptive behavior. We do not need to rewrite strictly penned rules each time that we find a new object that exhibits unexpected features and behaviors. We just need to adapt our previous distribution of objects in families with not-strictly boundaries around them. That is the thesis that Kuhn aims to convey, and that affects to the idea of "aura" here communicated, something that is strongly linked to his considerations on open texture.

To end this section, a final comment that Kuhn made at the end of the fifth Rockefeller session—which is the more interesting public presentation of these years as regards the issues discussed in this and the previous section—is in order. We have assumed that the objects are distributed in a machine's universe on the bases of a particular series of features, arranged in series of six, seven, eight, etc., such characteristics (or qualities). Kuhn, however, shows that it need not be that way. The first universe can be based on six features, but the second machine—the learner, so to speak—might receive a superior number; eight, for example. Of course, the learner should acquire the same universe and distribution of objects that the teacher tries to convey. To do so, it must select. For Kuhn, that process involves "learning a way of seeing."[159] This associates that selective process with a psychological thesis that Kuhn had often alluded to in *Structure*: our tendency to conform a diversity of appearances to something that we already know and even expect, a thesis that he obtained from the Gestalt psychologists. No wonder he calls it here, at Rockefeller, "the Gestalt function."[160] Kuhn is thus saying that the learner must, at some points, learn not to consider certain features that can introduce some extra diversity within the regularity. The regular universe is the main goal at that stage.

However, this well-behaved universe does not always work well. Anomalies are a case in point. He does not assign extra space to them beyond the following succinct commentary.[161] For him, these machines work well in such a kind of universe. Sometimes, however, the universe does not behave that way; sometimes, he says, an anomalous case of a "vertex for which the program doesn't work" occurs; it would be "[o]ne

---

[158] See Footnote 157.

[159] Rockefeller 5, 16.

[160] Rockefeller 5, 15.

[161] In fact, Kuhn's Rockefeller seminar is mainly the introduction of a critical doctrine against the logico-empiricist views on the meaning of scientific terms, based in turn on paradigm-based learning. It is, admittedly, an exploitation of the ideas already conveyed in *Structure*, and also a clarification—on the basis of new resources—of old points in the book that had been poorly explained or misunderstood. But it is also, perhaps mainly, a perspective that looks at some as yet undetermined future view. In that sense, it reflects Kuhn's current concerns very well. At the same time, it shows that he was no longer particularly concerned with notions like crises or anomalies. Oddly enough, incommensurability is not discussed at this point, either.

that falls equidistantly between two families."[162] In such cases, other than ascribing them to a family at random, a likely procedure would be to reconsider those extra dimensions that the Gestalt function had ignored. Though it changes the aura, and most probably the present grouping too, it is one of the best methods to face up to an anomalous case—something that spontaneously emerges. Kuhn calls it "the revolutions program," and it clearly explains scientific revolutions on the basis of this simple situation and this simulation.[163]

Given the role that interstices, or rather strips, between natural families that, being key for Kuhn in the successfulness of this learning process—the one only based, even for machines, on similarity relationships—play in this perspective, it would be prima facie reasonable if we considered Kuhn to be a subscriber of a realistic position. But he is trying to eschew the idea that a taxonomy of natural families is only an interpretation of already given families, just as he is also trying to avoid the idea that such a taxonomy is a pure creation of the human mind or group. Following the same argument that he was trying to convey to the historian of science David Joravsky in 1965, these taxonomies emerge from the collaboration of the human mind (and the group activity, we should not forget) together with the world. The way we should understand this collaboration according to him is not based on the opposition of a mind-independent world to the mind itself. It rather seems that the picture of the human mind—especially of the theoretical products of a socially organized group—without the nature it deals with makes as little sense to him as a representation, or image, of the world without the tool (the mind) that turns it into an image or representation itself. If we consider that this is a realistic position, then realism is deprived of all meaning, because for him there is no guarantee that a taxonomy can overcome the appearance of an anomaly, and that the outcome of dealing with it isn't a brand-new world.[164]

7.7 Historical Knowledge and the Size of Scientific Revolutions

We should pay attention to two more points, (f) and (g), as introduced earlier in this chapter. Point (f) concerns the size of revolutionary change; that is, as he said in New York in December 1966 at the Scientific Methods Colloquium, he is referring to the fact that, as already introduced in *Structure*, some revolutions only affect "particular specialties" and are therefore "micro-revolutions"; they are smaller than the greatest examples of their kind, "the big ones," as he says: the "Newtonian, [the] Darwinian, [or the] Einsteinian" revolutions.[165] The significance of this point is that it is related

[162] Rockefeller 5, 16.

[163] Rockefeller 5, 15; see also 16–17.

[164] See Hoyningen-Huene (1993, Chap. 2) for an already classic interpretation and analysis of this view.

[165] [Kuhn], "Scientific Methods Colloquium," 4 December 1966, TSKP 3.14, 3.

to the problem of the size of the scientific group, when this does not overlap an entire profession, as noted earlier. A few months later at Swarthmore, Kuhn mentioned that question as an introductory step to the isolation of the scientific community, which is more complex than merely finding out what the profession is or what the paradigm that they share is. For a group of scientists that are devoted to a very specific problem (or puzzle) that does not affect the rest of their colleagues greatly, their accomplishment in their particular area may satisfy the standards of a revolutionary step forward, though they still may look *normal* to the greater community.[166]

Kuhn does not go much further than this. The perception of normal and revolutionary change is thus context-relative, the context provided by the observer—whether the participant, the contemporary observers, who may be active scientists in a closely related field, or the historian of science, who approaches the case from the far future. For him, it is important to settle that historical relationship between the subject of inquiry—that piece of change—and the contexts in which such agents are located, from the discoverers' context to that of its recipients and the point of view of future inquirers. The fabric of science and its historical study is a complex, changing, and dynamic scenario, and Kuhn tries to convey that fact.

Finally, *historical knowledge* itself was a motive for reflection for Kuhn at Rockefeller—this is our point (g). After all, his own work is based on accepted historical knowledge, and he often produced samples of such knowledge, too. If the empirical basis for much of his criticism of the empiricist tradition is historical, what is his characterization of that achievement?

Kuhn approached that problem as an exercise of criticism against analyses of historical knowledge such as that provided by his colleague Hempel.[167] Kuhn's approach found a basis in Louis Mink's paper, "The Autonomy of Historical Understanding," published in 1966, and that Kuhn recommended to his audience at Rockefeller.[168] Kuhn's perspective is a confrontation of his own views about similarity-based relationship as a basis for acquiring already established knowledge, already examined, with the logical empiricism's account of scientific knowledge. His point of departure is the covering-law model, and the requirement that recourse to laws for prediction is needed in science. It would be difficult for history to provide new knowledge if that function is diminished by the proper nature of the subject. Hempel said that generalizations in history are mostly implicit and highly inexact, and that the result is not clearly a complete explanation, but rather an "explanation sketch."[169] Accordingly, prediction and testing of historical hypotheses, is limited.[170] For Kuhn, however, the idea that the covering-law model is universally applicable and that it must be the standard against which all kind of knowledge must be measured is wrong. Disciplinary particularities are paramount, and, from that point of view, the

[166] [Kuhn], "Paradigms and Theories in Scientific Research," 19 February 1967, TSKP 3.14, 3. See also *Structure*, 50.

[167] See Hempel (1942), and more generally, Hempel (1965).

[168] See the list of bibliography in TSKP 2.9. See also Mink (1966).

[169] See Hempel (1942, 237–238); Rockefeller 3, 8.

[170] Concerning this point, see Hempel (1942, 233–234, 239).

role that explicit laws truly play in a successful historical narrative is, if existent at all, very reduced. He also mentions that the difference between history and social sciences such as economics is not minor, and that the role laws may play in the latter is not as reduced as in history.[171] History is, thus, a different case. It is a case of empirical knowledge, and has virtues that we recognize in such a kind of knowledge. It is predictive, fruitful, fallible and refutable, liable to improvement, and produces understanding.[172]

If that is true, how is, therefore, historical knowledge achievable? For Kuhn, its basis is the communication of "recurrent types" and "developmental situations," which are gained and, he says, "recognize[d]" through the "direct experience" of human behavior, "institutions" and "situations."[173] He thus shows that, in historiography, we enlarge our experience of such elements in order to develop the proper expectations about other cases of them. These expectations do not involve laws, mainly because they are not so exactly and explicitly defined. Our expectations, Kuhn says, are based, as are the paradigm-based learning of mature disciplines such as physics, on *concrete* examples, and that basis is the most primitive. Obtaining laws may be a by-product of that kind of experience, but in history they are clearly not important. They would arise from an analysis of essential characteristics, or variables, but that is something that history does not handle. So, for Kuhn, history provides us with the experience that is the basis for creating expectations and for which concrete examples (or paradigms) serve as its ground. Echoes from Weber's perspective about "ideal types" might be discovered in this point of view, and especially from the first line of the following extract[174]:

> [In history, we l]earn to recognize recurrent types, developmental situations. [We w]ill recognize them again even though we can't say, or can't say fully, what it is that makes them typical or just what it is that *must* happen in [a] situation. [...] History thus enriches our *concrete* experience and prepares our expectations for similar situations. [...] Where, however, seeing things as similar, as in a recognizable pattern, is prior to saying characteristics with respect to which things are similar. [...] This is close to saying [that] history provides paradigms.[175]

For Kuhn, therefore, historical knowledge is based on finding similarity relationships among documentary evidence and previous experience with the historical past. In doing so, laws are not obtained, but they are not the basis of such knowledge. Discovering patterns, similarity relationships, and facts that commonly do not occur—which is for him the place for the "empty perceptual space[s]"[176] between families—provides the basis for an instance of paradigm-based knowledge. The historian thus provides a likely narrative that is empirically based and that avoids

[171] Rockefeller 3, 9–10.

[172] See Rockefeller 3, 7.

[173] Rockefeller 3, 10. The rest of my account in this paragraph is based on that page, too.

[174] Concerning the relationship between Kuhn's and Weber's works, see Mladenovic (2017, 43–53). See also our Chap. 2.

[175] Rockefeller 3, 10.

[176] Rockefeller 3, 12.

"implausible juxtapositions in place and time."[177] Assertions concerning the historical past remain fallible, and a part of the historian's interpretation is simply *open*; the facts are limits to a future narrative. "All possibilities with the facts are still open," he writes, and adds that "[t]hese specify only non-occurrence, empty perceptual space."[178]

The idea that historical knowledge is paradigm-based, and that it would follow on from his paradigm-based model for knowledge seems to be intended to support his criticism of previous accounts of scientific knowledge. Together with the model of learning that he had been pursuing, and with the recent reconsideration of the notion of paradigm (his primary source of concern), of community and even of revolution, he seems to have been trying to make *Structure* stronger. With such elements, Kuhn would soon publish his reactions to criticism to his position in a few papers that are commonly viewed as his own emendations to his philosophical position.

7.8 Reinforcing *Structure*, 1969–1974

All these points, (a) to (g), eventually appeared in a set of papers that were published in the first half of the 1970s. I am referring of course to the Postscript to *Structure*, to the two contributions to Lakatos and Musgrave's edition of the Kuhn–Popper confrontation at London in 1965, and to "Second Thoughts on Paradigms." Certainly, the earlier paper that Kuhn submitted was his first contribution to *Criticism and the Growth of Knowledge*. He had told Popper only ten days before meeting him and his wife at London in 1965 that his own paper was not yet ready. At that time, he had sent Popper a draft of the introduction and a very rough draft of the first section—a position that would probably change later, he told Popper[179]—on June 30, while the covering letter summarized the rest of the future paper, which was going to be Kuhn's chapter for P. A. Schilpp's volume on Popper, too. In that letter, Kuhn apologized for it and hoped that such lines together with *Structure* would help Popper to prepare his contribution.[180] Previously, on June 23, 1965, he still presaged that "Logic of Discovery or Psychology of Research?" would not be ready until some weeks later.[181] Yet, it seems that, in London, Kuhn had a full version very similar to the one that was later published.[182]

The rest of the papers that he published in the early 1970s came four years later. "Second Thoughts on Paradigms" was the earliest of all of them. It was ready in

---

[177] Rockefeller 3, 11.

[178] See Footnote 176. On the ideas presented in the second part of this section, see also Kuhn (1976c, 17–18).

[179] Kuhn to Popper, 30 June 1965, KPP 317.17, 2.

[180] Kuhn to Popper, 30 June 1965, KPP 317.17, 3.

[181] See Kuhn to Popper, 23 June 1965, TSKP 23.9; Kuhn to Popper, 30 June 1965, KPP 317.17, 2–3.

[182] Lakatos and Musgrave (eds.) (1970, vii).

late March 1969 for the meeting at Urbana (Illinois) to whose volume it contributed, and it was followed by his "Reflections on My Critics" for Lakatos and Musgrave's volume and by the Postscript to *Structure*, which Kuhn had not yet written when the two former ones were ready.[183] In August 1969, with "Reflections on My Critics" already finished, he did not expect the Postscript to be ready before October, though by November 1969 he was already making corrections to that new afterword.[184] On the relationship between both essays, he later wrote: "Much of the 'Postscript' can also be found somewhere or other in my 'Reflections on my Critics,' but it is there often hidden among unhappy answers to questions which I felt need never have been raised"; he went on saying that the Postscript was "a somewhat clearer and later work than the 'Reflections.'"[185] The Postscript was not a chapter long in the making, though the plan came up a bit earlier. In June 1968, he had planned to write "a new introduction" to *Structure*, now that he had "made some progress on the sorts of problems raised by" his book.[186] In March 1969, he met Barbara Babcock, from the University of Chicago Press, in Urbana and they talked about the Postscript—to which they referred at that point as "Second Thoughts."[187] But, as noted, nothing of that sort was ready until the following October, 1969.

These chapters have many themes and points in common because they are in essence the result of the work that he had been developing since 1962, and that we have examined in this and our previous chapter. In such chapters, Kuhn regrets that his message has so often been lost, and that his intentions have been misunderstood—especially by those who, like Popper, Lakatos, Feyerabend or Watkins, usually employ historical examples in their own writings.[188] However, the main problem is probably that such critics have often understood the importance of the history of science and of a sociological view of the epistemic agent in science—the community—but that they have misunderstood what the community *truly* represents in his theory. Kuhn is nevertheless aware that his account of the community, and its

183 See Kuhn to Wolfgang Stegmüller, 20 January 1975, TSKP 20.36, 1, where Kuhn describes the chronology of those papers for his German colleague.

184 See Kuhn to Lakatos, 24 March and 14 August, 1969, both in TSKP 11.37; Kuhn to Barbara Babcock, who was Science Editor at the University of Chicago Press, 20 August 1969, TSKP 25.55. "There is as yet no smell of the new chapter," he told her in that letter as regards the future "Postscript—1969." On November 16, 1969, however, Shigeru Nakayama had Kuhn's revisions to that afterword; see Nakayama to Kuhn, 16 November 1969, TSKP 25.55. The Japanese edition had prompted Kuhn's writing of that additional chapter, as he said in a footnote at the beginning of the Postscript; see *Structure*, 173, fn. 1.

185 Kuhn to Stegmüller, 20 January 1975, TSKP 20.36, 1.

186 Kuhn to P. D. Jones, Managing editor of the University of Chicago Press, 12 June 1968, TSKP 25.55.

187 Babcock to Kuhn, 2 April 1969, and Kuhn to Babcock, 15 April 1969, both in TSKP 25.55.

188 Kuhn (1970b, 125). Kuhn's disagreements with Feyerabend are surprising because of their coincidences in the defence of incommensurability—though from different points of view—and the use of the history of science as a basis, but Kuhn sees incommensurability under a less radical light and does not accept the link with irrationality and relativism that Feyerabend seems to associate with his position; see Kuhn (1970b, 124–127, 138–139, 140–141) for some of those disagreements.

relationship to paradigms, had not been as specific as was surely required.[189] That's why, just as he did in previous annotations and in his lectures and talks, he tried to find a proper method for identifying the community without recourse to either the full profession or the paradigm itself. He is clear that the community is a group of experts that the sociologists have taught us to identify in history without equating them with a discipline or a paradigm. The group of experts might be a full profession, of course—the practitioners of a full discipline. But significant changes in such big groups, especially those of a revolutionary kind, do not often happen. Transformative innovations, even normal, progressive ones, usually correspond to groups of lesser size with a reduced number of members, who are specialized around some themes. A paradigm is adopted and is replaced in such contexts more often than in the larger context of the full profession and discipline. Such groups are themselves less stable and sometimes even ephemeral. Transformations that result from their practice have an impact on the consciousness of the rest of the profession, and for an external observer as well, which must be studied in its proper context.[190]

Kuhn talks about this in the "Postscript—1969" and in "Reflections on My Critics," in which he tries to be clearer than in *Structure* about the true nature of the scientific community. This is related to our previous point (c). He also takes these chapters as opportunities to show to what extent this kind of group must be better understood in order to avoid the usual distrust about the irrationality of a group decision in the absence of clear and explicit rules or, as for Lakatos, recourse to a "psychology of [the] ideal mind."[191] No such "ideal minds" populate Kuhnian communities, but his account of the elements of *cohesion* and of *decision* within such groups does not depend either on the idea of community as an agglutination of subjectivities that have recourse to reasons but also to uncanny personal inclinations. Kuhn's account of the functioning of a community in a mature discipline around the articulation of a paradigm is not a matter of "mob psychology," as suggested by Lakatos.[192] Psychology and sociology are important for Kuhn, but their role in the study of the community, which is nevertheless central, does not lead to irrationality.[193] On the contrary, classical epistemic values—Kuhn mentions some in these chapters—contribute to the scientists' decision, though rather as limits than as determinants of the individual's decision.[194] In other words, complexity—the one that sociology and psychology of science may help to describe—does not amount to irrationality, but rather to a more exact account of the proper epistemic agent. Kuhn

---

[189] See Kuhn (1970b, 146–148); Postscript—1969, 174–177. The rest of the paragraph is also based on such passages.

[190] Haufe's (2022) is an interesting description of such transformations in an evolutionary key, though from a more traditional, less Kuhnian, perspective. For a critique, see Mayoral (2024a).

[191] Kuhn (1970b, 133–134).

[192] See Kuhn (1970b, 130–134; 155–158); for the expression "mob psychology," see Lakatos (1970, 178).

[193] See Kuhn (1970b, 130–134).

[194] See Kuhn (1970b, 155–158; see also his 1973); Postscript—1969, 198–199, 204, 208.

tries to show this in these chapters, and in his 1973 Machette Lecture, "Objectivity, Value Judgment, and Theory Choice."[195]

For him, the history of science is, just as sociology and psychology are—as he had countered to Reichenbach in his private notes—a source of insight into the processes of change. With such historical studies and perspective, the nature of processes within the community is clearer. His past record of historical studies, not many but significant, supports that view. Nevertheless, some good examples were still to come. *Black-Body Theory*, especially in Part II, Chapters VIII–IX, is a good contextualized report about the way in which a new theory—in fact, an emerging new mode of solving problems in the natural world—gains momentum in the experts' group. The inquiry into processes of change of this nature also shows a fair account of theory choice—he particularizes them to a few individuals in Chapter IX of *Black-Body Theory*—and verification compared to the standard, imaginary, abstract, uncontextualized version that logical empiricists set as the backdrop for exploring it.[196] As he said (see our previous points (a)–(b)), there is a perspective on rational reconstruction that he performs, which relies on such resources beyond an external, logically and abstractedly defined, account of scientific inference.

For him, this point of view is one that his many critics—even Toulmin, Feyerabend or Lakatos, in spite of their mastery of historical cases—have not come to understand properly, as noted. The reference to paradigms had been Kuhn's option in *Structure* for conveying this view of the community, but the book was less than appropriately explicit about them, which still awaited a more detailed account. Masterman's paper for *Criticism and the Growth of Knowledge*—the contribution that had done the most in that volume to help develop *Structure*—had showed Kuhn a proper point of view from which he should reconstruct a more exact vision of paradigms. Again, she had done so much less by pointing out the (too often mentioned) 21 senses of paradigm in the book than by emphasizing the importance of these pieces as autonomous with regard to theories, and by paying attention to the resources and "habits" involved in the group's research—Masterman's "sociological paradigm," eventually called the "disciplinary matrix"—as different from the very idea of a paradigm (now the "exemplar") as a representative concrete problem–solution.[197] (Our point (d) refers mainly to this question.)

Kuhn had also shown to what general point of view the notion of paradigm corresponds. During the years since *Structure*, he had been working hard to show how the acquisition of a language and a set of interrelated concepts also involved a continuous interaction with relevant objects and processes in nature. The discovery of one's own phenomenal world involved its construction in a continuous interaction (even a "dialogue") with nature as well as access to a community in which the agent also

---

[195] See Kuhn (1973).

[196] See *Black-Body Theory*, Chapters VIII–IX.

[197] On Masterman's contribution to Kuhn's work, according to these chapters, see Kuhn (1970b, 124, 127, 168; 1974, 294); Postscript—1969, 174. On sociological paradigms and their relationship with habits in Masterman, see Masterman (1970, 66 ff.).

acquired an identity as a member; as a researcher, in the case of scientific communities. The process looked like the way we—scientists or not—acquire our language and our classification of objects in the world on the basis of an adult's guidance, and without recourse to explicit rules (beyond those that we may come to produce by ourselves). Kuhn had investigated such a kind of training as a model for learning a scientific language and a way of practicing science on the basis of concrete examples, which hold priority over rules and definitions.[198] On the basis of such concrete examples, we acquire the family resemblances and the natural families that populate our phenomenal world without the need to answer questions like "*x* is similar to *y* with regard to *what*?"[199] His "Johnny example" and the computer simulation were auxiliary resources that he described in "Second Thoughts on Paradigms" (the paper he wrote first), and then mentioned in the "Postscript—1969" and in "Reflections on My Critics."[200] However, an interesting mention of paradigm-based learning appears in "Logic of Discovery or Psychology of Research?" as a basis for criticizing Popper's views on refutation instances.[201] The computer simulation was eventually left aside, though it showed him the way to the difficulties involved in such a model of paradigm-based learning, as we saw above. In that respect, his new chapters were a more or less complete report of what we have explored as point (e), probably the key result of all this phase. So, Kuhn's argument about perception and interpretation, similar to his brief points in Washington in 1967 on the significance of the process by which percepts are acquired ("Stage I" on that occasion), also appears in the Postscript to explain how perception is involuntary, and different from interpretation—the latter is not just a deliberate variety of the former.[202] That is a thesis that, as noted, traces back to his discussions with Kubie in 1955, and is a basis for his paradigm-based model of learning.

Such chapters from 1970 and 1974 display the aspects of Kuhn's development since the publication of the first edition of *Structure*. The interesting result of such development can be found in its application to the issues that most interest philosophers of science, that is, issues concerning the rationality of theory choice involved in episodes of revolutionary change and the ensuing effect of incommensurability between two distinct paradigm-based practices.[203] Kuhn discusses in depth how the processes of theory choice, whenever there is opportunity for them, involve a series of individually-based weighing of shared values. Together with the fact that different individuals pursue a decision at different moments in time—something that is again

---

[198] See Kuhn (1970b, 171–172).

[199] Postscript–1969, 191, 199. Kuhn's question is slightly different.

[200] For the whole model described in this paragraph, see Kuhn (1970b, 166 ff.; 1974, 307 ff.); Postscript—1969, 191–193, 196. See Kuhn (1970b, 171–172; 1974, 310), and Postscript—1969, 191, 196, for mentions of the computer program. The Johnny example can be found in Kuhn (1974, 307 ff.).

[201] See Kuhn (1970a, 16–19).

[202] See Postscript–1969, 191, 193–194.

[203] Kuhn also talks about progress and truth in the Postscript—1969, but it does not essentially change with regard to how he discusses the issue in *Structure*, first edition. See Postscript—1969, 203–205.

well illustrated in the above-mentioned chapters of *Black-Body Theory*—that application renders the transition to a new paradigm as an appropriately complex process. The role of unequivocal, general criteria of choice disappears from sight, and the process by which a new theory or paradigm emerges is treated as a socially based process. As Kuhn used to say, adopting a fresh solution amounts to running a risk on the part of a group. Individuals may bet on some new problem-solutions differently, and even reach different conclusions despite sharing the same system of values. However, the system of values works as a limit that helps them to make decisions that eventually converge. Although scientists may weigh their values differently, they cannot choose just *any* values whatsoever, nor may they opt for alternatives any way they please. The result of such activities, if extended to the community beyond, perhaps, a few eccentric members, would involve the disappearance of that science.[204]

Incommensurability is an understandable effect of science as seen in this way. The way the world in which scientific practice takes place is acquired involves, in turn, acquiring an interrelated set of terms and basic expressions, like laws, for example, which are adapted to more complex situations on the basis of more specialization for them.[205] If we add to that a common system of values, it is clear that differences between paradigms amount to unsurmountable difficulties in communication, which may turn out to be only *partial*,[206] among practitioners of different paradigms. This is a point that Kuhn had already made in *Structure* and that he now makes even more clearly on the basis of his paradigm-based perspective on perception, language, and practice. The basic understanding of incommensurability that results from it arises from that point of view, in which there is no conflict between specific aspects of different theories, but rather between whole points of view in which experience and normativity are intermingled. This makes the process of persuading others of the virtues of a newly produced problem–solution very difficult. The case of converts to a new viewpoint is a long process in which an auxiliary resource such as that described next as translation—later as apprenticeship of a different language, practice, and worldview—does not warrant any particular result.[207]

Overcoming communication breakdown between individuals or groups with mutually incommensurable practices, languages, and worldviews involves learning the other's paradigm. Kuhn speaks about a hard first step: *translating* between languages. As he says in 1993, speaking about translation is not appropriate here.[208] However, in 1969 he was already aware of the troubles involved in such a process,[209]

[204] See Kuhn (1970b, 157–159); Postscript—1969, 198–199.

[205] See Kuhn (1970b, 169–170); Postscript—1969, 187–188.

[206] On this important nuance—not always evident to his critics—see Kuhn (1970b, 124); Postscript—1969, 197–198.

[207] For that kind of apprenticeship in Kuhn's late philosophy, see our next chapter. *Black-Body Theory*, Chapters VIII–IX are again a good example of such kind of conversion. Kuhn talks about persuasion and conversion in these chapters in Kuhn (1970b, 174–175); Postscript—1969, 200–203.

[208] See Kuhn (1993, 237–238).

[209] See Kuhn (1970b, 163–165).

so overcoming incommensurability requires *relearning* how the world *might be* according to another point of view and from someone else's practices.[210] However, Kuhn would be more explicit about that process years later, when he substituted such a kind of translation-based relearning with the training of bilingual interpreters.[211]

In general, in these chapters, Kuhn answers some challenges that he had been observing in the years since publishing *Structure*. In doing so, he reinforced his criticism of the position that supporters of logical reconstruction of scientific languages—for short, the logical empiricist philosophers—had put forward, and that he had found faulty many years ago. His previous criticism of the existence of a purely observational language is reinforced by the idea that Stage I is particularly important in the process of acquiring percepts and is affected by our education and the world we live in—as he had already shown in 1951, in his Lowell Lectures. That perception is not just an unconscious form of interpretation is also part of that criticism of the perceptual view attached to the epistemology he wishes to disregard.[212] Once the former perceptual basis of the empiricist construal of knowledge is no longer reliable, his point of view about the nature of acquisition of a scientific language responds to another alternative viewpoint. He shows in these chapters that we acquire knowledge of a language on the basis of particular examples; he says, as we have seen, that *concrete models, facts, and objects*—something to be perceived in a given situation, normally as part of an interaction with other members of our linguistic community—are the pieces that helps us conform a full representation of the world. So, an example-based explanation of the configuration of the phenomenal world (Kuhn's behavioral world in the early 1950s) takes over the former role assumed by rules and definitions. In this new view, they are no longer needed.

We can take Kuhn's reinforcement of his perspective as an alternative to the logico-empiricist division of knowledge between an observational language and a theoretical language, and to its clarification on the basis of logical calculi and an empiricist account of perception. In their place, he puts forward a view in which the perception of models, facts, objects, etc., whatever their complexity, is not only the basis for providing a theory with a neutral, objective content, it is the material itself of which the theory and the underlying view of the world are composed. The congruence of the categorical structure thus formed in the minds of all individuals is the ground for the knowledge of language and of nature that serves as a bond for

---

[210] See Kuhn (1970b, 166); Postscript—1969, 200–201.

[211] On the idea of incommensurability revisited in this paragraph, see Kuhn (1970b, 162–166); Postscript—1969, 199–203. For more elaborate views after such works, see *The Road* and *The Last Writings*. Concerning his changes in view about such a concept, see Sankey (1993); Bird (2000, Chap. 5); Hoyningen-Huene (1993, Sect. 6.3); Oberheim and Hoyningen-Huene (2013); Marcum (2015b) Devlin (2021). There is an extensive literature on the concept, of course; for a representative list of relevant literature up to the early 1990s, see Hoyningen-Huene (1993, 207, fn. 58). That already long list increases in more recent times well beyond the reasonable extension of a footnote, so I will not attempt a more comprehensive recount. The references above will serve as a good starting point for a thorough exploration of the relevant literature.

[212] See Postscript—1969, 194.

the members of a community, and so for their subsequent activity. The reinforcement of this view makes the social and psychological nature of that perspective on scientific knowledge even clearer. Part of Kuhn's message is that this perspective is not just a naturalized explanation of the *development* of scientific knowledge, but a *philosophically* meaningful account of scientific *knowledge* that does not need the doctrinal basis of the epistemological tradition. He would spend much time and effort in the coming years to make this latter thesis credible. It was important, after all. The relevance of key aspects of his view—such as the theory-ladenness of observation, the denial of an essential role for application criteria of terms in the acquisition of a scientific language in real practice, or the defense of a crucial function in his theory for concrete examples (or paradigms), for family resemblances, and for statements that for previous empiricist thinkers belonged to a fruitless metaphysics—was clearer upon adopting that perspective. For the moment, however, and in these chapters, Kuhn shows that his controversial theses concerning incommensurability, the rationality of theory choice during scientific revolutions (according to his perspective on them), the role and nature of values in it, and his puzzle-solving-based account of progress are acceptable if the new point of view about paradigm-based learning and research is indeed adopted.

I shall examine the evolution of Kuhn's reinforced standpoint during the last twenty years of his life (roughly, 1975–1995) in the next chapter and in the Epilogue. Before doing so, however, I must recall that the decade of 1970s started for him with another kind of project in mind, whose origins trace back to the early 1960s, right after finishing his full draft of *Structure*. I am referring to his historiographical research on the origins and history of quantum theory. To that project, which Kuhn had been developing in parallel to his philosophical research, we now turn.[213]

7.9 Recovering the Past: The Quantum Physics Project, 1961–1972

We must return to the early 1960s for a while in order to understand, in the next section, what kind of research Kuhn was conducting right after that series of philosophical publications in the early 1970s. Kuhn had already spent almost five years in California as part of the faculty of two Departments at Berkeley, History and Philosophy, when, in March 1961, his former PhD advisor, theoretical physicist John Van Vleck, visited him. Actually, Kuhn was awaiting this visit. In the summer of 1960, solid-state physicist Charles Kittel, J. A. Wheeler, and historians Hunter Dupree, Harry Woolf, and Kuhn had met at Berkeley and had discussed the details of a plan, and drafted a proposal, for an ambitious but probably significant project of historiographical and archival research: gathering significant original documents from the pioneers of quantum physics and conducting interviews with them before all of them

[213] On that project, see te Heesen (2020).

disappeared.[214] During the rest of the year and the early months of 1961 that followed, Wheeler, who was very active in this matter, got an agreement between the American Philosophical Society and the American Physical Society with the help of the former society's librarian, Richard Shryock. Both societies formed a Joint Committee that included historian George Corner, Shryock, and Wheeler on the part of the former society and physicists Karl Darrow, George Uhlenbeck, and Van Vleck on the part of the latter. The Joint Committee was in charge of the project. The members of the committee took their time to choose the best candidate for chairing the researchers in charge of the project, and the chosen name was, of course, Kuhn.[215] On February 17, 1961, during the first committee meeting, its members called him by phone in order to find out if he was willing to chair the project. In mid-March, Van Vleck was traveling to Berkeley to have an in-person meeting and talk about this matter with him.[216]

Kuhn would lead a group of historians of science (all of them previously trained in physics, he himself included, of course) that, aided by practicing scientists, would gather the remnants of an exciting era in the recent history of physics: the development of quantum physics. That moment was timely. Just recently, some major figures in that development, such as Einstein, Schrödinger, Pauli, or von Neumann, had passed away in a period of five years (1955–1960), and many of their equals were very old.[217] In fact, the great Niels Bohr also died at the very beginning of the project.[218] It was important to preserve the documents behind the creation of such a new approach to physical reality—I mean, manuscripts, research notes, drafts, letters, etc. At that point in history, the proponents of this project were already aware that the uses of the telephone had replaced some epistolary relationships that leave a trace behind, so impeding the historians in tracking the, as Mara Beller later said, "dialogical" development of some ideas in history.[219] Despite that, the project itself revealed the

---

[214] See Corner et al., "Request for $202,996 for Period 1 July 1961–1 July 1964 to Record Interviews with Men Central in the 1913–1938 Quantum Revolution and to Retrieve Personal Documentary Material on Quantum Theory and Related Scientific Developments, 1898–1938," 7 June 1961, TSKP 6.12, 4–5.

[215] See Corner et al., "Request for $202,996 for Period 1 July 1961–1 July 1964 to Record Interviews with Men Central in the 1913–1938 Quantum Revolution and to Retrieve Personal Documentary Material on Quantum Theory and Related Scientific Developments, 1898–1938," 7 June 1961, TSKP 6.12, 4–6.

[216] See also Wheeler's preface to Kuhn et al. (1967, vi–ix) for the data on which this paragraph is based.

[217] See Corner et al., "Request for $202,996 for Period 1 July 1961–1 July 1964 to Record Interviews with Men Central in the 1913–1938 Quantum Revolution and to Retrieve Personal Documentary Material on Quantum Theory and Related Scientific Developments, 1898–1938," 7 June 1961, TSKP 6.12, 4. More on this below.

[218] As Anke te Heesen (2020, 86–87) notes. See also Wheeler's preface to Kuhn et al. (1967, ix).

[219] Beller's (1999) dialogical approach to the history of science is intended to be critical of Kuhn's views on normal science. It is a truly interesting approach that I do not see as completely detached from Kuhn's ideas, though. In any case, "dialogical" is a proper qualification for the kind of interaction behind the creation of such great theories in physics like quantum physics. In that sense, Beller's point, to which I am sympathetic, is really opportune for this creative process. Concerning the uses

great amount of documentary evidence of that kind dispersed throughout Europe and America, and perhaps elsewhere. It was also said that there was probably far more evidence from this phase in the history of physics than from any other before 1900. So, that was a good moment for gathering that evidence and deposit it in some specific places at the disposal of historians of science.

The documentary evidence was important for another reason. The Joint Committee had already remarked that the way results and research in physics were conveyed in the twentieth century differed from the way they had been made public before. The depersonalization of the public communication of research was significant at that point in history, and the process by which physicists achieved their results were far from evident after the historian read their main papers.[220] In fact, Kuhn's works on the history of science attested to that effect: the reconstruction of some lines of research are obstructed by the kind of (unintended and good-willed) hindsight reading of one's past that many physicists went through in their later years. Kuhn showed examples of that kind of reading in some of the physicists he interviewed during those years. Though innocent, that reading did not help historians of science to grasp the long creative process, often full of ideas that did not belong to the worldview that ultimately became the well-established setting for the new physics—Planck himself was a good case in point—and so it disguised the true nature of scientific inquiry and reasoning.[221] For Kuhn, even masterful accounts of the evolution of quantum physics like Max Jammer's *The Conceptual Development of Quantum Mechanics* was influenced by that kind of hindsight view.[222] In *Black-Body Theory*, Kuhn also talked about the kind of narrative that results from a hindsight reading of history, and exemplified the kind of effort that the historian must make to overcome it.[223] Only first-hand documentary evidence remedied it, but the relevant documents for that stage in the history of physics ran the risk, at that time, of disappearing, once dispersed among the families to which the pioneers of quantum physics bequeathed them and the private collectors, or simply being destroyed.

Perhaps for the same reason—the disguising effect of participant actors' recollections—the interviews that the members of the project conducted in the following

of new vehicles of communication in science such as the telephone, see Corner et al., "Request for $202,996 for Period 1 July 1961–1 July 1964 to Record Interviews with Men Central in the 1913–1938 Quantum Revolution and to Retrieve Personal Documentary Material on Quantum Theory and Related Scientific Developments, 1898–1938," 7 June 1961, TSKP 6.12, 3; Kuhn (1967, 418–419) makes a similar point.

[220] See Corner et al., "Request for $202,996 for Period 1 July 1961–1 July 1964 to Record Interviews with Men Central in the 1913–1938 Quantum Revolution and to Retrieve Personal Documentary Material on Quantum Theory and Related Scientific Developments, 1898–1938," 7 June 1961, TSKP 6.12, 3.

[221] The final report (Kuhn et al. 1967, esp. 4–5) shows the difficulties involved in the recovery of memories from the interviewees, and how, "prompted by an interviewer well acquainted with his work and its context, the subject of an interview will often produce his most revealing information about technical developments within a biographical framework" (Kuhn et al. 1967, 5).

[222] See Kuhn (1967, 416–417).

[223] See *Black-Body Theory*, 102–103; see also the following pages for Kuhn's account of Planck's derivation of his distribution law.

years were not very successful. Kuhn often regretted that the ultimate result was not satisfactory, and his first report, a year after the beginning of the project, revealed that, together with emotive recollections of some personal significance and historically significant details about the institutional context of physics, the older scientists that they first interviewed provided them with not only vague memories but also inconsistent reports of the process of their research from many years before.[224] To some extent, the members of the project had prepared themselves for such a possibility, and the preparatory stage for the project involved, especially for the chair and his direct collaborators, who ran the interviews, their own reconstruction of events, which was destined to boost the subjects' memories and to help fix their inconsistencies. Despite that, the result was not as satisfactory as they had been supposed it would be, though the many interviews conducted, they concluded, would probably serve historians nonetheless.[225] In his 1963 report for the Joint Committee, Kuhn said that Niels Bohr's participation, albeit helpful, would not have changed the overall result, because Bohr, despite "his first-hand acquaintance with almost the entire development of the field," Kuhn wrote, and many other virtues, "[...] displayed the same failures of memory and the same tendencies to linearize the record that have been shown by most of the physicists to whom we have talked." So, for Kuhn, Bohr's contribution would not have made "the difference between success and failure of an interviewing program."[226]

So, when Van Vleck visited his former student at Berkeley, he was inviting him to chair a difficult project that would occupy him for plenty of time during the forthcoming years, and that would require hard preparation—despite his already well-established acquaintance with the recent history of physics and with theoretical physics itself. No wonder, then, that Kuhn told the University of Chicago Press in 1961 that he intended to submit the final draft of *Structure* before the new project started.[227] He envisaged little additional time for philosophy of science during the next few years, especially for improving *Structure*, and also a period abroad. We have already seen that from 1961 to 1964 Kuhn devoted some spare moments to his former philosophical ideas—though they were indeed significant. Yet, he spent a great part of the first five years of the decade on this project, and the fact that it was also a productive philosophical time is a testimony to his intense activity in both fields, history and philosophy of science, and to his dedication. However, the project was also important for him personally. He intended to use the newly acquired knowledge of the history of quantum physics and of the pioneers' role in it, as well as the evidence gathered—he asked for copies of documents for Berkeley, and he

[224] In that respect, his 1963 report did not differ significantly from the previous one. See Kuhn, "Sources for History of Quantum Physics," Memorandum, 28 October 1963, TSKP 6.12, 2–5.

[225] See Kuhn, "Sources for the History of Quantum Physics," Memorandum, 8 August 1962, TSKP 6.12, 2–5; "Sources for History of Quantum Physics," Memorandum, 28 October 1963, TSKP 6.12, 2–5. Kuhn et al. (1967, 3) say that "the project staff conducted approximately 175 interviews with about 95 people".

[226] Kuhn, "Sources for History of Quantum Physics," Memorandum, 28 October 1963, TSKP 6.12, 5.

[227] See our Chap. 5 for details.

had access to transcripts of his conversations with scientists—to pursue his own line of investigation.[228] And so he did. His paper with John Heilbron, "The Genesis of the Bohr Atom," as well as, of course, *Black-Body Theory*, were born from this project.[229] It is no surprise that Kuhn asked for an NSF research grant after finishing the project, and that he extended it and asked for another one later, in 1973, in order to follow his own line of research.[230]

As noted, in meetings from early 1961 (17 February and 31 March), the project's plan was set up. Kuhn was at the second meeting. The project started on July 1, 1961, and the first stage was preliminary, mainly devoted to interviewing the oldest and most significant physicists on the list: Bohr and Born. The second stage started with the project itself and involved research at Berkeley on the part of the members of the project and of their research assistants, who focused on preparing biographical accounts of the physicists of interest and lists of relevant bibliographies. Besides running other interviews during this new stage, Kuhn and Heilbron focused on researching the literature. They also consulted experts in the field of the history of physics, such as Martin Klein and Gerald Holton, and other senior scientists in order to conduct the interviews, which always involved the chair (i.e. Kuhn) and one of the latter. The underlying plan was to create a positive atmosphere for the interviewee thanks to a former colleague, preferably a close friend, and then also a historian like Kuhn, who, together with the scientist, took care about the detail, quality and relevance of the information gathered.[231] The new interviews in this stage were conducted in 1962, and there were many of them. Kuhn reported in August 1962 having recorded interviews with 29 physicists, most of them in America.[232] In contrast, the gathering of documentation advanced to a lesser extent, but there were good contributions. Kuhn reported having obtained around 200 letters from Alfred Landé to such physicists as Heisenberg or Schrödinger, among others, which were a good example of the kind of substantial contribution obtained during this stage. Paul Forman became a member of the group, and helped Kuhn and Heilbron to study the literature, especially for the coming year in Europe, when more interviews were conducted.[233] So, in the period in which Kuhn was correcting the first full draft of *Structure*, and once promoted to full professor in July (some months since May when

---

[228] See Kuhn, "Proposal to the National Science Foundation for Support of Research in the History of Quantum Physics," n. d. (but see the "Effective Date: 1 July 1974" in the cover page), TSKP 26.16, 1.

[229] See Heilbron and Kuhn (1969).

[230] See Kuhn, "Proposal to the National Science Foundation for Support of Research in the History of Quantum Physics," n. d., TSKP 26.16, 1.

[231] See Corner et al., "Request for $202,996 for Period 1 July 1961–1 July 1964 to Record Interviews with Men Central in the 1913–1938 Quantum Revolution and to Retrieve Personal Documentary Material on Quantum Theory and Related Scientific Developments, 1898–1938," 7 June 1961, TSKP 6.12, 7–8.

[232] See Kuhn, "Sources for the History of Quantum Physics," Memorandum, 8 August 1962, TSKP 6.12, Attachment II.

[233] See Kuhn, "Sources for the History of Quantum Physics," Memorandum, 8 August 1962, TSKP 6.12, 1–2, 5.

he was detached from the Department of Philosophy), he started a period of work focused on the sources of quantum physics.

Stage 3 was different. It lasted a full academic year together with the summers that, respectively, preceded and succeeded it. So, it started in June 1962 and lasted until September 1963.[234] The Berkeley headquarters was transferred to Copenhagen from September 1962 to August 1963.[235] Kuhn received notice of the publication of *Structure* while at the Niels Bohr Institute, in Denmark. In his report, Kuhn expressed gratitude for the treatment in Copenhagen:

> During the summer before our arrival, Professor Bohr had undertaken, with the help of the Carlsberg Foundation, to expand his Secretariat so that it would provide office space for our large staff as well as his own. In most of these rooms, pleasantly set in the remodelled stable of the Carlsberg Brewery, handsome and comfortable furniture was provided as well as a modicum of office equipment. Our work was enhanced as well as our comfort, but we were badly spoiled and have had difficulty returning to the more confined quarters and the borrowed furniture which continue to characterize our Berkeley office. Furthermore, our Copenhagen quarters were only a single concrete manifestation of the immense cordiality and helpfulness shown to us by Professor Bohr, his family, associates, and staff.[236]

During that long period, many interviews were planned. They not only included major figures in the origins and development of quantum physics, but also others who could contribute significant information about those figures; Laura Fermi and Emilio Segrè, for example, were a case in point as regards Enrico Fermi. This period was devoted to interviews, and to getting copies from all the manuscripts, documents and letters from a variety of sources. Finally, Stage 4 lasted from September 1963 to July 1964, when the project was considered concluded. The Joint Committee planned to use this last year to organize the material, and to analyze gaps and inconsistencies. A follow-up of some interviews was also expected to be necessary in some cases.[237]

As Heilbron summarized in a survey article published four years later, "At the close of its operations in 1964, the project had conducted and transcribed some 200 interviews with 100 informants, and had arranged for the microfilming of about 100,000 frames of material."[238] Copies of all documents were submitted to Berkeley and the American Philosophical Society, and Copenhagen kept another set.[239] For Kuhn, the project was conceived to offset the ruthless passage of time. "The death

---

[234] Corner et al., "Request for $202,996 for Period 1 July 1961–1 July 1964 to Record Interviews with Men Central in the 1913–1938 Quantum Revolution and to Retrieve Personal Documentary Material on Quantum Theory and Related Scientific Developments, 1898–1938," 7 June 1961, TSKP 6.12, 8.

[235] Kuhn, "Sources for History of Quantum Physics," Memorandum, 28 October 1963, TSKP 6.12, 1.

[236] See Footnote 235.

[237] See Corner et al., "Request for $202,996 for Period 1 July 1961–1 July 1964 to Record Interviews with Men Central in the 1913–1938 Quantum Revolution and to Retrieve Personal Documentary Material on Quantum Theory and Related Scientific Developments, 1898–1938," 7 June 1961, TSKP 6.12, 8–9.

[238] Heilbron (1968, 98). The number of interviews and interviewees is a little greater, according to Heilbron, than the one that appears in the final report (Kuhn et al. 1967, 3), as we have seen.

[239] Heilbron (1968, 98–99).

during the last twelve years of men like Einstein, Schrödinger, Pauli and Bohr has been a particularly poignant reminder of the speed and completeness with which the profession is being separated from its past," he wrote.[240] Some contemporary physicists were thus concerned with a background of information that ran the risk of being lost, and which could be of interest for showing students the foundations, whether conceptual or institutional, of their disciplines. Kuhn summarized this idea, which, though ascribed to some physicists, he did not seem eager to counter—despite his usual critical zeal—which is significant.[241] As he said soon after the completion of the project: "An appropriate injection of historical materials into professional and preprofessional curricula may, a number of them [i.e. the physicists] feel, refine the student's grasp of basic concepts and increase his understanding of the nature of the scientific career."[242] It is unusual to see this suggested by Kuhn without a rebuttal, insofar as he supported the idea that a contextualized acquaintance with real historical cases of scientific research and of creative science need not be, and perhaps was not, pedagogically pertinent. Another reason was that recent advancements in physics—and the physicists themselves—had had a clear social influence, so this science had to be studied with the tools, and from the point of view, of the historians of science. The project would contribute to this goal with an important ingredient: original documents.[243]

Whatever the reasons for the project, for Kuhn it had fulfilled its initial aim of unearthing and recovering a collection of highly relevant documents in the recent revolution in physics, as well as some memories of its main characters. "The Turn to Recent Science," a 1967 essay review of six books on recent physics that had been published in the two previous years (a list that, he said, could have been extended to three more books), is a testimony to Kuhn's awareness of the public prominence that physics was gaining at that time.[244] The influence of that extended concern with the recent past in general, and with their project in particular, was not a subjective feeling on the part of Kuhn alone. As his correspondence from 1964 onwards shows, other people were interested in obtaining data from the repositories that they had created.

Let us see a few examples. In early 1965 an archivist from the University of Illinois Library consulted Wheeler in order to gain some information from the archives about a few physicists—Jakob Kunz, R. C. Tolman, F. W. Loomis and Polykarp Kusch. Wheeler, of course, referred to the forthcoming catalogue that would be published in 1967.[245] K. K. Darrow, which had been part of the Joint Committee, asked in 1964 for some references about the correspondence between Bohr and Raymond Birge,

---

[240] Kuhn (1967, 410).

[241] On Kuhn's views about the role of the history of science in scientific education, see Kindi (2005).

[242] Kuhn (1967, 409).

[243] See Kuhn (1967, 409–410).

[244] See Kuhn (1967, esp. 409–411).

[245] See Maynard Brichford to Wheeler, 24 February 1965, and Wheeler's response to Brichford, 12 March 1965, TSKP 6.23.

which Kuhn sent in his answer.[246] Heinrich G. Kuhn, the British physicist, asked Kuhn for information about, and interviews with, James Franck, and Kuhn referred him to the American Philosophical Society, where the interviews were deposited. Kuhn also told his British colleague that Franck's interviews were not particularly reliable nor useful, and that he would not find much more information about the matter of his interest, the Franck–Hertz experiments, than the published articles. Kuhn attached a copy of the biographical summary that the project's assistants had prepared in Stage 2 of the project. In any case, this petition also come in May 1965, soon after the completion of the project.[247] It is interesting to note that, for Heinrich Kuhn, Franck's memories were not reliable, either, but that he had found reading the transcripts interesting, and that this had made him aware of the importance of the project, as Kuhn had noted at the beginning.[248] Precisely concerning Franck, his daughter, Herman Lisco, asked him to see the interviews with her father. Kuhn referred her to the American Philosophical Society once again, though this time he used very moving words about her father.[249]

This will do as examples of the kind of queries that Kuhn received from physicists, historians, and even families, as regards the kind of information that they had gathered, either from manuscripts, letters and so forth, or from the interviews that they had conducted. The importance granted at that time to the recent history and the living memories of a recent past was not minor at all. Kuhn and Wheeler were right when they thought and said that an action such as the one pursued by the project was particularly relevant at that time, when the recent revolution in physics remained in the memories of many disciples and descendants of direct participants, but the danger of losing all of them was more evident than ever before.

This was an ongoing project, which some grants from the National Science Foundation helped to keep alive until 1972.[250] Though the first installment, with Kuhn in charge, only lasted until the summer of 1964, further efforts were made to bring other collections of documents to the extant repository. Kuhn was also active during those stages, but his efforts were now mainly devoted to his new duties at Princeton, to his constant philosophical activity, and, of course, to getting the best from his newly gained knowledge of the revolutionary episode of the rise of quantum physics and his control over original sources—which is also revealed in the correspondence with some individuals asking for advice about the sources collected in the project. The outcome of the project in which he invested most of his time during his last years at Berkeley (1961–1964) was a short but significant series of talks and publications

[246] See Darrow to Kuhn, 4 December 1964, and Kuhn to Darrow, 11 December 1964, TSKP 6.23.

[247] See H. G. Kuhn to T. S. Kuhn, 22 May 1965, and the latter's reply, 27 May 1965, TSKP 6.23.

[248] See H. G. Kuhn to T. S. Kuhn, 6 June 1965, TSKP 6.23.

[249] See the letters from E. Merzbacher (University of Carolina at Chapel Hill) to Kuhn, 23 May 1966, from Kuhn to Lisco, 31 May 1966, and from Lisco to Kuhn, 27 June 1966, all of them in TSKP 6.23.

[250] See Wheeler, "Final Report of the Joint Committee of the American Physical Society and the American Philosophical Society on the History of Theoretical Physics in the Twentieth Century," 29 December 1972 and 12 January 1973 (there are two reports by Wheeler with the same title), TSKP 6.8.

from 1965 to 1978 on the history of quantum physics. Examining that phase shall bring this chapter to an end.

7.10 The Road to *Black-Body Theory*

During the eleven years from 1964 to 1975, Kuhn investigated the history of quantum theory beyond the point at which he had stopped in his previous research for the archival project. He worked on that history (roughly 1894–1928) in reverse fashion with regard to the true historical development of quantum theory, at least as his main results show. In April 1966, before the American Philosophical Society, he gave a talk about "The Crisis of the Old Quantum Theory, 1922–1925," which dealt with the challenges that eventually led to matrix mechanics and the electron spin. Between that date and 1969, he wrote "The Genesis of the Bohr Atom" with his close collaborator John Heilbron, and then published the paper the latter year in the new journal *Historical Studies in the Physical Sciences*, vol. 1, edited by Russell McCormmach. From the quantum revolution at an advanced stage, he was progressing to previous stages, such as Bohr's quantized model for the atom. The beginning of the 1970s brought him back to the history of quantum theory, this time to its very earliest steps: Planck's groundbreaking work on the black body and the derivation of his radiation law, and its acceptance by the community of physicists. In 1972 and 1973, Kuhn was working hard on the mathematical details and on the list of publications that mentioned Planck's innovation right after it was published—which would later become a key element in Chapter IX of his book.

That research in a reverse fashion responds, however, to a plan that, in 1973 or 1974, was already pretty clear—though it may have been conceived before. Kuhn had developed a plan for a multi-volume series of books on the subject, which he exhibited before the National Science Foundation in a proposal for a research grant that, if awarded, would begin in July 1974. It was not the first grant obtained for such a project. The NSF had awarded a grant for two years after 1964, and had extended it, first until 1968, and then until 1972. His university duties, and probably also his parallel research on the philosophy of science, left him only a fraction of his time for this history. Those intermediate steps had probably been spent in preparing "The Crisis of the Old Quantum Theory, 1922–1925," which, as I mentioned, was conceived as a short book, and his collaboration with Heilbron.[251] But now, Kuhn had started to work hard on the early days of quantum theory, and a "complete draft" of *Black-Body Theory*, he said, was already prepared.[252] The series of books that

[251] "The Crisis of the Old Quantum Theory, 1922–1925" is mentioned in Heilbron and Kuhn (1969, 283, fn. 162) as a forthcoming title, though its length is not announced. Interestingly, the mention sets a connection between both works by Kuhn—the kind of connection that the four-volume set he proposed would have exploited.

[252] See Kuhn, "Proposal to the National Science Foundation for Support of Research in the History of Quantum Physics," n. d., TSKP 26.16, 2. Data included in this paragraph have been obtained from that detailed document, especially from pp. 1–2 and Appendix II.

he had planned included *Black-Body Theory* as the first one, and another one on *The Crisis of the Old Quantum Theory, 1922–1925* as the third volume, but it also included two more books, *The Evolution of the Quantum Conditions, 1911–1923* as the second volume, and *Light Particles and Wave Mechanics, 1905–1928* as the fourth one.[253] Detailed abstracts were provided for those volumes in Appendix II of the proposal, and a chapter-by-chapter description of the former was included as Appendix I. Clearly, Kuhn never accomplished such an ambitious project, and *Black-Body Theory* was the only outcome of this plan. Although "The Crisis of the Old Quantum Theory, 1922–1925," cited by Paul Forman in 1971, was sometimes requested, Kuhn said years later that he "had very nearly forgotten" it, and that its preparation as a book was unlikely to happen.[254]

In *Black-Body Theory*, Kuhn reconstructed how Planck obtained his now well-known radiation law for the black body, which was the basis for the introduction of the quantum view of energy in physics. Kuhn shows the long road that Planck followed, from his early work on irreversibility and his debate with Boltzmann, in order to derive his own law from the original distribution law developed by Wien, to the different views that Planck had about the nature of his own innovation.[255] In obtaining a proof for that law, Planck had to divide the continuum of energy among frequency-dependent elements of size $h\nu$, where h stands for the Planck well-known constant, and ν for the frequency of radiation. Planck initially referred to such elements as "quanta of action."[256] Despite that division into quanta, Kuhn argued that Planck did not see the energy as divided accordingly. It was rather a sort of mathematical device assumed in order to make his own parallel with Boltzmann's combinatorial approach to thermodynamics. Kuhn was thus advancing a new view about the revolutionary nature of Planck's contribution.[257] Planck obtained his expression for the law in October 1900, but it was many years later, in the period 1906–1909, when Einstein and Ehrenfest pointed out that the division truly involved a quantization of energy and, therefore, a non-classical treatment of the problem of black-body radiation.[258]

A first result of Kuhn's *Black-Body Theory* was the reinterpretation of Planck's innovation; a second interesting result was the description of its acceptance as a revolutionary step in physics among practicing physicists. Kuhn shows that the acceptance of Planck's approach was reticent at first, but that it was more widespread later.

---

[253] See Kuhn, "Proposal to the National Science Foundation for Support of Research in the History of Quantum Physics," n. d., TSKP 26.16, Appendix II (pp. 14–15).

[254] See Kuhn to Peter Kraft, 8 September 1980, who had commented about it; see also Daniel H. Deutsch to Kuhn, 15 March 1987, with a similar commentary on the paper; both of them are located in TSKP 5.2. See also Forman (1971, 61, fn. 141).

[255] See *Black-Body Theory*, Chapter IV, esp. 92–97, 102–110. On Planck's previous works on irreversibility, see *Black-Body Theory*, Chapters I–III.

[256] See *Black-Body Theory*, 130–131.

[257] This argument is developed in *Black-Body Theory*, Chapter V, esp. 115 ff., 125–130. On the further debate about Kuhn's interpretation, see Seth (2013, 818–819, and 815–823 as a good summary of this historical episode).

[258] On Ehrenfest, see *Black-Body Theory*, Chapter VI, and on Einstein's contribution and an overall perspective, see *Black-Body Theory*, Chapter VII, esp. 170–171; 182–187.

However, Kuhn emphasizes that the problem that truly helped to spread Planck's views in the wider community of physicists was not the black-body problem, but rather the problem of specific heats, to which Einstein had also made an essential contribution.[259] As in "The Crisis of the Old Quantum Theory, 1922–1925," Kuhn uses his inquiry into the published treatments of both problems during the years 1905–1914, and shows that the application of the new quantum view was extended among those that dealt with the problem of specific heats; by contrast, the interest in the black-body problem reduced in the same span.[260] The book did not close with the narrative about acceptance on the part of the community, but with Planck's "second theory" about the black-body radiation, "in which Planck found a place for discontinuity of any sort," in Kuhn's words.[261]

His notes on Planck between February 1972 and late January 1973 show the evolution of his research on Planck's 1900 paper and the subsequent history that he was about to share in a book. In fact, as he wrote in the preface for the book, *Black-Body Theory* grew from another kind of project. In early 1972, he set out to investigate the historical background of the crisis of the old quantum theory, and the appearance of such theories as the electron spin and matrix and wave mechanics.[262] At that point, Kuhn had just begun a research leave from Princeton, was no longer chairing Princeton's HPS program, and had become a part-time member of the IAS, which he would be for the following five years.[263] In the summer of 1971, the crisis period in the history of quantum theory still seemed to be his main target.[264] His early notes from mid-February 1972 still showed a different period of initial interest for him: around 1910–1912. In a group of notes titled "Evol[ution] of Quant[um] Cond[itions]," he wrote, under the headline "Emergence of the Problem," that "[c]learly it's there by 1910–12. Bjerrum, Lorentz, Planck, Einstein, Ehrenfest (?)," and added that "[p]reliminary look is suggesting that it's not there before 1908." Despite that, he also noted that he "[n]eeded careful study of Planck's papers and lectures."[265] Among the authors that he planned to deal with was Planck, of course, but also Sommerfeld and Bohr.[266] Yet, notes from ten days later show he was already focused on Planck and even on his acquaintance with Boltzmann's work; by March, Kuhn was already wondering about Planck's early work on black-body radiation.[267]

---

259 See *Black-Body Theory*, Chapters VIII–IX.

260 See *Black-Body Theory*, 216–217, 228–229.

261 *Black-Body Theory*, x; see also Chapter X (Epilogue), esp. 235–237. The influence of Planck's second theory is examined at the end of Kuhn's epilogue (244–254).

262 *Black-Body Theory*, vii–viii.

263 See Kuhn, "Proposal to the National Science Foundation for Support of Research in the History of Quantum Physics," n. d., TSKP 26.16, 2; *Black-Body Theory*, vii.

264 A note attached to a questionnaire for Princeton's faculty, and dated 19 August 1971, attests to that fact. See TSKP, 26.16.

265 All the previous extracts come from Kuhn's handwritten notes "Evol[ution] of Quant[um] Cond[itions]," 16 February 1972, TSKP 5.78, 1.

266 See [Kuhn], "Evol[ution] of Quant[um] Cond[itions]," 16 February 1972, TSKP 5.78, 2.

267 See Kuhn's notes from 26 February and 6–13 March 1972, TSKP 5.78.

The rest of his notes from that moment until late January 1973 show how he focused on the transition to the quantum view starting with Planck's 1900 work on that problem. In late March, for example, he was already studying Ehrenfest 1905, 1906 and 1911 papers—those he explores in *Black-Body Theory*.[268] Between March and the following summer, Kuhn explored many papers and ideas. Some (unfortunately undated) notes show purely mathematical reconstructions, such as that on "Planck's First Derivation,"[269] while others are explorations of bibliography, some of which refer to the reception of Planck's work in the physics community, an essential part of his arguments in Chapter IX.[270] By the end of the summer, some dated notes show that he was still thinking about the role Ehrenfest and Einstein had played in this episode. In a document titled "Where from: Energy Quant[ization]?", dated late August 1972, he states the overall significance of Ehrenfest's 1905 and 1906 papers, but he seems more doubtful about their role in "point[ing] towards energy quantization as a way out." Concerning Einstein, it seems that he still had more work to do; "[m]ust read Einstein['s] papers in full carefully," he annotates. Although this section was later crossed out, it shows that, by the end of the summer, Kuhn was still fighting with an important part of what was going to be one of his essential theses—the role Ehrenfest and Einstein played in the establishment of energy quantization.[271] Overall, these notes show Kuhn on the long road that took him from the conviction that Hans Kangro and Martin Klein had sufficiently dealt with this episode in the early history of quantum physics to the conviction that there were some unnoticed aspects to highlight, and eventually a new view of the details of the development that started with Planck's work.[272]

Despite a deepening interest in the origins of the quantum, Kuhn's account of Planck was part of a greater plan, as previously noted. Most probably, Kuhn planned to, so to speak, set the record straight as regards the history of quantum theory. This was an episode in the history of science that had "been little studied," as Trevor Pinch wrote, and whose "material," he went on "although unusually accessible, is technically complex." He also shows that Kuhn has the necessary expertise in this area of history to accomplish that task.[273] Pinch was right in that sense. However, Kuhn's aim was not only to explore that field in depth, but also to show *exactly* how the many historical stages through which quantum physics had been established had happened. As Peter Galison said when reviewing this book, the "quest for coherence

[268] See Kuhn's annotations on Ehrenfest, 31 March 1972, TSKP 5.78, 1.

[269] Undated, TSKP 5.78.

[270] Some of the notes relevant to the study of the reception of Planck's works are from early September; see Kuhn's "Planck Citations," 8 September 1972, TSKP 5.78.

[271] See Kuhn's "Where from: Energy Quant[ization]?", 26 August 1972, TSKP 5.78, 1. In further notes from early September, Kuhn is still reassuring himself about the role Ehrenfest played. See Kuhn's notes on "Misc[ellaneous] Thoughts," 5 September 1972, TSKP 5.78.

[272] On the role he attributes to Kangro and Klein and on his departure from their work, see *Black-Body Theory*, vii–ix, 256 (fn. 1). I shall give further details about the steps that followed until the submission of the final draft of the book at the beginning of the next chapter, when examining his years as a part-time member of the IAS. See our Chap. 8.

[273] See Pinch's review of *Black-Body Theory* in Klein et al. (1979, 437).

in the history of science" was a key motive for his work, and *Black-Body Theory* is probably the best example of that goal.[274]

I also agree with Galison on *Black-Body Theory* when he writes about it as follows:

> I cannot [...] accept his repeated protests that his history and philosophy are the products of completely separate concerns. [... T]hough at the beginning of this review I said Kuhn's philosophy is never explicitly presented, it is deeply embedded implicitly in his view of the history of science. Moreover, Kuhn's work is perhaps best seen as just this sort of continuous interaction between the history and philosophy of science.[275]

Galison's statement is based on the nature of scientific change as expounded in *Structure* and as visible in his historiographical works, too. I will argue for a similar connection on the basis of the methodological aspects of his historiographical practice (more on this below). In any case, Galison is right in pointing to the *continuity* between Kuhn's philosophical and historical works (he includes not only *Black-Body Theory*, but also *The Copernican Revolutions* in the latter group), and in pointing out that *coherence* is a goal he pursues when exploring "the work of pivotal scientists, a coherence within their works and a coherence with past exemplary problem solutions."[276] It is indeed those two kinds of coherence that led Kuhn throughout 1972 from the quantum physics during the second decade of the twentieth century to December 1900 (and even before, to Planck's early years and Boltzmann's times), and to doubting previous accounts, especially Klein's views.

Martin Klein did not accept Kuhn's interpretation of Planck. He disagreed in a now well-known review symposium that he shared with Pinch and with Abner Shimony ("Paradigm Lost? A Review Symposium").[277] In turn, Kuhn did not receive Klein's rebuttal very well either. "I have now seen proof for the *Isis* 'Review Symposium' about my book," Kuhn wrote in a long and detailed letter, "and I'm of course disappointed and disturbed by your contribution to it."[278] The exchange on the book and the review included another long reply by letter from Klein and a further long letter from Kuhn, but it finished when Klein accepted that they would never agree with each other.[279] Perhaps the worst part was that it damaged a "friendship [that] began more than twenty years" before, in Kuhn's words.[280] They both attempted to restore that friendship, but their relationship was no longer the same. In fact, new problems emerged in a meeting they had in Yale in January 1984. In his letter after the meeting, Kuhn said: "We were both pretty anxious and tense at our meeting in New Haven. I left it feeling extremely pleased that none of the things I'd most feared had

274 See Galison (1981, 83).

275 Galison (1981, 84).

276 See Footnote 275.

277 Klein et al. (1979). The exchange between Klein and Kuhn that I am about to examine has already been explored in depth, with reference to the same letters that I describe here (though in that case from Klein's papers deposited at Yale University), by van Dongen (2020, 11–19).

278 Kuhn to Klein, 1 August 1979, TSKP 21.59, 1.

279 Klein to Kuhn, 7 December 1979, TSKP 21.59.

280 Kuhn to Klein, 1 August 1979, TSKP 21.59, 5.

occurred."[281] However, "during the week that followed," Kuhn went from relief to "anger."[282] He believed that Klein's conciliatory attitude was unacceptable unless Klein clarified the difference between his new opinion about *Black-Body Theory* and his original opinion in his review, but Klein had not done it yet. Klein's negative reaction to Kuhn's book was—Kuhn believed—the origin of a negative interpretation of *Black-Body Theory*, and that interpretation, in Kuhn's words, "has done me great harm."[283] Definitely, Kuhn could not accept it. I am not sure if their friendship was restored after 1984. Apparently, Kuhn refused to attend the celebration of Klein's seventieth birthday in 1994, arranged by the Department of Physics at Yale.[284]

Leaving aside the emotional burden of this debate, the historical and the methodological aspects of this dispute are remarkable for those of us interested in Kuhn's views on the philosophy of science. I shall close this chapter by exploring them briefly. Concerning the historical aspects, the main point against Kuhn's book remained the attribution to Planck's December 1900 work of a classical, not a quantum nature. For Kuhn,

> Planck's [...] view of the radiation problem is still, in the *Lectures* [*on the Theory of Thermal Radiation*] of 1906, fully classical. [...] Both in his original derivation papers and, far more clearly, in the *Lectures*, Planck's radiation theory is incompatible with the quantization of resonator energy. That theory does require fixing the size of the small intervals into which the energy continuum is subdivided for purposes of combinatorial computation, and the restriction to a fixed size does isolate the main respect in which Planck's theory diverges from Boltzmann's. But the divergence does not [...] make radiation theory less classical than gas theory, for it does not of itself demand that the values of resonator energy be limited to a restricted set. On the contrary [...], any such restriction would conflict both with the global structure and with multiple details of Planck's argument.[285]

Kuhn went on to argue that the restriction on the resonators' energies is not present in Planck's thought until the arrival of Ehrenfest's and Einstein's works after 1906.[286] There is evidence on the contrary, which Kuhn admits, in passages from Planck's original 1900–1901 papers.[287] For Klein, that is sufficient evidence of that restriction, but Kuhn explains at that point why such passages "need not be read literally, and important sections of the *Lectures* show that they should not be."[288] While for Kuhn the 1906 *Lectures* help to explain Planck's original intention, they are for Klein an "attempt to get around the discreteness that others [like Hendrik Lorentz,

---

[281] Kuhn to Klein, 31 January 1984, TSKP 21.59, 1.

[282] See Footnote 281.

[283] See Footnote 281.

[284] See Asger Aaboe and Michael E. Zellner to Kuhn, 25 April 1994, TSKP 21.59. On the left margin of Aaboe and Zellner's invitation (both from the Department of Physics), an annotation suggests that Kuhn turned it down.

[285] *Black-Body Theory*, 125. For the original, German edition (and title) of Planck's book, see *Black-Body Theory*, 342.

[286] *Black-Body Theory*, 126.

[287] See *Black-Body Theory*, 127–128.

[288] *Black-Body Theory*, 128 (his explanation runs until p. 130); Klein et al. (1979, 432) also quotes such passage in part.

Ehrenfest and James Jeans] were taking more seriously than he had intended."[289] In Klein's hands, Planck "was not always completely clear about what he was doing."[290] In Kuhn's view, meanwhile, Planck was clear about the solution he was pursuing from 1900 to 1906, and only other physicists let him see the consequences of that solution.[291]

Apart from other minor critical corrections, Klein finished his review by means of a more personal complaint. Despite having cited 11 of his publications—there is no other author with a higher number of publications mentioned among the "Secondary Sources"—Klein still felt that Kuhn had not been fair to him; in particular, when referring to Ehrenfest's 1905–1906 papers.[292] Kuhn's letter did not mention this criticism either. He devoted most of his time to the discussion of the previous point, the most central of all of them. This new point was key for Klein, though, and he ascribed to Kuhn's apparent neglect of his expertise on this episode of the history of physics the origin of a growing tension between them that, as agreed by both, had begun around six years before.[293]

Indeed, as Galison says in his review, Kuhn's discussion of secondary literature is truly reduced.[294] In that sense, Klein's work is mentioned in the Preface, indeed, but it was not enough for him. In fact, Klein believed that his papers and his book on Ehrenfest merited a deeper discussion.[295] He did not even think that he should had been mentioned together with Kangro in the Preface, insofar as his work on Planck was prior to Kangro's, who, at some points, had based his accounts on his

[289] Klein et al. (1979, 432).

[290] See Footnote 289.

[291] Klein devotes three paragraphs to what those other readers—Lorentz, Ehrenfest and Einstein—had seen in Planck's papers from 1903 to 1905, and contrasts his own perspective, in which all of them recognize Planck's innovative quantization of the resonators' energy, with Kuhn's perspective, in which they do not. In the case of Ehrenfest and Lorentz, Kuhn's explanation of the reading of Planck's solution is, indeed, somewhat speculative at some points. In his first letter to Klein, however, Kuhn adds some more evidence concerning his interpretation of Lorentz's reactions (Ehrenfest's views remain a matter of interpretation, however). See Klein et al. (1979, 432–433); *Black-Body Theory*, 135, 138–140, Chapter VII (on Einstein and Planck); Kuhn to Klein, 1 August 1979, TSKP 21.59, 2–3 (as regards Kuhn's latter reply).

[292] Klein et al. (1979, 433).

[293] See Klein to Kuhn, 4 September 1979, 2, and Kuhn to Klein, 1 August 1979, 5; both in TSKP 21.59. See also van Dongen (2020, 17).

[294] See Galison (1981, 73).

[295] See also the review itself, when he alludes to the scarce commentary on his book on Ehrenfest (Klein 1970). See Klein et al. (1979, 433).

own work.[296] Kuhn's answer depends on a sort of methodological motto, and is worth being quoted almost in full, at least as regards his overall attitude towards citation:

> I believe that what bothers you is primarily a result of deliberate policy on my part [...]. When someone else had previously described the contents of papers I was discussing, I tried to point out the existence of that earlier work in my notes. Where interpretation as well as description was involved, I tried also to point out whether I was following the previous position or departing from it. But I deliberately refrained from itemizing points of agreement, indicating where either descriptions or narratives differed, and discussing the reasons why. A different policy would, I felt, have resulted in persistent confrontation, primarily but by no means only with you, and I wanted to avoid that. I think you sometimes misinterpret my attempts to do so.[297]

The "deliberate policy" helps to explain those omissions that were for Klein indicative of Kuhn's apparent disregard for his work. Just in case, Kuhn acknowledges in his last long letter (the second reply) his debt to Klein, and the virtues of his scholarship.[298]

It is therefore clear that, in historical terms, Kuhn and Klein did not agree on some key points about the internal history of Planck's innovation. This historical aspect of the debate is related to a methodological remark. Kuhn added to his two letters some details concerning the practice of the history of science. Despite its significance in theoretical terms, Kuhn's somewhat aggressive tone, at least in his initial reply, was considered by Klein as another sign of disrespect towards his expertise.[299]

Kuhn shows in his first letter that Klein has no reason to assume a change in mind in Planck's case from the papers from 1900–1901 to the 1906 *Lectures*. Kuhn claims to have handled evidence that goes well beyond the *Lectures* themselves—despite Klein's claim on the contrary—and refers to Chapter V and VIII in the book. Another argument is more properly Kuhnian: the use of terms by Planck, from "resonator" to "oscillator," for instance, only changes after 1908.[300] Beyond purely terminological differences—though, as we know well, that would be important

[296] Kuhn has often been accused of insufficient reference to other authors, as in the case of Fleck, Polanyi, and some others. If that is true, that problem seems to repeat in this case. Although he acknowledges a debt or a connection, he does not seem generous enough with references to the author(s) in question within the text. The extract below may be helpful in understanding that practice, though it only seems applicable to *Black-Body Theory*. As van Dongen (2020, 16, 24, nn. 44–45) shows, it was not the first time that Klein had felt that Kuhn had not given credit to his work. When Kuhn presented his paper on "The Quantum Theory of Specific Heats: A Problem in Professional Recognition" (Kuhn 1975b; see also TSKP 25.27) in August 1974, in a congress in Japan, he did not give credit to Klein for the interpretation that Kuhn had given in his own paper, even though Klein had given it in a previous paper. According to van Dongen, Kuhn felt "greatly embarrassed" (van Dongen's words; 2020, 16), and acknowledged Klein's work by means of an additional note (van Dongen 2020, 17). Years later, in a letter to Klein (18 October 1979, TSKP 21.59, 1), Kuhn still felt that he "had erred badly, and" that he "continue[d] to be disturbed by having done so." Yet, Kuhn believed that "nothing of the same sort occur[ed] in" *Black-Body Theory* (Kuhn to Klein, 18 October 1979, TSKP 21.59, 1).

[297] Kuhn to Klein, 18 October 1979, TSKP 21.59, 1–2.

[298] Kuhn to Klein, 18 October 1979, TSKP 21.59, 2–3.

[299] Klein to Kuhn, 4 September 1979, TSKP 21.59, 3.

[300] See Kuhn to Klein, 1 August 1979, TSKP 21.59, 1.

for Kuhn even philosophically in a subsequent phase of his thought[301]—Planck's accounts from the 1900–1901 papers to the *Lectures* is scarcely different: the *coherent* way of reading those texts is based on continuity of thought and there is no evidence on the contrary, neither textually nor purely historical. In this latter sense, Klein had argued that "Planck [had] devoted much effort in those intervening years to trying to deepen his understanding of statistical mechanics," which was supposed to be key in his account in the *Lectures*.[302] Although Kuhn found this criticism to be "secondary," he nonetheless stated that "I know of no evidence that Planck attempted to improve his grasp of statistical theory between 1901 and 1906, and we do know that his published research in those years dealt with a non-statistical topic also plausibly related to the puzzles of his black-body theory."[303] In short, Kuhn said to Klein:

> You are the first person I know of to suggest that Planck not only clarified and extended but also fundamentally altered his ideas between 1901 and 1906, and you've developed that viewpoint only in responding to my book. Given the almost complete parallelism of the relevant passages in Planck's texts of 1900, 1901, and 1906, the papers and lectures must, I think, be read in the same way. Your new account of Planck's development makes no sense.[304]

As noted above, this difference between their perspectives on this historical case leads to a methodological problem.[305] Kuhn was sure that this division was very deep, and in his second letter he points to a difference between historiographical schools:

> people who know your work and mine [...] are likely also to see that the divergences between our readings, especially of Planck and Boltzmann, relate closely to aspects of my work about which you have shown yourself to be at least uneasy: the influence of Koyré and the closely related views developed in *Structure*. [...] People who reach conclusions of this sort will be right, and there is nothing I could have done to prevent their doing so short of suppressing the book. In certain fundamental respects, you and I disagree about the nature and function of the history of science.[306]

Almost at the end of his review, Galison says that both historians are looking to be fair to Planck, and that Kuhn is following his own lemma concerning the search for coherence.[307] I agree, of course, but Kuhn would have disagreed with the idea

---

301 See Kuhn (1987, 25 ff.); *The Last Writings* (41 ff., 147 ff.) provides similar versions of the historical case.

302 See Footnote 289.

303 Kuhn to Klein, 1 August 1979, TSKP 21.59, 2. Klein argued (in his letter to Kuhn, 4 September 1979, TSKP 21.59, 3) that one of his papers cited in Kuhn's book (Klein 1963) dealt with that question. Kuhn replied in his second letter (Kuhn to Klein, 18 October 1979, TSKP 21.59, 2) that he knew that paper by Klein and that, nevertheless, "Planck's understanding of statistical theory in 1900 was very nearly the same as the one he demonstrated in 1906." For Kuhn, *Black-Body Theory* sufficiently deals with that issue, which he was unwilling to reiterate once again, either in his first letter or in the book itself (recall his "policy" regarding citation).

304 Kuhn to Klein, 1 August 1979, TSKP 21.59, 2.

305 This has also been explored by van Dongen (2020, 17–19).

306 Kuhn to Klein, 18 October 1979, TSKP 21.59, 2.

307 Galison (1981, 82–83).

that "coherence" means the same in both cases. A different sense of the word seems to underlie the disparity in methodological terms that he observes in their respective practices. His first letter to Klein devotes some paragraphs to make that point.

For Kuhn, there is a difference in the ways in which they both, he and Klein, use words such as "confused," or "confusion," and "contradiction" to refer to the state of mind of a historical agent. (The relationship of this term usage to the search for coherence in history is straightforward, of course.) For Kuhn, one can attribute confusion or contradiction once texts show that there is some. As we know, his anti-Whig attitude demands a comprehensive reading of past texts and an initial tolerance towards the seemingly unaccountable stances that the interpreter sometimes finds in historical agents.[308] Attributing confusion or contradiction to the historical agent without exploring what the texts say adds to the historical narrative "a sort of mysticism that's anathema to most historians"—and Kuhn ascribes that vacuous solution to Klein.[309] As Kuhn says, "I do insist—and this is a key point—that confusion and contradiction are always between two or more things (e.g., Bohr's two models; Boltzmann's 'molecular' and 'molar' disorder) and that both must be demonstrated to exist *within the texts* written by the person who is said to be confused."[310] In contrast, for Kuhn, Klein attributes confusion or contradiction to the agent's *mind* from the vantage historical point at which he is located as a historian. For Kuhn, the attribution of lack of clarity that Klein makes in his review has these roots.[311] Kuhn argued that Klein assigned to Planck (or anyone else), in a still early stage, the "viewpoint that emerged from" his later research. Of course, Klein does so with hindsight, and his interpretive result is that, at first, the historical agent "glimpsed [that viewpoint] only vaguely, or intuitively, or in a confused manner." Kuhn calls this "the sleepwalker view of scientific development," and he of course rejects it out of hand.[312]

Kuhn also defends against the idea that, in his book, he is trying "too hard to establish the internal consistency of Planck's position."[313] "My mode of operation," he says in his first letter of reply, "has, however, no tendency to force clarity," and he refers to his article on Bohr and his model of the atom, written with Heilbron, where they showed how Bohr had used "two incompatible models at once," despite his reports to the contrary.[314] So, while Klein assumes that he is going to find "contradictions *within* the thought patterns reconstructed by" him, Kuhn only finds them at the end of a long inquiry into original texts, after having sought for coherence *in* them—a process in which he always has in mind the effects of incommensurability, that is, the difference between the historian's reading of history from his or her

---

308 Again, Galison (1981, 83) echoes this fact.

309 Kuhn to Klein, 1 August 1979, TSKP 21.59, 4.

310 See Footnote 309.

311 See Footnote 289.

312 See Footnote 309.

313 See Footnote 289

314 See Footnote 309.

own background and the agent's own perspective.[315] Indeed, for Kuhn, the difference between them was not only a matter of historiography, but also a question of the *methodology* of that historiography—a question that ultimately leads to Kuhn's arguments in *Structure*. Maybe he was right, after all, that Klein did not accept the messages of that philosophical book and the influence of Koyré on Kuhn's practice of the history of science.

I must say that I am not sure if this methodological criticism is entirely fair to Klein as a professional historian, but I shall not pursue that suspicion beyond this point. I am interested in this debate for two main reasons. The first reason is the more general. As we saw that Galison had already pointed out, Kuhn's historiography and philosophy are intermingled, despite his usual testimony to the contrary. This debate shows it very clearly. The second reason points to a more particular aspect within that relationship between the history and the philosophy of science that can be observed in Kuhn's work. For him, the search for coherence in past scientists (the one that Galison points out) relies on a specific method of exploring the original texts and their background, and depends on immersing oneself, as interpreter, in a different scientific culture and language, speaking and thinking from a pretty alien point of view. In order for Kuhn to speak about Planck the way he did in *Black-Body Theory* he had, indeed, to learn to "live in another world," to use his famous phrase from *Structure*. His historiography of science has always been philosophically shaped, though not necessarily by the reference to paradigms.[316] In 1975, when the essential work for the book was almost ready, he turned to the philosophical background that makes such a method meaningful. Let us proceed to explore that last stage in Kuhn's philosophical development in our last chapter and in the Epilogue.

---

[315] Kuhn to Klein, 1 August 1979, TSKP 21.59, 3.

[316] As he says in Kuhn (1984), for example. For the previous phrase of *Structure*, see its Sect. X.

Chapter 8
Kuhn's Late Philosophy of Science

8.1 The Years at the Institute of Advanced Study

In the academic year 1971–1972, Kuhn was granted a new leave of absence from Princeton University and began the spring term of 1972 at the IAS.[1] In April 1972, he received a formal invitation from the IAS's director, Carl Kaysen, to become a member of the School of Social Science for the next five years. Kuhn, of course, accepted.[2] He could not have foreseen that the School, and Kaysen in particular, would become the target of serious criticism and controversy from a group of natural scientists by the end of that year, and that in early 1973 he would be asked to take a stand for or against Kaysen. I shall return to this point in a further section below. For the moment, it is important to remark that this problem did not seem as important for Kuhn as the development of his own work on the history of quantum physics. In turn, the latter was not as significant for him, especially from 1975 onwards, as his return to *Structure*. By the time he left Princeton and also the IAS, where he had spent a term per year for the last seven years, for the MIT in 1979, he was devoted full-time to his old problems in the philosophy of science, though from the new vistas that his recent work had opened for him. I shall talk about all this in the next pages.

As had happened often during his lifetime, he was travelling from physics to the philosophy of science, from the history of his discipline to the reflections on it, and that journey had always been truly attractive to him. Unforeseen for him, this last journey was going to be long, though, and unfortunately, he did not get to see the end of it. From the end of 1975, he had started to reconsider some of the themes of *Structure* along the lines that we have been exploring in the last two chapters, and he had started to consider the possibility of preparing a new book on the philosophy of science.[3] During the next twenty years of his life, and with his characteristic

[1] See Kuhn's full working life at Princeton, undated, PUTK, 2.

[2] Kaysen to Kuhn, 4 April 1972, and Kuhn to Kaysen, 5 April 1972, both in TSKP 14.13.

[3] See Kuhn to Kaysen, 14 May 1975, TSKP 14.13.

J. V. Mayoral, *Thomas Kuhn*, Springer Biographies,
https://doi.org/10.1007/978-3-031-86512-1_8

zeal, Kuhn explored his philosophical theses and the possibilities of a new book; first timidly, then more focused on its very idea. In that period, he considered on successive occasions the grounds from which his particular approach to the history of science and the pattern that he had revealed in *Structure*, which he *knew* were unified, could be built on and expounded in a more systematic way. Of course, the basis for that unification had been explored in depth during the 1960s. We examined that work in the last two chapters. The book was not easy to construct, though. After successive attempts to show the essentials of his well-grounded position in some lecture series (which I shall examine in this chapter), the scheme that he would partially reveal in *The Plurality of Worlds* started to emerge in the early 1990s. Sadly, Kuhn's life ended before he could see the work concluded. His last twenty years show how he climbed the staircase that seemed to lead to his full view on scientific development. I still wonder whether that view would have been ultimately realizable, because he rarely seemed totally satisfied with the result. In any case, he zealously pursued it to the end.

As noted, in spring 1972, he was pursuing his own research at the IAS, and enjoying a leave of absence from Princeton. On March 10, 1972, he made an agreement with R. A. Lester, the Dean of the Faculty, that would provide him with enough time for his own research in the next five years, by shifting from full-time to part-time.[4] The conditions for a long-term arrangement could be studied once it had finished.[5] The agreement that he proposed, and ultimately obtained, involved that part-time, five-year period with two-thirds of his salary and one additional paid leave every six years. During that period, Kuhn would keep his own office, together with the usual assistantship, but he would also enjoy the one at the IAS; if this latter was unavailable, he would obtain a second office at Princeton's library. In exchange, he would teach one term per year, and while on leave at the IAS he would be available for usual student supervision and some administrative duties. These latter had nevertheless been reduced as he had resigned as Director of the HPS Program in January 1972.[6] Moreover, after the offer of the Benjamin Franklin chair from the University of Pennsylvania, his salary increased more than expected in the same agreement.[7] Besides all that, an inheritance he had just received made his life, both in terms of time for research and of economic well-being, much easier. It was actually the key for negotiating this part-time status.[8]

We saw in the last chapter that Kuhn had begun to work on the black-body problem as a brief preliminary step towards the background of the crisis of the old quantum

[4] See Kuhn to Lester, 10 March 1972, PUTK. The terms of agreement can be found in Lester's "Proposal for Two-Thirds Time Status," sent to Kuhn on 8 March 1972, PUTK.

[5] Kuhn to Lester, 25 January 1972; Lester to Kuhn, 11 February 1972, PUTK.

[6] See Lester to Goheen and Bowen, 7 September 1971, PUTK. For the details of Kuhn's agreement, see Lester's "Proposal for Two-Thirds Time Status," 8 March 1972, PUTK.

[7] See Lester, "Proposal for Two-Thirds Time Status," 8 March 1972, PUTK, 1. See also Lester to Goheen and Bowen, 3 March 1972, PUTK, 1.

[8] Concerning that inheritance, see Lester to Goheen and Bowen, 28 January and 3 March 1972, PUTK. Not everyone in the faculty was happy with some aspects of Kuhn's arrangement, as evident from the tone of Ricardo Mestres's letter to Lester, 1 February 1972, PUTK.

theory. From a period of interest in the years 1910–1912, he went back in time to 1900 and before. Such a change took place in February 1972, during the early days at the IAS that spring term. By the following summer, he was focusing on the themes that would eventually be the core of *Black-Body Theory*. The years that he spent at the IAS were highly significant for the book, which he nonetheless started as a paper.[9] In June 1972, a first draft of an article on Planck was in sight, and he planned to continue that work in the next fall term.[10] The following year, he reported that the paper had turned out to be a book, and that he had been talking with Oxford University Press to publish it there, as eventually happened.[11] In July 1974, the first half of *Black-Body Theory* was written and he expected to finish the full draft before Christmas. Indeed, by May 1975, he was already revising the full manuscript with the help of Oxford University Press. Revisions, however, were not a minor task, just as making it clearer and adding endnotes had not been the year before. Moreover, the book was no longer an exciting project for him because he was just improving it for publication. In short, he wished to finish it, but there was still more work to do for a while—the kind of work that Kuhn no longer enjoyed.[12] In September 1976, he reported to Kaysen again that the book still needed revisions. He had returned to the first chapters and had made substantial changes. Although the book was turning out to be hard work, he seemed satisfied with the result. He wished to submit the final manuscript to the press by February 1977, but in the end it seems to have been submitted in late October 1977.[13] An important part of his days at the IAS had been spent in finishing his book on the history of the quantum theory.

During his years at the IAS, he had been pursuing other works, such as his famous paper, "Mathematical versus Experimental Traditions in the Development of Physical Science."[14] He also reported in May 1975 to have rewritten the paper, formerly a lecture he had given in 1972: the George Sarton Memorial Lecture in the History and Philosophy of Science (AAAS) in Washington, DC.[15] He published it first in French in 1975 and then revised it for the English 1976 version.[16] Similarly, he had been working on an improved version of his 1973 Machette Lecture at Furman University, "Objectivity, Value Judgment, and Theory Choice," in order to publish it.

---

[9] See *Black-Body Theory*, ix; Kuhn to Kaysen, 22 June 1972, TSKP 14.13.

[10] Kuhn to Kaysen, 22 June 1972, TSKP 14.13. Kuhn felt that, with the benefits of the HPS Program in mind, fall terms would be better for his leaves during the five-year period; see Kuhn to Lester, 25 January 1972, PUTK, 2.

[11] Kuhn to Kaysen, 28 June 1973, TSKP 14.13.

[12] See Kuhn's reports to Kaysen, 8 July 1974 and 14 May 1975, TSKP 14.13.

[13] See Kuhn to Kaysen, 30 September 1976; Kuhn to Harry Woolf, 15 November 1977; both in TSKP 14.13.

[14] See Kuhn (1976a); its French version was Kuhn (1975).

[15] See Kuhn to Kaysen, 28 June 1973, TSKP 14.13; Rossiter (ed.) (1999, Appendix B, S329).

[16] See Kuhn to Kaysen, 14 May 1975 and 30 September 1976, TSKP 14.13; Kuhn (1975a; 1976a). See also the exchange between Kuhn and Merton, January–February 1976, where the paper, in both versions, is mentioned. See Merton to Kuhn, 19 January and 18 February 1976, and Kuhn to Merton, 26 January and 24 February 1976, all of them in TSKP 22.3.

The latter was only published in *The Essential Tension*, in 1977, though.[17] Likewise, the preparation of the latter volume, *The Essential Tension*, took place during those years, as did the translation of Fleck's book to English, to which he contributed a foreword.

8.2 *The Essential Tension* and the Return to Fleck

Kuhn had been thinking about a book of essays (published and unpublished) just as *Essential Tension* would be, since at least mid-1969, around the time he presented "Second Thoughts on Paradigms" in Illinois. Barbara Babcock, from the University of Chicago Press, had shown an interest in the project. Kuhn had the intention of including that paper in the volume, too.[18] Five years later, the idea returned in conversation again with Chicago, now with Philip D. Jones. At that point, Kuhn was struggling with *Black-Body Theory* in its early stages, though he said, somewhat optimistically: "I am at the moment deeply engaged in trying to finish revisions of a volume promised to Clarendon and Oxford on a part of the history of the quantum theory. Until that is finished," he added, "I want to avoid all other writing commitments."[19] A few months later, in June 1974, he started talking about the likely translation to German of some of his essays and, although he had initially left aside any plan until late that year, he had started to think about both volumes: the German-language one and its equivalent in English.[20] In July, Kuhn did not think that the plan for the book would be ready before the summer of 1976 or so.[21]

The book, which in June 1976 was still tentatively titled *Tradition and Innovation: Selected Studies of Scientific Change*, had nevertheless started to be discussed in 1974. Kuhn made something clear in 1976, and probably long before: "I am not prepared to push that on you," he told John Ryden, the editor-in-chief of the Chicago Press, "but the suggestion may at least indicate that I refuse to have 'Paradigm' in the title and simultaneously give you some notion of the sort of banner that is on my mind."[22] In the book, whose title, in the end, was *The Essential Tension: Selected Studies in Scientific Tradition and Change*, the concept of a paradigm had the advantage of being shown together with an essay on the extended background that supported their usage, and with the corrections that emerged from his post-*Structure* work.[23]

[17] Kuhn to Kaysen, 8 July 1974, TSKP 14.13.

[18] See Babcock to Kuhn, 2 and 28 April 1969, and Kuhn to Babcock, 15 April 1969, all of them in TSKP 25.55.

[19] Kuhn to Jones, 19 February 1974, TSKP 26.6.

[20] Kuhn to Jones, 19 June 1974, TSKP 26.6, 1.

[21] Kuhn to Jones, 16 July 1974, TSKP 26.6, 1.

[22] Kuhn to Ryden, 14 June 1976, TSKP 26.6, 2. On the title, see also Kuhn to Ryden, 12 October 1976, TSKP 26.6, 1. "I'm not enchanted by it, but have not been able to come up with a better version; nor have a few friends I've consulted here," he told Ryden in this second letter, p. 1.

[23] See our previous two chapters.

"Second Thoughts on Paradigms," an already significant paper for Kuhn, gave a glimpse (but not the whole) of his broader vision, and in 1975 it had become a hinge, or a catalyst for his entire point of view (more on this below). That was important for Kuhn; keeping on the same old track was not. Our Sect. 8.3 shows the connection he observed in 1975 between that paper and the 1964 (actually from 1958) essay, "A Function for Thought Experiments."[24] It is easy to understand that this was, for him, the kind of core idea that had to be presented in the volume.

I would like to summarize the varied number of events that happened in Kuhn's life in 1976, either intellectually or emotionally. I will not dwell on them at this point, because they are examined in more detail elsewhere—either before or after this point in my examination. However, I think it worth emphasizing their simultaneity here. For most of 1976, he was still working hard on *Black-Body Theory*, as we have just seen, while, at the same time, he wrote a foreword for the translation of Fleck's book to English, and started to write a preface for *The Essential Tension*. In 1975, he had just seen—he confessed this thought to Wolfgang Stegmüller, as we shall see later—that there was a coherence in his system of thought from "A Function for Thought Experiments," originally written in 1958, to the 1969 draft of "Second Thoughts on Paradigms." He seemed excited about the connection. Moreover, part of his musings on the theory of knowledge were rendered in an apparently promising lecture, his Foerster Lecture at Berkeley the following November, titled "Does Knowledge 'Grow'?", which seemed to fit his new view on paradigms, as expounded in the 1970–1974 chapters.[25] I shall also talk about that lecture later. This amount of work reflected that the several layers of his thinking were not only evolving in parallel, but also that something fresh on the old ways of thought, perhaps a new view, was emerging. I explore this in the next pages. So, *The Essential Tension* reflects that solid and varied structure without being explicit about it. The volume reflects a summary of his views so far, but it also represents a productive moment of *transition* in a couple of senses: philosophical, first of all, but also the move from the history of science to the philosophy of science. Unfortunately, 1976 was also a difficult year for Kuhn in personal terms, which we will look at a little later, and it had consequences for his life, too.

Let us now return to *The Essential Tension*. From the beginning, it was important for him to have a paperback edition of the new book available quickly in bookshops, because he intended to offer material that could be useful for students and scholars. "If we cannot put together a volume which will have significant current use, in classrooms and elsewhere," he told Jones in 1974, "I have other things to do."[26] The book would combine already published papers together with unpublished ones, and Kuhn aimed to include new material from his recent work on the history of quantum theory, which would make at least a part of it interesting for the specialist.[27] A good example was "The Genesis of the Bohr's Atom," written with John Heilbron. For Kuhn, it was a

---

[24] See Kuhn (1964).

[25] See our Chap. 7 on these chapters.

[26] Kuhn to Jones, 10 June 1974, TSKP 26.6, 1.

[27] Kuhn to Jones, 10 June 1974, TSKP 26.6, 1–2.

good example of the kind of writing that could be "picked up for assigned reading in some physics courses, though that estimate may be excessively optimistic."[28] Jones, however, seemed to have been thinking about a paperback edition of less technical, even less historically-laden, essays, which could be complemented by a hardback edition with more technical, harder-to-understand papers, not initially addressed to a wide audience. As noted, Kuhn had something different in mind. His papers, technical or not, could be of some use for different groups of students and scholars, historians of recent science in particular, and these latter usually had access to the original editions. With the physicists it was different. This was a group that was likely interested in papers such as that on the Bohr's atom. For them, however, a hardback edition would make little difference according to Kuhn. This group would access a hardback edition as little as they did original publications in the history of science. Meanwhile, they and their students would find his more historically-laden and more technical papers useful and available in a paperback edition. There was no role for a hardback extended version of the book that was mainly addressed to specialists.[29]

As we know, Kuhn decided not to reprint "The Genesis of the Bohr's Atom" in *The Essential Tension*. Yet, it helps to show Kuhn's first goal for the volume: to reach a wide audience. A second goal, which was also important for him, was to make his history of science available to an audience mainly devoted to his philosophical disquisitions. As he told Jones:

> Ignoring a few early articles in straight physics, which I have no thought of reprinting, the only papers of mine properly described as professional are in history of science. Most of them can, however, be read by a relatively general audience, and one of my objects in proposing a volume of selected articles is to persuade the relatively large number of people who have been looking at my more philosophical stuff that they ought to be seeing the historical materials from which my philosophical viewpoints derive. They will not often go looking for them, but they could read a good many of them if they were willing to try.[30]

For Kuhn, those interested in his philosophical views were often little familiarized with his papers with a historical content and nature. *The Essential Tension* was also aimed at this readership. This new book thus had a philosophical role to play, too. When he criticizes the idea of a pure epistemologically-based reconstruction of scientific work on the basis of the history of science, the practice of this latter discipline is essential in understanding his philosophical views. He is not just flagging a likely point of interest that can be neglected in a philosophical analysis of his work. On the contrary, understanding how *a historian* acts is essential in order to grasp the basics of his own perspective *in philosophy*.[31]

Kuhn also declined to include another likely item in the book. Initially, the presence of Merton's views was not limited to the relevant passages in "The History of

---

[28] Kuhn to Jones, 16 July 1974, TSKP 26.6, 3.

[29] Kuhn to Jones, 16 July 1974, TSKP 26.6, 3–4.

[30] Kuhn to Jones, 16 July 1974, TSKP 26.6, 2.

[31] Some of this overall attitude is also visible in the Preface to *Essential Tension*, xiii–xiv.

Science," the fifth essay in the book.[32] Merton was included in 1976 as a contributor to the volume with some kind of piece, such as an afterword.[33] There was a reason for them to do so. Kuhn explained this to his editor as follows:

> Robert K. Merton of Columbia is both an old friend and the world's most famous sociologist of science. For the last half dozen years a number of members of the new wave in sociology of science have been writing about the conflict between Merton's views and mine, not infrequently implying that we are locked in a struggle to the death. We see no conflict between our views and find this recurrent line in the sociological literature disturbing, both intellectually and emotionally. Since I first approached you about this volume, Merton and I have been in touch about one recent example of this distasteful genre, and it suddenly occurred to me that my volume would provide a natural and an easy podium from which to make a statement on the subject. [...] He is delighted by the idea and would very much like to supply such a piece.[34]

This was written in May 1976; by June, Chicago had accepted Merton's contribution and the deadline for his piece was initially October of the same year.[35]

Merton's piece was not yet ready by that date, though.[36] Merton agreed to contribute that piece because he also agreed that the interpretation of Kuhn's views on science as representing a position against Merton was forced. Merton explained in a February 1976 letter to Kuhn that he saw that interpretation as a recent trend originating mainly in the United Kingdom and which was profitable for its promoters, who were creating an opposition where there was not any. He, similarly to Kuhn, shows himself somewhat bored with it.[37] However, they saw the benefit in the kind of reply that a conjoint publication would provide in order to set the record straight. *The Essential Tension* would have helped in that sense. Kuhn introduced Merton to his plan for an afterword in the book in precisely this way:

> As you know, I share your distress over the attempts to create a Kuhn–Merton confrontation. If my feelings are less strong than yours (I am not sure that they are) that is only because I see less of the literature involved than you do. We ought, I think, to try to do something about it. [...] Doubtless, we shall find some [...] podium at some point. It occurs to me [...] that this [afterword] might be an easy and useful one, which is why I raise the subject at all.[38]

---

[32] See Kuhn (1968, 115–118). He also mentions Merton in Kuhn (1971b, 136–137), reprinted in *Essential Tension* (just as Kuhn 1968).

[33] Kuhn to Ryden, 3 May 1976, TSKP 26.6, 2.

[34] Kuhn to Ryden, 3 May 1976, TSKP 26.6, 2. Kuhn made a similar statement against that view in the preface to *Essential Tension*, xxi–xxii, as he had told Ryden (in the same letter) that he intended to do. A classical paper that represents the Kuhn–Merton confrontation that they did not like was Barnes and Dolby (1970), cit. by Kuhn, *Essential Tension*, xxi, fn. 9, where, in effect, Kuhn is presented as an alternative to Merton's seemingly outdated views.

[35] Kuhn to Ryden, 14 June 1976, TSKP 26.6, 1.

[36] Kuhn to Ryden, 12 October 1976, TSKP 26.6, 1–2.

[37] See Merton to Kuhn, 18 February 1976, TSKP 22.3, 2.

[38] Kuhn to Merton, 24 February 1976, TSKP 22.3, 1–2. Another possible occasion to confront their views as they pleased would have been a volume that Kuhn seems to have been planning about his "Mathematical versus Experimental Traditions in the Development of Physical Science," which never came to fruition, either; see Kuhn to Merton, 24 February 1976, TSKP 22.3, 2.

Before mid-October 1976, Kuhn had sent Merton a copy of his Preface to *The Essential Tension*.[39] Kuhn, however, had not had news from Merton, which he excused on the basis of Merton's heavy agenda. Yet, he told Ryden that, if Merton could not submit his afterword in time (or not at all), they could "go ahead without it. [...] I hope we do get this piece," he added, "but I don't want to delay the volume much for that purpose."[40] As we know, Merton's afterword was never a part of *The Essential Tension*, which was a pity in any case.

In the same February 1976 letter in which Merton expressed his frustrations with the new trend in the sociology of science, he also talked about the translation of Fleck's book and invited Kuhn to contribute a foreword. Merton argued that Kuhn had been the ultimate cause of his and T. J. Trenn's plan to translate Fleck's book into English, because of Kuhn's mention of the Polish physician's monography in the Preface to *Structure*. Merton could not think of a better option for a foreword.[41] Kuhn responded that, "Under the circumstances, I do not quite know how to refuse to do at least the very brief introduction." He made suggestions as to other collaborators, either to contribute or to replace him in that role, such as the Polish mathematician Mark Kac, at Rockefeller University, who "actually knew Fleck."[42] In June 1976, Kuhn had written the foreword and submitted it to Merton, and was already suggesting changes in the translation. One of the suggestions was to substitute "genesis" in the title for a previous option, "origin." Kuhn argued that "genesis" was more appropriate. He said that "'Origin' can too easily be read in this context as equivalent to 'Discovery,' and the German original cannot."[43] He went on as follows:

> When you have looked at my account of how I first discovered the volume, you will see why I particularly press the difference. Incidentally, if you do not change the title, I shall have to do something about my text, probably simply quoting your form but including a parenthetical or footnote indication of what seems to me misleading about it.[44]

Fortunately, Kuhn did not need to take this latter step. Merton and Trenn accepted "genesis" as the proper option for the title.[45] Kuhn finished his June 1976 letter by giving them "[t]hanks for bringing me back to Fleck. It has been exciting to reread it."[46]

Merton's reply did not change things much in his early draft of the June Foreword, even though Trenn did not seem to like how Kuhn had dealt with the terms

---

[39] Kuhn to Ryden, 12 October 1976, TSKP 26.6, 2.

[40] See Footnote 39.

[41] Merton to Kuhn, 18 February 1976, TSKP 22.3, 3.

[42] Kuhn to Merton, 24 February 1976, TSKP 22.3, 1. Kac and his personal contact with Fleck are also mentioned in Kuhn's Foreword; see Kuhn (1979c, vii).

[43] Kuhn to Merton, 9 June 1976, TSKP 22.3, 1.

[44] See Footnote 43.

[45] Merton to Kuhn, 29 June 1976, TSKP 25.31, 2.

[46] Kuhn to Merton, 9 June 1976, TSKP 22.3, 2. The discussion of details of the translation with Trenn went on. In 1979, Kuhn received some explanations from Margaret Flack, from the University of Chicago Press (not to be confused with Fleck himself), concerning some decisions between them and Trenn about some terms in German and their translation. See Flack to Kuhn, 31 January 1979, and Kuhn to Flack, 6 February 1979, both in TSKP 25.31.

"thought-collective" and "thought-style" in this text. For Trenn, Kuhn had attributed features to the thought-collective that belonged to the thought-style. Trenn suggests that the thought-collective equals categorization for Kuhn in the last paragraphs of the foreword; worse still, a categorization of a Kantian nature. In that case, Kuhn should rather be talking about the thought-style, not the thought-collective.[47] However, Kuhn was not saying that; "Thad [Trenn]," he replied, "has totally misunderstood my point (or else is defending the aspect of Fleck to which I'm objecting)."[48] Although Kuhn did not extend his reply against Trenn beyond that point in that letter, the foreword itself clearly shows that Trenn's misunderstanding was real, insofar as Kuhn says there that, once the individual belongs to a thought-collective, this latter is a *source* of such categorization. There is a component of an epistemological nature on individual thought once (or at the same time) that the subject acquires membership, and that, "like the Kantian categories, [it is] prerequisite to any thought at all."[49] But he is *not* saying that such a prerequisite *is* the thought-collective *itself*. In any case, Kuhn did not wish to spend much more time detailing his foreword, and he opted for suggesting the elimination of the last paragraphs. As they are still there, it seems that Trenn and Merton accepted his explanation.[50]

8.3 Troubles and Connections in the Ivory Tower

The years that brought the publication of *The Essential Tension* and Kuhn's collaboration on the translation of Fleck's book to English were also those in which Kuhn spent most of his time in a sort of ivory tower in search of his own thoughts about the history of science and his closely related philosophy of science. The ivory tower that he wilfully sought was not, however, free of more worldly concerns. As soon as Kuhn entered the School of Social Sciences, he found himself dealing with the complications that the appointment of sociologist Robert N. Bellah at the School (itself a motive for a previous debate) had produced.[51]

Carl Kaysen, to whom Kuhn addressed almost all of his annual reports during his stay at the IAS, starting in 1972, promoted the creation of the School of Social Sciences that Kuhn joined during his five-year part-time status at Princeton. As a part of that creation, which was polemic for some of the already full-time members, such as the mathematician André Weil (among others), Kaysen also got a full-time

47 Merton to Kuhn, 29 June 1976, TSKP 25.31, 2. Trenn's commentary is copied in that part of the letter. See also Kuhn (1979c, x–xi).

48 Kuhn to Merton, 22 July 1976, TSKP 25.31.

49 Kuhn (1979c, xi).

50 See Kuhn to Merton, 22 July 1976, TSKP 25.31. Compare Kuhn's Foreword (Kuhn 1979c) with his draft, "On Encountering Fleck: Some Personal Reminiscences," dated June 1976 (TSKP 25.31), which, despite showing many minor corrections, remains essentially the same as that which was later published.

51 This controversy over Bellah's nomination for a permanent appointment at the IAS has been described and analyzed in detail by Bortolini (2011).

appointment for anthropologist Clifford Geertz—who was very close to Kuhn, too—in 1970, and later on, he tried to achieve the same for a former colleague of Geertz: Bellah.[52] Members from the IAS's Schools of Mathematics and History disagreed with this latter possibility upon reading Bellah's scholarly publications, which they considered to be of low value. As journalist Ed Regis says in his book about the IAS, this conclusion was not only drawn by a mathematician like Weil, but also by an expert in the history of Greek philosophy like Harold Cherniss.[53] Kaysen, the School itself, and of course Bellah, seem to have been part of the same problem for a part of the Schools of Mathematics and History within the IAS, so they demonstrated clear disagreement against that new step.[54]

The subsequent history of the affair has often been told. An ad hoc committee, which included some of Kuhn's closest academic friends (Merton and Cavell were among them), met on December 3, 1972, in order to discuss Bellah's permanent appointment.[55] However, it adjourned with attendees showing mixed feelings about this appointment.[56] Or so Weil informed Kuhn in a letter from January 29, 1973, at least. In contrast, in a memorandum from a few days later, February 2, 1973, Geertz reported that an agreement in support of Bellah's appointment had been reached.[57] In any case, the reports coming from the ad hoc committee were used in a meeting of the IAS faculty on February 15, 1973, in which the appointment was considered, and ultimately rejected. Kaysen, in turn, who still wanted Bellah to become a permanent member of the IAS, took his proposal to the Board of Trustees, who approved it on January 20, 1973.[58] Weil also says that, even though some faculty members told the Board of Trustees about their misgivings, these were not taken into account.[59] Bellah got his appointment (although he ended up going to UC Berkeley), but Kaysen's clash with the faculty even reached the newspapers.[60]

Soon after the faculty vote took place, Kuhn was invited by Weil to take a stand on the conflict.[61] Weil felt that the faculty had lost its legitimate grip on matters such as the appointment of new permanent members of the IAS.[62] Kuhn, however, did not feel the same as Weil. It is probable that this issue brought back bitter memories of his old days at UC Berkeley, when the Philosophy Department's senior faculty voted against him and rather for his promotion in the Department of History alone.[63] In

[52] See Bortolini (2011, 5–7). On Geertz and the IAS, see also Gross (2008, 211–212).

[53] Regis (1987, 204–205).

[54] Bortolini (2011, 7).

[55] Bortolini (2011, 9).

[56] See Footnote 55.

[57] See Weil to Kuhn, 29 January 1973, and Geertz, "Memorandum to All Members in the Social Science Program," 2 February 1973; both in TSKP 14.13.

[58] Bortolini (2011, 9).

[59] Weil to Kuhn, 29 January 1973, TSKP 14.13.

[60] See Bortolini (2011, 9 ff.) for a description of the following events.

[61] See Weil to Kuhn, 1 February 1973, TSKP 14.13.

[62] Weil to Kuhn, 29 January 1973, TSKP 14.13.

[63] See the beginning of our Chap. 6.

his reply to Weil, he actually mentioned Berkeley as a place where departmental life had degraded, and he saw the intervention of outside authorities as more convenient for the appointment of new faculty. In this and other cases, faculties and departments should propose new candidates, but the decision, for Kuhn, should remain in the hands of a different authority. That way, he said, personal bias could be avoided—though it was probably difficult to suppress them entirely. The idea that *all* the faculty, whatever the field of expertise, had a share in such matters was unusual for him, and he did not seem to like it.[64] As expected, Weil's reply showed disagreement, and not in a wholly friendly manner. Kuhn had offered to discuss the issue further at the IAS, but Weil did not seem particularly keen on the idea.[65] So, at the very beginning of his stay at the IAS, Kuhn could have considered himself as part of a somewhat turbulent ivory tower. It was probably nothing that he had not experienced before—except for the press coverage, of course.

At the School of Social Sciences, Kuhn gave two talks in different years, 1974 and 1976, and alongside Geertz took part in the organization of a seminar at the School in the 1976–1977 academic year, to which sociologists that he used to read were invited, such as N. C. Mullins, Diana Crane, and Jerome Ravetz (whose *Scientific Knowledge and Its Social Problems* he had just read), together with Barry Barnes, who represented the view of Kuhn's opposition to Merton that he considered misleading.[66] Concerning his contributions for the IAS's School of Social Sciences, he gave a talk on "Puzzles versus Problems in Scientific Development" on January 7, 1974. The Lunch Group, as it was called, had a variety of members; Kaysen and Geertz, of course, but also physicists such as John Bahcall and Freeman Dyson.[67] "Puzzles versus Problems" was often repeated during the rest of the year: at Washington University, St. Louis, as the A. H. Compton Lecture, for example, and at Vassar College.[68]

The lecture on "Puzzles versus Problems" at Vassar College is worth emphasizing, not because of the contents of the lecture itself, but because of Kuhn's reaction to the audience's reception of it.[69] In this paper, Kuhn reintroduced the difference between puzzle-solving and problem-solving, and explored the consequences for our views about the sciences and the differences between scientific disciplines. That is the argument that he presented at Vassar and elsewhere. Basically, it was not particularly special for common readers of *Structure*. Before the lecture, Kuhn seems to have been particularly excited about the prospect of lecturing at the Alma Mater of his

---

64 Kuhn to Weil, 2 February 1973, TSKP 14.13. Talcott Parsons had a similar reaction in 1973, as Bortolini (2011, 14) shows.

65 Weil to Kuhn, 4 February 1973, TSKP 14.13.

66 See Gross (2008, 212). Kuhn had read Ravetz's 1971 book, *Scientific Knowledge and Its Social Problems*, in June 1972, and had almost finished it at that point. He nevertheless sent the sociologist a relatively long letter with some comments and criticism, though he said that his "reaction [...] is unqualified enthusiasm and admiration" (Kuhn to Ravetz, 21 June 1972, TSKP 12.1, 1). See, of course, Ravetz (1971).

67 See Geertz, "Memorandum Re the Social Science Lunch Group," 7 January 1974, TSKP 5.8.

68 Texts for these lectures can be found at TSKP 5.8–10, together with relevant correspondence.

69 See Marcum (2015a, 118–119) for a useful summary of the lecture.

mother, of his wife, and of his daughter in the near future.[70] The result, however, was not what he had expected it to be. Kuhn told Stephen Rousseas—the economics professor in charge of the Science, Technology and Society (STS) Program there—that "the Vassar trip was for me a nightmare."[71] He did not spare his comments on the reasons for that conclusion, which Rousseas definitely found offensive.[72]

Apparently, it wasn't just the student audience asking questions that for Kuhn involved a serious misunderstanding of his views, but also the faculty members from the STS Program. He told Rousseas that "it bothered me a great deal to find a program with a title like yours [STS] led by people who appear to know nothing about either science or technology and to be totally uninterested in learning more."[73] A student told him that he "was no prophet," which he also found striking.[74] Yet, he was particularly astonished by the overall attitude about his role as the author of *Structure*:

> During my visit you repeatedly pointed out to me that it made no difference if what you and your colleagues took from my book was something very different from what I had intended. You could well be right and I wrong. With all of that I agreed then, and I still do. Those thoughts are not new ones, and they have helped to prevent my responding to a number of misrepresentations of the book in print. But my agreement does not at all relieve me of responsibility. After all, what you said to me must be rather like what Truman and Stimson said to James Franck, and other atomic physicists concerning the decision about dropping the bomb. You have done your job; leave the rest to us; and don't rock the boat. No one has been very happy since about the outcome of that exchange, and I am not very happy about the outcome of this one.[75]

Kuhn was clearly annoyed by that attitude—probably as much as Rousseas was upon receiving his missive, for that matter—because of his perpetual wish to be properly understood as well as to properly understand others.[76] His increasing popularity as the author of *Structure* probably made that wish difficult (if not impossible) to satisfy. Yet, things were worse in some contexts; for instance, this one, to which he was emotionally so close. He actually made that clear in that letter, too: Vassar was significant for him.[77]

As noted, he gave a second talk at the School of Social Sciences in 1976. His notes for the talk show some points of interest, which I shall mention in a further section. However, the cover letter that he submitted to the members of the Symbolic Anthropology Seminar for which he talked is perhaps more significant. Moreover,

---

[70] [Kuhn], "Puzzles vs. Problems in Scientific Development," Vassar College, 4 November 1974, TSKP 5.9, 1. See also Kuhn to Rousseas, 20 June 1973, TSKP 5.9.

[71] Kuhn to Rousseas, 9 December 1974, TSKP 5.9, 1.

[72] Rousseas to Kuhn, 20 December 1974, TSKP 5.9, 1.

[73] See Footnote 71.

[74] See Footnote 71.

[75] Kuhn to Rousseas, 9 December 1974, TSKP 5.9, 2.

[76] Rousseas to Kuhn, 20 December 1974, TSKP 5.9. Rousseas' reply (formal, blunt, and occasionally elegant) emphasized Kuhn's lack of interest in their program and their ideas during his visit.

[77] See Footnote 75.

the full context for the interesting point in that letter involves an exchange he had with the German philosopher Wolfgang Stegmüller more than a year before, in January 1975. I will examine Kuhn's letter to Stegmüller first, and then will proceed to the letter to the School's anthropologists.

The second part of the second volume of *Probleme und Resultate der Wissenschaftstheorie und analytischen Philosophie* by Stegmüller had been published in 1973. That part was *Theorie und Erfahrung, Pt. 2: Theorienstrukturen und Theoriendynamik*, and Kuhn read it carefully.[78] In April 1974, he had not finished it yet, he told Jaakko Hintikka,[79] the Finnish philosopher, but in January 1975 he told Stegmüller that he had "forced [himself] to set aside other tasks in order to finish the remaining pages of [Stegmüller's] book and then put together the long letter I promised you last August."[80] Actually, "long letter" was an understatement. The letter extends over 33 pages, so Kuhn extracted its central part and constructed an independent piece out of it—though he pointed out that the piece was "not a separate piece designed for separate distribution."[81] The 30 extracted pages received the long and somewhat archaic title of "Excerpt from a Letter of 20 January 1975 by T. S. Kuhn to W. Stegmüller Concerning *Theorie und Erfahrung*."[82] It is a very good summary of Kuhn's state of mind, not only with regard to the structuralist viewpoint, but also with regard to his philosophy of science at large, and it will help us later in a further section. At this moment, it is only going to serve as a testimony for a significant but simpler point.

The long extract of his letter was going to serve as a basis for Kuhn's contribution to a symposium which would deal with Stegmüller's application of Sneed's views to Kuhn's theory of science in *Structure*, and that would take place in the International Congress of Logic, Methodology, and Philosophy of Science, to be held in London, Ontario, the following summer of 1975. Hintikka, at the suggestion of Hempel, had started to arrange the terms of the symposium in March 1974, and Kuhn accepted the invitation in June 1974. Joseph Sneed would be the third member of the symposium.[83] As is well known, Kuhn's paper, titled "Theory Change as Structure Change: Comments on the Sneed Formalism," subsequently published in *Erkenntnis*, was the result of that amiable discussion.[84] And it was indeed amiable. Hintikka told Kuhn that he was not going to suggest any of Popper's followers as a commentator, or any other usual opponent to his views.[85] For his part, Kuhn liked Stegmüller's and

---

[78] See Stegmüller (1976)—its translation to English.

[79] See Kuhn to Hintikka, 11 April 1974, TSKP 23.15.

[80] Kuhn to Stegmüller, 20 January 1975, TSKP 20.36, 1.

[81] Kuhn to Stegmüller, 20 January 1975, TSKP 20.36, 2.

[82] See Kuhn, "Excerpt from a Letter of 20 January 1975 by T. S. Kuhn to W. Stegmüller Concerning *Theorie und Erfahrung*," attached to Kuhn's letter to Stegmüller, 20 January 1975, TSKP 20.36.

[83] See Kuhn to Stegmüller, 20 January 1975, TSKP 20.36, 2–3; Hintikka to Kuhn, 26 March and 13 June (two letters) 1974, and Kuhn to Hintikka 11 April and 27 June 1974, all of them in TSKP 23.15.

[84] Kuhn (1976b). The paper was submitted to the journal by November 1975, perhaps before; see Hintikka to Kuhn, 10 November 1976, TSKP 23.15.

[85] Hintikka to Kuhn, 26 March 1974, TSKP 23.15.

Sneed's work. About the German philosopher, he told Hintikka that "he is one of the relatively few philosophers writing about my work who does see clearly what I am trying to say."[86] To Stegmüller himself, he wrote: "I am certain that I shall learn from the exchange, as I have not from attempts to communicate with the Popperians, Scheffler, or, with occasional exceptions, Shapere."[87] Kuhn, who was put more at ease by Hintikka's caution concerning the additional commentator on that occasion, only had reservations about the likely excessive use (for him, surely rather an abuse) of formalism in the symposium. He told Hintikka, before accepting the invitation, "I do not mean at all to be devaluing formalism but only to be expressing the special problems which a response to that sort of *presentation* would confront me."[88] After all, Kuhn told Stegmüller that he still had some problems with his book because of his lack of expertise with set theory.[89]

Midway through the "Excerpt" of his letter, Kuhn confessed that he had seen a connection that seemed significant for him. In his letter, Kuhn struggles with the distinction between the normal and the revolutionary in order to shed some light on the difference between a "core" and an "expanded core," in the vocabulary of Sneed and Stegmüller.[90] Of course, a way to characterize revolutionary change is by appealing to shifts in term-meaning. In revolutionary change, these shifts are different from the "refinements, reductions of vagueness" that unproblematically populate change in normal-science periods.[91] The former often depend, in turn, on previous changes in the statements that, as in the case of some scientific laws in some contexts, play a tautological or a constitutive role, thus giving meaning to some key terms, or changing that meaning as such statements change. Kuhn mentions roles like these for Newton's second law of motion, or Ohm's law, at some points in their history. In the latter case, the change in the meaning of "resistance" during its revolutionary introduction is a good example of the kind of revolutionary change produced by a law that plays such a definitory role.[92]

Kuhn found that "A Function for Thought Experiments" clearly showed the resources that demonstrated the need and the pathway for that kind of change. Mentioning in the "Excerpt" Einstein's revolutionary step to show "the relativity of simultaneity," he says that the "[c]onsideration of the train-struck-by-lightning was primarily a device to unlock its tautological force, make an alternative sort of behavior possible by permitting a small but all important redefinition of the concepts

[86] Kuhn to Hintikka, 11 April 1974, TSKP 23.15.

[87] Kuhn to Stegmüller, 20 January 1975, TSKP 20.36, 1.

[88] See Footnote 86.

[89] Kuhn to Stegmüller, 20 January 1975, TSKP 20.36, 3. Kuhn said something similar at the beginning of his paper; see Kuhn (1976b, 176).

[90] See Kuhn, "Excerpt from a Letter of 20 January 1975 by T. S. Kuhn to W. Stegmüller Concerning *Theorie und Erfahrung*," TSKP 20.36, 9–17; also, Kuhn (1976b, 181 ff.); Sneed (1971, 171).

[91] Kuhn, "Excerpt from a Letter of 20 January 1975 by T. S. Kuhn to W. Stegmüller Concerning *Theorie und Erfahrung*," TSKP 20.36, 13.

[92] Kuhn, "Excerpt from a Letter of 20 January 1975 by T. S. Kuhn to W. Stegmüller Concerning *Theorie und Erfahrung*," TSKP 20.36, 14–15.

involved."[93] Kuhn did not mention at this moment the case of the Aristotelian account of motion and its relationship with Piaget's experiments on children's perception of speed, as is shown in "A Function for Thought Experiments." There is, however, no need for it. He shows that this discussion in the former paper was later clearer after the vision conveyed in "Second Thoughts on Paradigms," which helps to explain change, while, at the same time, "simultaneously eliminating the term 'definition.'"[94] For Kuhn, in other words, his newly developed view on paradigm-based learning, and the way a theory is acquired without resources to rules and definitions, explained the mechanism of change, and thus the normal/revolutionary difference, fairly well.[95]

We shall return to this letter a little later. Now, however, we must go back to the day, more than a year later, on May 5, 1976, when, through another letter (not as long in this case), he suggested the subject of his talk to the Symbolic Anthropology Seminar. At that point, Kuhn proposed a sort of puzzle formed by the two papers previously mentioned—namely, "A Function for Thought Experiments" and "Second Thoughts on Paradigms." It was a puzzle that he swiftly solved, though. Kuhn shows in that letter that "A Function for Thought Experiments" emerged from discussions at the Berkeley Philosophy Department in which ordinary-language philosophy, and more particularly the work of J. L. Austin, was debated.[96] Kuhn disagreed with the idea, expounded in these debates, that a philosopher exploring ordinary language must support his or her statements on empirically-based premises. Fieldwork was thus required for ordinary-language philosophy. Kuhn disagreed with the idea that this requirement was indispensable, and he found similarities between that problem in ordinary language and the role thought experiments may play in revolutionary changes in science. Shortly after, he wrote that paper, whose earliest version is dated late 1958.[97]

A draft of a paper with the title "The Status of Thought Experiments," unfortunately undated, remains in Kuhn's archives, and this might be that early version. As Kuhn says in his letter, "'Thought Experiments' was written a few months before I came upon the notion of a paradigm in its, for me, original sense of a concrete

---

93 Kuhn, "Excerpt from a Letter of 20 January 1975 by T. S. Kuhn to W. Stegmüller Concerning *Theorie und Erfahrung*," TSKP 20.36, 15.

94 See Footnote 93.

95 See our Chap. 7 for further details about this new view.

96 As Kuhn mentions a discussion about the application of the term "voluntary," it seems that a part of the discussion may have centred around "A Plea for Excuses," by the Oxford philosopher (see Austin 1956–1957, 191). See Kuhn's letter to the members of the Symbolic Anthropology Seminar, 5 May 1976, TSKP 5.16, 1–2. A reference to "responsibility" in the intimately related paper "The Status of Thought Experiments" (on which I shall talk below) also refers to that essay (see Austin 1956–1957, 181). See [Kuhn], "The Status of Thought Experiments," n. d., TSKP 25.14, 28. Of course, it need not be the only paper by Austin that they explored, though, and we know that Kuhn read at least another of Austin's papers (Austin 1946: "Other Minds"; see below), but it gives us an idea as to which of Austin's papers Kuhn did read. It shall be of some value later, when exploring "Does Knowledge 'Grow'?", his Foerster Lecture at UC Berkeley in 1976, and Kuhn's connections with Austin's work.

97 Kuhn's letter to the members of the Symbolic Anthropology Seminar, 5 May 1976, TSKP 5.16, 1.

and accepted problem solution illustrating scientific terms and laws in use."[98] And, actually, though the paper uses the term "paradigm," it is referring to examples of psychological experimentation or to actual thought experiments, not to the concept that plays such a central role in *Structure.* Moreover, in the published version of 1964 and 1977 (that is, in *The Essential Tension* in that last date), the term "paradigm" has been deleted, and there is no trace of it at all.[99] That reflects the former usage of the term on Kuhn's part (just as, for instance, he used in the Lowell Lectures) rather than the post-1959 usage, committed to the theory of *Structure.* In any case, that paper also describes the same situation that Kuhn shows in his letter, and the same questions: Do we need fieldwork in order to make a philosophical claim concerning ordinary language? And, would that tell us anything about the relationship of our terms and concepts with the world?[100]

Seen together, Kuhn says in his letter, both papers show the starting point for clarifying the relationship between the scientific language and the world. That clarification is sometimes normal, to use Kuhn's term, and sometimes extraordinary. In the latter case, changes in language made in order to better grasp our real knowledge of the world lead to revolutionary change, as noted above, which potentially affects either of both dimensions, language and knowledge, or both of them at the same time. Knowledge and language are acquired together, but it does not mean, he says, that they always match completely.[101] If we wish to understand how that could be so, remember Kuhn's post-*Structure* learning model. In that model, we learn to arrange the inhabitants (in the end, components) of our phenomenal world in categories at the same time that, first, we create a categorical space and, second, learn how to use kind terms according to this arrangement. These processes are not necessarily always well matched insofar as such knowledge and language is not learned on the basis of specific definitions or rules. From that point of view—visible in Kuhn's notes from 1962 to 1967, and then visible in "Second Thoughts on Paradigms" and other texts from 1970–1974—the process of change can be understood as a progressive improvement of language and of its matching the arrangement of our phenomenal world.

So, yes, thought experiments, not only actual experimentation, help to make scientific concepts clearer, just as they sometimes provide us with a new acquaintance with the world—with our knowledge of it. Yet, they sometimes upset our language and our knowledge, too, and lead to revolutionary change.[102] In this letter, Kuhn talks about other aspects of his views in "Second Thoughts on Paradigms" and its relationship to "A Function for Thought Experiments." I shall return to this later. But I would like to

[98] See Footnote 97.

[99] Contrast "The Status of Thought Experiments," n. d., TSKP 25.14, 2, 3, 20, 24–26, 28, with Kuhn (1964). The second does no longer include the last section, in which we find such description of the situation repeated years later in his 1976 letter.

[100] See [Kuhn], "The Status of Thought Experiments," n. d., TSKP 25.14, 28.

[101] See Kuhn's letter to the members of the Symbolic Anthropology Seminar, 5 May 1976, TSKP 5.16, 2.

[102] See Footnote 101.

reemphasize that the main point was the connection that both papers helped to establish. The whole of his theory—the set of ideas that underlies *Structure* and that was not completely understood by Kuhn in that former full presentation—kept gaining more and more layers of meaning, and the last one acquired was this connection. As he says in the letter:

> In writing the paper ["Second Thoughts on Paradigms"], it did not occur to me that I was supplying apparatus that might be used to explicate the intuitions developed in "[A Function for] Thought Experiments" a decade before. Nor did I imagine that the pre-*Structure* essay might, if thus explicated, be the most fundamental paper I had written about paradigms and the nature of revolutionary change. But I now believe that both those characterizations hold.[103]

Kuhn's newly grasped relationship between both papers is a beautiful connection between parts of his own thinking. It should not be surprising that he wished to convey that connection to his colleagues at the IAS's School of Social Sciences.

I said before that 1976 was a year full of key events in Kuhn's life—intellectually and personally. We have already covered some of these, but in the next section, we will look at the events before the end of this decade. In 1979, he would leave Princeton, both the University and the IAS, for the MIT, and with it any trace of dedication to the history of science. His new work in philosophy took priority, and new aspects of his daily life would become different, too. Let us explore the second half of the 1970s in the next few pages.

8.4 Rethinking the Growth of Scientific Knowledge

At the beginning of the fall of 1976, Kuhn confessed to his former student, Roger Hahn, Professor of History at the University of California, Berkeley, that the previous year and a half had been particularly difficult for him. At that moment, he was very worried about his father's frail health—in fact, Samuel would die just a month later, in October. Besides that, one of his daughters' marriage had just broken down, and his own problems with Kay Kuhn had become worse to the point that they had decided to separate from the previous July. He had just moved to an apartment at the IAS. In November, he was going to visit Berkeley in order to give the Foerster Lecture, to which he had been invited, and in August he still hoped that the problems with Kay could be resolved, but that possibility now looked hopeless. So, Kuhn seems understandably depressed in his letter to Hahn, and he tells him that he knows his forthcoming visit to Berkeley will end up being a particularly difficult time. In his words, "though I very much want a chance to visit with friends, I fear that I am going to find a return to Berkeley at this time pretty tough, and I have not felt up to deciding on my length of stay or other plans." He went on to say that though he was not going

[103] Kuhn's letter to the members of the Symbolic Anthropology Seminar, 5 May 1976, TSKP 5.16, 2.

to return to Princeton immediately, he still did not know what he was going to do, and that he would "fly out with [his] return ticket open."[104]

Concerning his relationship to Princeton University, in the academic year 1976–1977 his five-year trial of a part-time appointment as professor there and a part-time appointment at the IAS was about to finish, and he planned to make that arrangement permanent, because his time for research had increased considerably.[105] The extension of the plan was accepted by the dean and the program.[106] His full switch to philosophy, not only in his research but also in his teaching, was by now evident. We saw that he was no longer enjoying the last stages of his correction of the *Black-Body Theory* manuscript. In fact, except for a couple of papers that are fully dependent on his previous work on the history of the quantum theory—and even more, like "Revisiting Planck," dependent on the book itself—he did not publish or do any further research on the history of science itself.[107] Gillispie told Lemonick in January 1977 that Kuhn's teaching was now more focused on philosophy, and that he would replace Hempel in Philosophy 204, a course traditionally taught by the renowned German philosopher.[108] In fact, in the spring term of 1976, he had given a course on the history of physics from the mid-eighteenth century to the beginnings of the old quantum theory, but he did not like the results—the students, in contrast, had enjoyed it.[109] The following spring term, in 1977, he started his course on the philosophy of science, but a heart problem of pericarditis on March 31 prevented him from finishing it; Hempel took over the course until the end of the year.[110] In any case, his transition to philosophy was fully made at that time.

We have just seen that, at least from 1975, his thinking about the connections between older and newer parts of his philosophy of science was much clearer, and that he seemed excited about such links. *Structure* had been written as a report about the dynamics of change without a clear relationship to the fundamentals of this dynamics—especially in terms of his analysis of the nature and origins of conceptual systems and the languages attached to them. In the Lowell Lectures, he had started to explore such fundamentals, but once paradigms became the central parts of his explanation eight years later, a new move to locate their proper role and nature in his theory had started upon publishing *Structure*. From then, Kuhn published some revisions of his early theory (1970–1974) in which that extended treatment of paradigms with a view to the fundamentals—how a theory is truly learned without rules, how it is improved and enhanced and how it changes later—was a bit clearer. However, the moment for pursuing the whole theory finally came in 1975–1976,

[104] Kuhn to Hahn, 21 September 1976, TSKP 5.13, 1.

[105] See Kuhn to Lemonick, 8 November 1976, PUTK.

[106] See Lemonick to Kuhn, 11 January 1977, PUTK. Other documents around this one in the same place also inform us about the details of the arrangement, especially as concerns salary.

[107] See Kuhn (1980a; 1984).

[108] Gillispie to Lemonick, 28 January 1977, PUTK.

[109] See Gillispie, "Faculty Salary Recommendation Form," 28 January 1977, PUTK, 1.

[110] See Lemonick to Gillispie, 7 April 1977; Gillispie, "Faculty Salary Recommendation Form," 30 January 1978, PUTK, 1.

when the internal connections of his theory were even clearer. Kuhn's full theory seems to have never been completely finished, but in its continuous evolution, this seems to be a decisive turning point.

At this turning point, a new view about the progress of knowledge came to Kuhn's mind, and he pursued it at a difficult moment in his life: the months in which he moved to an apartment at the IAS, and now alone, looked to his forthcoming return to Berkeley with bitter regret. During that time, however, he prepared an interesting document: his Foerster Lecture, "Does Knowledge 'Grow'?" It was unpublished, though a year later he told Harry Woolf, at that time director of the IAS, that he wished to (and surely would) publish it once he had solved some difficulties in the conclusion which he did not know how to deal with at that moment.[111] Kuhn never did publish it, though, so it seems that he did not find out how to solve these issues. We shall examine the lecture in what follows.[112]

The Agnes A. and Constantine E. A. Foerster Lecture is an annual series of lectures at UC Berkeley. In Kuhn's year, he shared it with the Polish philosopher Stefan Amsterdamski, whom Kuhn had had contact with in advance. Amsterdamski had visited the United States previously, but Kuhn had not had the chance to meet him. He wrote in order to show his regret for this lost opportunity, and to talk about Amsterdamski's book, which he had just read. It was June 1976. At that point, he talked in passing about the planned text for his lecture, which had seemingly not been written down yet. In that regard, he says that "I shall probably use as my title the phrase, 'Does Knowledge Grow?', and I hope, among other things, to say something about the resemblance between some of my views and those expressed by Wittgenstein in *On Certainty*."[113] That is a good summary, in effect, of some of the points Kuhn made in the lecture, but it now seems to be, perhaps, too condensed. There are some features of the lecture that must be highlighted in order to give a more faithful representation. As a result, we will also see why this text—which Kuhn repeated on several occasions from 1976 to, at least, 1980—is relevant for his interpreters.

A first feature of this lecture was the arrangement of the argument, starting with a properly described case study, in this case devoted to Aristotle's physics.[114] It would not be the only time that Kuhn would use this case. "What Are Scientific Revolutions?" is the best-known example of its reuse in a much-improved, extended version; and the first chapter of *The Plurality of Worlds* shows an even more improved version of that case, which has even changed in comparison to the two accompanying cases from the earlier version in "What Are Scientific Revolutions?", which remain

[111] Kuhn to Woolf, 15 November 1977, TSKP 14.13.

[112] Kuhn's Foerster Lecture has just been published as Kuhn (2024), with the kind collaboration of Yafeng Shan as editor of the corresponding volume. Mayoral (2024b), in the same volume, is a study on that lecture; see also Marcum (2015a, 119–122). My next paragraphs in this chapter give a different, shorter version, but see that essay for further details and reflections. See also a brief account in Mayoral (2023a, 109–111).

[113] Kuhn to Amsterdamski, 29 June 1976, TSKP 5.13, 2.

[114] See [Kuhn], "Does Knowledge 'Grow'?", Foerster Lecture, UC Berkeley, 29 November 1976, TSKP 5.13, 5–11.

almost the same.[115] The attention to detail is measured with a clear function in mind: to show the central elements of a doctrine that should be learned all at once in order to be properly understood. Kuhn called those elements "the CORE of Aristotelian physics or of the Aristotelian worldview."[116] It is important for him to show to what extent such elements "lock together, lending each other mutual support."[117] Such a core and its composing elements are the basis for showing how Aristotle's physics is a fairly consistent theory, but also, on the basis of this example, to show a case that has philosophical meaning, as it helps to illustrate a central notion in Kuhn's theory: incommensurability.[118] So, the example comes first, as an exploratory step of historical evidence. It is not an unusual step in Kuhn's usual way of proceeding. The Lowell Lectures, we should recall, are similarly arranged, and even *The Essential Tension* presents the historical cases first. Owing to the structure of the argument, as we know, *Structure* is not arranged that way.

A second feature of this text is that he shows its relevance through an autobiographical introduction. After having read those autobiographical prefacing remarks very often, those that have read his writings are accustomed to seeing the importance of the "Aristotle experience" for him. As noted in Chap. 2, for his daughter, Sarah Kuhn, the significance of that experience for her father cannot be minimized.[119] I have explored that experience in Chap. 2, but I would now like to emphasize that it starts to play an explicit key philosophical role in this lecture, just as in the Preface to *The Essential Tension*. Kuhn shows how the historian's experience with incommensurability not only leads to applying the historical method of contextual understanding, of hermeneutical inquiry, properly, but he also shows that a certain philosophical perspective is required in order to *explain* why those resources must be applied, which paves the way for his own account of incommensurability and scientific revolutions, as expounded in *Structure*. So, the Aristotle case is a source of insight about how a core's elements are inextricably linked, and about how their terms and concepts form a rigid structure that gives shape to (that *is*, in fact) the core itself. At the same time, though, that net of terms and concepts have epistemological consequences: our *experience* of the world around us—its constitution itself—is based on such a kind of core. Incommensurability is thus not only a relationship, or rather a partial lack of it, among two different theories taken in abstract, but also, perhaps mainly, a somewhat contradictory, paradoxical kind of experience that takes place in the historian's mind; in fact, it is the same kind of experience that the creator of a thought experiment wishes to force onto those who opt to listen to it, as Galileo does with the Aristotelians in the seventeenth century, or as Einstein does with twentieth century physicists, or as Piaget does with the children that are the subjects of his experimentation. The Aristotle case and the Aristotle experience are, in Kuhn's

[115] See Kuhn (1987; 2022).

[116] [Kuhn], "Does Knowledge 'Grow'?", 29 November 1976, TSKP 5.13, 9.

[117] See Footnote 116.

[118] See [Kuhn], "Does Knowledge 'Grow'?", 29 November 1976, TSKP 5.13, 5.

[119] See her testimony in Andresen (1999, S56). Wray (2021, Chap. 1, esp. 16 ff.) has explored the relationship between that experience and Kuhn's philosophy of science that I comment on below.

hands, a central part of *the core of his own theory*—essential elements in it, and virtually inseparable ones, too.

A third feature of this lecture is the fact that, despite introducing the necessary elements for talking properly about incommensurability, Kuhn does not address it as the main concept to examine.[120] Rather, he assumes that incommensurability happens in order to show the problem that it poses for the application of the term "growth," and that, as he told Woolf even a year later, he did not know how to handle very well. The rest of Kuhn's research in the following years, especially during the 1980s, would consist of an increasingly refined treatment of incommensurability. The lexical theory would soon be born, and it would develop until incommensurability and its implications for truth and relativism were more or less clear. However, his conclusions as to the nature of knowledge, about the way it develops, and concerning the application or not of the term "growth" to it would remain largely untouched after the Foerster Lecture and later versions (which are virtually the same as the one he wrote in 1976).

A fourth and last feature that I wish to emphasize is Kuhn's treatment of elements of the philosophical tradition that he had not dealt with openly before. I am referring to the classical account of knowledge, the one based on the triad of conditions often summarized as "knowledge as justified, true belief," or the JTB analysis, and the way it is, to Kuhn's mind, insufficient.[121] Once Kuhn did so, he opened up the possibility for discussing such matters with philosophers. So, as we shall see later, Rorty showed his discrepancy with Kuhn's (for Rorty unfair) criticism of the JTB analysis. His Princeton colleague showed him some ideas in line with the kind of view that would shortly lead Rorty to write *Philosophy and the Mirror of Nature*.[122] Part of this feature is the alternative that Kuhn suggests to the JTB analysis: a view of his own based on the works of Wittgenstein, as he told Amsterdamski, and of Austin.[123] In that respect, we return again to ordinary-language philosophy as a source of insight into scientific language and, in this case, into the nature of knowledge itself. The Foerster Lecture is, therefore, a testimony to Kuhn's inspiration by key authors of that current. Kuhn is therefore dealing with more traditional aspects of philosophy from a philosophical point of view, though one that was not common in the philosophy of science.

Once we take these features into account, what does the lecture itself says? The lecture relates his experience of the rupture between Aristotle's and post-Newtonian physics with the idea of the core, which the example helps to illustrate. It then locates the notion of knowledge in the core, and shows how the only way in which knowledge can be said to grow is when it is understood as "know-how." A core, understood as a

---

[120] See Footnote 118.

[121] See [Kuhn], "Does Knowledge 'Grow'?", 29 November 1976, TSKP 5.13, 11–13, for this discussion.

[122] See Rorty to Kuhn, 5 January 1977, TSKP 5.13; see also Mayoral (2023a, 111–113) on that correspondence.

[123] See [Kuhn], "Does Knowledge 'Grow'?", 29 November 1976, TSKP 5.13, 14–18. For a comparison of Kuhn's views with Wittgenstein's views, see Kindi (1995; 2021), and Read (2012, Part I). Cedarbaum (1983) takes a biographical view about the relationship between Kuhn's philosophy and Wittgenstein's works.

set of intimately interconnected pieces of theory, cannot *grow*, properly speaking. Our practical abilities may grow continuously, and we can easily see that they are *better* than before through the historical stages of science. By contrast, our knowledge *that* phenomena in the world are a specific way is something that cannot be said to grow. In Kuhn's words:

> If by "knowledge" we mean knowing how—if, that is, we take a purely instrumental view of knowledge—then knowledge clearly does grow. We know how to do many things that Newton and his contemporaries could not do, and they knew how to do many things that the Greeks could not. On the other hand, if we mean by "knowledge" the more usual "knowledge that": That stones fall because their natural qualities are fully realized at the center, or instead *that* they fall because of gravitational attraction to the earth, or, still instead, that they fall because their path is a geodesic in 4-d curved spacetime. If that's the sense of "knowledge" that we have in mind, then I think the answer must be that it does not grow. Surely there is a change. But I see no evidence at all of growth or even of some asymptotic approach to a final state.[124]

Kuhn's answer to the query that is the title of the essay, "Does Knowledge 'Grow'?", is, in the end, that "the question is badly formed" and that it is rather an "*aporia*."[125] (He would delete this latter qualification in further versions of the lecture.)[126] His main premise for concluding this is that both senses of "know"—know-that and know-how—are "inextricably connected."[127] So, we must conclude, we cannot say that knowledge grows in one sense and, at the same time, it does not grow in the other. Both senses cannot be separated from each other. Kuhn, however, seems to have remained unsatisfied with this answer, which probably stopped him from publishing the paper in the end.

Explaining how Kuhn comes to that premise involves referring to his critique of the JTB analysis, and the proposal of his own alternative based on Austin's and Wittgenstein's works. Briefly, Kuhn devotes some passages of his lecture to show that the "standard doctrine," the JTB analysis, is, in his words, "ultimately unilluminating, or too little illuminating, with respect to the difference between the circumstances under which one may properly make a belief and a knowledge claim."[128] For Kuhn, the difference in our discourse between a knowledge claim and a belief claim is evident. I make a belief claim like "I have left my laptop on the desk" on the basis of my (maybe vague) memories of a recent event, an unfortunate oversight on my part, for example, and the confirmation that it is on the desk right now, once I have

---

124 [Kuhn], "Does Knowledge 'Grow'?", 29 November 1976, TSKP 5.13, 19. This lecture originally has the aspect of a series of notes with a list of statements (the ones that form the paragraph extracted) preceded by an alphanumeric reference. As in the edition (Kuhn 2024), I have avoided that arrangement and have presented the extract as a paragraph. Kuhn's corrections are included in the text. The same applies to further extracts from Kuhn's Foerster Lecture and of his 1980 revision. See Kuhn (2024) for further details.

125 [Kuhn], "Does Knowledge 'Grow'?", 29 November 1976, TSKP 5.13, 19.

126 See, e.g., [Kuhn], "Does Knowledge 'Grow'?", "Fall 1980: Revision of 1976 Foerster Lecture," TSKP 5.13, 19.

127 [Kuhn], "Does Knowledge 'Grow'?", 29 November 1976, TSKP 5.13, 19.

128 [Kuhn], "Does Knowledge 'Grow'?", 29 November 1976, TSKP 5.13, 13.

returned to the spot I was occupying a minute ago, and thereby I may be said to be making such a belief claim justifiably. For Kuhn, however, the content in whose truth I believe may turn out to be false—that's not my laptop, for example, but rather someone else's laptop, though it is in the same spot in which I was working before. Although I thought I was, I was not actually justified in believing that. My expression after my failure is, Kuhn says, something like "I used to believe that such and such, but I was wrong"[129]; in my example, I would say something like "I thought that was my laptop, but I was wrong."

Insofar as the contents I state are false because of my imperfect memories, for example, or because a friend is playing tricks on me, I cannot claim to have *known* that. Kuhn shows that, evidently, we are clearly talking about a different kind of propositional attitude, and that our reaction to a wrong knowledge claim cannot be, for example, "I *knew* that such and such, but I was wrong"; the most I can say is that "I *believed* that I knew that I had left my computer on the desk, but I did not."[130] So, there is a change in the way we express our failures, and Kuhn wonders what the difference between both kinds of claim is. For him, in the belief case, I am wrong about the *content* of my belief—that is not *my* laptop, so my laptop is not located at that spot, and (in the case that I did it at all) I did not *forget* it *there*, and so on; in the knowledge case, by contrast, I am wrong about something else, which Kuhn says is a *relationship.* In his words: "In one case your mistake has been about 'such and such,' the object of belief; in the other it's been about what I'll gloss over as your relation to the object of belief.—You didn't really know it at all."[131] In a further 1980 version, Kuhn expounds that relationship differently; he says: "in the other [case, that is, false knowledge-claims, your mistake …] was about your state of mind; you did not really know 'such-and-such' at all."[132] So, for Kuhn there is a *sufficient* amount of evidence that allows us to make a belief claim justifiably—*insufficient* in case it becomes a belief in a false content—which helps to turn that doxastic stance into a state of knowledge (a "state of mind" in his 1980 formulation), which involves a *relationship* between the subject and the content of his or her belief, and which for some philosophers would be a correspondence relationship.

However, what criteria must be satisfied for the amount of evidence gathered to be *sufficient* so that our belief turns out to be a legitimate state of knowledge? Kuhn speculates whether checking our memories in search of the real itinerary that justifies our beliefs (in my example, that I left the laptop on the desk) would be sufficient; should I also check, for instance, that no friend of mine is playing tricks on me? Perhaps I should check that and also some still more disturbing and imaginative scenarios. Kuhn's answer is that it is not completely clear if it would be sufficient

---

[129] [Kuhn], "Does Knowledge 'Grow'?", 29 November 1976, TSKP 5.13, 12. I am adapting an example by Kuhn on p. 13, in that case based on a raincoat and his garage.

[130] See [Kuhn], "Does Knowledge 'Grow'?", 29 November 1976, TSKP 5.13, 12. The last two statements between quotation marks in this paragraph are not a quotation of Kuhn's own words; they are my own, though they describe Kuhn's thoughts and adapt his example, as previously noted.

[131] [Kuhn], "Does Knowledge 'Grow'?", 29 November 1976, TSKP 5.13, 12.

[132] [Kuhn], "Does Knowledge 'Grow'?", "Fall 1980: Revision of 1976 Foerster Lecture," TSKP 5.13, 12.

or not.[133] In his 1980 version, an additional statement follows to that of uncertainty: there is probably not a *uniform* answer for everyone that knows how to apply the distinction belief/knowledge. It is a typically Kuhnian reflection. For him, we know how to apply both terms, "knowledge" and "belief" because we have *learned* how to do it, but we do not follow instructions or apply definitions when doing so. As he says, "people, all of whom, use the knowledge/belief distinction in the same way, would give different answers to questions about the requisite amount of evidence."[134] In a second (and also typically Kuhnian) reflection, he says:

> Be clear, children do learn how to do this at a relatively early age. After a few corrections they distinguish easily between circumstances when they may say they know and those under which they are restricted to statements about belief. What have they learned?[135]

Of course, he does not mention any specific case of Piagetian experimentation on children, but rather makes a common-sense claim. But it is the same kind of statement that he makes in other texts, such as those in which he appeals to the "Johnny example" to show that our learned behavior does not require the kind of rules that a great part of the philosophical tradition assumes and requires in order to obtain a result.

For him, questions such as these on evidence and its relationship to the belief/knowledge distinction show genuine problems that the JTB analysis leads us to neglect on the basis of its standard thesis, namely, that, together with justification, truth makes all the difference between belief and knowledge.[136] As he says,

> Truth may appear to provide the missing distinction between knowledge and belief. But since we can't ever tell—except perhaps in mathematics and direct reports of subjective observations—whether the object of a knowledge claim is in fact true, we're left as puzzled as ever about the nature of the circumstances under which we may appropriately claim knowledge. We clearly do use the distinction without absolute assurance about truth. The notion of truth is being asked to bear inappropriate weight. It can't, that is, be the cue that tells us whether to say "know" or "believe."[137]

Appealing to truth looks like an impractical solution to the problem whether we are (or must be) making a knowledge or a belief claim. For him, this problem is more successfully approached from another point of view, in that case inherited from Wittgenstein's and Austin's philosophies. Biographically, Kuhn says, Wittgenstein's influence is more recent than Austin's, and not only because *On Certainty*, the book that serves Kuhn as a foundation for a part of his argument, had only recently been

---

[133] [Kuhn], "Does Knowledge 'Grow'?", 29 November 1976, TSKP 5.13, 13; see also his "Does Knowledge 'Grow'?", "Fall 1980: Revision of 1976 Foerster Lecture," TSKP 5.13, 13.

[134] [Kuhn], "Does Knowledge 'Grow'?", "Fall 1980: Revision of 1976 Foerster Lecture," TSKP 5.13, 13.

[135] See Footnote 134.

[136] See [Kuhn], "Does Knowledge 'Grow'?", "Fall 1980: Revision of 1976 Foerster Lecture," TSKP 5.13, 13.

[137] [Kuhn], "Does Knowledge 'Grow'?", 29 November 1976, TSKP 5.13, 13.

published, but also because of his contact with Austin's work long ago, as we have already seen.[138]

In fact, in his letter to Stegmüller the previous year (indeed, very early in 1975), Kuhn already said that his account of the relationship between the distinction knowledge/belief and the problem of distinguishing the elements within the core from those that do not belong to it traces back to his acquaintance with Austin and, in particular, to his approach to the former distinction, which I examine below. Later—or so it seems, as Kuhn's acquaintance with Austin seems to come from his days with the ordinary-language philosophers at Berkeley—he read *On Certainty* by Wittgenstein, and saw a relationship with Austin's views that was highly pertinent to his own views. "Ever since [G. H.] von Wright called its existence to my attention a few years ago," Kuhn recalled, "I have been astounded by the completeness with which it reproduces in *aperçus* those aspects of my position that have most disturbed my critics."[139] In fact, in "Does Knowledge 'Grow'?" Kuhn uses a series of passages from *On Certainty*, also reproduced in that letter, in connection with Austin's views in order to show what he, himself, has in mind.[140] Let us see that alternative point of view and how it leads to the premise that I mentioned above.

For Kuhn, the distinction between knowledge and belief has to do with the behavior deployed around each of those concepts, which is not the same in each case. He had already shown in the 1980 version of the lecture that individuals learn to behave that way (that is, to apply verbs like "to believe" and "to know" consistently) from childhood and without resource to explicit rules—as is the case with the application of other terms—and that their explanation about that use might not be uniform. However, their behavior *is* consistent throughout. The way we exhibit our behavior around knowledge claims is not the same way in which we deploy it around belief claims.

Kuhn turns to the way Austin shows that we do not ask about each case identically. In a passage from "Other Minds," Austin says that "[t]here is a singular difference between the two forms of challenge: '*How* do you know?' and '*Why* do you believe?' We seem never to ask '*Why* do you know?' or '*How* do you believe?'"[141] Each challenge seems to lead to an assumption that differs from the one underlying the other challenge. So, Austin goes on, "'How do you know?' suggests that perhaps you *don't* know it at all, whereas 'Why do you believe?' suggests that perhaps you *oughtn't* believe it."[142] Neither assumption can be associated with the alternative question, he adds. If we refer to beliefs, for instance, defective evidence does not lead the challenger to conclude that you *do not* believe in such content, but rather

[138] See [Kuhn], "Does Knowledge 'Grow'?", 29 November 1976, TSKP 5.13, 15.

[139] Kuhn, "Excerpt from a Letter of 20 January 1975 by T. S. Kuhn to W. Stegmüller Concerning *Theorie und Erfahrung*," TSKP 20.36, 16. For a comparison of Kuhn's views and Wittgenstein's *On Certainty*, see von Wright (1972, 59–60), Kindi (1995, 84 ff.), and Williams (2004, 274).

[140] See Kuhn, "Excerpt from a Letter of 20 January 1975 by T. S. Kuhn to W. Stegmüller Concerning *Theorie und Erfahrung*," TSKP 20.36, 16–17.

[141] Austin (1946, 78).

[142] See Footnote 141.

that you *should not*. The contrary applies to the uses of "to know," of course. Austin also says that "[t]he 'existence' of your alleged belief is not challenged, but the 'existence' of your alleged knowledge *is* challenged. If we like to say that 'I believe,' and likewise 'I am sure' and 'I am certain,' are descriptions of subjective mental or cognitive states or attitudes, or what not, then 'I know' is not that, or at least not merely that: it functions differently in talking."[143]

Recall that Kuhn said that the kind of mistake an individual, S, makes when saying that S *believes* in the truth of a specific content that is actually false, and the kind of mistake committed when S claims to *know* something, though S does not, is not the same. In the former case, S's mistake is about the object of S's belief; in the latter case, the mistake is about S's (epistemic, or cognitive) relationship with the object. In the first case, S should *not* believe such a thing; in the second case, the relationship between S and the object does not *exist*—as Austin says, too. Accordingly, their respective positions seem to agree that "S knows" is wrongly applied in the second case (or its application is doubted in the case of a challenge, as in the previous paragraph), whereas "S believes" is correctly applied in the former case (and also assumed when challenging S's belief) but the truth of the content in which S believes is the target of our challenge, or of our conclusion of incorrectness.

Kuhn is also inspired by Austin's idea that the way such challenges are answered is by appealing to different statements. In the first case (that is, as regards statements like "S believes that *p*"), we enter the game of exchanging evidence concerning the content that S believes in. In the second case, we may do so as well, or, alternatively, we may appeal to S's "credentials," as Austin says.[144] Kuhn is particularly interested in the latter option. "This second answer—the one in terms of credentials—is likely, I think to prove the more basic," he says.[145] He adds that:

> [T]he credentials answer is a standard and, under appropriate circumstances, a fully satisfactory one. When it is, then the answer in terms of credentials simply specifies a special sort of training, a special sort of initiation procedure, after which one is a member of a group which, under ordinary circumstances can simply recognize the standard birds on sight.[146]

Credentials are, after all, acquired once a specific training process is completed, from learning a language as a native speaker (as with those children that learn to apply the verb "to know" properly) to being trained as a theoretical physicist.[147] Interestingly, that kind of training does not involve the acquisition of a set of rules, as Kuhn had emphasized in *Structure* and afterwards, during the 1960s. That kind of training is, he says, more often referring to the acquisition of "special skills," such as, for example, riding a bike or playing a guitar, and those are cases of "knowing how" rather than of "knowing that."[148] However, Austin shows that the recourse to

---

[143] Austin (1946, 78–79).

[144] Austin (1946, 86).

[145] [Kuhn], "Does Knowledge 'Grow'?", 29 November 1976, TSKP 5.13, 15.

[146] [Kuhn], "Does Knowledge 'Grow'?", 29 November 1976, TSKP 5.13, 15. Of course, the mention of a bird is related to Austin's well-known example, about which I shall say more below.

[147] See [Kuhn], "Does Knowledge 'Grow'?", 29 November 1976, TSKP 5.13, 17.

[148] [Kuhn], "Does Knowledge 'Grow'?", 29 November 1976, TSKP 5.13, 15.

credentials is as legitimate as an answer to questions concerning whether a given subject knows or not in the case of knowing *that* something is the case (for instance, that a given bird under observation is an instance of "goldfinch")[149] as it is to questions about whether S knows how to play a guitar or not. In Kuhn's words:

> Austin brings the two uses of the term ["to know"] back together by pointing out that the answer to a question about "knowing that" is often simply the specification of the credentials, the special training that permits one to know—to know on sight, without judging, weighing evidence.[150]

In short, Austin's "Other Minds" was, biographically, the vehicle for Kuhn to assume that the verb "to know" had associated a practical dimension, whether we are talking about skills or about linguistically formulable knowledge. Once that association made "knowledge" easier for Kuhn to apply to his own views, his vision of the nature of progress was also easier to understand. The reason for finding the question concerning its growth problematic was also easier to understand.

He further considers that the elements composing the core are those that correspond to knowledge claims. "I'd describe the core as the region of knowledge claims," he says.[151] That region has a *holistic* nature, as the close relationship among the elements involved reveals. His example of Aristotle's physics—described through his view of motion as a kind of change, the qualitative explanation of physical phenomena based on a limited, hierarchized cosmos in which there is the absence of a void—is a good case in point.[152] The elements leading to this explanation of physical phenomena form a core of interrelated theses that are hard to describe and understand without referring to the rest of them. On the other hand, insofar as the core overlaps knowledge for Kuhn, its elements, according to the tradition in epistemology, should be appropriately justified and true beliefs. Kuhn, however, finds support in Wittgenstein, and again in Austin, in order to show that such elements in the core are not expected to play this precise epistemological game as expounded in the JTB analysis.

Kuhn follows Wittgenstein in stating that basic statements concerning commonsense situations, like G. E. Moore's "This is a hand," are not those that require justificatory evidence; as Kuhn says, "[o]ne can simply look and see."[153] For him, the foundations of a scientific discipline have the same character and its main statements form a core with a similar status (though less related, of course, to immediate perceptual experience).[154] As regards elements in the core, where knowledge belongs, "evidence and its evaluation are not quite relevant," he adds.[155] He recalls that such elements may turn out to be false, though they are usually neither doubted

---

[149] A classic example by Austin (1946).

[150] [Kuhn], "Does Knowledge 'Grow'?", 29 November 1976, TSKP 5.13, 16.

[151] [Kuhn], "Does Knowledge 'Grow'?", 29 November 1976, TSKP 5.13, 18.

[152] [Kuhn], "Does Knowledge 'Grow'?", 29 November 1976, TSKP 5.13, 6–9.

[153] [Kuhn], "Does Knowledge 'Grow'?", 29 November 1976, TSKP 5.13, 14. see also Moore (1925, 53 ff.; 1939, 145 ff.); Wittgenstein (1969, §1).

[154] See [Kuhn], "Does Knowledge 'Grow'?", 29 November 1976, TSKP 5.13, 17.

[155] [Kuhn], "Does Knowledge 'Grow'?", 29 November 1976, TSKP 5.13, 18.

nor judged; they are rather assumed. As Wittgenstein says in *On Certainty*, propositions such as Moore's "This is a hand" "are all of such a kind that it is difficult to imagine *why* anyone should believe the contrary."[156] For Kuhn, however, it is *not only difficult* to imagine that: "it would be *perverse* to try to imagine *why* anyone should believe the contrary," he says.[157] In short, core pieces lie outside the epistemological game. After all, they form the setting against which the epistemological game itself takes place. Citing Wittgenstein's words again, Kuhn says that "the core is 'the inherited background against which I distinguish between true and false' by the evaluation of evidence."[158] There is a clear difference between our beliefs, which "are *justified* by reference to evidence that is itself made relevant by a prior system of thought," he told Stegmüller in January 1975 by letter, and knowledge claims, which "are *explained* by identifying the system of thought itself."[159] Austin remained his inspiration for that just as Wittgenstein did.

It is no wonder that these two philosophers, Wittgenstein and Austin, as well as Kuhn himself, of course, think that an experience that upsets elements in the core causes disturbance for those whose credentials are based on it, or else are simply ignored. Kuhn presents an extensive argument on behalf of both cases in *Structure* when he talks about the emergence of anomalies and the resistance of the profession to it—or to the solutions that involve a theoretical upheaval. Austin says in "Other Minds" that in confronting an unforeseeable situation such as that in which a bird "explodes," we run out of words: "Words literally fail us," Austin says, which Kuhn also quotes in his lecture.[160] In other words, the resources—both conceptual and linguistic—from our core are useless before such a situation. Instead of instantaneously recognizing an item in our experience that ipso facto resembles a past item, something that Austin relates to the behavior around the use of "to know,"[161] we have no resources at our disposal for classifying the incoming information, and it leaves us speechless.

On the other hand, the emergence of an alternative core may be the only answer to the challenge. Again, Kuhn's *Structure* is a source of historical evidence about this, in the form of cases of scientific revolutions. In substituting one core for another, the epistemological game does not take place either. Wittgenstein shows the kind of situation that Kuhn sees as the more reasonable outcome. In fact, in both his philosophical and in his historical work Kuhn has examples of that outcome. As Wittgenstein says:

---

[156] Wittgenstein (1969 §93); see also [Kuhn], "Does Knowledge 'Grow'?", 29 November 1976, TSKP 5.13, 14.

[157] [Kuhn], "Does Knowledge 'Grow'?", 29 November 1976, TSKP 5.13, 14. Italics for "perverse" are mine; italics for "why" are Kuhn's.

[158] [Kuhn], "Does Knowledge 'Grow'?", 29 November 1976, TSKP 5.13, 18. The extract within simple quotation marks is from Wittgenstein (1969, §94).

[159] Kuhn, "Excerpt from a Letter of 20 January 1975 by T. S. Kuhn to W. Stegmüller Concerning *Theorie und Erfahrung*," TSKP 20.36, 16.

[160] Austin (1946, 88); [Kuhn], "Does Knowledge 'Grow'?", 29 November 1976, TSKP 5.13, 16.

[161] See Austin (1946, 84–85).

Fig. 8.1 Kuhn on "core" and "periphery." Reproduced from Mayoral (2024b, 301)

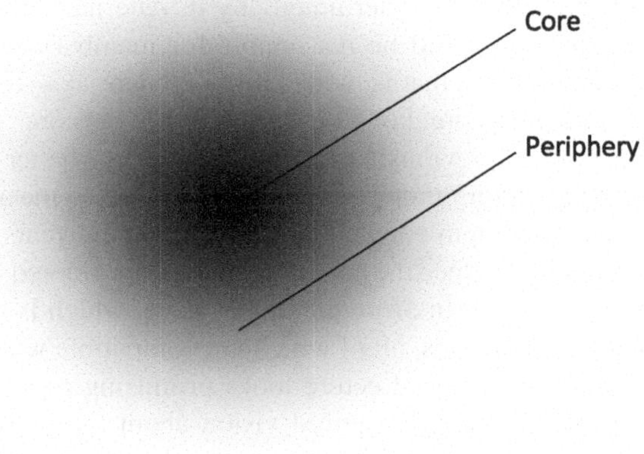

However, we can ask: May someone have telling ground for believing that the earth has only existed for a short time, say since his own birth?—Suppose he has always been told that,—would he have any good reason to doubt it? Men have believed that they could make rain; why should not a king be brought up in the belief that the world began with him? And if Moore and this king were to meet and discuss, could Moore really prove his belief to be the right one? I do not say that Moore could not convert the king to his view, but it would be a conversion of a special kind; the king would be brought to look at the world in a different way.

Remember that one is sometimes convinced of the *correctness* of a view by its *simplicity* or *symmetry*, i.e., these are what induce one to go over to this point of view. One then simply says something like: "*That's* how it must be."[162]

In his works, Kuhn has presented many examples of this kind of change that affects the core. He did so in *Structure*, and later, he did so in *Black-Body Theory* as well.

To bring this (already long) section on Kuhn's Foerster Lecture to a close, we should ask what happens with *beliefs*. Beliefs are those propositional attitudes that *do* play the epistemological game. If we *believe* that some propositional content is true, we and our equals often evaluate that content and weigh the evidence for and against it. As they are not as certain as the elements in the core, they are, figuratively speaking, part of a surrounding area that Kuhn calls "periphery."[163] If we try to illustrate the relationship Kuhn has in mind between the core and the periphery, our best approximation is probably a continuous degradation of a dark tone from the core to the periphery (see an example in Fig. 8.1). As he says, "[i]t's not clear to me just how precise this notion can be made," so he opts for representing the core as "a continuum from the center out."[164]

---

[162] Wittgenstein (1969, §92). See Kuhn's quotation in "Does Knowledge 'Grow'?", 29 November 1976, TSKP 5.13, 17.

[163] [Kuhn], "Does Knowledge 'Grow'?", 29 November 1976, TSKP 5.13, 9–11, 18.

[164] [Kuhn], "Does Knowledge 'Grow'?", 29 November 1976, TSKP 5.13, 9.

Kuhn shows that this region contains those elements that, though compatible with the core, are "not determined" by it. Alternatives to elements in the periphery are available. It would be reasonable for members of the same community, or background—those, that is, with similar credentials—to pursue such alternatives that do not affect the core. For Kuhn, therefore, if the core "points to the aspect of a scientific theory that I previously tried to cover with the better known of two distinct uses of the term paradigm," then "periphery [...] points to the area in which people with a given core or paradigm conduct what I've for some years called normal or puzzle-solving research."[165] Once the paradigm, or core, is granted, the rest of the exploration, called normal science in *Structure*, takes place, which is here a consideration of solutions and of alternatives, all of which are compatible with the extant core or paradigm.[166]

Kuhn's Foerster Lecture looks promising as a way to adjust his older views on change to a newly acquired vision about the theory of knowledge. With it, he is exploring the idea of scientific *knowledge* that may come to behave according to the discontinuous pattern visible in *Structure*. As with some other aspects of Kuhn's theory, this one remained unfinished, though this lecture, at least, leaves us with some glimpses as to the kind of direction he seemed to envision. As previously noted, his future work would turn to the problem of incommensurability, as we shall see in the rest of this chapter.

8.5 A New Decade

Princeton philosopher Richard Rorty, soon to become the author of a landmark book himself, *Philosophy and the Mirror of Nature*, was fond of quoting Kuhn's writings; indeed, he famously would do so again in his book. By the winter that followed Kuhn's Foerster Lecture, in January 1977, Rorty had read his Princeton colleague's typescript and written a letter highlighting some discrepancies.

For Rorty, Kuhn had gone too far in considering that justification does not envision reference to credentials, and much more besides. We can understand Rorty's point by saying that, despite Kuhn's alleged association with evidence, justification could be extended to include aspects of our social behavior, from credentials obtained in a community of experts or scientists to membership in a culture or religion at large. The JTB account is fine as it is, Rorty suggested, except for the view that some philosophers and, unfortunately, Kuhn in this lecture give to that concept of justification. Kuhn's contribution to that tradition did not please Rorty insofar as Kuhn himself—like Wittgenstein, Dewey, Sellars, and others—had precisely contributed to that extended usage of the term "justification." Even if Kuhn had launched an attack on some aspects of an epistemological tradition that traced back to Descartes

[165] [Kuhn], "Does Knowledge 'Grow'?", 29 November 1976, TSKP 5.13, 10.

[166] See Footnote 165.

(Rorty mentions it in his letter), he was now assuming some points of the analytical current in epistemology that used to give support to that tradition.[167]

Rorty's sympathetic series of criticisms in his letter also included other minor points. For instance, he also talked about Austin's true message, to wit, that he did not point to the analytical tradition in epistemology as a complete failure; he rather showed its ineffectiveness, its limits, as a theory of knowledge. He also pointed out that, even if core-based expectancies are unfulfilled and we are surprised by an anomaly, we can handle the situation, and we need not run out of words. For him, there is some flexibility in the kind of knowledge that grants us credentials. This point signals a motive for the discrepancy between them that would never disappear, as Kuhn emphasizes the importance of a kind of rigidity in the core that allows for the normal/revolutionary distinction. I shall return to this later. Finally, Rorty liked the analysis of growth on the basis of the difference between know-how and know-that; yet, the association of the former, of know-how, with "growth" should have led to a broader meaning and not reduced to a quantitative progression.[168]

Although all of these points were insightful, Kuhn does not seem to have followed them in his revision of the Foerster Lecture in the next few years. This letter is a good beginning for a series of exchanges that Kuhn valued highly, and that endured for many years to come, as we shall see. Kuhn was not fond of his sometimes bitter discrepancies with Rorty, which shows that he esteemed his Princeton colleague highly. However, at this moment, Kuhn did not seem to be ready to make many changes along Rorty's suggested guidelines.

At the end of the 1970s, Kuhn was still struggling with personal issues. In May 1979, he told his friend, the philosopher Emmanuel ("Manny") Mesthene, as follows: "I find writing sufficiently difficult these days to restrict myself entirely to projects that grow directly from the development of my own work."[169] Kuhn was actually trying to avoid a commitment to the new journal promoted by Mesthene—the *Journal of Policy Modeling*—but his lament was nevertheless sincere.[170] In 1982, he would confess the University of Chicago Press Science Editor and long-time friend, Susan Abrams, that he had scarcely written anything since 1977 or 1978. His personal crises, described at the beginning of the previous section, explained his low mood, which seems to have affected the progress of his writing. He was already referring to a likely future book, of whose contents he had not yet written anything at all (more on this below).[171] Accordingly, a review that he had agreed with John Watkins in 1976, which they both knew was going to take Kuhn some time to prepare, was long delayed. It was a review of *Method and Appraisal in the Physical Sciences*, published in 1976, and which included contributions by several authors, including Lakatos and his former colleague at Berkeley, Feyerabend. As Kuhn said at the beginning of his,

---

[167] Rorty to Kuhn, 5 January 1977, TSKP 5.13, 1–2. On this correspondence, see Mayoral (2023a, 111–113).

[168] Rorty to Kuhn, 5 January 1977, TSKP 5.13, 2–3.

[169] Kuhn to Mesthene, 17 May 1979, TSKP 5.26, 1.

[170] See Footnote 169.

[171] See Kuhn to Abrams, 15 October 1982, TSKP 25.54.

in the end, lengthy essay review, "The Halt and the Blind: Philosophy and History of Science," Lakatos's paper had already been the target of his own commentary in 1970, and it had been their last exchange.[172] The book was interesting for him, but, during the long process of preparation, he told Watkins at least twice that some problems of a personal nature were an obstacle for him in writing the review—which in May 1979 was not ready yet, and was only published in 1980.[173] In January 1978, Kuhn told Watkins that, owing to his personal (and also professional) problems, he felt "unprepared" for his new teaching load in the new term, and that the review was, of course, still pending.[174] Sixteen months later, in May 1979, he referred back to his personal problems in 1976, and now to some new ones—in that case, concerning his health. He had had to stop smoking, which had affected his writing, too. He had left Princeton for New York City in his last leave of absence. He told his friend Mesthene that they could have lunch together, but that it had to be in New York, and that he was at his apartment every morning (so easy to contact by phone). He was going to occupy a new post at MIT the following academic year, in whose Department of Linguistics and Philosophy he would spend the rest of his professional life.[175] He seems to have rejected the Albert Schweitzer Chair in the Humanities at New York University in 1979, owing to "the difficulty of finding intellectual seclusion in New York."[176] Most probably, Kuhn expected that the new decade could bring some relief to his life, and the kind of solitude that he had found at the IAS. To judge from his words to Abrams in 1982, however, he would be waiting a while longer for this.

At MIT, Kuhn was going to teach in two programs: in courses in the Department of Linguistics and Philosophy, and in those of the program in Science, Technology and Society, (STS) which was beginning that year 1979–1980. So, together with philosophers such as Ned Block, Paul Horwich, George Boolos, Jerry Fodor, Sylvain Bromberger (with whom he eventually taught a course), and others—including of course Noam Chomsky—he shared a program with historians of science like Loren Graham and Gerald Holton (he knew the latter very well), and also with his long-time friend, Carl Kaysen. Kaysen had left the IAS in 1976 and had been, since then, Professor of Political Economy at MIT, though on leave from 1977. In 1979, they were both going to have teaching duties in the new program in STS. In 1979–1980, Kuhn would teach an undergraduate course on "Topics in the History of Physical Science," as well as a graduate course on the "Philosophy of Scientific Development."[177]

---

[172] See Kuhn (1980b, 181). See also Kuhn (1971a); Howson (ed.) (1976); Lakatos (1971).

[173] See Kuhn to Watkins, 24 January 1978 and 2 May 1979, both in TSKP 25.32.

[174] Kuhn to Watkins, 24 January 1978, TSKP 25.32.

[175] Kuhn to Watkins, 2 May 1979, TSKP 25.32. Kuhn to Mesthene, 17 May 1979, TSKP 5.26, 1.

[176] Kuhn to Feyerabend, 20 April 1983, TSKP 21.28, 1. Feyerabend also said to have rejected such chair four years later; see Feyerabend to Kuhn, 4 April 1983, TSKP 21.28. These letters are available in Collodel and Oberheim (eds.) (2024, 314–317; see the quotation on p. 316).

[177] Data for this paragraph were obtained from MIT's General Catalogue Issue, 1979–80, *MIT Bulletin*, 115, September 1979, esp. 234–236, 254–256. See also the "Report of the President and the Chancellor, 1979–80, Massachusetts Institute of Technology," 226–230, 244–247. The university newspaper *MIT Tech Talk*, in the 23 April 1979 issue, esp. 1, 8, also reflects that Kuhn joined the

The new teaching activity was surely attractive for Kuhn, but we must keep in mind that, more than that, he expected to enjoy more time for his research in philosophy. The first notes for a future book are dated from 1980, the same year in which he not only revised (very lightly) the Foerster Lecture, but also gave the Perspective Lectures at Notre Dame University.

8.6 Rethinking the Philosophical Traditions

In his letter to Mesthene, right after repeating that his work was progressing too slowly, he tried to contribute to an open conversation between them with a brief commentary and some literature. "You ask," he told his colleague, "how to go about finding a language (let us admit that the term 'language' here is being used in part metaphorically) suitable for fruitful discussion of the technology–society relationship."[178] Before proceeding to submit the commentary, however, he said as follows:

> I am terribly protective of my time these days. The difficulty is to know how to do it. For personal reasons of no present relevance, my work has advanced far less during the past year than I had expected, and I have substantially nothing new to send you. Equally to the point, I suspect that if I did it would prove too abstract for your purposes. I think I am moving towards a series of confrontations with Quine, Kripke, and Putnam. That is a sort of discussion likely to leave most members of your group [...] quite cold. I shall make a very few suggestions, but they are directed in the first instance to you. I am very dubious that most of them are suitable for transmission.[179]

Most probably, the subject matter of Mesthene's group would be related to an STS program. Kuhn, in a later letter, referred to "a quite successful STS group" led by geographer Robert W. Kates at Clark University, which was, for him, a good example of the kind of contribution that could be of help for Mesthene at Rutgers University.[180] Moreover, his recommendations had to do with aspects of the field (STS). Yet, it is clear that though Kuhn was eager to contribute and would do so in the near future as part of the STS program at MIT, his mind was already focused on something else. After discussing the epistemological tradition in his Foerster Lecture in the previous few years, Kuhn—as he had already made clear at the beginning of the lecture—was eager to deal with problems emerging from incommensurability. In his words, less than three years before, he said of such problems that "they're the most likely focus for the next round of my own work"; and he added: "Quine on radical translation and Kripke and Putnam on reference do provide clues that may

MIT and the courses that he was going to offer. All of these documents are available through the MIT online archival resources (<https://dome.mit.edu>).

[178] Kuhn to Mesthene, 17 May 1979, TSKP 5.26, 1.

[179] See Footnote 178.

[180] Kuhn to Mesthene, 4 June 1979, TSKP 5.26.

permit more sense to be made of the notion"—incommensurability, of course; that is, the notion he was not going to examine in his Foerster Lecture.[181]

Kuhn went further than that. In his approach to such authors (among which, for him, Quine was a well-known source of insight), Kuhn seems to have planned to re-examine both the tradition which his entire philosophy of science opposed, and the one that he had helped to found. Logical empiricism was important as a target for his line of *criticism*, but his colleagues on the historicist side were, too. At some points, as we shall see later, even Hempel, an old friend but also a philosopher of logical empiricism, was closer to his true interests than some of his colleagues. Kuhn announced his two focal points of critical re-examination in the form of such traditions to Kaysen in a memorandum written two years later, in 1981, as a member of MIT's faculty, as part of his presentation of a manuscript of a paper whose final version is well-known today: "Scientific Knowledge as Historical Product," whose first full draft was ready in the fall of 1981.[182] This paper seems to be key in these years, as are a series of lectures that he repeated in the following few years: *The Natures of Conceptual Change*, his Perspective Lectures, which he delivered at Notre Dame University on 17, 19 and 21 November, 1980. The first two of these were also the content for his James and David Orr Lecture on Culture and Religion at Dartmouth College (Hanover, New Hampshire), also in 1980.[183]

As Bojana Mladenovic says, there are different versions of "Scientific Knowledge as Historical Product." The first of them was ready for an STS symposium at MIT in the fall term of 1981—the one I mentioned before. I am aware of at least another version, which Mladenovic also refers to, presented at Brandeis University in 1984, and then in Tokyo in 1986, eventually published in Japanese and posthumously in English in 2022.[184] Kuhn introduced some variations from 1981 to 1984, though there is a core of theses that is already mentioned in the previous exposition for Kaysen: the opposition of two traditions and the additional message for the new one in the form of Kuhn's particular attention to languages and vocabularies. As in other cases during these years, these theses are often repeated, and from these repetitions a kind of template for his ideas in this decade becomes visible.[185] In the series of lectures in 1984 and 1987, as well as in the last unfinished book, *The Plurality of Worlds*, these arguments commenting on the tradition to overcome, and the reasons for doing so, preface the core of his presentation about his model of language. Further reflections usually ensued from that model. The model itself also requires a previous introduction to some philosophically moulded examples of scientific revolution. His

[181] [Kuhn], "Does Knowledge 'Grow'?", 29 November 1976, TSKP 5.13, 5.

[182] See Kuhn's "Memorandum," addressed to Kaysen, 27 July 1981, TSKP 5.42; see also Kuhn, "Scientific Knowledge as Historical Product," fall 1981, TSKP 5.42.

[183] See the copies of the two first lectures with an added note as providing a different presentation of the first one in TSKP 5.33. Marcum (2015a, 122–123) summarizes the main theses of the Perspective Lectures.

[184] A copy of this latter is available in TSKP 13.10. Mladenovic's notes on it can be found in her edition of Kuhn's *Last Writings*, xiv, 276 (nn. 13–15).

[185] A reconstruction of the relationships between the series of lectures in the 1980s can be found in Mayoral (2023b), and in Melogno (2023).

historical case on Aristotle holds primacy over the rest—on Volta's battery and on Planck's quantization of energy. It is, in some sense, an appropriate way of coming full circle in his case, insofar as his experience of the problems of interpretation of Aristotle's thinking as a physicist led him to the problems about incommensurability. So, at that point, with a pretty complete theory about incommensurability and scientific change at large almost ready, it seemed particularly appropriate that Kuhn explained the problems of interpretation involved in *this* particular case with the explanatory resources developed by him thus far. No wonder Kuhn began his series of lectures and the draft of his book with a case for the kind of change that was most significant for him in many senses—personally, biographically, idiosyncratically—followed by two that were also useful and well known for him. That's why the triple case-history that is visible nowadays in "What Are Scientific Revolutions?" reappears in 1984, 1987 and 1994–1995 as it was first presented in the Perspective Lectures and, reduced to the "Aristotle experience," in "Does Knowledge 'Grow'?", too. That rough template includes an account of his lexicon-based model of learning, practice, scientific change, and incommensurability. The template usually begins with the introduction to the two traditions, their open issues, and their consequences.[186]

If we read both "Scientific Knowledge as Historical Product" and "Does Knowledge 'Grow'?" carefully, we can observe connections between them. I'm not referring, of course, to the fact that they must be there insofar as they were written by Kuhn and so they reflect a defence of his own brand of discontinuism. That is true but far too obvious. We can also see that his revolt against the main tradition in philosophy of science on the basis of a revision of the latter's theses in epistemology gains momentum. In the Foerster Lecture, he attempts to question some elements in the epistemological tradition that he had not yet examined in *Structure*, but he is more focused on the proposal of a new perspective more akin to his own discontinuist convictions. Without entering into the problem of incommensurability, his conclusions concerning the problems of talking about the growth of knowledge without further qualification eventually and implicitly lead to that concept, incommensurability, from a viewpoint based on the views of knowledge that he supports. The next step would involve leaving, for a short time, issues concerning the theory of knowledge, and returning to problems about scientific languages. This led him, in turn, to a return to his previous views from the 1960s on paradigm-based learning, now with an eye to the lexical structure that underlies scientific languages. The Perspective Lectures are a first stop in this long journey, in which Kuhn would be engaged for the rest of his life. However, at some points in the period 1981–1984, at the same time he was starting that journey, Kuhn returned to epistemology, as associated to traditional (that is logico-empiricist) philosophy of science, and showed the inefficiency

[186] Melogno (2023, esp. 99 and the previous discussion) helps us to see the documentary evidence for this kind of template on the basis of a brilliant schematic comparison of unpublished texts. Key theoretical common points are also expounded in Mayoral (2023b) as well as below, in this same chapter. See also Mayoral (2024c) for a few details about Kuhn's *Plurality of Worlds* and his previous work.

of some of the main features of that tradition and the point of view from which a new view, the one he propounded, was likely to emerge.

In his 1981 version of "Scientific Knowledge as Historical Product," Kuhn talks about some constants of the tradition in philosophy of science and epistemology that he is discussing. He mentions its fixation with the justification of beliefs, and the emphasis in the almost exclusive relevance for philosophy of science of the context in which that justification is clearly, timelessly obtained and exhibited. There is a second fixation of that tradition with the related idea that knowledge claims (and so knowledge at large) are understood in propositional terms alone. These two features are accompanied by a third one: foundationalism. This is a truly traditional feature that shows that, for an empirical discipline to be science at all, it must have strong foundations for its claims in the form of a descriptive observational language that underlies any theoretical statement. We can grant that science has strong foundations that are exhibited on paper but also in practice, but Kuhn questions whether ignoring the process by which such guaranteed results are gained helps us to perceive the source of their genuine justification and, as he often says, the grounds of their authority.[187]

In re-examining the problems of confirmation, for instance, a classic theme as regards the validations of scientific hypotheses, and so concerning the epistemology of scientific belief, Kuhn had already tried to make his own perspective on confirmation theory clearer for his colleague Stegmüller some years before. For Kuhn, as he told Stegmüller in his 1975 letter, there is not any claim for his part against the idea of the "proof" for a theory. He recounted in that letter that he was surprised, even by Stegmüller's book, that he could be considered a *denier* of the key role for a confirmation theory in the process of scientific research. "Nothing I have ever written *was intended* to deny the need for confirmation theory," he wrote in that letter, in a section he devoted to confirmation theory exclusively.[188] Actually, confirmation was present in "normal science," he said, with the proviso that we do not consider there to be a "definite fixed boundary between confirmed and unconfirmed, or between falsified and still possible" hypotheses.[189] Hypothesis evaluation is, however, different from theory choice, that is from making a decision between two theories—if theories, moreover, are understood the way he, Stegmüller and Sneed see them, as Kuhn points out, as a "core plus applications."[190] The conditions for making a theory choice require an exploration that is not limited to proof, or even in which proof itself does

[187] I am summarizing here some of the main points in, roughly, the first half of Kuhn's "Scientific Knowledge as Historical Product," fall 1981, TSKP 5.42, 4–15. Similar points can be observed in the version from *The Last Writings*.

[188] See Kuhn, "Excerpt from a Letter of 20 January 1975 by T. S. Kuhn to W. Stegmüller Concerning *Theorie und Erfahrung*," TSKP 20.36, 17. The section extends to p. 21.

[189] See Kuhn, "Excerpt from a Letter of 20 January 1975 by T. S. Kuhn to W. Stegmüller Concerning *Theorie und Erfahrung*," TSKP 20.36, 17.

[190] See Kuhn, "Excerpt from a Letter of 20 January 1975 by T. S. Kuhn to W. Stegmüller Concerning *Theorie und Erfahrung*," TSKP 20.36, 19. Admittedly, Kuhn only mentions that association with the Sneed–Stegmüller view on theories a little below the passages in which he is making that distinction between evaluating a theory and choosing between theoretical alternatives (see 17 on that point),

not play any role at all. We should pause to think about this position in a little more detail.

Despite the many views to the contrary, he did not disagree with the idea that there were *reasons*—moreover, "good reasons," as he says—for theory choice.[191] The tradition that considers that theory choice can be eventually explained in a context of justification on the basis of reasons is not one that Kuhn denies; moreover, he says that there are a number of additional factors to be considered, which enrich that perspective without undermining it. As we know, he had just written "Objectivity, Value Judgment, and Theory Choice" as a reinforcement for the passages in *Structure* in which he talks about such a vision of theory choice, and he refers to that paper in his letter.[192] However, Kuhn assumes and argues for the idea, here and elsewhere, that the *process* by which a theoretical choice is made is much more complex than a situation in which a proof is involved. Theory choice is thus a composite process, extended in (historical) time, that is more properly studied in a broader context than the useful, but ultimately too idealized context of justification that logical empiricism so eagerly supported. We would be better off in such a study if we erased the frontier between the context of discovery and the context of justification, and we showed how the development of a new theory and its gradual acceptance take place in parallel; both contexts interpenetrate each other.[193] The process is *social*, of course, and it is not properly understood as the result of an idealized individual mind's option, easily formulated along algorithmic lines. Perhaps this latter might be done, but it says almost nothing about true theory choice at all. Moreover, perhaps exploring the role that some individuals have played in some theory-choice processes—I am thinking of examples from *The Copernican Revolution* to *Black-Body Theory*—surely shows how that process takes place, particularly once we have in mind the true role that experimentation and observation plays in a community of people trained according to Kuhn's description. However, the way a theory is considered "chosen" to the detriment of a competitor is something only properly accounted for in social terms, as Kuhn shows in "Objectivity, Value Judgment, and Theory Choice."

Many critics of Kuhn from the previous tradition will consider this an opportunity for "externalism" to take hold of the philosophical account of theory choice.[194] However, Kuhn is just thinking of the kind of processes that only involve *scientific* resources, strictly speaking. I have just said that *Black-Body Theory* is a good example of the historical complexity of a theory-choice process—a genuine revolutionary process—that is not easy to accommodate to a pure context of justification

---

but there is no doubt that his view of theories is the one he shares with Sneed and Stegmüller, and that he states two pages below.

[191] See Kuhn, "Excerpt from a Letter of 20 January 1975 by T. S. Kuhn to W. Stegmüller Concerning *Theorie und Erfahrung*," TSKP 20.36, 17, 21.

[192] See Kuhn, "Excerpt from a Letter of 20 January 1975 by T. S. Kuhn to W. Stegmüller Concerning *Theorie und Erfahrung*," TSKP 20.36, 18.

[193] It is only fair to say, Rorty was referring to precisely this broader perspective on justification in his 1976 letter to Kuhn. He was right, but Kuhn surely agreed with that point, as is easy to see here.

[194] Haufe's (2022, esp. 29 ff., 190 ff.) recent account of Kuhn on theory choice and theory change is a good example of that traditional position.

without losing essential information about when and how Planck's radiation theory was ultimately established. It is possible to understand such an account of that process as not purely philosophical because of its social component, and so imbued with an "externalist" point of view. At the same time, in other quarters, among historians and sociologists of science, that book is considered "internalist" because it does not take into account what we can consider participant, extra-scientific, causal factors of physical research and theory choice.[195] Leaving aside that the internalist/externalist difference is somewhat outdated nowadays, this difference in views shows, first, a striking conceptual disagreement between disciplines, the history and the philosophy of science, that should agree a little more because of their common interests of understanding science. But secondly, it also shows that Kuhn is probably trying to say something different for both of them than a mere addition of irrational or subjective factors to theory choice. "Social" and "historical" are not synonymous of "irrational." Kuhn suggests crossing the line between contexts more easily and definitely, and leaving aside previous associated categorizations. As in other respects, he is trying to forge a new vocabulary for the philosophy of science, with all the conceptual transformations that it involves.

That transformation starts with the assumption that theory choice, on the one hand, and confirmation and proof, on the other, are not paired as if the latter explained the former. At some point in his argument for Stegmüller, he says that maybe there can a be a second kind of context of confirmation in theory choice, but he soon gives up the alternative. He alludes to a passage of Wittgenstein in *On Certainty*, once again, in order to show what his thoughts on proof and confirmation are. Wittgenstein, as Kuhn says, wrote (I quote the whole paragraph):

> 105. All testing, all confirmation and disconfirmation of a hypothesis takes place already within a system. And this system is not a more or less arbitrary and doubtful point of departure for all our arguments: no, it belongs to the essence of what we call an argument. The system is not so much the point of departure, as the element in which arguments have their life.[196]

For Kuhn, theory choice involves an option between such systems, between such groups of essential elements in which the rest of arguments are embodied and are evaluated. If we take into account all the processes that Kuhn, in 1975, still deems apt for describing a revolutionary change—from conversion processes, as illustrated in his other books on the history of science, to the less desirable expression of "gestalt switches," which he would later criticize—confirmation and proof are not the proper starting point to understand theory choice.[197]

So, when in 1981 he shows his disagreement with the epistemological tradition that underlies the logico-empiricist vision of philosophy of science, he is only rehearsing

[195] See Wray (2021, 128–132) in that regard.

[196] Wittgenstein (1969, §105); see also Kuhn, "Excerpt from a Letter of 20 January 1975 by T. S. Kuhn to W. Stegmüller Concerning *Theorie und Erfahrung*," TSKP 20.36, 20. Kuhn does not include the last sentence in his quotation. He quotes the German version of Wittgenstein's words, not this English translation.

[197] See Kuhn, "Excerpt from a Letter of 20 January 1975 by T. S. Kuhn to W. Stegmüller Concerning *Theorie und Erfahrung*," TSKP 20.36, 18–21.

the same old argument in a new key, this time more focused on the groundwork for an epistemology of science. The emphasis on justification has been taken too far, taking into account that the essentials of a system are scarcely doubted and their grounds, at least in science, could be argued for if asked—which does not happen often. It is, Kuhn says, the justification of abandoning a certain statement on behalf of another, particularly if that statement is key—a law or definition, for example—that is, the substitution of an element of a theory core for another, which normally brings down others with it, that has to be provided. The justification of change, not a piecemeal justification of the whole system, is what must be discovered.[198] As he says in his 1981 version of "Scientific Knowledge as Historical Product,"

> Justification of an entire current body of laws and theories, which was *the* justificatory problem for the older tradition, can be achieved only by justifying the entire historical series of local decisions that brought scientific knowledge to its present state of development. The reasons to choose quantum theory over Newtonian theory are clear, but the basis on which the choice is made depends not only on the quantum theory but on the demonstrated strengths and weaknesses of Newtonian theory as well. The same is true of the choice of Newtonian physics over Aristotelian. But there is no similar clear-cut way in which to confront quantum theory with Aristotelian unless one employs Newtonian theory to find out what is common about their concerns.[199]

The last lines of this passage are noteworthy for two main reasons. First, they are significant because they indirectly show that comparison of theories, as usually understood in the logico-empiricist tradition, are difficult to make if they only try to compare theories that are not clearly historically linked. It is, moreover, superfluous to do so, if we think about it a little. Second, they are remarkable because this emphasis on the kind of belief-change that requires exploration and comparative justification involves a historical situation throughout, which cannot be avoided. When, in his successive trials in the 1980s, he uses lexical theory in order to show how language-change successfully explains scientific change and incommensurability, that kind of historically-situated process of debate and replacement is imperative. I shall deal with that mature theory in what follows, but it must now be plain to everyone that this lexical theory, and its evolution in *The Plurality of Worlds*, serves as a foundation for the kind of reform of the tradition in the philosophy of science that uses that limited view on knowledge and belief as the only ground for explaining rational theory choice. "[T]he developmental approach to philosophy of science," as he already called his perspective in 1981, would be established in the coming years, and its foundations concerning epistemological matters were already in place. Let us see the groundwork for explaining change and incommensurability in the rest of work that Kuhn would develop in the following years. Unsurprisingly, the background for paradigm-based learning that he developed in the 1960s would act as the foundations for these new views.

---

[198] See Kuhn, "Scientific Knowledge as Historical Product," fall 1981, TSKP 5.42, 9–15.

[199] Kuhn, "Scientific Knowledge as Historical Product," fall 1981, TSKP 5.42, 11–12.

8.7 Kuhn's Research Activity in the 1980s

"Scientific Knowledge as Historical Product" is a statement of Kuhn's views on the dialogue between the traditions that he sees as representative of the second half of the twentieth century. I say "dialogue," though at some points we should talk about mere opposition. His genuine dialogue with Hempel, though, or his collaboration at MIT with Bromberger when teaching a course on the theory of reference in 1987 show that there were steps that could be taken in order for both traditions to get closer to each other. "Scientific Knowledge as Historical Product" sets the scene just as "Does Knowledge 'Grow'?" helps him to show what the problem of the growth of knowledge is from the point of view of his own tradition, and what his preferred approach is. It is more than that, though. In approaching the idea of knowledge from resources from ordinary language philosophy, the Foerster Lecture is also a statement concerning his view about the nature of scientific language—to wit, as the kind of vehicle for the theory *and the practice* of science whose essential pragmatic features a strict logical reconstruction can only help to hide. The *historical* perspective for the reconstruction of the past is more effective in the philosophy of science than the logical one. For Kuhn, this is true of his own developmental view at least, though he shows that it should be of any perspective on science if our philosophy of science is intended to speak about science at all.

It has often been said that Kuhn undertook a "linguistic turn" in his latter phase during these years.[200] I agree more with George Reisch when he says that this was rather a "return," but I am not sure whether this is all that remains to be said about Kuhn's late phase.[201] His alleged "linguistic turn" seems to point to an intention to translate his theses to a vocabulary that even analytic philosophers could understand. That is true insofar as, as we have just seen, he confessed that intention to Emmanuel Mesthene. Yet, his theses were more akin to the vision of language that Wittgenstein aimed to exhibit, and that the latter and Austin used to probe. Moreover, as we shall also see, Kuhn's use of the insight into language and cognition obtained from psychologists and cognitive scientists did not stop from the Lowell Lectures and *Structure* up to this point. The 1960s were an experiment in examining language and learning on the basis of models and attempts at computer simulations. In other words, Kuhn may have been examining scientific language in more detail in those years in his published writings, and he may have been pursuing a discussion of his views with Quine's, Putnam's or Kripke's, but he never shared either their philosophical vision or their philosophical resources, or methods, to pursue that task. Methodologically and philosophically, he was still far from them. If "linguistic turn" means anything for interpreting this last phase of Kuhn's thinking, it should include the view that Kuhn was not doing anything very different from what he had been doing in 1949–1951 and later on in the 1960s, and that this *is* different from what the expression "linguistic turn" means for the history of philosophy at large. In that case, maybe the expression is used in a misleading way.

---

200 See Gattei (2008, Chap. 4). On that phrase, see also Irzik and Grünberg (1998).

201 See Reisch (2019, 355).

In the 1980s, Kuhn was exploring all these views on the scientific traditions, on knowledge and on language, at the same time. When, in the fall of 1984, the young philosopher and upcoming canonical interpreter of Kuhn's philosophy, Paul Hoyningen-Huene, arrived in Cambridge, Massachusetts, in order to spend a whole year pursuing his own Habilitation Thesis on Kuhn with Kuhn himself, they read and commented on "Scientific Knowledge as Historical Product," now in its 1984 version (that is, the one now published, or a version very close to it), and discussed some points about the contents of "Does Knowledge 'Grow'?"—those referring to the difference between belief and knowledge.[202] Of course, they did much more than that. They discuss the point of view about language that Kuhn was pursuing in those years, and whose results led to a significant series of lectures during the 1980s. One of those series, the Thalheimer Lectures, in November 1984, took place in the same fall term in which Hoyningen-Huene was pursuing his research at Cambridge, working closely with Kuhn. I shall talk about this and the rest of the lectures in this and the next few sections.

This stage in our journey starts again where we left it in the previous section, in mid-1981, when he was overcoming a period in which he had been in low spirits and unable to work much. In October 1982, he married the British writer and artist, Jehane Barton (then Jehane Kuhn). Incidentally, she had known Margaret Masterman since her studies at Cambridge, where she studied with Richard Braithwaite, and where she obtained her degree.[203] After that difficult period from the late 1970s, things in Kuhn's life now seemed to change substantially, at least in personal terms. In February 1987, he wrote to Hempel: "Life continues to go well for me. The transformation effected by Jehane in my sense of self has stayed with me—even the euphoria not infrequently recurs—and I revel in it."[204] In 1985, he told June Z. Fullmer, the historian of chemistry, that he had "not been getting much into print lately," though he sent her a copy of the Thalheimer Lectures.[205] That summer, however, Kuhn travelled to Paris with Jehane in order to lecture at L'École des hautes études en sciences sociales. There, he gave the Thalheimer Lectures once again—he did so on some occasions during these years.[206] In late 1987, Kuhn would travel to London to deliver a new lecture series, today published in *The Last Writings*, titled *The Presence of Past Science*. These lectures, the Shearman Memorial Lectures at University College London, were, like the former, a bridge towards a complete account of his ideas that could find its way into a new book.[207] In that sense, Kuhn told Hempel in early 1987 that "my work moves along in ways that excite me, and I still hope to bring it all together in a book, though I'm not clear how soon. Things go more slowly these

[202] I thank Paul Hoyningen-Huene for his generosity in letting me study his notes (PHA) from the discussions with Kuhn, which have served me as a basis for the statements in this paragraph.

[203] See Kuhn to Abrams, 15 October 1982, TSKP 25.54; Jehane Kuhn (2011).

[204] Kuhn to Hempel, 2 February 1987, TSKP 21.42.

[205] Kuhn to Fullmer, 23 May 1985, TSKP 13.18.

[206] See Kuhn to François Furet and to Pierre Jacob, 9 November 1984, and Kuhn to Pierre Jacob and Gérard Lemaine, 21 December 1984, all of them in TSKP 13.16.

[207] Or so he seems to suggest to Fullmer; see Kuhn to Fullmer, 23 May 1985, TSKP 13.18.

days, but I continue to learn."[208] Kuhn seems to be in a better mood during these years. His publications might have been more modest, which is debatable, but the ideas were advancing at a good pace.

The book was, indeed, quite another thing. The earliest notes and comments I have had access to on its *possibility* (that is, not about *The Plurality of Worlds*, or similar) are from 1980. However, in 1987 he had not yet prepared his lecture series as a worked-out basis for a book, and the first clues that, around 1989–1990, he had started to prepare it seriously comes from an application grant that James Marcum has described, and from a letter to Abrams in May 1990.[209] Let me comment on these steps briefly.

Concerning the 1980 notes, Kuhn had a rough vision of *a* possible book based on his recent work. These notes were written after delivering the Perspective Lectures in Notre Dame, in December 1980, and at the same time or before he obtained feedback from colleagues at MIT, Fodor and James Higginbotham among them.[210] The notes show that the first Perspective Lecture, based on the historical cases on the transition from Aristotelian to Newtonian physics, on Volta's battery, and on Planck's innovation, was already part of the plan, which is visible nowadays in "What Are Scientific Revolutions?" as well as in *The Last Writings*, both in the Shearman Memorial Lectures and in *The Plurality of Worlds*.[211] This part of the plan did not change in essence; it was only improved in the details given in each historical case. At that moment, though, in late 1980, Kuhn was considering suppressing Planck's case, possibly because of its inherent difficulties, and showing something about the revolution in chemistry. A second chapter would deal with "Language and Language Change," where he mentions the priority of entities over features, a key point of view that we saw in his views about paradigm-based learning. Kuhn also added other elements from old research in the 1960s, such as the "role of models, i.e. metaphors in language learning," which is also part of the first Perspective Lecture and related versions, and the attention to—surely the discussion of—"[r]eduction sentences."[212] The rest of the extant scant notes give a limited indication of the contents of four more chapters. The first of them (Chapter III) was going to deal with "Explanation: The Holistic Part," and seems to have included a discussion of Newton's second law and, probably, its constitutive role, as already discussed with Stegmüller—though this reconstruction is highly speculative on my part—and a reference to the distinction between knowledge and belief, as usual from "Does Knowledge 'Grow'?" The last three chapters (IV–VI) refer to his discussion about Quine on translatability, on the

---

[208] Kuhn to Hempel, 2 February 1987, TSKP 21.42.

[209] See Marcum (2015a, 135–137).

[210] These notes are available at TSKP 12.49. His "First Notes t[o]w[a]rds the Book: Conceptual Change in Science," 25 December 1980, are also deposited there.

[211] Again, Melogno (2023) compares the texts of such lecture and the published texts with the three examples carefully. The part that deals with Aristotle in *The Plurality of Worlds* is extended and improved; see *The Last Writings*, 135–141, 159–166.

[212] Concerning the role of models and metaphors in the first Perspective Lecture, see Kuhn (1987, 30–31). On the role of metaphor in Kuhn's thinking, see, of course, Kuhn (1979b).

causal theory of reference, and on scientific realism.[213] In the next few notes, he writes that "Chapter III is background for all that follows, thus like I and II," and he seems to want to discuss open issues on the role of verification and persuasion during scientific change, as well as on the communication between paradigms.[214]

From all this, we can see that, at this point, Kuhn was thinking about a new book on the themes of *Structure* as based on his previous research—which was also apparent in his Perspective Lectures. Visibly, the work that we explored in Chaps. 6 and 7 of the present book is still central in devising his new view on the themes of *Structure*, but so is his new perspective on knowledge and belief explored in the 1976 Foerster Lecture. He also takes Quine's, Putnam's, and Kripke's recent work as point of departure for his own views, just as he told Mesthene the previous year. The Perspective Lectures show this set of concerns very clearly.

However, as the 1980s advanced, Kuhn's vision evolved significantly, and his preferences on the themes to be examined changed accordingly. In the second Perspective lecture,[215] Kuhn combines his old account of paradigm-based learning with the idea that this learning conveys a classification of the world, a taxonomy, as he starts to say at that point. That learning is, as he says very often, of a language and simultaneously of a world—the phenomenal world in question. That way of expressing how a scientific discipline is constructed starts to be standard for him. So, for example, in his second letter to Mesthene, he said that Robert Kates's STS group could be a model for his own group, insofar as Mesthene was looking for "a language [...] suitable for fruitful discussion of the technology–society relationship."[216] Interestingly, Kuhn talked about the writings that he had received from Kates's group as follows:

> I was not originally much impressed with the written material, for it all seemed pretty obvious. But Kates was fascinating, and after listening to him and participating in discussion, I felt differently also about the reading. What makes the material in the papers seem obvious is that the group that produced them has found a language and a taxonomy with which to describe the situations that concern them. It did not pre-exist, and it was found by deliberate and self conscious effort. Finding something of that sort is what I apparently suggested to your group.[217]

Clearly, Kuhn told Mesthene that, in order to pursue his goal in STS, he should try to develop a language and its corresponding taxonomy in order to deal with the phenomena of interest. Any other scientific discipline would take some such step in its origins—probably adopting and transforming a previous taxonomy and language. His Perspective Lectures, which he prepared shortly after these letters, generalize that point, which for Kuhn seems to be more than a manner of speaking, because

---

213 All the previous reconstruction, extracted phrases into quotation marks included, come from Kuhn's "First Notes t[o]w[a]rds the Book: Conceptual Change in Science," 25 December 1980, TSKP 12.49, 1.

214 See [Kuhn], "First Notes t[o]w[a]rds the Book: Conceptual Change in Science," 25 December 1980, TSKP 12.49, 2. The statement in quotation marks comes from that page as well, of course.

215 I will return to this lecture below.

216 Kuhn to Mesthene, 17 May 1979, TSKP 5.26, 1.

217 Kuhn to Mesthene, 4 June 1979, TSKP 5.26.

his way of showing what happens to Johnny when adopting a *language* about nature from his father—when obtaining *knowledge* about nature, *credentials* as a speaker of English, if we also return to the views in "Does Knowledge 'Grow'?"—in those lectures is a *theory* of what underlies that way of speaking.

The origins of this standard date to his previous work. In 1976, in a talk to the IAS Symbolic Anthropology Seminar, he had already stated that a renewed point of view embodied in his new way of talking has to do with substituting reference to terms for reference to concepts:

> I think that in all the episodes I want to call revolutions, something happens to language, i.e., [the] professional language of the specialists. Most of it says the same, but there's some network of related terms that attaches to experiences, or cuts up the world differently. If I were rewriting the "Thought Experiments" paper now, I'd not talk about concepts but just about language. It's the application of words like "faster," "speed," "motion" that are being altered by my exemplary thought experiment. They fit nature differently. Language–nature fit becomes my central problem. Recourse to "concepts" disguises what's going on.[218]

And a few months later, in "Does Knowledge 'Grow'?" for example, he combines both forms of expression when he is referring to his own Aristotle experience and to his interpretation of what had been happening; he says: "I'd been trying to read Aristotle with a vocabulary and a set of associated concepts that worked very well for Newton and that could be used, though with somewhat deceptive results, for Galileo."[219] And shortly after, he says that:

> within the field on which Aristotle was working a number of terms—like motion, matter, space, and speed—functioned rather differently from the way they would when they later recurred within Newtonian mechanics. By and large, the problem wasn't translation. After Newton there were no better English terms into which to render Aristotle's Greek. But somehow the terms related to each other and also to the phenomenal world in different ways when they occurred in Aristotle than they did when encountered in Newton.[220]

The particularities of what that theory of language says—of what that "network of related terms that attaches to experiences" is, in his words at the IAS—evolved in the forthcoming years, and in 1984 and 1987, mainly, he developed two more versions of that theory. For him, it is crucial to say how language works, because the idea of a scientific theory—what a scientific theory *is*—depends on it. He does not forget that there is a mental, or cognitive, counterpart for language; in fact, such an assumption is part of his theory of what a language is, or rather, what it works for. He does not forget the ontological counterpart of languages, either, insofar as it is equally fundamental in order to show how the language is learned and how we learn about the world at the same time, as noted. Yet, knowing about language itself is the sure pathway to find out how a scientific theory is built and how it evolves. As he told his colleague Feyerabend three years later, in 1983,

---

[218] [Kuhn], "Symbolic Anthropology Seminar, IAS," 18 May 1976, TSKP 5.16, 10. In the original, handwritten text, the sentences "They fit nature differently. Language–nature fit becomes my central problem" were later added to the original text. "Thought Experiments" refers, of course, to Kuhn (1964).

[219] [Kuhn], "Does Knowledge 'Grow'?", 29 November 1976, TSKP 5.13, 5.

[220] See Footnote 219.

> I think, as you do, that a language embodies cognition, and that it may therefore be necessary to change language here and there when one makes the transition to a new theory. On the other hand, I am reluctant to speak of language as embodying a theory. It seems to me better to reserve the term "theory" for an explicitly articulated structure (it may be a structure of statements or one more like Stegmüller's, indifferently), and to acknowledge that there are other ways of embodying cognition. One advantage of this mode of speech is that it permits one to acknowledge that, given a language, there are an infinity of possible theories that might be stated in it, even though it may also be the case that that infinity does not include the theory that is going to emerge next in science.[221]

"Language embodies cognition," as he says at the beginning of the extract, but language is not exactly a theory; it is rather its predecessor, and some changes of theory involve a previous change of language. No wonder Kuhn starts to pay more attention to language, as the way it is established needs simultaneous information about the world to which it is attached, and to the categories, or taxonomies, that we can assume its speakers have in mind. Most of what Kuhn has to say about his developmental philosophy of science, and also about his practice of the history of science, is explained well by a theory of languages and their modes of change. Even his theory of meaning depends on it. Accordingly, the 1984 and 1987 series of lectures give us new glimpses of his new perspective in full. I shall examine them in what follows, starting with the Perspective Lectures themselves.

Just a brief note before finishing this section. Despite those series of lectures, the book, as noted, was not ready in 1987. However, in 1989, as Marcum has shown, Kuhn had clear plans that were based on the work that we are examining in this chapter.[222] In an exchange of letters with Abrams in May 1990, Kuhn is clear about the book: "there have been fundamental further developments in my thinking since the Shearman lectures" he said, and added,

> I am at last writing the book (no longer lectures, etc., pointing towards it) and am greatly excited about the way it's going. I'm doubtless at least two or three years from a full draft manuscript, but the pieces seem at last to be in hand, and they feel near enough right to go on with. (An indiscretion for your eyes only: I think repeatedly of the time when—also after many years of beating my head against a wall—my conception of *Structure* began to feel this way. The parallel is risky, and I may be inviting a bad letdown. But I can't think of a better way to convey the reality of my excitement. [...]).[223]

There are many elements in this extract that are typical of Kuhn's way of proceeding: the considerable delay in putting the pieces of his arguments together, but also the enthusiasm over the promise of the whole set arranged and properly synthesized. In that letter, he shows his hope that, by focusing on the book alone and by leaving other activities aside, and also with his retirement the following year, he would be able to finish the book.[224] In the end he was not able to do so, as we now

[221] Kuhn to Feyerabend, 20 April 1983, TSKP 21.28, 1. I have omitted an extra punctuation mark. See an edited version of this letter in Collodel and Oberheim (eds) (2024, 316–317; the quotation is on p. 316).

[222] See Marcum (2015a, 135–137).

[223] Kuhn to Abrams, 29 May 1990, TSKP 21.1, 2.

[224] See Footnote 223.

know. But the process of preparation—those lecture series that "pointed towards" the book, in his words—are worth examining before coming to Kuhn's last days and last thoughts.

8.8 Johnny Comes of Age

As I already mentioned in the previous section, the first Perspective Lecture develops the three-cases example of scientific change that illustrates Kuhn's characteristic description of his own encounter with incommensurable worldviews. He also introduces three key aspects of such changes: the holistic nature of change; the change in language and the terms that are attached to the phenomenal world when this latter changes as well; and the parallel changes in the model or metaphor that paradigmatically help to build the basic ontology and the explanation that comes from the theory in question.[225] A characteristic of his perspective is evident here: the *developmental* nature of the view of science he is trying to convey, in which language works as a sort of "living," "organic" whole that is engaged in continuous change, and that affects the way we relate to the world around us—our phenomenal world—and the way we model phenomena in order to have control over them via our theoretical and practical resources.[226]

The second Perspective Lecture, develops his own view about language, which is now the core of his main work on the philosophy of science. His view is based on the remainders of the formal model that he had developed in the 1960s and then expounded in "Second Thoughts on Paradigms" and the other texts from that period 1970–1974 that we examined in our previous chapter.[227] As we know, the main public result of that model was his well-known "Johnny example," developed in the previously mentioned 1974 paper. Here, Kuhn prefaces the presentation of that example with a justification of his critical attitude against the descriptive tradition about meaning in the philosophy of language from Frege to Searle and Strawson. For Kuhn, neither classical descriptivism nor the improved cluster theory of meaning will do; though, for the sake of argument, he shows some promising features of the latter.[228] Of course, in the Lowell Lectures, as we know, and *Structure*, Kuhn set himself against the idea that capturing the meaning of term—of a natural-kind term

[225] See Kuhn, *The Natures of Conceptual Change*, Perspective Lectures, University of Notre Dame, 17, 19, 21 November 1980, TSKP 5.36, Lecture 1, 17–20.

[226] "Living" and "organic" are my own adjectives, not Kuhn's.

[227] See our Chaps. 6 and 7 for this background.

[228] I would like to emphasize that Kuhn does so *for the sake of argument*, only. I disagree with Melogno and Giri when they state that there is a span of time, whose length is not clear to me, in which Kuhn temporarily assumed the cluster theory. As evident from the lectures that we are examining, Kuhn made it clear that such theory only improves a little on its, for him, not very compelling predecessor, but that it is, in any case, affected by similar problems. See, however, the alternative interpretation in Melogno (2023, 91, 93–96, 100), and Melogno and Giri (2023, 387, 394–398, 401). My point of view can be read in the rest of this paragraph and in the next one.

in his case—depends on getting to know the necessary and sufficient conditions (that is, the definition) for the applicability of that term. The development of that view that he pursued in the 1960s followed that line of thinking as well. An alternative, as suggested by Searle and Strawson, said that, if an appropriate number of features that characterize a class are true of a possible instance of that class, we are entitled to apply the corresponding kind-term that we usually apply to members of that class. For Kuhn, that alternative would be promising, indeed, in order to overcome the problem of definitions. But he uses that alternative to show what is lacking in such a vision of meaning. For him, we do not know what would be an appropriate number of features; in his words: "an object is a swan if and only if it has *enough* of the features listed for the 'swan' cluster."[229] However, how much is enough? How many features must the new instance share with the members of the class? A positive answer to this kind of question, Kuhn says, presents the same issue as the traditional response.[230] What issue am I referring to?

It is interesting to note that not many things have changed since 1949–1951—the period of time in which he composed the Lowell Lectures and preparatory work for them—or since the 1960s, when he was led to his formal model of paradigm-based learning. One of the things that had not changed in the early 1980s was that, whatever our vision of language, its consequences for behavior must be taken into account. And that's the case with the classical theory of meaning, whether Frege's, Russell's, or the cluster theory. For Kuhn, the idea that meaning equates to a list of necessary or sufficient conditions means that some features must be considered to be on the list or out of it. If they are, they are definitional; otherwise, they are part of mere fallible, empirical generalizations. Our linguistic behavior, Kuhn says, does not usually depend on such a list, but if it depended on it, then our behavior would be different if a new likely sample of a class or kind did not fulfil the expectations based on the definitional list and the associated generalizations. The classification of features in both groups in a language based on instructions for term-application is fundamental, whether the theory involves a definite list, or one based on a certain average, or number, of features. What will happen next, Kuhn wonders, if the new instance does not meet the expectations? If we consider that the new item is just an anomalous case of a given class, on the basis of a failed generalization, we might be willing to ascribe it to that class, and to abandon, or correct, the generalization itself; if we believe that it *cannot* be part of that class because it fails to meet a definitional expectation, then our next decision is different: the definition must be altered in order to correct its problems, which probably affects other aspects of our categories. In any case, our future behavior, and probably our life itself, would be affected. The real case that Kuhn uses to exemplify his argument—the reaction of Europeans towards an example of waterfowl that looked like a swan but was black—does not seem as urgent as one in which the animal in front of us is a previously unobserved snake that looks like some kind of venomous serpent but lacks a regular or definitional feature or more. Kuhn does not see that, in the case of the cluster concepts, things go much

[229] See Kuhn, *The Natures of Conceptual Change*, TSKP 5.36, Lecture 2, 25.

[230] See Footnote 229.

better. What if the new likely sample of a class exhibits a lower number of features than the amount expected for applying the corresponding kind-term to it?[231]

These descriptive theories, in short, are not good guides for the use of terms in our common behavior. Kuhn's account of scientific language must be applied in real-life situations. Language is an adaptive vehicle, conditioned by our behavior, which helps to develop such behavior in positive ways for us. As we have seen, he pays attention to *ordinary* scientific language as the medium in which theories are developed, and, in that case, term-application and the theory of meaning have, in his hands, a different nature. Language, is not, from that point of view, a vehicle to be used in *any possible* landscape, as he says here. The tradition, conversely, asks for something more general, versatile in every context, whether next to the learning-context or not, and applicable by whosoever. Kuhn thinks otherwise:

> Why should a language that has evolved from experience in dealing with a particular environment function unproblematically when transferred to another one? Why, to borrow a phrase from Austin, should we even know what to say if placed in an environment that violates age-old expectations? Is it not more likely that such expectations, some of them already embodied in our physical endowment, be embodied in our language as well?[232]

Kuhn's alternative is rendered on the basis of the Johnny example, which shows how language is acquired in the field, and how the lack of definitions and rules does not make it a less effective adaptive tool—rather to the contrary. I need not repeat the example once again. Kuhn develops it after his criticism of the descriptive theories of meaning.[233] It is important to remark, though, that many of the main features of his position are now clear here. Let me list them once more.

First, Kuhn emphasizes that we acquire knowledge *of language* at the same time as we acquire knowledge *of nature*. This simultaneous acquisition is properly illustrated by Kuhn in the way Johnny arranges his clusters of entities—waterfowl, in his case—in taxonomic fashion.[234] Second, this arrangement does not involve the acquisition of definitions or rules for applying terms.[235] They can be obtained later, of course, but, as was plain through his argument in "Second Thoughts on Paradigms," they are neither useful nor convenient. Ever since the Lowell Lectures and before, in his 1949 notes, we know that Kuhn considered such semantic closure around naturally emerging clusters of entities a drawback for the kind of (again) tool that the resulting language is for the subject that acquires it. Third, learning that language, at least learning how to apply natural-kind terms to items in our perceptual field, takes place in a *holistic* manner, by discerning how to apply two or more terms at the same time. For him, the learner is acquiring a set of terms and categories on the basis of similarity and difference relationships, which usually involve an iteration of samples to be located

[231] My account of Kuhn's criticism in this and the previous paragraph is based on *The Natures of Conceptual Change*, TSKP 5.36, Lecture 2, 22–27.

[232] Kuhn, *The Natures of Conceptual Change*, TSKP 5.36, Lecture 2, 26.

[233] Kuhn, *The Natures of Conceptual Change*, TSKP 5.36, Lecture 2, 28–35. See our Chaps. 6 and 7 for details.

[234] See Kuhn, *The Natures of Conceptual Change*, TSKP 5.36, Lecture 2, 31–32.

[235] See Kuhn, *The Natures of Conceptual Change*, TSKP 5.36, Lecture 2, 32.

in one of two or more different kinds.[236] Fourth, that membership is obtained on the basis of *approximative* similarities to other members of the same kind—and the same happens to the differences with members of more distant kinds.[237] Kuhn's formal model of paradigm-based learning, that we examined in previous chapters, is a fair representation of that mode of gradual formation of categories and of future enlargement with new instances of each kind. Finally, insofar as we are talking about natural kinds—about natural families, in a more recognizable denomination in Kuhn's case—we *must* also expect that *empty spaces* between categories help to a smooth use of language without the use of rules and definitions.[238]

The phenomenal world and the language ascribed to it are liable to change according to the emergent novelty in the field of stimuli. Kuhn meets that challenge and shows that, at some points, an existing categorization cannot simply be changed by adding one more kind to the ones already available; sometimes—as in the case of a revolution—the already arranged entities must be redistributed. The paradigm-based model helps us to find out what happens in that case. We had distributed entities among already existent groups that had been formed by the distance that we could find, by virtue of their features, among the entities themselves. A new entity that is equidistant to two such groups breaks into a previously empty space, and makes the individual, or the scientist, think about the resulting behavior. Change of the local part of the categorization affected by the new, initially unclassifiable item is an option, though we know that, in the case of categories handled by a scientific community, the decision to pursue that kind of change is sometimes slow; anomalies are not immediately accepted as such. In these lectures, Kuhn shows that this is an option, the *revolutionary* option, in any case.[239]

In these lectures, Kuhn combines evidence from several fields, a common practice in many of his arguments since the Lowell Lectures. His research about paradigm-based learning in the 1960s followed that same path. For him, the same kind of change that we find in science, and which was examined in the first lecture, reappears when approaching natural language. Language-learning and language-change are, first of all, *holistic*. In addition, they use *metaphors* as a means of discovering similarities and differences between entities. He mentions the juxtaposition of samples of waterfowl in order to discover the differences between different kinds of them. Finally, *language* itself changes as our *knowledge* of the phenomenal world changes as well.[240] But, of course, ordinary language and scientific language are not the same thing. Usually, a scientific language means a theory and, as he told Feyerabend, they are not the same, and the latter is structured in a specific, sometimes formal, fashion. For Kuhn, however, all the sciences involve such a categorical step. To put it in other words, every scientific language involves classification. He had already made it clear in the 1960s that classification, or taxonomy, as it often appears in some disciplines and

[236] See Kuhn, *The Natures of Conceptual Change*, TSKP 5.36, Lecture 2, 30–31.

[237] See Kuhn, *The Natures of Conceptual Change*, TSKP 5.36, Lecture 2, 32–33.

[238] See Kuhn, *The Natures of Conceptual Change*, TSKP 5.36, Lecture 2, 33.

[239] See Kuhn, *The Natures of Conceptual Change*, TSKP 5.36, Lecture 2, 34–35.

[240] See Kuhn, *The Natures of Conceptual Change*, TSKP 5.36, Lecture 2, 30–33.

in our ordinary activities, is explained according to his paradigm-based model of learning.[241] He returns here to that point and remarks that "[e]very scientific theory presupposes a taxonomy, whether the one that groups the falling stone with the growing oak, or the one that groups an atom with a solar system for the purpose of writing an appropriate Schrödinger equation."[242] So, as Kuhn says, if anyone is concerned about the relevance of the Johnny example for the philosophy of science, the answer is that this thought example serves for both fields of human activity at the same time—science and everyday life. Johnny's learning is strong enough for both, and it shows how the basic vocabulary for any science, the one that radically changes in revolutions, is built in the same way as the basic scheme of our ordinary language.

That is the kind of language, involving a taxonomy, that Mesthene had to find for his own projects. It is also the kind of language that is implicit in Kuhn's historical cases and works, and the one that any interpreter of a past language, or a philosopher of science not exclusively involved in formal models, must have in mind, seek, and investigate. In the following series of lectures, Kuhn would be explicit about this requirement, which involves an ethnographic phase for every philosophical inquiry (see next section). Let us examine this penultimate stage of Kuhn's thinking, and let us see what he has to say about some (at that time) recent philosophical theories.

8.9 The New Developmental Philosophy of Science

The Perspective Lectures are not only a defense of Kuhn's view of scientific change as examined through the lens of his own theory of language. They also add, in a third lecture, a criticism of the causal theory of reference and of Putnam's internal realism. However, Kuhn's critical stance towards Kripke's and Putnam's perspectives on reference can be interestingly examined against the background that the further development of Kuhn's views provide. I would like, therefore, to show Kuhn's core lexical theory first, and then deal with his critical points against this and other aspects of the philosophy of language.

Two lecture series are relevant in providing such a background. As we know, one of them, the Thalheimer Lectures, dates from late 1984, and the other, the Shearman Memorial Lectures, from late 1987.[243] These are definitely the main focus of interest for Kuhn before engaging in the preparation of the new book in the early 1990s. That's why he delivered them more than once in each case. That interest is perhaps

241 See our Chap. 7, especially Sect. 7.5.

242 Kuhn, *The Natures of Conceptual Change*, TSKP 5.36, Lecture 2, 38.

243 Kuhn, *Scientific Development and Lexical Change*, The Thalheimer Lectures, The Johns Hopkins University, 12–19 November 1984, TSKP 23.21; *The Presence of Past Science*, The Shearman Memorial Lectures, University College London, TSKP 23.32; these were edited and published by Bojana Mladenovic in *The Last Writings*, 23–89. Marcum (2015a, 130–132) is a study of the Thalheimer Lectures.

the reason why the silence of a friend like Susan Abrams about *both* of them was so deafening for him. Kuhn had sent her copies of both lecture series, and had had difficulties obtaining a response in each case. In the case of the Thalheimer Lectures, Abrams simply told him, according to Kuhn, that she "didn't quite see what [he] was up to," and suggested that he do something else, which she detailed a little. However, he missed that frankness—which had nevertheless distressed him—after the second case, when he submitted the Shearman Lectures. At that point, Abrams did not answer for a year, and, according to Kuhn's description, upon asking him "on the phone to see something else," he had "remarked somewhat testily that [she was] in arrears on a previous mailing." Despite that, another year passed before he obtained a response from Abrams. At that time, Kuhn recalled, "I was in a good deal of a state about it," and he, given his character and his respect for Abrams, "found it impossible simply to let go."[244] In the covering letter with her notes about the Shearman lectures, Abrams expressed better feelings after reading them, as compared to those she had had when reading the Johns Hopkins lectures of 1984.[245] To tell the truth, the Shearman lectures show more clearly where Kuhn is heading. Abrams, however, had more difficulties in seeing the point in each of the lecture series individually, which is rather understandable. These lecture series are better understood comparatively.[246]

Rather than complementing each other, such series show some progress through the years during the 1980s. The progress from the 1980 Perspective Lectures is clear, too. This older series, *The Natures of Conceptual Change*, devotes a full lecture, the third one, to the causal theory of reference, and Kuhn demonstrates that he is clearly eager to give a response to that theory. In the Thalheimer Lectures, in contrast, he devotes a paragraph to the theory.[247] That interest for this theory disappears in the London series, though he pays attention to it in "Possible Worlds in the History of Science," prepared in 1986, and in the related "Dubbing and Redubbing: The Vulnerability of Rigid Designation."[248] Shortly later, in the spring term of 1987, he and Sylvain Bromberger offered a course in philosophy at MIT on natural kinds, with clear emphasis on Kripke and Putnam's work, and commenting on David Wiggins's *Sameness and Substance*, which seems to have left a lasting impression on Kuhn.[249] So, the theory might not have been a central problem in the lectures, but in Kuhn's mind it remained a subject of interest for him. Besides that, the Johnny example, which in the 1980 series helps to support his vision of language—as opposed to the

---

[244] Kuhn to Abrams, 29 May 1990, TSKP 21.1, 2. All extracts in these paragraphs come from it.

[245] Abrams to Kuhn, 1 February 1990, TSKP 21.1.

[246] See Mayoral (2023b) and Melogno (2023) for two examples of that comparative approach.

[247] Compare Kuhn's *The Natures of Conceptual Change*, TSKP 5.36, Lecture 3, with his *Scientific Development and Lexical Change*, TSKP 32.21, Lecture III, 77–78. See Mayoral (2023b, 62–64) for other details on his criticism of the causal theory of reference. On Kuhn and the causal theory of reference, see Kuukkanen (2010).

[248] Kuhn (1989; 1990).

[249] See Wiggins (1980), as well as Kuhn's notes for "Topics in Philosophy of Science," a course taught at MIT in Spring 1987, TSKP 23.1–2 (available online at: <https://dome.mit.edu/handle/1721.3/189545>).

descriptive theories of meaning—is replaced in 1984, as the core of his language-based explanation of revolutionary change and incommensurability, by the lexical theory, which he develops in depth in 1987. The systematic exposition was clearer in the last series, which surely appealed to his Chicago Press editor more than the previous one. Finally, while the first lecture of the former and the last, which is the second of the Thalheimer Lectures—that is, the ones devoted to the three historical cases on the transition from Aristotle to Newton, on Volta's battery and on Planck's quantum—are quite similar, the last lecture of the Thalheimer and the Shearman series, though often present some points in common, are dissimilar, and the latter confronts the usual challenges that Kuhn had to deal with since *Structure*: the questions about world-changes, truth, relativism, realism and so on. To sum up, these lecture series present some differences and evolution. The contrast between them may help us to better understand Kuhn's evolution in those years (roughly, 1984–1987). Let us examine them in turn in this and the next section.

Scientific Development and Lexical Change starts with a version of "Scientific Knowledge as Historical Product" which is very similar to the 1984–1986 draft that was published in *The Last Writings*.[250] There are differences with regard to the 1981 version I previously commented on. The new version also introduced the virtues of a developmental philosophy of science. He also talks in both versions about the main features of the previous tradition, the "static" one, like its emphasis on the justification of beliefs, instead of the justification of belief-change, propositionalism (his denomination) and foundationalism. He does, however, add two more features now: methodological solipsism and the effect of holism on evaluations. In the first case, he refers to the "coercive" nature of the results of the evaluation of a hypothesis, as in the conclusion of a mathematical proof, and without recourse to judgment, which is an inseparable part of his own vision about evaluation.[251] As to the second feature, he alludes to the disadvantageous properties of the Duhem–Quine thesis, according to which a whole body of knowledge claims—and not one or a specific subgroup of them alone—may be considered, with certainty, responsible for a predictive or explanatory failure.

Kuhn's developmental alternative emphasizes the historical point of view, according to which the emphasis is made on the belief-claims that must be changed as a result of scientific research, and on the evidence that the process of change takes time and a whole community of researchers, with different views, judgments and decisions concerning the change itself—which is not an entry point for irrationalism and subjectivism in any case.[252] The attention paid to the beliefs that change also helps to avoid the problem of general holism, insofar as the focus of attention are those beliefs that truly change. Kuhn complements this criticism, which we can

[250] Compare, however, the differences between the version in *Scientific Development and Lexical Change*, TSKP 32.21, Lecture I, 20 ff., and *The Last Writings*, 14 ff. Melogno (2019) is a study of the first Thalheimer lecture.

[251] Kuhn, *Scientific Development and Lexical Change*, TSKP 32.21, Lecture I, 8–10. Quoted terms come from those pages. See Kuhn (1973) as regards his position concerning evaluation, as well as Lecture I of this series.

[252] See Kuhn, *Scientific Development and Lexical Change*, TSKP 32.21, Lecture I, 9, 18–19.

also see in the version published in *The Last Writings*, with a comparative form of historical accounts of the same scientific innovation: that of Torricelli's barometer. Although not dealt with in the *Last Writings* version, it was included in Chap. 1 of *The Plurality of Worlds* in far more detail.[253] In the Thalheimer version, Kuhn contrasts the historical account of that case that a textbook offers with one in which the developmental details of belief-change are emphasized. The historical record reveals a point of view about science that the developmental philosophy of science takes into account in its study of the dynamic structure of science.[254]

The first Thalheimer lecture helps Kuhn to emphasize the virtues that the developmental viewpoint in philosophy of science can offer, and the principles from which its practice comes, which involve a kind of historiographic viewpoint of science, as far as it is explaining and evaluating *change* that holds primacy over the more classical epistemological dimensions of the static tradition.[255] In the Shearman Memorial Lectures, Kuhn starts his Lecture I with a similar perspective, in that case referring to the "ethnographic" stage of the interpreter's inquiry.[256]

The ethnographic stage involves a form of reconstruction that is not the one practiced by the static tradition. In Kuhn's own brand of historical reconstruction, interpreters must "regain for themselves and their audience the past from which their narrative sets out," he says.[257] It is no longer a static set of statements, but rather a "body of knowledge" that uses a limited language that serves as a way of speaking and as a communication vehicle in which ordinary beliefs based on such a body are ordinarily conveyed.[258] Kuhn's previous work in the history of science is an example of the kind of reconstruction he is seeking. I am not just referring to the three examples that usually precede philosophical discussion in these lecture series. In the first three chapters of *The Copernican Revolution* and of *Black-Body Theory*, we can see how he sets the scene for understanding further developments. In them, as he says in the later series, *The Presence of Past Science*, the Shearman Memorial Lectures in London, we see how the historian (that is, the interpreter at large) "reestablish[es] both an older body of knowledge claims and also the nature of its appeal."[259] In *The Copernican Revolution*, that historical, developmental relationship explored between a former body of knowledge—astronomical, physical, cosmological, philosophical—and its replacement serves him as a background to account for the philosophical interpretation of the process of change and its consequences.[260]

Once again, the body of knowledge to be reconstructed is not a mere static context in which a given statement or its contrary is considered meaningful, justified, or simply declared true or false. Such a body supplies the standard of change, that is,

---

253 See *The Last Writings*, 116–123.

254 See Kuhn, *Scientific Development and Lexical Change*, TSKP 32.21, Lecture I, 22–27.

255 See Kuhn, *Scientific Development and Lexical Change*, TSKP 32.21, Lecture I, 11, 17.

256 See Kuhn, *The Presence of Past Science*, TSKP 23.32, Lecture I, 3–4; *The Last Writings*, 29.

257 Kuhn, *The Presence of Past Science*, TSKP 23.32, Lecture I, 3; *The Last Writings*, 29.

258 See Kuhn, *The Presence of Past Science*, TSKP 23.32, Lecture I, 3–4; *The Last Writings*, 28–29.

259 See Footnote 256.

260 See our Chap. 4 for details.

the basis for showing if a new result is justified or not, how it can be considered meaningful or not, true or not, and sometimes also, whether its meaningfulness, justification and truth require a foundational change in the body of knowledge and in the categorical organization of the phenomenal world. It is no wonder that, for him, knowing the body of knowledge that is going to change afterwards is required both for questions concerning the so-called context of discovery—the usual question for this current, that is, that concerning the road to a given result, heuristic for example, previous to justificatory matters—and for those other questions concerning the context of justification, a less usual kind of question for the developmental, historicist trend.[261] From "Scientific Knowledge as Historical Product" (1981) onwards, the traditional context distinction is increasingly blurred.[262] In *Presence*, the developmental approach is already ingrained as the point of departure for showing how the categorical relationships that are affected by incommensurability are better explored from a historical, dynamical perspective.

So far, I have been talking, as Kuhn does, about bodies of knowledge, the relationship between which helps to show what new beliefs are justifiable and accepted as replacements for older, possibly false predecessors—the former then probably being declared as true. The sentences conveying such beliefs, which form such bodies of knowledge, could be taken to be holistically interrelated, a post-empiricist viewpoint that shows that, either epistemologically or semantically, they obtain their justification and meaning from the whole body of knowledge. When one or several such beliefs are considered false, or simply unjustified, and some competitors are granted greater reliability and later deemed true, the former are simply eliminated from the group and, at most, preserved as former false statements that some predecessors of our current, standard viewpoint about nature used to believe in. They count as historical relics, in short, useful only in order to provide a museum-like demonstration of past science.

There are problems with this view about such bodies of knowledge. Pessimistic views about knowledge, according to which all previous statements about nature have turned out to be false, and so will happen to our more reliable ones, opposed a second alternative: the alethic relativism alternative. This second view shows that every statement that *is* false in our conceptual scheme, *is* true—not simply "was deemed" true—in another, perhaps past, conceptual scheme. So, a statement like "The sun is a planet" (s_A) conveys a false belief that the Aristotelian philosophers used to hold, but whose falseness we are today well aware of. The problem with this viewpoint, that of the former view, is that of pessimism: how many of our current beliefs are merely *deemed* true as well? Perhaps, (even more, *surely*) there are many of them. It is probable that many of them will turn out to be false eventually, too. The alternative view, the relativistic mode, argues that s_A is true in a certain, in this case

[261] Kuhn, *The Presence of Past Science*, TSKP 23.32, Lecture I, 2; *The Last Writings*, 28.

[262] See Kuhn, "Scientific Knowledge as Historical Product," fall 1981, TSKP 5.42, 12; *The Last Writings* (for the 1986 version), 9. See also his *Scientific Development and Lexical Change*, TSKP 32.21, Lecture I, 13.

Aristotelian, conceptual scheme, and that its denial is true in ours, where the sun is a star at the center of our Solar System.[263]

In the former case, the pessimistic construal, the coherence of our whole list of belief-claims—our body of knowledge—involves every belief that we can formulate and its truth value. In the latter case, global holism around a specific set of belief-claims restricts the coherence to those that are taken as true within the conceptual scheme. For the relativistic thinker, there is no external judge or platform outside the world envisaged by the conceptual scheme that allows for belief evaluation through diverse conceptual schemes. So, although the same belief-claims are often formulable in alternative conceptual schemes, some statements like s_{A} are true in some of them and false in other ones.

If introduced to each of these philosophical alternatives, which here represent currents that Kuhn only partially agrees with, he would say, particularly after *Presence*, that a main problem in both is that they assume that there is no variation in the statement that is subject to truth evaluation. Of course, for him, and for such currents too, the meaning of "planet" in s_{A}, which is true in the Aristotelian worldview, has changed to, say, the concept a Newtonian subject holds when using "planet" from the point of view of Newton's three laws of motion, gravitation, and its celestial mechanics. From that point of view, once we discover what planets truly are, the list of properties "planet" has in intensional relationship has changed from one theory to the next. So, either we have come to know better what planets are, so we declare s_{A} as false, or we have changed the conceptual scheme from which we evaluate statements concerning planets, and we state that the truth of s_{A} is limited to the Aristotelian conceptual scheme.

For Kuhn, however, the diagnosis of this situation is different. For him, even if there was a clear, simple translation of the Greek expression for s_{A} in seventeenth-century (or later) English, so that both linguistic renderings of s_{A} would be grammatically identical, the Newtonian speaker would not be uttering the same statement that the Aristotelian speaker did. As Kuhn shows in *Presence*, once "planet" has changed its meaning, the resulting statement is not simply another version of the same original statement, s_{A}, but rather another statement, perhaps s_{N}, which is altogether different. In short, s_{A} has not changed its truth-value over time and with meaning-change for its kind-term, or—as the relativistic philosopher would say—because of a change in the conceptual scheme from which the statement is valued. Plainly, s_{A} is no longer expressible for a Newtonian physicist (with a caveat, which I shall talk about below), and the latter expresses a quite different statement, s_{N}, which could also have a

[263] My usage of the term "pessimism" and its application in this case is inspired by Godfrey-Smith (2003, 98, 177–178). I am fairly sure that my depictions here of a pessimistic view and a relativistic alternative do not represent fully formed views of specific philosophers. However, these viewpoints are basic parts of specific philosophies, and represent some assumptions that Kuhn opposes on the basis of his own viewpoint, expounded below. In the pessimistic case, the point of view could be termed "objectivist," though its realistic bias is mitigated by inductive skepticism (on the latter, see again Godfrey-Smith 2003, 20, 40); in the relativistic case, this version is too rough, but it is stated in some general views of relativism. In any case, Kuhn does not introduce the main thesis that I am examining in this way. This is part of my own reconstruction and interpretation.

similar, maybe identical, grammatical appearance, "The sun is a planet," and its own truth value—in this case, the statement is evidently false.

Kuhn's alternative, in short, makes s_A and s_N different and is explained through the theory of lexicons and taxonomies that he develops in the Thalheimer and the Shearman lecture series.[264] It is also present in the papers that were posthumously published in *The Road since Structure*, which covers his post-*Structure* thinking. However, the Shearman Lectures may be the most finished view on that theory that he bequeathed to those interested in this theory and its consequences. In the next section, I shall examine the lexical theory in some detail.

8.10 Lexicons and Taxonomies

In his own view, Kuhn had elements of both alternatives, pessimistic and relativistic, previously mentioned. However, his own view about the fate of truth evaluations through scientific change is different from both, and the way he explains incommensurability does not fit squarely in either alternative.

In order to see how to understand Kuhn's perspective, as developed mostly between 1984 and 1987, we must turn to the theory about the change in meaning and reference of kind-terms that produced his own explanation of incommensurability. To do that, during these years Kuhn developed a vision of lexicons as nets of internally related and hierarchized rigid structures of kind-terms. Lexicons, in turn, are attached to a rigid structure of kinds, or taxonomy, which only vary in those events that we know as scientific revolutions. "Taxonomy," though, had been a term he had used for some years before this, as we have already seen. In *Presence*, Kuhn provided a more systematic treatment of such structures than in the previous Thalheimer series, although his vision turned out more systematic, of course, in *The Plurality of Worlds*. In any case, in *Scientific Development* the basic features of such structures were already clear.[265]

The background model on which the idea of a lexicon is projected as a covering theory is the one we examined when, in Chaps. 6 and 7, we spent some time looking at how Kuhn developed the paradigm-based model of learning. It is sometimes said that Kuhn abandoned the idea of paradigm, or of exemplar-plus-disciplinary-matrix, altogether when advancing toward the lexical theory, but that is not, properly speaking, true. Admittedly, his references to paradigms in further writings after 1974 do seem to be references to something located in the past. However, insofar as his lexical model finds support in his previous paradigm-based model for learning and research, its presence is nevertheless permanent, though indeed implicit. His recourse to such a

[264] My account in this and the last four paragraphs is mainly based on Kuhn, *The Presence of Past Science*, TSKP 23.32, Lecture I, 7 ff.; Lecture II, 40–41; Lecture III, 72–74; *The Last Writings*, 31 ff., 55, 75–77. See also Kuhn (1999, 34–36); Mayoral (2023b, 58, 70–71).

[265] See Andersen, Barker and Chen (2006) for a now classic synthesis of Kuhn's views—including his published writings from the 1980s and 1990s—from the point of view of cognitive science.

model remains. He still insists that there is no need for rules and definitions when learning a language, because we learn how to apply terms to new experienced items in perception according to a previous categorization of the world, founded on the creation and recognition of a net of categories created, as we already saw, on the basis of similarity/difference relationships. The categorization of the world (what in different though compatible words he used to call a behavioral and a phenomenal world) is gained through a process of coherent distribution of items in perception into different groups mediated by empty interstices, and then settled by ostensive recognition of membership into extant categories for further new items. Recall, briefly, the *n*-cube model, how it was built and, despite its problems with stability, how it showed the way an initial categorization of the world is gained and fed by the following experiences.

In the third Thalheimer lecture, he starts by showing what a feature space is.[266] In short, such a space is easily understandable if we turn to the background model that I am mentioning again. Although the feature space actually works for constructing each individual instantiation of a lexicon that is shared throughout the community, it is useful if we connect it to a simulation in terms of the old geometrical representation that Kuhn developed in the 1960s. He also does so in a paragraph that I would like to reproduce here almost in full for the sake of this useful connection. Almost at the end of Lecture III, he wrote:

> I have just spoken of a feature space as providing a similarity/difference metric that permits objects and situations to be compared. That metric, in turn, provides the lexicon with a structure determined by the relative distances between the nodes at which the referents of terms cluster and to which the names of those referents are attached. Think of the structure, if you will, as a multi-dimensional lattice of nodal points, each labelled by a referring term and all interconnected by lines of different but determinate length. Like any other geometric structure, this one is independent of the particular space or set of coordinates used to specify its vertices and nodes. In the literally geometric case, homology of structure is preserved by any coordinate system that preserves both congruence, the relative distances between corresponding vertices, and also chirality or handedness. In the lexical case homology of structure requires preservation of relative similarity relations and also, though for this requirement I have not prepared the way, of the hierarchical relations among terms. Doubtless, other aspects of the lexicon also require preservation, but the list of features with which individuals specify the structure they share is not one of them. Shared lexical structure alone is what characterizes a language community; a feature space able to support that structure is what characterizes each community member.[267]

As he says shortly before, talking about the feature space as "a similarity/difference metric" is definitely metaphoric: "That way of speaking is metaphor"; however, he adds, "it has for some time provided me a source of valued suggestions and challenges."[268] And we saw in previous chapters how that model played that part in the development of his philosophical views after *Structure*.

In this way, we can explain how a lexicon and its related feature space—the latter being speaker-relative and context-relative—are formed in every individual

---

[266] See Kuhn *Scientific Development and Lexical Change*, TSKP 32.21, Lecture III, 64–68.

[267] Kuhn, *Scientific Development and Lexical Change*, TSKP 32.21, Lecture III, 91–92.

[268] Kuhn, *Scientific Development and Lexical Change*, TSKP 32.21, Lecture III, 90.

that belongs to a certain community. Owning that lexicon, and having formed their own individual feature space in doing so, scientists acquire the kind of membership, of credentials, that Kuhn mentioned in 1976. Thus knowledge, both of nature and of language, the very core of the scientist's activity, is thereby acquired. The *shared* part of the basic vocabulary of science (the lexicon), and also the shared knowledge that allows for the practice of science is based on a feature space that each individual composes and that depends on the experience and the context that serves as a basis for it.[269]

If we wish to explain how the lexicon works, we can say that the latter is made up of natural-kind terms that label classes of objects and substances arranged in natural families and forming a taxonomy for the phenomenal, discipline-specific world. Features of objects and substances are aids for locating newly perceived objects in the corresponding group, or taxon (or, at the beginning of our training or education, for the formation of the initial groups, just as in the computer simulation Kuhn tried to run in the late 1960s). As we saw above, he does not believe that there is a minimum number of such features that permits us to construct the lexicon, and that are therefore necessary for them. Insofar as he abhorred the idea of a lexicon created by rules and definitions—recall his criticism of the descriptive theories of meaning, which he repeats here—the role of features is to *help* to create such lexicon in each speaker, the way the speaker is able to do it.[270] But each individual constructs *the same* lexicon on the basis of *different* experiences, different training routes, and so having access to different groups of features. He shows that even the exemplars that permit the construction of a functional lexicon do not always provide the speaker with a full list of features—which is a perspective he already held in the Lowell Lectures, 33 years before. In his own 1984 example:

> Undulating tails have nowhere been singled out as a defining feature of cats. The tail of my cat, Tabitha, has in any case been bobbed; the statement is not true of her. But the generalization about undulating tails cannot be purely contingent either, for it is a significant criterion by virtue of which the term 'cat' is applied.[271]

If a child has access for the first time to a bobbed-tail cat, this latter feature, the bobbed tail, is only one among the many (just *many*) of the features that can be of help in forming the local part of child's lexicon devoted to the application of the words "cat," "dog," and those for other pets. That feature helps them, and need not be removed from the child's acquaintance with the natural family of cats, just because it does not belong to the core features usually ascribed to members of the group of cats.

A lexicon, thus, corresponds to the phenomenal world, with access to exemplars (that is, to those paradigms that have served as samples of each category, or taxon) as "tie points for that matching process," in Kuhn's words, while the speaker acquires

---

[269] Kuhn, *Scientific Development and Lexical Change*, TSKP 32.21, Lecture II, 59; Lecture III, 92.

[270] See Kuhn, *Scientific Development and Lexical Change*, TSKP 32.21, Lecture III, 67, 79. "Talk of definitions and the meaning of terms supplies no paraphrase of what I am saying," he says there (79).

[271] Kuhn, *Scientific Development and Lexical Change*, TSKP 32.21, Lecture III, 68.

the lexicon. This latter, the lexicon, "is applied to the world," he also says, "as a whole or in chunks, by virtue of the structural homology between them."[272]

Other aspects of this model of lexicon are also visible in the Thalheimer series. To name three of them, first, Kuhn refers to the alternative ways in which a lexicon such as that of Newtonian mechanics can be acquired; this is an example that he famously repeated in 1989 in "Possible Worlds in the History of Science" as part of his rebuttal of the causal theory of reference for natural-kind terms.[273] This example demonstrates a language-learning model based on the relationship between lexicons and feature spaces.

A second aspect concerns the definition/generalization distinction, and the application in his model of the term "constitutive." Kuhn shows that learning a language involves having access to *many* generalizations about the extension of a given kind, which are part of its intension. The whole series of generalizations need, however, not be known by every speaker that applies the corresponding kind-term well, and some of them are doubtfully applied to each member of that kind, though they are often found among some, or even many, of their members. For Kuhn, learning a language, especially for some of their members—depending, of course, on the learning situation which they become a part of—involves applying the term on the basis of some such generalizations, fallible though they are. Once again, we can locate such generalizations in the fringe of vague meaning he talked about in the Lowell Lectures.[274] This point is important for him, because he is aware that we surely assume that some statements are more than this kind of generalization because of the essential association of the kind term with some features. Such statements are rather "*constitutive* of a community's practice," he says, and they seem to be part of the definitions of language terms, that is, they form that part of language without which no speaker could apply its vocabulary properly. But he insists that this belief is not true. Members of the same language do not exhibit such behavior. They do not depend on the *same* group of criteria, even though they communicate with each other successfully, and so there is no line surrounding the essential and constitutive, while leaving the contingent outside. In his words, "[t]here is no conceivable decision procedure which would permit attributing to each and every shared belief either the label 'constitutive' or else the label 'contingent'."[275] The absence of frontiers of that sort, however, does not mean that there are no requirements for a proper learning of the lexicon and that "some generalizations [are not] more central than others."[276] The previously mentioned example of the alternative ways in which the Newtonian lexicon can be acquired is a good example. Yet, that proviso does not mean that fixing frontiers around the extension of terms—as in the Johnny example—is realistic, or even convenient for good use of language. This is a very typical point in Kuhn's view

[272] Kuhn, *Scientific Development and Lexical Change*, TSKP 32.21, Lecture III, 79.

[273] See Kuhn, *Scientific Development and Lexical Change*, TSKP 32.21, Lecture III, 70–77.

[274] See our Chap. 3.

[275] Kuhn, *Scientific Development and Lexical Change*, TSKP 32.21, Lecture III, 69; see also 68 for what had been said in this paragraph.

[276] Kuhn, *Scientific Development and Lexical Change*, TSKP 32.21, Lecture III, 69.

that returns in these lectures, and that distinguishes him from other philosophers such as Reichenbach.

Third, the success of such a kind of learning model depends on a further characteristic of Kuhn's theory: that, insofar as we are talking about the learning of natural-kind terms, that is, of terms attached to a distribution of objects and substances *in natural families*, we expect of the world that it does not exhibit a gradation of instances that could ambiguously belong to two families at the same time. The existence of empty spaces between families is, as we know, a related characteristic of this kind of taxonomy.[277]

This list does not exhaust the high number of features that Kuhn's lexical model exhibits in these lectures, but it will do for the sake of this presentation. When compared to the Shearman Memorial Lectures, it is possible to see that the information about the model is also there, but that the way it is presented is a little different. To begin with, in *Presence*, Kuhn already reduces his introduction to the new developmental perspective in philosophy of science. Although it has not disappeared altogether, and its brief account is reduced to the essentials, it occupies a much more reduced space at the beginning of Lecture I.[278] The end of it, moreover, is not as full of details about its position as in the previous series. The function for the ethnographic phase that necessarily precedes the philosophical evaluation is repeated, and the recourse to lexicons as the forthcoming basis for understanding the past is immediately presented to the audience. A few passing comments about the relationship between belief and terms are also provided, but they do not stop the progress towards Lecture II for long.[279]

The second lecture of *Presence* is different and also more focused on displaying the lexical model. Kuhn introduces us to the problem of defining incommensurability and its more accurate association with the notion of "untranslatability" from the point of view of Quine.[280] Right after that, Kuhn shows what his own brand of relativism truly is—"linguistic" relativism—which makes not truth, but "effability," lexicon-relative. Kuhn returns to this point later, in Lecture III, once the lexicon-model is in place and already used to explain how revolutions happen, what their consequences are and what incommensurability consists in.[281] Terms in the lexicon are natural-kind terms, on the basis of John Stuart Mill's perspective. Later, in the 1990s, he would change this latter perspective to something more general, first in his "Afterwords" and then in *The Plurality of Worlds*. At this point, however, he talks about natural-kind terms alone.[282] Section II of Lecture II displays the basics of the lexical model in a fairly

---

[277] See Kuhn, *Scientific Development and Lexical Change*, TSKP 32.21, Lecture III, 67.

[278] See Kuhn, *The Presence of Past Science*, TSKP 23.32, Lecture I, 1–4; *The Last Writings*, 27–30.

[279] See Kuhn, *The Presence of Past Science*, TSKP 23.32, Lecture I, 29–33; *The Last Writings*, 46–49.

[280] See Kuhn, *The Presence of Past Science*, TSKP 23.32, Lecture II, 34–37; *The Last Writings*, 50–53. See Quine (1960).

[281] See Kuhn, *The Presence of Past Science*, TSKP 23.32, Lecture II, 38–39, 50–51; Lecture III, 72; *The Last Writings*, 53–54, 62, 75–76.

[282] See Kuhn, *The Presence of Past Science*, TSKP 23.32, Lecture I, 30; Lecture II, 40; *The Last Writings*, 47, 55. See also Buchwald (1992, 39) and Kuhn (1993, 22). See also, of course, his last

orderly manner.[283] Kuhn explains again how natural-kind terms are hierarchically related and how feature spaces, as developed by each individual, must reproduce such arrangements, which form identical lexical structures. After all, as he says, "the meaning of a term is associated, not with any particular set of features, but with what I shall henceforth refer to as the lexicon's *structure*, the hierarchical and similarity–difference relations that it embodies."[284] There is, of course, convergence among those feature spaces created individually (by training in the discipline), but some divergence is allowed, and there is, as he says, "some play in the system, some room for adjustment."[285] The convergence and divergence in origin—that is, the commonalities among the reduced diversity of routes through which a part of the Newtonian lexicon is obtained in scientific training—is once again explored by way of an example that had already been used in the previous series: that of the Newtonian lexicon.[286]

The second lecture emphasizes once again the essential idea that the lexicon is not based on definitions, but on the relationship of the phenomenal world with natural-kind terms and other property terms that convey the features that form the feature space. "Let me underscore [...]," he says towards the last paragraphs of the lecture, "the very large role played by the world in the lexical acquisition process and the correspondingly small role played by anything quite like definition."[287] Kuhn shows how some categories named by terms in the lexicon are more or less general, some others much less. We could even say that, at some points in the development of a discipline like ancient astronomy for instance, "a pure observational language," to use Kuhn's terms in this part of his argument, *seems* to become an important part of the lexicon, and actually, something like the kind of teaching waterfowl to a young boy, as in the Johnny example, seems to be available. The language, he emphasizes right after that, "was not pure: the division between stars, planets, and meteors can be made in other ways; the particular taxonomy implemented by the Greeks was not dictated by observation alone."[288] In any case, it shows similarities to the Johnny example, because the access to the relevant referents by direct observation is fairly

book in *The Last Writings*, 176–179, 261, 264; see also Mladenovic's introduction to the volume, xxxiii–xxxv, as well as Mayoral (2024c). "Artifactual kinds," one of the new key kind of terms in the last book, are also passingly mentioned in the Shearman Lectures (Lecture II, 45; *The Last Writings*, 58). I shall return to all this in my Epilogue.

283 See esp. Kuhn, *The Presence of Past Science*, TSKP 23.32, Lecture II, 42–47, 49–50; see also *The Last Writings*, 55–59, 61–62.

284 Kuhn, *The Presence of Past Science*, TSKP 23.32, Lecture II, 50; *The Last Writings*, 61–62. "Structure" is in italics in the edited versions of these lectures. His opposition to the cluster theory of meaning is again present in Lecture II, 46–50 (*The Last Writings*, 59–61).

285 Kuhn, *The Presence of Past Science*, TSKP 23.32, Lecture II, 63; see also 55; *The Last Writings*, 70; see also 65. On the convergence of feature spaces, see also Lecture II, 50; *The Last Writings*, 62–63.

286 Kuhn, *The Presence of Past Science*, TSKP 23.32, Lecture II, 57–60; *The Last Writings*, 66–68. Kuhn refers to his 1989 paper (Kuhn 1989) from which the example, here abridged, had been obtained (see Lecture II, 60; *The Last Writings*, 68).

287 Kuhn, *The Presence of Past Science*, TSKP 23.32, Lecture II, 61; *The Last Writings*, 68.

288 Kuhn, *The Presence of Past Science*, TSKP 23.32, Lecture II, 52; *The Last Writings*, 63.

direct. As science develops, however, categorizations become more specific, and the lexicon involves more than pure direct inspection of the phenomenal world: previously available language plus, at times, some instrumentation is required.[289] This point in this lecture is crucial for what is to come later. Some of our categories, he says near the beginning of Section II of this lecture, "must be innate," while some others, "though not innate, are likely to be species-universal by virtue of shared aspects of the natural environment." And he goes on, significantly, "[i]t is hard to imagine a language which has no word for the sun, no words for night and day, no word whose referents include the stars."[290] The way our taxonomies evolve and in which the lexicon is inextricably intermingled with aspects of the phenomenal world often grows more complex throughout history. But Kuhn seems to be interested in obtaining, by exploring different points in such development, key details about this vision of the lexical model. By contrast, we will soon see that this descent to the lower levels of generality is a way to overcome the challenges of this point of view, tainted with the dangers of relativism. In the next section, we see how Kuhn confronts such challenges, mainly in the Shearman Memorial Lectures but also in other texts, some of which lead the way to the drafts of *The Plurality of Worlds*, which I examine in the Epilogue of this book.

8.11 Relationships with the Past

At the beginning of Lecture II of *Presence*, Kuhn shows the problems that the historian must meet in order to convey past thoughts and statements to a modern reader. If the lexical structure is rigid and permanent while the language in question is used, the possibilities of transmitting its content to the present on the part of the historian are reduced. The lexical terms are characteristically those that populate key statements in our language, particularly laws. Terms such as "mass," "force," "resistance," "planet," and so on, are those in use in those sentences that figure in the predictions and explanations that we formulate in using our scientific theories. Kuhn recalls that these terms have some specific characteristics. One of them is based on Nelson Goodman's concept of *projectability*.[291] Such terms are "projectible," Kuhn says, so they "appear in laws of nature, counterfactual conditionals, or candidates for inductive generalization."[292] Besides that, a second characteristic that is important for him is that they satisfy the *no-overlap principle*. He defined this principle often in his late writings such as "The Road since *Structure*," for example, where he says that "no two kind terms, no two terms with the kind label, may overlap in their referents

[289] See Kuhn, *The Presence of Past Science*, TSKP 23.32, Lecture II, 53 ff.; *The Last Writings*, 63 ff.

[290] Kuhn, *The Presence of Past Science*, TSKP 23.32, Lecture II, 43; *The Last Writings*, 56.

[291] See Kuhn, *The Presence of Past Science*, TSKP 23.32, Lecture II, 40; Lecture III, 80; *The Last Writings*, 54–55, 81. See also Goodman (1954).

[292] Kuhn, *The Presence of Past Science*, TSKP 23.32, Lecture II, 40; *The Last Writings*, 55.

unless they are related as species to genus."[293] It is a reasonable principle in order, first, to avoid inconsistencies in the use of such key terms in our language, and also to meet the requirement that these terms name natural families, which usually have an interstice between them, an "empty space," as he typically says.[294] In Lecture II, he says that such terms "cannot overlap in their referents unless one of the kinds entirely contains the other." These two characteristics also make a third one reasonable, to wit, that they, and only they, hold the kind of "label" that makes them "projectible" and "suitable for occurrence in natural laws." If there were other similar candidates, only those that fulfill such conditions truly belong to the lexicon.[295]

A historian of science like Kuhn, who tries to convey the contents of a past theory, faces precisely that situation. That historian must use, as Kuhn does—he says—with his examples in Lecture I, terms that, though in use in our own language, must represent vicariously those no longer present in our current lexicon. Even if we ignore the many problems of translation from, say, ancient Greek language to modern English, which Kuhn mentions in Lecture I, we should isolate the substitute from its current use in our language, and we should consider it as signaling a relic from a previous, outdated lexicon that no longer provides the rigid net of terms, the structure, that applies to the current taxonomy for our phenomenal world.[296] The lexicon they belong to has no structural homology with our own; the no-overlap principle is violated when applying terms both our way and that of, say, the ancient Greeks; the empty spaces and the similarity/difference relationships are no longer preserved; the vocabulary is incommensurable with our own. The only solution is thus to "freeze" (if my metaphor is allowed) such terms only for the historical reconstruction. In doing so, they no longer hold the label that turns them into natural-kind terms for us and key elements of our lexical structure.[297]

Let us assume, for the sake of argument, that the historian is achieving such a task for us, and that he or she is transmitting a whole different phenomenal world and its science through such careful explanations. To do so, the historian of science has to "become *bilingual*," in Kuhn's words.[298] In a paper for a symposium on his work at the Philosophy of Science Association in 1982, "Commensurability, Comparability, Communicability," a fairly well-known paper for Kuhn's readers, he made it clear that the result of acquiring a different lexicon is to turn oneself into an *interpreter*,

[293] Kuhn (1991a, 92).

[294] Kuhn, *The Presence of Past Science*, TSKP 23.32, Lecture II, 46; *The Last Writings*, 59.

[295] Kuhn, *The Presence of Past Science*, TSKP 23.32, Lecture II, 41; *The Last Writings*, 55. Previous quotations belong to those pages. On this vision, see also Mayoral (2023b, 58–59).

[296] For the problems of translation of Aristotle's writings, see, for example, Kuhn, *The Presence of Past Science*, TSKP 23.32, Lecture I, 7, n. 5; *The Last Writings*, 33. In *The Plurality of Worlds*, he comments on even more examples about that case than in these lectures; see *The Last Writings*, Chap. 2, Sects. I, IV, from the book, 135–141, 153–166.

[297] For the ideas described in this paragraph, see Kuhn, *The Presence of Past Science*, TSKP 23.32, Lecture II, 40; *The Last Writings*, 54. The notion of "homology of structure," if we recall, appears in his *Scientific Development and Lexical Change*, TSKP 32.21, Lecture III, 91–92.

[298] Kuhn (1989, 61); my italics.

not to provide the audience with a (highly difficult) translation.[299] In Lecture III of *Presence*, he says that "past beliefs are recaptured by language learning, or rather by acquiring the lexicon of kind terms in which those beliefs were stated." Then, he adds: "That process, when successful, produces bilinguals, but it need not produce translators."[300] Bilingualism is key in his view, and a result of the lexical-model explanation of incommensurability: the only way for the historian to overcome the difficulties of understanding that it produces.[301]

The idea of interpretation that Kuhn develops in this paper finds support in thought experiments such as that of Johnny, of course, and in the lexical model itself, which still seems to be in the making—the terms "lexical structure" and "taxonomy" are already employed in this paper, though, and with identical functions to those in these lecture series.[302] The paper, as with the lectures we are examining here, shows the difficulties of overcoming incommensurability and holism on the part of the historian. How does he or she become bilingual? The question is crucial. The historian of science shows what the body of knowledge that is superseded by a new one was before, so the justification of change depends on that activity of bringing the past to the present. However, that evaluative activity is not the only one of interest. Our own identity as scientists may depend on our sharing our activity with past researchers as well. Kuhn is concerned with that problem at the end of Lecture IV in the Thalheimer series.[303] Although the lexicons and the world they match set the limits for our evaluative activity and for our understanding of the past, he showed, there are possibilities of linking the past and the present. Incommensurability, holism, and even relativism are acceptable consequences of his model of change—rupturist as they nevertheless are. Let us examine how.

A learning process is, in Kuhn's hands, always a process of acquisition of the kind of categorical space and lexicon that we have already seen in such a basic example as that of Johnny learning to sort waterfowl, and is better explained by means of Kuhn's *n*-cube model. Learning a second lexicon and its corresponding taxonomy is a little different. I am not talking about substituting it for one's own, or about enriching our particular view of nature with a more specialized, though compatible one. A mixture of both is probably what the convert to a new paradigm is bound to do by the requirement of living in a new specialized, expert world. Here, I am talking about holding two world-views and lexicons in parallel, one of them non-functional nowadays but completely functional and useful for the historian of science to explain and predict phenomena on its basis *as if* such a historian was

[299] See Kuhn (1983a, 44–53).

[300] Kuhn, *The Presence of Past Science*, TSKP 23.32, Lecture III, 65–66; *The Last Writings*, 72. Recall also Kuhn's amendments to his own account about the way a different paradigm can be properly understood by a historian of science, a process that, in the Postscript—1969 is still expounded as translation and then, in these years, considered as the result of learning a second language. Compare Postscript—1969, 202–203, with Kuhn (1993, 237–238).

[301] On bilingualism, see also Kuhn (1991a, 93; 1993, 238; both in *The Road*).

[302] See Kuhn (1983a, 52–53). See also Kuhn (1983a, 51) for a passing mention of the Johnny example.

[303] See Kuhn, *Scientific Development and Lexical Change*, TSKP 32.21, Lecture IV, 112 ff.

living in the past historical time of the actors that he or she explores. For Kuhn, the typical historian of science—and this is important for him—must come to know how the historical actor must have thought or even act (when there is no record of that), given the historian's learning of both preconditions of past scientific knowledge—the lexicon and the taxonomy.

Kuhn often said that, despite the similarities, and even the apparent symmetry, between the journey that the convert to a new theory makes, and the one that the historian of science that travels back in time to a past theory must undertake, there are differences. The process, in short, is not truly symmetric. It *cannot* be, because historians of science travel back in time alone, and the change that such a journey produces in them is not shared by a whole group of close companions, as happens in the conversion to a new theory. Kuhn often shows in this part of his life that he had been applying the idea of Gestalt switch as representative of the experience of change for a whole community. He had, that is, "modelled the experience of scientists moving forward in time too closely on the experience of the historian trying to move back."[304] However, the latter is an individual experience to which the idea of a Gestalt switch, carefully applied, would do as a metaphor; the "experience" of the group is more difficult to express for three reasons: first, groups simply don't have experiences—hence my scare quotes; second, the experience of change can be understood, at least for external observers, from an examination of changes in language, more particularly in lexicons; and third, conversions of group members are extended in time.[305]

Given those conditions, the historian must be travelling in time alone to another world, and he or she must be ready to apply the new lexicon-based theory to that new phenomenal world completely alone—perhaps, with the companion of some scholarly texts written before about the same theme. For Kuhn, either in this task or in the attempt to convey the findings to an audience, the historian can find support in some resources. The first one is the stability of the world of stimuli that impinges upon both the historical actors and us, with which we all form our respective phenomenal world. Recall here his musings on the formation of percepts around 1967.[306] The *assumed* uniformity of stimuli gives shape to a plurality of percepts arranged, in turn, in a "plurality of phenomenal worlds" and, we would say now, with their own taxonomies and lexicons attached in each case.[307] However, that assumed commonality of stimuli is useful for him—a "minimal hypothesis," as he says.[308] Hoyningen-Huene, who was already at MIT at the time Kuhn presented these arguments at Johns Hopkins in the Thalheimer Lectures, provided an interesting discussion of the "stimulus ontology"

---

[304] Kuhn, *The Presence of Past Science*, TSKP 23.32, Lecture II, 35; *The Last Writings*, 51.

[305] See Kuhn, *Scientific Development and Lexical Change*, TSKP 32.21, Lecture II, 61–62; *The Presence of Past Science*, TSKP 23.32, Lecture II, 35; *The Last Writings*, 51.

[306] See our Chap. 7, Sect. 7.5.

[307] For the expression "plurality of phenomenal worlds," see Hoyningen-Huene (1993, Chap. 2, 31ff.).

[308] See Kuhn, *Scientific Development and Lexical Change*, TSKP 32.21, Lecture IV, 115–116.

that Kuhn assumes here in his book.[309] This minimal hypothesis is good starting point for a reconstruction of a world that is, at some points, alien to us.

In the Thalheimer series, Kuhn showed that we must find support in the points in common with "our ancestors"—his words; at this moment, they seem to have counted for him as such. We must avoid seeing past science as, in his words, "an alien island culture." To do so, we must build on a common link with the past: "The cognitive ties that bind us to our for[e]bears are as strong as the biological ties, and they place rigid constraint on what we can or should accept as science or as history."[310] This is a second assumption that leads us to use a second set of resources to understand the past: our own lexicon and our own theories in use. For him, it is in our own lexicon, our second resource, that we discover the anomalous passages in which the initially apparent structural homology between the historical agents' lexicon and ours is broken, in which incommensurability manifests, which leads us to explore local variations between both languages. However, it is clear for Kuhn that points in common between the two respective lexicons are a good starting point. In the Shearman Lectures, Kuhn comments on a possibility expressed by Martin Hollis, that of finding "bridgeheads," which is very similar to this option. Bridgeheads involve a partial overlap of two different taxonomies, and the use of two similar terms linked to partially, locally, dissimilar categorizations is useful in finding the way to learn a different language.[311]

If linguistic commonalities help to reconstruct a past lexicon, current theories are a third resource, expounded in the Thalheimer series, to find out what past agents were looking at when making statements that seem strange to us. "It is not anachronistic for a historian," Kuhn says, "to make use of [...] modern eclipse computations, chemical or spectroscopic examinations of pottery, or other techniques developed by modern science," which improve the empirical basis from "which the historian's narrative will grow." Similarly, Kuhn says that his current physics knowledge helps him to "understand what [...] Planck can have meant when [using ...] 'energy element'. Without such knowledge," he says, "I would not have been able to discover the relevant parts of their lexicons, understand what they said and why they said it."[312] Points in common and a common biological and cognitive equipment, reinforced by our current knowledge of the world, fallible and lexicon-relative though this latter is, help us to reconstruct a different lexicon, and to set ourselves in front of (or surrounded by) a different world in which we must compose a functional lexicon and phenomenal world useful enough to bring the past, as complete as it can be for us, to our minds.

On the basis of this procedure, a past lexicon can be brought to life once again and conveyed to the historian's audience. The connection with the past is established,

---

309 See Hoyningen-Huene (1993, 42–63).

310 See Kuhn, *Scientific Development and Lexical Change*, TSKP 32.21, Lecture IV, 112.

311 See Kuhn, *Scientific Development and Lexical Change*, TSKP 32.21, Lecture IV, 113; the idea of structural homology reappears there. See also *The Presence of Past Science*, TSKP 23.32, Lecture III, 67–68; *The Last Writings*, 72–73. See Hollis (1970, 21 ff.); Mayoral (2023b, 70).

312 Kuhn, *Scientific Development and Lexical Change*, TSKP 32.21, Lecture IV, 114–115.

and we could say that our identity as scientists is guaranteed thanks to this. *Local holism*, the key for Kuhn for his own account of learning, justification, and incommensurability, is also crucial for exhibiting the nature of our link with the historical past of science. Albeit different from more global species of holism, mostly devoted to examining scientific theories in a more static way, Kuhn's localist brand helps to show how it is possible to exhibit the paradigm-based model of learning successfully, how justification is a belief-change problem, and in the Thalheimer Lectures, how our scientific *identity* through past scientific revolutions and the occurrence of incommensurability (which is also *local*) is nevertheless possible.[313]

And yet, this apparent optimism concerning our historical link with the past seems broken in the Shearman Lectures. In these latter series, Kuhn is somewhat pessimistic about the chances of showing connections with the past. The kind of historical reconstruction of past science that he supports, of an anti-Whig nature since he was a young undergraduate student at Harvard College, and which he now vindicates once again (and perhaps more forcefully), transmits to us the history of science "as the story of an alien tribe."[314] The "alien-island-culture" account of past science that he straightforwardly tried to avoid in 1984 recurs now, three years later, in 1987, and seems to intend to stay on the basis of a successful, economic primer to the lexical theory such as the one expounded in Lecture II of the latter series. Pessimism regarding that connection with the past—not about *knowing* it, but rather about finding the true, genuine, *kinship relationship* with the groups that seem to be our forebears—emerges in this Kuhn, who shifts responsibility for that task onto Whiggish history, which is definitely less faithful in bringing the past to present, but more practical in forging a non-disturbing identity for practicing scientists.[315]

8.12 Vindicating the Essential Tension

Sciences, all over history and both synchronically and diachronically, seem to be partially accessible island cultures, to which we travel with some effort. If we leave the metaphor aside, it requires us to become interpreters of a different culture, and thus to learn a new language, thereby becoming bilingual. The perspective Kuhn finally puts forward around 1987 is a lexicon-based perspective of his theory of revolutions, which, all things considered, amounts to vindicating the old idea that science requires moving towards convergence, and also often towards divergence, on the part of a community.

This vindication reappears in Kuhn's writings in this mature period of his philosophy with the same force that it exhibited when, in late 1958, he started to talk

---

313 On local incommensurability, see Kuhn (1983a, 35–36).

314 Kuhn, *The Presence of Past Science*, TSKP 23.32, Lecture III, 89; *The Last Writings*, 87.

315 Kuhn, *The Presence of Past Science*, TSKP 23.32, Lecture III, 89–91; *The Last Writings*, 87–88. On these kinds of history and their applications in education from Kuhn's point of view, see Kindi (2005).

about the essential tension—a moment in which paradigms became a term of the art. Paradigms then turned out to be a meaningful sample of a kind, and this latter part of a locally holistic net of categories usually named by kind-terms. The view ultimately evolved until generalizing J. S. Mill's *infima species* to cover natural kinds and artifactual kinds, as we shall see in the Epilogue of this book.[316] Such kind-terms were arranged as lexicons, hierarchized structures that were stable for a community, whose terms played central roles in key statements of that community. For instance, when that community developed a scientific theory about some natural phenomena on the basis of such a language, its laws took them as central pieces of their statements. Such terms did not overlap in their referents; they were, so to speak, consistent in the way they helped the members of such a community to express true statements about nature. We have seen this in the previous section. The lexicon confers stability on a whole way of approaching nature epistemically and also practically. For Kuhn, moreover, the lexicon opens up "a set of possible worlds," with slight alternatives that converge to a common lexical core. The practice of normal science is, as a result, the discovery "of the actual world among the members of that set."[317] The difficulties for exploring anomalous novelties, when the community is aware of them, or for suppressing inconsistencies in the conceptual background to science, and sometimes the initial disintegrations of the community (what Kuhn used to call "crises"), lead to a new lexicon, and to a new set of possible worlds—a new practice of that science entirely. A scientific revolution happens, in which local incommensurability between the two lexicons, the old and the new, takes place.

The essential tension is intrinsic to Kuhn's perspective, and it took some effort on his part to vindicate it once and again. His perspective on the causal theory of reference from 1980 to 1989 shows how Kuhn defends the idea of scientific revolutions and its consequences against the views that either ignore or attack the phenomenon of incommensurability. In the same period, he would show, contrary to Rorty, how the idea of a permanent medium for change, in a sort of permanent form of revolutionary advancement, which is really an oxymoron, does not make sense, and that the equilibrium that the idea of essential tension represents is the one that need be emphasized and explored in depth.[318]

Concerning Kripke's and Putnam's causal theory of reference, Kuhn devotes Lecture 3 of his Perspective Lectures to showing in what respects he disagrees with such a theory. In the first part of the lecture, however, he rather expounds the causal theory of proper names according to Kripke's *Naming and Necessity*, and, in most of this, he agrees with it. He says that, "[t]he encounter with [the] causal theory has been especially fruitful, and I'm by no means certain that the harvest is yet complete." Despite that, he says earlier, "[i]n the long run it may be possible to say all that's on

---

[316] See Kuhn, *The Presence of Past Science*, TSKP 23.32, Lecture II, 41, n. 8; *The Last Writings*, 55.

[317] Kuhn, *The Presence of Past Science*, TSKP 23.32, Lecture II, 63; *The Last Writings*, 70.

[318] See Kuhn to Rorty, 19 September 1986, TSKP 22.14, 3. I have examined this correspondence in more detail in Mayoral (2023a, esp. 115–117).

my mind without mentioning the causal theory of reference."[319] His main goal there is to show that Kripke's perspective works for periods of stability, but that the theory does not account for some kinds of change in which the language involved is not able to express the variations that a lexical change involves. Kuhn shows that there are two requisites for the causal theory: (1) that individuals exist, and (2) that we can "trace their lifelines."[320] Concerning (1), it is possible to question the existence of individualities in every world; among Kuhn's examples are also those that belong to our actual world, but not to our common-sense world, as in the case of the "quantum–mechanical world," whose "English description [...] embraces paradox."[321] In such "worlds," he says, "the concept of individuals seem[s] forced."[322] Now, concerning (2), it is sometimes difficult to trace the lifeline of a given object. If that is true, alternative scenarios make it difficult to accept that some a posteriori identity true statements are truly necessary.[323]

In order to describe Kuhn's argument about (2), it is advisable to speak briefly about the idea of "lifeline" that he is handling here. Individuals, he says, are characterized by "lifelines, a trajectory through time [...] extending continuously from birth to death, from coming-to-be until passing-away." A second important characteristic is that these lifelines do not split into two or more branches, and that two or more of them do not merge into one alone.[324] It is not the same case, incidentally, with natural families: "natural kinds have no lifelines," he says.[325] For him, there must be "conditions of continuity and singularity" for those lifelines, and they "must be preserved." His argument against (2) is that they cannot be granted in all cases.[326] He describes a scenario in which an identity statement that, according to Kripke, is necessarily true, would actually be false. First of all, let's consider that the real conditions for the identity "Hesperus = Phosphorus" depend on astronomical theories developed since antiquity. They are fairly accurate, Kuhn says; they help us to see that, if we "extrapolate the lifeline of Phosphorus during the time of its disappearance, then" we see how that heavenly object "reaches just about the position that Hesperus is at when it appears."[327] Yet, we can imagine an alternative scenario, in which a crew of individuals (of unknown origin in the example) put a replica of the planet in the place and at the precise moment in which Hesperus disappears, and instead of the latter when it stays out of sight for the alternative-earth amateur and professional

[319] Kuhn, *The Natures of Conceptual Change*, TSKP 5.36, Lecture 3, 55.

[320] Kuhn, *The Natures of Conceptual Change*, TSKP 5.36, Lecture 3, 50.

[321] Kuhn, *The Natures of Conceptual Change*, TSKP 5.36, Lecture 3, 52.

[322] See Footnote 320.

[323] See Kuhn, *The Natures of Conceptual Change*, TSKP 5.36, Lecture 3, 51–52.

[324] See Kuhn, *The Natures of Conceptual Change*, TSKP 5.36, Lecture 3, 47. The previous quotation comes from that page as well.

[325] Kuhn, *The Natures of Conceptual Change*, TSKP 5.36, Lecture 3, 56; see also Kuhn (1979b, 198–199, 205).

[326] Kuhn, *The Natures of Conceptual Change*, TSKP 5.36, Lecture 3, 48.

[327] Kuhn, *The Natures of Conceptual Change*, TSKP 5.36, Lecture 3, 49.

astronomers. The old Hesperus is melted down, Kuhn says, "for subsequent refabrication." With this imaginative scenario, Kuhn intends to show that the lifeline may be broken, that the conditions of singularity and continuity would not be fulfilled in that case, and that our identity in question, "Hesperus = Phosphorus," would not be a necessary truth anymore, for it would not be true in at least this possible world.[328]

Kuhn is aware of the kind of criticism that a proposal like this would raise. As he says,

> At this point I'm sure to be told that I have misunderstood a central feature of the causal theory. [... O]ne may not legitimately make up a world in which Hesperus is not Phosphorus and use it to claim that the causal theory breaks down. Perhaps even Kripke would respond to what I've said in this way. But I doubt it, for the response would indicate that there's something about his argument he's not seeing.[329]

Indeed, Kuhn's intention goes much further. He is more interested in showing that in such examples in which neither requisite (1) nor (2) are fulfilled, we are talking about worlds in whose description the usual functioning of our language is interrupted. They are possible worlds in which *many* changes are required before our descriptions can be made, not only the truth of a statement. For Kuhn, "'Hesperus is phosphorus' is [...] true in every possible world. But the class of possible worlds [Kripke] considers is strictly limited."[330] In Kuhn's view, Kripke talks about sets of possible worlds in which the "gradual," as Kuhn says, piecemeal transition to another counterfactual scenario is smooth, easily describable in terms of the current language. In contrast, "[n]o transitions so gradual are possible if one includes in the itinerary of one's journey one or more of the worlds I've been imagining," he says, and adds:

> If Hesperus is replaced by Phosphorus during a passage near the sun, then Hesperus need not even be Hesperus for more than a single interval of a few months, nor need Mars remain Mars, Saturn remain Saturn, or the Zodiacal stars remain themselves. In that world the criteria of astronomical individuation fail and a great deal is put up for grabs with them.[331]

Kuhn calls the sets of worlds in which the Kripkean smooth transition is respected, "Kripke worlds," and these are, of course, different from those in which holistic changes of language are necessary for accessing them, and in which the similarity–difference relationships that ground the lexical categorization (based on the corresponding lifelines) are broken in the process of doing so. To describe it in terms of the recently expounded version of 1987, the set of Kripkean worlds are those in which

---

[328] See Kuhn, *The Natures of Conceptual Change*, TSKP 5.36, Lecture 3, 51–52.

[329] Kuhn, *The Natures of Conceptual Change*, TSKP 5.36, Lecture 3, 52.

[330] Kuhn, *The Natures of Conceptual Change*, TSKP 5.36, Lecture 3, 52. Kuhn also says, just below the previous line, that: "Kripke himself emphasizes that, whatever language is spoken by the natives of a possible world, *we* must be able to describe what occur[s] there in our own language."

[331] Kuhn, *The Natures of Conceptual Change*, TSKP 5.36, Lecture 3, 53; see also 54. These statements are arranged as a list, the latest separate from the rest, but I have suppressed that arrangement for the convenience of the reader as in other cases. This applies also to other extracts of the same lectures in this section.

normal science takes place, and in which the no-overlap principle is respected; incommensurability, when transiting from the actual world to a counterfactual setting, does not take place. By contrast, when travelling from the current world to a "Kuhn world," as he calls his own scenarios—which in his case are not mere imaginative alternatives—that holistic modification of key parts of the lexicon takes place, the no-overlap principle is violated, and incommensurability ensues.[332]

Kuhn then deals, in the rest of the lecture, with the problem of applying to natural-kind terms this theory about the reference of proper names. If Kuhn was more optimistic about the former, he finds the problem of individuation more difficult to overcome in the second case.[333] The situations that made identity statements false in Kripke's theory of proper names were counterfactual scenarios in which reidentifying individual lifelines was difficult or even impossible. Now, the situations that lead to such interruption, when we are talking about natural kinds, are recurrent through scientific revolutions. The focus of his criticism is, as he says, not only Kripke's *Naming and Necessity*, but also Putnam's texts, "The Meaning of 'Meaning'" in particular. Kuhn worked this out in his well-known paper "Possible Worlds in the History of Science" a few years later.[334] By now, however, the main idea I wanted to illustrate at first is clear. Kuhn, in these years, from 1980 on, has not abandoned the notion that both the normal practice of science and the notion of revolutionary change are necessary to understand the development of science, and the causal theory of reference, despite being a successful theory of reference (at large), only provides a part of the story. In Kripke's and Putnam's cases, the idea of changes to another, quite different—moreover, incommensurable—world is not even taken into account. For Kuhn, therefore, it is not powerful enough, and his work in the 1980s—in particular the *late* 1980s, with the Shearman series and the following steps towards the book—is Kuhn's attempt to show how that full account of the structure of scientific advancement must be built.

One last discussion by Kuhn, in this case rather an almost face to face debate, is interesting to reflect on here. It was a debate he had with Rorty for a few years; the most significant interactions for us took place after Kuhn's arrival at MIT. Sometimes, it was a very lively debate, and it was not about Rorty's well-known 1979 book, but rather about some exchanges of papers and letters that made Kuhn aware that his former Princeton colleague was grasping, once again, only a part of the full picture. Let us examine it briefly to close this section and this chapter.

The relationship between both colleagues was always cordial. After commenting on "Does Knowledge 'Grow'?" by Kuhn, Rorty had the opportunity to confront Kuhn's commentaries on a paper of his in a meeting that took place from 28 to 31 March, 1984, at the University of Iowa.[335] Kuhn presented a short commentary titled "Rhetoric and Liberation," which was intended to be a response to Rorty's initial

[332] See Kuhn, *The Natures of Conceptual Change*, TSKP 5.36, Lecture 3, 53–55.

[333] See Kuhn, *The Natures of Conceptual Change*, TSKP 5.36, Lecture 3, 44a.

[334] See Kuhn, *The Natures of Conceptual Change*, TSKP 5.36, Lecture 3, 57. See also Kripke (1972), Putnam (1975), and Kuhn (1989).

[335] Nelson, Megill and McCloskey (eds.) (1987, x).

submission: "Solidarity or Objectivity?" Rorty, however, had submitted a different paper in the end: "Science as Solidarity."[336] Although this setback did not seem to have upset Kuhn, he did not submit his commentary for publication in the volume that ultimately appeared with the contribution three years later. In fact, Rorty did his best to fix the situation. His former paper had already been committed to another publication and he corrected the second so as to introduce Kuhn's excerpts of the original text in the second one. It was not enough, though. For Kuhn, it was already something different, and he told John Nelson, one of the editors of the volume, that he had at that moment no time to make the kind of complete revision that the occasion required.[337] However, as noted, it had no serious consequences for the relationship between them.

There was some friction between them, though, and precisely after he withdrew from the volume. Kuhn replied to Rorty's submission of his second paper with a brief letter, in which he said:

> The twin concepts of "revolution" and "incommensurability" are the linchpins of my view of scientific development. Whig history has, throughout my career, provided the anvil against which that view has been forged. Your use of my work requires that you systematically omit all such considerations, and the views of both science and history you then attribute to me strike me as not even bowdlerized versions of the ones I actually hold.[338]

Rorty's reply was short and not very enthusiastic. He seems to be throwing in the towel as regards a mutual understanding.[339] Kuhn replied, in his words, "a good deal distressed." Yet, he seemed more worried about the possibility of having misunderstood his colleague: "we have talked quite a lot, especially at Iowa, and I had not previously had a hint that you felt that I misinterpreted you." He added that "[a]s one hermeneuticist to another, it hurts to hear it in this way."[340] Shortly after, the same year, Rorty sent a more reconciliatory letter, which nevertheless showed surprise at Kuhn's reaction towards his second paper. Rorty felt that he had given an accurate depiction of Kuhn's views in that text, and seemed frustrated after realizing that Kuhn did not feel the same way. In any case, Rorty confirmed that incommensurability was not the most important concept for him, and that, in his view, it should not be Kuhn's main concern either.[341] Kuhn and Rorty did not ever seem to have agreed on this point.

In an interesting exchange between them in September and October 1986, Kuhn explained what he had concluded to be a key difference between them: the defense

---

[336] See Kuhn, "Rhetoric and Liberation," University of Iowa, 28 March 1984, TSKP 13.9; this commentary was edited as Kuhn (2014); see also Schiappa (2014) on that text and its edition. See also Rorty (1985; 1987), and Mayoral (2023a) on this exchange.

[337] See Kuhn to Nelson, 10 May 1984; Nelson to Kuhn, 15 May 1984; Rorty to Kuhn, 2 November 1984; Rorty to Nelson, 2 November 1984; Nelson to Kuhn, 19 November 1984; Kuhn to Nelson 20 August 1985; all of them in TSKP 13.13.

[338] Kuhn to Rorty, 20 August 1985, TSKP 22.14, 1.

[339] Rorty to Kuhn, 2 September 1985, TSKP 22.14.

[340] Kuhn to Rorty, 17 September 1985, TSKP 22.14.

[341] Rorty to Kuhn, 11 October 1985, TSKP 22.14.

he (Kuhn) used to make of the necessity of normal science. Accordingly, incommensurability, as a relationship (or local lack of it) between stable though mutually incompatible forms of language, follows. For him, Rorty focuses mainly on the powerful transformative effects that new metaphorical forms of discourse produce on literal speech. Rorty, however, does not seem to realize that there is no metaphor, no new proposal of *reorganization* of our taxonomic arrangement of objects and substances in the phenomenal world, without an *organization* to modify. Individuals are not members of a community, and so of a culture, reluctantly; on the contrary: "Individuals," Kuhn says, "are individuals only with respect to a culture, and what it is to be an individual depends upon what culture the individual is in."[342] For him, Rorty was doing the inverse of the job that Michel Foucault used to do. Talking about the latter, Kuhn says: "He restricts himself to 'epistemes', archaeological strata, language-games, and refuses to say anything about the passage from one to the next. You turn his pattern inside out."[343] Similarly, for him, Rorty did not seem to be particularly interested in a sympathetic reconstruction of past philosophy as an element that belongs to a culture and its language, especially in his criticism of epistemology as a tradition. Kuhn writes:

> When I complained to you in Iowa that *The Mirror of Nature* did not even try to make plausible that earlier thinkers had held the beliefs you ascribed to them, I was complaining that you didn't describe their language-game and therefore could not provide a plausible (because causal) narrative. When I remained unconvinced that epistemology was dead, it was because epistemology (and the knowledge for which it provides theories) is constituted by and within language-games. To say that epistemology is culture-bound is not to say that there could be such a thing as an epistemology-free culture.[344]

Rorty's attitude towards the way language changes rings in Kuhn's ears like Popper's and his "revolution in permanence," which Kuhn finds (as it indeed is) "a contradiction in terms." Rorty is thus somebody who does not understand that normal science is as important a part of the whole picture as revolutionary change is. When reading some of Rorty's recent writings—his Northcliffe Lectures at London in particular—Kuhn told him, he had been "looking for [...] some acknowledgement and examination of the role of the group from whom the individual inherits, of the essential role of inherited, literal, usage in making metaphor possible, of the essential tension (the echo is *not* accidental) between the need to innovate and the need to communicate."[345] However, Kuhn was frustrated by the vain search for such a picture in Rorty's recent writings.

For Kuhn, as noted at the beginning of this section, that dynamic-equilibrium picture of the dialectic of normal and revolutionary science motivates his entire search throughout these last years of a way to show how language, and its attachment

---

[342] Kuhn to Rorty, 19 September 1986, TSKP 22.14, 3.

[343] Kuhn to Rorty, 19 September 1986, TSKP 22.14, 4.

[344] Kuhn to Rorty, 19 September 1986, TSKP 22.14, 4. Kuhn is, of course, referring to *Philosophy and the Mirror of Nature* (Rorty 1979).

[345] See Kuhn to Rorty, 19 September 1986, TSKP 22.14, 3, for all excerpts in this paragraph. Such lectures by Rorty were later Part I of *Contingency, Irony, and Solidarity* (Rorty 1989).

to a phenomenal world, grounds our approach to nature—and who we are—and also how its limits produce drastic changes in the language itself and in the group that uses it. There is no way for him to escape from the *full* picture, even though many philosophers around him had interesting interpretations of some aspects of that picture. The global structure was still to be found, and Kuhn made a last serious effort, after this long decade of successive attempts, to write a full book synthesizing it. The Epilogue shall guide us through those last steps.

Chapter 9
Epilogue

In this last, much shorter chapter of this book, I would like to briefly re-examine what we have seen so far in order to situate Kuhn's last work in a context of continuous development in which his last philosophical steps are easily understandable. Such steps were contained in his last book, *The Plurality of Worlds*, which remained an unfinished manuscript at the time of his death. In May 1991, Kuhn told his good friend, Paul Hoyningen-Huene, that he had just begun to write this text. Unfortunately, in fall 1995, deteriorating health forced him to stop writing.[1] In the five completed chapters (plus some pages of the sixth), he started to suggestively and informatively restructure the views that had brought him to that point. Are such chapters a summary of his preceding views from *Structure*? Definitely not. In fact, Kuhn does not return to his masterpiece again. However, their ideas can only be explained on the basis of this earlier book and of the biographical itinerary that had led him to that point. The journey that we started in Chap. 1 of this book, which finishes in this Epilogue, helps us to understand his itinerary, and so the slowly improved philosophical vision of science that lies behind the theses that appear in the extant pages of *The Plurality of Worlds*. It will also help us to grasp what remained absent from them for lack of time—mainly his theses about relativism, realism, and truth. Accordingly, this is the proper moment to remind ourselves of what we have examined so far, step by step.

If we return again to Kuhn's early days at Harvard College, we remember that he was a young, talented, and passionate student with a good scholarly record, who graduated in three years with a somewhat unusual concentration of courses in electronics for a theoretical physicist, and who was ready to enter the wartime Radio Research Laboratory at Harvard. The war was the reason for this, just as it surely changed his mind, too, in the more ideological and moral aspects of his life—his decision to abandon pacifism. Kuhn was not an aspiring philosopher at that time yet, but his inquiring, reflective mind did not let the consequences of such options pass without further investigation. The vision of an individual whose beliefs and decisions

[1] See Kuhn to Hoyningen-Huene, 13 May 1991, and 10 April 1996, both in PHA.

J. V. Mayoral, *Thomas Kuhn*, Springer Biographies,
https://doi.org/10.1007/978-3-031-86512-1_9

are inextricably linked to a group, and that of the autonomy of stages in individual development, are two interesting intuitive ideas that he obtained from reflecting about his abandonment of pacifism. The idea that some foundational principles of action, rationalism, and liberalism in the main, may be, somewhat ironically, foundationless and pragmatically warranted, arouse from that reflection. His other vital choice—to follow his vocational option for physics to the limits—also showed him how the narrowing of his training and a dedication to applied physics were not compatible with the kind of critical attitude he used to have about most aspects of his intellectual life.

This contrast must have been in the origin of the component of "convergent thinking" that he added to normal scientific research in his essay, "The Essential Tension," a component whose partial origins he confessed to be in his "own experience in scientific research."[2] From his graduate studies in physics under the supervision of Van Vleck he also obtained a pragmatic feature of scientific behavior: that physicists usually work on problems they know how to solve, not on a more uncertain kind of problem—a point of view that his supervisor himself suggested.[3] Critics have seen in Kuhn's normal science a result of his lack of contact with the kind of science that physicists practice in the later stages of their careers. For such critics, the debate and the search for alternatives populate the usual practice of physics beyond the picture of normal science.[4] Convergent thinking would thus be an incorrect component that was obtained, in Kuhn's case, from his own experience of physics research as only a pre-doctoral student. However, that Van Vleck, a senior physicist nearing his fifties, confirmed (or perhaps was in part the origin of) that impression seems to show that Kuhn was not the only one that thought that way in those days. He seems to have witnessed a certain structure of research—which is, in any case, a central source of contribution to physical knowledge.

On the other hand, Kuhn never denied the creativity component of normal science.[5] He must have been able to appreciate the necessary creativity, not only hard work, involved in his own doctoral work. The difference was probably rather a matter of perspective. His doctoral research did not aim to transform the broad landscape of physics. His essays in philosophy and literature show that he was fond of exploring broad-scope consequences of looking into specific problems in other fields—such as those shown above. It seems that he opted for a career in a field, philosophy of science, in which these latter were the order of the day. However, he

[2] Kuhn (1959a, 226).

[3] See [Kuhn], "Puzzles versus Problems in Scientific Development," Talk at Vassar College, 4 November 1974, TSKP 5.9, 11.

[4] See Beller (1999, 306); Hamlin (2016, 303). I mentioned Beller's point in Chap. 1.

[5] See Kuhn (1976b, 179, n. 3). See again our Chap. 5, Sect. 5.6, on this point about the creativity involved in normal science in the early drafts of *Structure*, Chapter II.

also took advantage of his first-hand experience of physics, not only of his second-hand experience of the past as a historian of science,[6] in showing important details for his theory of science, with long-range consequences for our view of science.

We have also seen that Kuhn exhibited his critical attitude in literature essays and in his many extra-academic activities. Some further important ideas are also visible in his essays on literature, as the contingency of evaluative standards and the almost ethnographic study of the past previous to any evaluative activity. His praise of Browning's poetry against the charges of Santayana are a good case in point. In those essays, Kuhn exhibits a critical attitude against Whig historiography and against a universalist philosophical account of values that is present in his mind and in his writings over the years, though later transplanted to science. Even among his opportunities to explore philosophical questions about science, he showed his hope to explore the limits of scientific realism at that time. He did not have much confidence in the empiricist views of senior members of the Department of Physics, nor did he assume that realism itself is, after all, settled. His critical attitude was also at work in this field.

I could be accused of reading the early days of Kuhn's biography with the help of hindsight; however, note that I am not claiming that he *related* such ideas to one another yet, which is the truly crucial and original step. He would do so later, from 1947, and, in fact, he would present his thesis in the original Chapter I of *Structure* in precisely that way, that is, as a series of resemblances and differences between the sciences and the arts. At this moment, though, there is a visible divide between them, and also between the kinds of philosophical problems that each of them elicit. For Kuhn, the philosophical problems of science are mostly epistemic, whereas moral and aesthetic issues are related to a different kind of problem in which the historical contingency of standards and the irrationality of belief might be involved. Kuhn would need his "Aristotle experience," and the long journey to *Structure* and beyond to figure out to what extent such ideas are intermingled and, especially, *how*. But the pieces *were* already there at this point, and that cannot be denied, either.

The separate presence of the pieces that were already there in those years (roughly, 1940–1947) should be discussed further here. Andresen calls our attention to the presence of the word "crisis" in his essay on war and on his changes of mind about pacifism and intervention. At some other points, she shrewdly notes how Kuhn recalls how he discovered the geometrical root of "incommensurability" as a term. There were, she also says, some experiences in "The War and My Crisis" that unequivocally fall under that denomination.[7] Although Andresen is careful when linking such terms and the specialized meanings in Kuhn's own vocabulary, we must emphasize that this link *took time*. His perspectives on the sciences, on the one hand, and on the arts, on the other, had to be combined first, and the attachment of such terms and

[6] Such information for the history of science is an important source of that attention to the convergent-thinking component, he says in "The Essential Tension." See Kuhn (1959a, 226).

[7] See Andresen (1999, S47–S49, S67). There are many similar interesting passages in this pioneer work.

the phenomena they named had still to be made. "Crisis," "paradigm," or "incommensurable" were later useful in a theory that was steadily discovered and improved over the years. Metaphorically speaking, Kuhn's "lexicon" was gradually attached to his own phenomenal world, although some of the resources were already old for him, and sometimes had passed unnoticed—or so it seems. The word "paradigm" is a case in point, as we saw in Chap. 5. Some central generalizations and foundational beliefs were in place at an initial stage, but *unconnected*, dispersed across his many fields of interest.

Connecting all of them, composing Kuhn's vocabulary, was a long process that I have tried to explore in Chaps. 2–5. The highlights of these chapters are, firstly, the 1951 Lowell Lectures, and secondly *Structure*. One would expect that *The Quest for Physical Theory* (the Lowell series) served as a blueprint for *Structure*, and that Kuhn worked out the latter from the former. Thematically, there are evident connections, and many older ideas reappear there. However, there is a remarkable lack of interplay between them. The Lowell Lectures are arranged as a plan for a book from around 1949 or 1950. They focus on scientific revolutions as problem-solving devices in cases in which an already established viewpoint is unable to do its job. There is no sociological approach to the periods between revolutions, but it is left pretty clear that, psychologically and semantically, this is the period when the version of a scientific theory in the minds of the practicing scientists is closer to the one that appears in books. In *Structure*, there is greater emphasis on the way the practice of science within communities, whether the training of new scientists or their professional job, need not be based on rules; paradigms fulfill that function. Paradigms structure the practice of science in communities sociologically. The Lowell Lectures, meanwhile, focus on the role our psychological constructions of nature play in order for our minds to control nature. And *Structure* does its job by means of a different script, which was not even what was originally planned. It is a book born out of frustration with previous plans, which found its center on the basis of a marvelous key concept, the paradigm, which Kuhn dusted off from his stockpile of frequent expressions.

Despite these differences, some points in common are visible. Three of them are, in my view, remarkable. First, Kuhn shows that the practice of science is not well represented by the simple, pedagogical structure of a textbook, and that there is a more complex, less structured relationship between the main expressions of a science and the world to which they are applied. This world, on the other hand, is no longer simply "the world," so to speak, but rather a cognitive scenario, a liaison with nature that, as a Kantian phenomenal world, simplifies and regularizes the flow of stimuli coming from the environment. He had designed that vision in previous years, while in the Society of Fellows, on the basis of many readings—mainly of Jean Piaget. Our way of dealing with the environment depends on that intermediary and its language-based relationship. A science is a specialized version of that language (which includes the behavior deployed around it) and, in the periods of stability in science, that expert version trades with the environment without interruptions or doubts. The practicing scientist knows to what extent that version is imperfect, but that knowledge is avoided in textbooks. So, although they are good training devices, textbooks conceal the authentic nature of science, and philosophers, and the public in general, should approach science on the basis of this inadvertently hidden truism. This point of view is common to *Structure* and to *Quest*.

Second, Kuhn's crusade against empiricism was born from his own training and experiences, but it is also emphasized by Conant's own critical perspective against philosophers. A defender himself of a view like Bridgman's—which shares features with the logical empiricism of their day—Conant is, however, against some naivetes coming from that point of view, specifically when contrasting its depiction of science with the practice he knows. Kuhn shares that view, too. His vision of science from his undergraduate studies is more characterized by a skeptical attitude than by the empiricist convictions of his elders.[8] From that point on, he tried to find out what is behind the simple arrangement based on a theory and its relationship with observation. The difference between textbook science and creative science responds to that conviction. That critical perspective against empiricism is common to the Lowell Lectures and to *Structure*. It is a long-standing attitude which is again visible in his developmental alternative to static philosophy of science.

Finally, Kuhn's perspective adopts a form of dynamic functionalism, in which basic components, such as paradigms in *Structure* or the basic scientific orientations that affect the creation of theories and problem-solving in *Quest*, are accounted for not in terms of a logical architecture for the theory plus a pretty traditional epistemology, but rather on the basis of a language that shares many features with ordinary language—the latter would be increasingly important for Kuhn in his more mature years—and on a revised version of the philosophical vision of knowledge. This latter element is pursued with persistence both in the Lowell Lectures, where the criticism of classic epistemology is visible, and in *Structure*, whose Section X is a core argument for Kuhn. Even in his mature years, in other pieces of research such as the Foerster Lecture of 1976, this search for an alternative to traditional epistemology—without considering it to be a dead tradition, like Rorty—remains visible as a primary concern for him.

Despite all that, *Structure* is still very different. The Lowell Lectures are a pretty Conantian text. The historical case studies that Kuhn handles in Lectures II–IV are precisely that—i.e. case studies, examples of science. With due caution, we can consider them paradigms for the public to learn what science is without dealing with the intricacies of modern chemistry and physics.[9] *Structure*, meanwhile, has a historical perspective embodied in its master argument from Section I. There are no case studies as in the Lowell Lectures—or in *The Copernican Revolution*, for that matter, which is a longish case study with an illuminating philosophical subtext. In *Structure*, the argument does not simply visit the past: it refers to science at large, which *has* by itself a historical nature. It is here that Kuhn's old ideas as a student recur. I am referring to the ideas concerning the relationship between the

[8] So, he might be right when he suggests that such skepticism is not a good ingredient in the training of a good normal scientist. After all it promoted his own exit from practicing science.

[9] Indeed, they are not *pure* case histories of science: they show orientations in operation. But, after all, Conant also shows his own cases as something arranged according to his own vocabulary of conceptual schemes.

individual's beliefs and decisions and the group to which the subject belongs; to the autonomy of developmental stages, whether for an individual or for a group of them; to the contingency of evaluative standards—a point that is already visible in the Lowell Lectures; and, of course, to his attacks on everything Whiggish. No, these latter elements are not born from his relationship with Conant, and the president would have difficulties in accepting them or their consequences. When Kuhn stopped proposing a Conantian view, they parted ways, and that had not yet happened in the Lowell Lectures nor in *The Copernican Revolution*. It would take place in *Structure*.

After *Structure*, Kuhn started to worry about some decisions made on behalf of its argument. We should distinguish between the improvements that the new concepts in *Structure* required and those developments of older views that Kuhn undertook. From the first point of view, the notion of paradigm received most of Kuhn's attention from the beginning. Masterman helped in that regard, but Kuhn's revision of the concept was well underway when the interaction between them started. He had already been worried about the monolithic nature of the notion as expounded in *Structure*, and about the relationship with the group and with the kind of change exhibited there. More precisely, the kind of "object," actually a model (more or less, as he told Boring), that he called a "paradigm" could change within the community as the same solution was reinterpreted, or articulated, in a different key; for instance, with a different mathematical interpretation. Individuals at different moments in time within the same group might consider two apparently different, though related, models as paradigms. He started to be concerned with the difference between articulation and revolution because it was possibly more complicated than had been described in *Structure*. Similarly, if a revolution is felt as such depends on the effects that a given change has on a group (or subgroup) of scientists; so for Kuhn, the concept of revolution seemed to have the features of a subject-relative phenomenon (where the subject remains, of course, a group). What a scientific revolution is, if and when it is different from mere articulation, what paradigms are, how they evolve, and how communities do not overlap with paradigms, were still open issues that he would deal with during the 1960s. In contrast, he did not yet seem to be very concerned about the concept of incommensurability; at least, not as much as with other questions that I shall mention later. This concept, together with the idea of revolution, would start to be more precisely delimited on the basis of the lexical theory in the 1980s.

From the second point of view (that is, those developments of older ideas that Kuhn resumed after *Structure*), his crusade against empiricism continued, and he attempted to show with precision to what extent the idea of reconstructing the past, his own perspective, was detached from the idea of rational reconstruction that he observed among the empiricist philosophers of his time. This idea took up again his old criticism both of that current and of the Whig trends in the way philosophers viewed the past. In the 1980s, that would again be a central foundation of his perspective as it was in *Structure*. However, perhaps the most decisive work, especially for further developments of old notions in the future, comes from an old theme that was already visible in the Lowell Lectures, which he re-examined in more detail. It is rooted in his years in the Society of Fellows, though unrelated to Conant himself, and first displayed in 1951, in *Quest*. I dealt with this in Chaps. 6–7 in this book, and it has two dimensions: how a categorical arrangement of the phenomenal world is obtained without recourse to rules and definitions, and how a language is attached to it. In *Structure*, this older theme was linked to paradigms in the new Section V of the final version of the book. Then, Kuhn returned to that perspective in order to make it clearer. As he says in the second edition of *Structure*, in the "Postscript—1969," we don't need to avoid the word "intuition" when talking about science. It need not be replaced by rules, definitions, and criteria, but it does not correspond to a mysterious capacity in human beings, which cannot be subject to scrutiny, either.[10] Showing how intuition is trained is the basis for Kuhn to display a model for arranging the phenomenal world and to attach a language to that cognitive scenario.

As regards the categorical arrangement of the phenomenal world, in 1951, in the Lowell Lectures, Kuhn had already demonstrated a preliminary view that referred to a perceptual or phenomenal world that every subject develops and that is conditioned by behavior. In fact, it is the scenario in which behavior is displayed, and its objects are those that play a behavioral key role. In *Structure*, this overall scenario is transplanted to the scientific case, but Kuhn no longer pays attention to the scenario itself and to its metaphysical and methodological components, as he did in *Quest*, but rather to some particular objects within it: the paradigms. Such paradigms are successful models of solved problem that, in order to learn the profession, the scientist learns to duplicate and then to apply to new situations, thereby articulating the original example—i.e. the paradigm. These paradigms are elementary inhabitants of the scientist's behavioral world, just like dogs, or ducks, or trucks are in our own common-sense world.

After *Structure*, Kuhn tried his luck with a more general perspective, one that worked for both the common-sense world and for the more specialized practice of science. In Chaps. 6 and 7, we saw how he developed a paradigm-based model of formation and learning of a categorical space in every individual in order (and owing) to his or her acquisition of membership to a group. These are based in turn on learned similarity and difference relationships, whose effectivity is based on a correct distribution of incoming information about the world, in particular, about its objects, which in this case act as paradigms, among the extant categories, usually learned from other members of the group. When such categories are of natural families, the interstices between them, the "empty spaces" between such families, are correctly respected.[11] With this model, there is no need for rules and definitions, and the world is composed of behavioral necessities, whether those of our ordinary, common world, or of specialized activities like science. Kuhn tried hard to offer a mathematical, geometrical analogy for such a model, and even tried simulating it in computers, but complications in this latter endeavor forced him to stop. In any case, a well-known intuitive example of paradigm-based learning emerged out of this research around 1967, the "Johnny example," which helped him to demonstrate the likeliness of that point of view.

Concerning the corresponding language, that is, the one applied to a kind of world comprising a well-arranged set of natural families, Kuhn showed that there was no

[10] See Postscript–1969, 190–191; see also our Chap. 6, Sect. 6.6.

[11] See our Chap. 6, esp. Sects. 6.5–6.6; Kuhn (1974, 316–318).

need for rules and definitions for the application of natural-kind terms in this case, either. This also traces a long way back to 1951 and before (his earliest notes in that respect are from 1949), when he showed how we attach terms to inhabitants in our world on the basis of our limited experience, not always identical to that of our peers, and with a clear target in our behavioral necessities. In those years, he considered the idea of a vagueness fringe around the core semantical aspects of a term. He already had the conviction that such a fringe can only be reduced to those senses known with certainty by the whole group at the risk of losing semantic details located at the fringe of vagueness that present a positive function for adaptation to environment. Once again, Kuhn's picture is attached to adaptation, to behavior, and to the practice world. In the 1960s and later, on the basis of the previously examined model of paradigm-based learning and, as in *Structure*, by way of Wittgenstein's considerations on the meaning of a term like "game," Kuhn displayed a model for the learning of natural-kind terms that worked pretty well with the previous model of categorization of the world picture. By that time, Kuhn was already doubting the appropriateness of considering that meaning should be vague, properly speaking, as if lacking a definiteness that it could gain if language could be devoid of the variability of experience, something that a formalized language could achieve.[12]

As noted, old perspectives on the world and language can be discerned here. Firstly, the idea that we configure a world—the *psychological* world, as he used to call it—by means of a simplification of the flow of stimuli is already visible in his notes from the early days in the Society of Fellows. Secondly, the idea that the meaning of our terms in everyday use (like "reason," or "liberalism") is captured from the uses to which we are accustomed in the group to which we belong is as old as his 1941 essay "The War and My Crisis." There are echoes of views cherished by Kuhn for a long time, which were probably part of his mental setup from the very beginnings of his career, and even before.

This conjoined model of learning a world and a language at the same time locates us in the anteroom of *The Plurality of Worlds*. During the years 1962–1978, Kuhn had worked hard not only in philosophy of science (and, lest we forget, for the Princeton's Program in History and Philosophy of Science), but also in the history of quantum physics. By 1976, however, he had started to reconsider some ideas in *Structure* from a point of view that emphasized the model above from the side of the language attached to the phenomenal world and its categorization. For Kuhn, from the mid-1970s, the idea of conceptual scheme was not as important as the idea of a structure of terms attached to categorizations. During the rest of the 1970s and in the 1980s, he developed two central ideas: the notion of a *developmental* philosophy of science, and the notion of a *lexical* structure, shared by every speaker of a language, attached to a taxonomy of inhabitants of the phenomenal world that is also respected by the communications within the community. We have already seen the details of both up to 1987, with the version of both that two now well-known series of lectures transmitted: the Thalheimer Lectures (Johns Hopkins University, 1984) and

[12] See in Chap. 6, esp. in Sect. 6.5, Kuhn's discussion of Braithwaite's and Waismann's perspectives on the meaning of theoretical terms.

the Shearman Memorial Lectures (University College London, 1987). Both lecture series show how the alternative to a traditional, static version of the epistemology of science should be replaced with a perspective that emphasizes the description and evaluation of belief-change, and the way a version of incommensurability based on untranslatability of statements from a lexicon to another is the best way to understand the kind of rupture that scientific revolutions produce. *The Plurality of Worlds* was intended to explain both global ideas in detail from solid foundations. Kuhn, however, did not have time to accomplish this task.

In our last chapter, we left Kuhn around 1987, when he had delivered the Shearman Lectures in London. In the second half of the 1980s, he travelled a lot with Jehane, talking about his work and trying to tour the places he used to visit. They went together to Paris in 1985, though she later had to stay at home because of her own work, while he visited Minnesota for some days upon returning from Europe in order to give a seminar, invited by his former student, Philip Kitcher. Wesley Salmon joined them on that occasion, which followed a previous one in which they both had been commenting Hempel's paper on rationality two years before. In 1986, Kuhn and Jehane visited Japan and Stockholm. Their trip to Japan was arranged by Shigeru Nakayama. They really enjoyed the experience, and they hoped (rather, they planned) to repeat it. The following year, he went to the Pittsburgh University's Center for Philosophy of Science, where Hoyningen-Huene was enjoying a period of research as visitor, and presented a paper in late October, 1987. In 1989, he travelled to Russia.[13]

As the new decade began, his interest in travelling around the world diminished. He told Greek philosopher of science, Kostas Gavroglu, that he would not visit Greece at that time. "I travel badly," he said, "and will not undertake a significant change in time zones unless I can spend more than a week in the place to which I'm going." Kuhn and his wife moved to an apartment in Cambridge and decided to sell their Boston house, so he had to spend time in getting rid of some "possessions drastically," which was "a consuming task, both temporally and emotionally," he told his Greek colleague.[14]

However, in 1990, he produced many interesting pieces. In January 1990, he commented on a paper by Abner Shimony at the Boston University Colloquium for Philosophy of Science, and in April he went to the west coast in order to deliver a couple of lectures at the University of California, Los Angeles, one of them to the Cognitive Science Colloquium, 26 April, and another one to the Philosophy Colloquium the following day, 27 April. In the former, "An Historian's Theory of Meaning," he talked again about the nature of incommensurability on the basis of lexical theory; in the latter, he examined the problem of incommensurability itself with some mentions of the problem of truth and of his Darwinian-based view of scientific progress. On both occasions, he mentioned the new book. Just like his most significant text of that year, "The Road since *Structure*," his presidential address for the Philosophy of Science Association, given in October 1990, those papers provided

[13] This itinerary is based on Kuhn's letters to Hoyningen-Huene, 14 October 1986, 17 September 1989, both from PHA.

[14] Kuhn to Gavroglu, 29 March 1991, TSKP 21.33.

good summaries of the book he was starting to prepare. Some additional ideas were also delivered months before in his paper for the conference that was organized in his honor at MIT in May—a paper that was published as "Afterwords" in the volume of contributions, *World Changes*.[15] For the Independent Activities Period at MIT, Kuhn joined Bromberger in a new common session in January 1991, and he talked about incommensurability and its relation to his theory of kind-terms and taxonomies (he no longer mentioned lexicons in that lecture).

In early 1991, Kuhn told Hempel that "The Road since *Structure*" was a good piece of work. It was "a sketch of the project with which I'm at last quite pleased," and, reasonably, it was a "fuller and better balanced" paper than the group of comments that would end up forming his "Afterwords."[16] In his talk at the Philosophy Colloquium at UCLA, he introduced his new book the following way: "The book as a whole is the outcome of attempts to return to the philosophical problems raised in *The Structure of Scientific Revolutions*." Among such issues, "a key obstacle," he said, "has been the unsatisfactory state of the notion of incommensurability."[17] The close connection between both books was important for Kuhn. Hempel had told him that he felt the position in "The Road since *Structure*" was quite far off that of his former book.[18] Kuhn reacted to that comment as follows:

> I have been puzzling over your remark that the position I'm developing seems so very different from that in *Structure* and some of my papers. I would have said that I'd made great progress and many significant consequent adjustments, but that my general direction was precisely the same. Then I remembered that someone else had made a remark that might be related, and putting the two together I realized that, at the point where I introduce speciation, I took for granted its relation to the concept of revolution as change of lexical structure. The incommensurability that obtains between old and new theories in the same domain also characterizes specialties with different domains at one and the same time.[19]

Clearly, the differences he is starting to note are already visible in the final version of the paper. Kuhn told Hempel that he had changed it a little in some specific points, and he must be referring to the passages in which he makes the parallel between specialization in the sciences and speciation in biology, including the branching process that revolutions bring about in the tree of existing specialties, and the degree of incommensurability involved between the resulting branches.[20]

In spite of that significant difference, some connections between both books are apparent. I shall comment on them below, but I would like to call attention to one in particular, of a more methodological nature. During the preceding decade, Kuhn had been paying increasing attention to lexicons (also in the paper just mentioned).

---

[15] See Horwich (ed.) (1993, 1).

[16] Kuhn to Hempel, 1 February 1991, TSKP 21.42.

[17] [Kuhn], "A Function for Incommensurability," 27 April 1990, TSKP 24.8, 1. On this talk, see Marcum (2015a, 138–140).

[18] Hempel to Kuhn, 6 February 1991, TSKP 21.42, 1.

[19] Kuhn to Hempel, 18 February 1991, TSKP 21.42, 1.

[20] See Kuhn (1991a, 97–99). On this parallelism, see Gattei (2008, 160–163); Wray (2011, Chap. 7); Kuukkanen (2012); Haufe (2022, 81 ff.).

In *The Plurality of Worlds*, however, he works out the structure of kinds, not specifically of kind-terms (and other terms), and the basic psychological configuration that undergirds such a structure. In doing so, he finds support from psychological work on children, such as that carried out by his new acquaintance at MIT, Susan Carey, and some of her colleagues. Having recourse to current experimentation in psychology is reminiscent of *Structure*, which had been based on Piaget's works on children in some essential points, as we saw in previous chapters.[21] In short, in *Plurality* he is revisiting old theses in *Structure* and changing some of his former arguments and making some notions clearer, but he also seems to be using his own plural, naturalistic perspective of the good old times to do so. No methodological decision seems to limit that naturalistic stance.[22]

It is time to take a look at the contents of that book, which he seems to have been ready to write in 1991, as he told Hoyningen-Huene and other colleagues. The arrangement of the written chapters is roughly as follows. The first two chapters show, first of all, Kuhn's own brand of developmental philosophy of science as a comparative alternative to the static tradition, as already seen in our Chap. 8. Two bodies of knowledge are compared by paying attention to some local groups of beliefs in conflict whose choice must be properly justified. Kuhn's vision of the nature of knowledge from 1976 onwards had shown him that this kind of choice was the focus for the search of justification and judgement in philosophy of science, and that it avoided some problems such as those related to a global holism and to the quest for a basic observational language, which was no longer needed.[23] Interestingly, Kuhn briefly explores the traditional sources of such problems in the footnotes of Chap. 1.[24] The kind of comparison that Kuhn's view involves for the historian of science (for the analyst of the scientific past at large) is properly illustrated here by means of the Torricelli case, which in Chap. 1 helps to show the differences between historiographic traditions, and in Chap. 2 is reinforced by examining, once again, Aristotle's arguments against a vacuum once his physics is reconstructed as usual, together with the examples of Volta and Planck.[25]

There seems to be nothing particularly striking thus far after examining the lectures that led Kuhn to this point throughout the 1980s. This is understandable once we have examined his journey from early in the decade to this point in time, and especially the lectures that usually conform his introductions to lecture series. He had been carefully constructing that part of the argument and he holds on to it as it is a good starting point, which is based on two main ideas: first, the characteristics of his alternative to traditional philosophy of science, and second, the elements of change through historical cases, which a more elaborated theory than that of *Structure* helps to explain. In arriving at Chaps. 3–5 and the first pages of Chap. 6 that were written,

21 On Kuhn's interaction with Carey at MIT, see Carey (2022, 11–12).

22 See Mladenovic (2022, xvii, xxvii); Mayoral (2024c, 174).

23 See *The Last Writings*, 126–123. All references to *The Plurality of Worlds* shall be hereafter through the reference to Mladenovic's volume.

24 See *The Last Writings*, 125, fn. 17, 18.

25 See *The Last Writings*, 160–165, and compare it with 116–122 in Chap. 1.

things change and we find many elements of renewed interest, because the previous recourse to lexical theory finds now an improved theory. That is the explanatory theory, more powerful than that in *Structure*, that explains the facts of change that he exhibits through historical cases. No mention of lexicons is made. Surely the fact that the categorical arrangement of objects in the world around us is common to linguistic and non-linguistic animals makes the reference to lexicons a bit strained.[26] He had already pointed to that forced situation in "The Road since *Structure*."[27]

In Chap. 3, he presents his own "replacement theory of meaning."[28] There is little mention of his criticism of the causal theory of reference, to which he devotes a footnote, but the descriptive theory of meaning receives greater attention as the traditional choice that he typically rejects.[29] His own view is devoted to showing the way our phenomenal worlds are arranged on the basis of kinds that are now generalized, and of at least two types: taxonomic kinds and singletons. These latter mainly refer to substances that are not learned on the basis of similarity/difference relationships (or rather by virtue of "differentiae," as he often underscores in the book), as usual for individuals.[30] He now emphasizes the importance in such cases of the simultaneous function of some of them in contexts like logically related groups of scientific laws. In that respect, there are clear relationships between this Chap. 3 and the recourse to the more general set of artefactual kinds in Chap. 6. Such new sets of kinds are not formed on the basis of ostensive signaling and similarity/difference relationships, but rather by virtue of a functional context and something more than immediate observation by pointing to objects (normally a more elaborated laboratory setting or descriptions). Singletons and material kinds are examples of these.[31] The distinction between taxonomic kinds and singletons (artifactual kinds in general) is related to the normic/nomic difference, which he had already demonstrated in his "Afterwords" and that reappears here. The specific nature of singletons as different from taxonomic kinds is, in addition, well-illustrated by the learning of the meaning of terms such as "mass," "weight," or "force" in the vocabulary, or lexicon, of the Newtonian physics in a context combining law statements and some laboratory devices exhibiting the phenomena described by the laws. This latter example was displayed in "Possible Worlds in the History of Science," and Kuhn refers to it again in this book, without developing it.[32]

The relationship of this chapter with the contents of Chap. 5 is evident as soon as one examines it. Kuhn expounds his vision of the acquisition of a structured kind

---

[26] On Kuhn's assumption about the structured kind set as shared in communities of non-linguistic animals, see *The Last Writings*, 168.

[27] See Kuhn (1991a, 94).

[28] *The Last Writings*, 176.

[29] See *The Last Writings*, 169–174; see 174, fn. 6, on the causal theory of reference.

[30] On "differentiae," to which I shall return in what follows, see *The Last Writings*, 210 ff.

[31] See *The Last Writings*, 176–180 of Chap. 3 for singletons, 252–256 of Chap. 5 for material kinds. On artifactual kinds, see the few extant pages of Chap. 6.

[32] See *The Last Writings*, 220, 240, 243, 251, and 180 for reference to the example of terms in the Newtonian lexicon. See also Kuhn (1989, 65–75) for that example, and Kuhn (1993, 230–231) for the normic/nomic distinction; see also Mayoral (2017a, 125–126) on the latter.

set on the basis of a feature space that varies from individual to individual a little, but eventually converges to the same structured, hierarchized set that is shared by the members of the group. As he says much earlier, in Chap. 3, the role paradigms played in *Structure* is now played by these kind sets—although, of course, we are not talking about the very same kind of concept at all.[33] In Chap. 5, the acquisition of the structured kind set goes through the steps that Johnny followed with his father in learning how to identify kinds of waterfowl. The connection, then, with the result of Kuhn's work since the late 1960s is evident. Indeed, that work, which was slow and that at some points (as in the case of the computer simulation of paradigm-based, and similarity-based, learning and training) led to a dead end, is the very basis for the idea of structured kind sets that appears here—leaving aside the evident differences that result from his evolution since those years. When it comes to the relationship between the feature space and the *common* kind set as a necessary condition for communication, Kuhn is returning home; he is visiting a space that has been configuring during the middle years of the past decade.[34] Incommensurability is thus well-explained on that basis, as he sought. When that commonality, the structured kind set, is not present in a situation whose participants try to develop overlapping behaviors, incompatible conducts emerge. For instance, when the language is part of the behavior, and the common kind set is not present or has been dropped out, and there are points of incongruence among speakers, a communication breakdown ensues and incommensurability is the resulting state.[35]

Possibly, however, one of the richest chapters in this new book is Chap. 4. Kuhn told Hoyningen-Huene in June 1995 that he had a soft spot for that chapter—which, according to that letter and from the date that appears in the published version, was in its final form at that date.[36] I understand (and share) that feeling. Chapter 4 returns to the basic categories needed for forming a given taxonomy, those to which "object," "space," "time," or "kind" apply, and he shows through a series of real psychological experiments how such basic categories are slowly formed and evolve in babies who are just a few months old. In these pages, he extends the application of the no-overlap principle to the formation of the basic category of objecthood, which is essentially founded on the principles that he had already expounded in 1980 when displaying the casual theory of reference in the third Perspective lecture. Those references, he said, have lifelines, that is, they undergo changes, which we track.[37] Those lifelines, as applied of course to objects only, cannot split into two branches or more at any moment, nor can they mix with others. Kuhn states that two forms of the no-overlap principle, the "principle of impenetrability" and the "no-branch-point principle,"

---

33 See *The Last Writings*, 181.

34 See our Chap. 8 for details. In *The Last Writings* (i.e., in *Plurality*), this appears on 230–237, 240–244. This is the main basis for what has been said in this paragraph.

35 See *The Last Writings*, 256–259.

36 Kuhn to Hoyningen-Huene, 25 June 1995, PHA, 1.

37 See Kuhn, *The Natures of Conceptual Change*, Perspective Lectures, University of Notre Dame, 17, 19, 21 November 1980, TSKP 5.36, Lecture 3, 47; see our Chap. 8, esp. Sect. 8.12.

apply in the process of acquisition of that basic category.[38] He then explores the formation of the basic category of kinds.[39]

A second important block of the chapter shows how, with that equipment, which may evolve in adulthood in different forms, our cognitive activity depends on our creation—surely by learning, as in the Johnny case—of a hierarchized space of categories, from objects themselves to kinds.[40] Our main resource for doing so does not depend on a group of similarities, but on the differences created among the samples (the paradigms) that we face in different situations. The kind set appears out of that situated group of differentiae that allow us to recognize the same object or a new instance of a kind to which it belongs. Though asking for similarity conditions may sound less prone to error in reidentification, Kuhn believes the contrary: differentiae-based recognition is a better adaptive strategy, itself less prone to error.[41]

Two characteristics of this statement are consequential for our examination. First, this scheme shows the way the basic cognitive equipment out of which structured kind sets emerge is naturalistically explained, as Mladenovic notes.[42] This is not new in Kuhn's thinking. His vision of Kant was limited in *Structure* by the kind of psychological explanation based on Piaget's work as it is now based on the works of more recent psychologists (such as Carey, for instance).[43] This newly formed explanation helps to substantiate examples like that of Johnny, on whose basis Kuhn shows how the structured kind set is learned and becomes functional through its use. A model for that kind set, similar to that of a geometric nature that he developed in the 1960s, would be a good analogy for that cognitive equipment based on recognition by differentiae. We can see that clearly in the following passage from Chap. 4:

> This account of recognition does, however, depend critically on the assertion that, in a space of suitable chosen differentiae, different presentations of the same individual or [the same] kind form a cluster at a distance from those formed by presentations of other individuals or other kinds.[44]

It is not difficult to see that the geometrical *n*-cube model would fit, with its distances between clusters of natural families, this abstractly presented space, and even its echoes behind Kuhn's assertion. Just as the former model fitted the Johnny case, so does this newer one.

Second, for him, the recognition-by-differentiae strategy is better in adaptive terms. Just as early in his career, the perspective for constructing a model for cognition, and even for meaning, is born from a behavioral perspective of the functions

---

[38] See *The Last Writings*, 200–204.

[39] See *The Last Writings*, Chap. 4, Section II.

[40] As he says elsewhere in the same book, not every type of kind has objects as the most basic of its categories. Material kinds, for instance, the result of an analysis of individuals in terms, for instance, of their composition—a table in the kind of wooden material of which it is made, and so on—have kinds, not objects, at that basic point. See *The Last Writings*, 253–256.

[41] See *The Last Writings*, 210–214.

[42] On Kuhn and naturalism, see Mladenovic (2022, xxvii).

[43] See *The Last Writings*, 200–201.

[44] *The Last Writings*, 213. Words within square brackets are Mladenovic's.

cognition and language must play for human beings. This perspective also belongs to the biological view that he adopts in this chapter, in which the overall naturalization of Kuhn's views achieves maximum expression. It is not only the evolutionary features of *Structure* and later improvements that count for that naturalization, the foundations for cognition and language *in behavior* (and in society, lest we forget) are also an important part of that coherent set of steps of that point of view.[45] I shall return to this below.

A problem in *Plurality* is nothing to fault Kuhn for. He did not get to write those chapters in which he planned to discuss some elements that, traditionally, had been the main concern of philosophers of science when it came to discussing *Structure* and further writings. Is theory-choice based on good reasons, or does the Kuhnian depiction of scientific revolutions involve the mediation of drives of a less rational nature? That is an often-formulated question that Kuhn had tried to answer to the best of his abilities since *Structure*. Another issue is familiar. The nature of truth in Kuhn's view does not lead to a correspondence theory of truth, but his final alternative is not completely clear. Besides that, and returning to another classic epistemological theme, his developmental view of philosophy of science says that belief-justification takes place within a paradigm or lexicon, and it is sort of circular, without recourse to any extra-phenomenal-world background that makes belief warranted non-relatively. On these premises, Kuhn's perspective on scientific beliefs and hypotheses is that they are epistemically and alethically relative. Their justification and truth, and so the very possibility for them to be elements in our knowledge, are context-relative, where the framework for the evaluation of any of them is given by the paradigm (from the point of view of *Structure*), or, more recently, by the lexicon or the structured kind set to which the categories of the phenomenal world are relative.

In the book, however, these answers are not available, and—as Mladenovic, its editor, says—we must turn to previous texts, particularly the Shearman Lectures (and other papers as well), to find them.[46] As regards the problem of relativizing knowledge, Kuhn would not have had problems in showing that the epistemological game is played within a context that is marked by the hegemony of a paradigm (or a set of them), with a world stabilized around a structured kind set and on the basis of a common lexicon. I am not sure whether he would have held on to the knowledge/belief distinction on the basis of his Wittgenstein-like reconstruction or not, but I am fairly sure that the basic model would be there, because it plays a role that goes beyond the strict epistemic evaluation of beliefs, as classically studied by epistemologists. For him, the parts of knowledge that are taken as certain are those that are gained through training in the group, and that way of acquiring it—which is coherent with his

[45] It is not hard to compare this position with proposals of classic American philosophers like John Dewey, as Mladenovic aptly points out (see Mladenovic 2017, Part III). Like Mladenovic (2017, 181–184), I consider the connections between Kuhn's and Dewey's works around this set of themes pretty illuminating both in interpreting the former's theory and in finding new ways of developing it. For more details on this, see Mayoral (2022); other connections with pragmatism are explored in Mayoral (2017b).

[46] See Mladenovic (2022, xvii–xix, xli–xliii). For her reconstructions of the rest of the book, see *The Last Writings*, 97 ff.

classic model of learning—provides the individual with credentials both for owning that knowledge and for membership in the group. So, the epistemological activity of searching for guaranteed beliefs remains important for him. We shall not find, he told Rorty, anything like a culture without epistemology.[47] But that activity is significant as *part* of a denser depiction of knowledge that has more layers than the proposition-based representation. It embraces the *social* dimension of knowledge, which is not just a mere complement to the proposition-based representation—which would be the one with truly philosophical significance, as has often been assumed. Accordingly, as knowledge and language are inextricably connected, as he used to say, and they both have limits, knowledge, like language, can only be context-bound.

Kuhn never denied that our contact with nature is formed through an intermediary of a cognitive nature that was termed differently depending on the stage of his thinking: psychological world, behavioral world, perceptual world, or phenomenal world. His vision of these cognitive intermediaries involves two main things: first, they are endorsed (and, so to speak, validated) by practice—and the reference to *behavioral* worlds in 1951 suggests the importance for him of such an assumption—and second, they are formed on the basis of an interaction with nature. We do not form such scenarios on the basis of socially shared beliefs; at least, not *only* on such foundations in any case. Some subgroups of such beliefs are the guides for acquiring features that, relative to the subclass of people to which we belong, give access to a commonly shared kind set, to explain it through the more mature view in this book.[48] But, as Kuhn untiringly showed, this is something we form in interaction with the natural objects that are present to our senses. Although our basic categories are not permanent, as we have just learned, they are there for deploying our behavior in a world that is not completely prepared for it, nor designed by us. So, the intermediary helps us to gain control (practically and cognitively) over the space around us, but it is formed, first and foremost, from that environment over which we must exert control, and, of course, on the basis of individually formed and socially shared sets of beliefs. Only thus are we able to deal with the spontaneous emergence of novelties, on which Kuhn counted, and so that a phenomenal world can become such an *intermediary* that helps us to deal with nature.

So, when he suggests that the real problem for epistemology depends on studying belief-change and its justification—not the timeless justification of beliefs themselves—he sees such beliefs as piecemeal, artificially unconnected parts *only* from the point of view of traditional epistemology. He rather opts for studying them as parts of the functional intermediary that, in his view, must be observed as an integrated whole, with such beliefs as its integrating parts. At the same time, their real function and the problems of justification that they truly present are easier to grasp when observed in their "moments of motion," so to speak; that is, either as they change

---

[47] Kuhn to Rorty, 19 September 1986, TSKP 22.14, 4. See our Chap. 8, esp. Section 8.12.

[48] I apologize for jumping from a phase of Kuhn's thought to another, but the jump is justified by an underlying coherence throughout that philosophy, despite evident differences, and it has explanatory virtues. Evidence of the differences between phases remains, and this book testifies to them. In any case, an opportunely synchronic view is allowed for the purposes of the explanation of a point, as in this case.

with regard to the increasingly better understanding of a phenomenal world, or—in the most interesting case—when one phenomenal world is being replaced by another. What justifies their change in that enriched context tells us the story of the role they play and the reasons for adopting them. The alleged individual justification of each of them alone, though intended to be timeless and eventually leading to truth, ultimately refers us to the context "in motion" that their intermediary provides. On that account, Kuhn may appear to be a sort of "Galilean" thinker (if this rough analogy is allowed), who is interested in details that for a more "Aristotelian" philosopher would not be central—as acceleration was differently considered by Galileo and Aristotle, for example. For Kuhn, it is *change*, in brief, that really matters, not its results. It is what really conveys the essential details of the grounds for holding a given belief in science. He delves into the apparent accidents of the *passage* from one set of beliefs to another rather than in the initial and final positions that they occupy in each stable historical stage of belief. Such apparent accidents are, of course, no longer mere accidents but truly crucial aspects of the proper epistemological evaluation. What happens during the passage from a set of beliefs about the world to a new one truly explains the status of our scientific beliefs: their legitimacy and the duration of their validity.

At this point, the concern with the assumption that the elements of our knowledge are context-relative, and that there is some form of epistemic relativism involved, seems futile. There *is* indeed a form of relativism, but not of the kind that the epistemological tradition considers harmful. It is that tradition that forms a somewhat artificial context in which relativism, thus exposed, *is* harmful. With Kuhn, in contrast, we are not in a position to regret the lack of a permanent background for evaluating every belief by itself as justified or true, because such a situation is a meaningless artificial analysis-context. That situation is artificial and says nothing about the true nature of beliefs—or about the role of *believing* itself. It does not mean that saying whether a belief is justified or not does not play any role in our common discourse and as a basis for our behavior. Once again, epistemology seems linked to our cultures in some essential way. It is just that the relative, context-bound nature of evaluations is inescapable, and there is no harm done by assuming that.

Finally, with regard to truth, Kuhn's position is similar. He says in the Shearman Lectures, and in some other writings from his last half-decade, that the contents whose truth or falseness we claim are context-bound, as is evident from all of we have just learned. The sentences we use to convey our scientific hypotheses are formulated on the basis of our lexicons in use, whose terms—nouns, natural-kind terms, taxonomic terms, singletons, material-kind terms, artificial terms at large, and of course property terms—name individuals, properties, and kinds that are arranged in a specific, stable, and hierarchized form in the taxonomy for our common world. Those contents change from one kind set to another, we have just seen, and their change produces incommensurability. If we wish to figure out by ourselves what true statements we can find in Aristotle's time and for Aristotle and his followers, we must not only learn classical Greek, but must also find out to what categorized phenomenal world that language applies—or, at least, a vicarious form of it. Finding *truths* somewhere else requires that we become bilingual, as Kuhn used to say, but we can do it. Finding *the truth* in the past is as meaningless as it is today, for Kuhn, because predicating the truth of a belief is something we need a kind-set and a lexicon for. As he says in the Shearman Memorial Lectures, asking whether a certain statement by Ptolemy or Aristotle concerning whether the sun is a planet is true or not is an ill-formulated question; it is actually senseless. That statement might be true or false then. An historical inquiry would help us to find that out. About its permanent truth, we can say nothing at all. We lack the proper words and the proper world to apply them.

Kuhn knew that the classic epistemologist would keep questioning whether this means that *the same* proposition that both Ptolemy and we can formulate may therefore change its truth-value. Kuhn, however, firmly believed that propositions do not change their truth value in any way. If a certain proposition *is* true (not that we *believe* it true, but that it *is* true), it remains true. It is just that *that* specific statement cannot be formulated as it is in a different language (and by "different" I mean a language based on a different lexicon and kind set, and so relative to a different phenomenal world). We might be fooled by the grammatical structure of two sentences, apparently similar and easy to translate, that are formulated from two different kind-sets, and thus with different lexicons as well. However, for Kuhn, we must avoid that deception if we pay attention to the *content* they convey. There are, Kuhn recalls, two propositions involved. *Truth*, he says, is not relative. Rather, I would say, *truths are* relative for him. They are lexicon-relative, kind-set-relative. Certainly, for him, it is the *content* itself that is truly relative. *Effability* ("speakability") is language-relative. So, not every content, not every thought we have—which depends on the categories by virtue of which we perceive and talk about the phenomenal world we live in—can be communicated in every language. The *universality* of language, he says, is an illusion. In this position we find the explanation for the apparent change in truth-value that many found so disturbing in Kuhn's perspective.[49]

As to the positive statements concerning a theory of truth, Kuhn was not explicit. He stated the necessity of a kind of discourse in which the evaluation of truth is not absent, just as he defended the role of epistemology. In his Thalheimer Lectures, he opposed Rorty's views on truth and epistemology by making the following statement: "Truth values and the legitimacy of epistemology will not be set aside [...] without banishing language and community as well."[50] And, as he says in the Shearman Lectures, "Human communities, I am suggesting, are discourse communities, and the truth-value game is essential to them."[51] For fulfilling that function, at least a minimum commitment with a non-contradiction law is required, he says, something

[49] See Kuhn (1999, 35); *The Last Writings*, 53–54, 75–76. Note that in the first reference (which is later in time than the second one), he is referring to propositions when explaining his form of relativism, and that he refers to statements (and propositions, too; see 53, fn. 7) in the former. See Mayoral (2017b, 97; 2023b, 75).

[50] Kuhn, *Scientific Development and Lexical Change*, The Thalheimer Lectures, The Johns Hopkins University, 12–19 November 1984, TSKP 23.21, Lecture IV, 120.

[51] Kuhn, *The Presence of Past Science*, The Shearman Memorial Lectures, University College London, TSKP 23.32, Lecture III, 76; *The Last Writings*, 79.

that avoids making contradictory statements.[52] But he did not seem ready, as yet, for more positive views on the kind of theory of truth that should be a substitute for the classical correspondence theory; and alternative views, as Mladenovic points out, were, in his opinion, equally inadequate, except for redundancy theories of truth.[53]

Similarly to the problems which these previous solutions already answered, the oft-repeated question of his critics about the rationality and nature of theory change in Kuhn's theory was not discussed in *Plurality*. As regards theory choice, Kuhn's answers and criticism of the traditional epistemological view in *Structure*, in "Objectivity, Value Judgment, and Theory Choice," and in other recent papers like "Scientific Knowledge as Historical Product" provide a background from which his new replies can be reconstructed.[54] However, we can recall how Kuhn explored the complicated question of theory choice in a revolutionary context both philosophically and historically—*Black-Body Theory* is a good example of the latter. Kuhn seems to have followed this same pathway in *Plurality*, for he says, in closing Chap. 5:

> If one cannot state two competing beliefs (or sets of belief[s]) in the same language, then one cannot compare them directly with observational evidence. [... I]t should suggest that the standard conception of a *choice* between the two on the basis of observational evidence cannot be quite right. Comparison requires simultaneous access to the things being compared, and that is here barred by the no-overlap principle.[55]

The situation is no different from the one usually explored by him. His model for the structured kind sets seems to have reinforced the basis for the kind of answers he had already provided.

Kuhn's views about the problems of truth, relativism, and rationality in his own theory can be obtained from a core doctrine that we have seen in development throughout this entire book. In my view, that is the very center of his lifetime's work: the improvement in detail of his broad perspective on human learning and cognition; his inquiry into their relationship with scientific theorization and research in society; and his continual attempt to figure out how that perspective explains the gaps in understanding on the part of the historically advanced interpreter that tries to approach an episode of change in a scientific discipline. Kuhn was successful in producing good reports of the problems and sketches of the theory, the most famous of which was *Structure*. That book was enough for many readers to become interested in further developments. Yet, the specific theory and vocabulary of *Structure*, especially the appealing central concept of paradigm, made it difficult for many readers to look for (that is, to become interested in) the broader view he was constructing. Even for himself, the particularly useful notion of a paradigm was a major concern

---

[52] Kuhn (1991a, 95, 99; 1992, 115).

[53] See Mladenovic (2022, xlii–xliii); Kuhn (1991a, 99). An interesting study about the concept of truth in the context of Kuhn's evolutionary views and a proposal for a theory of truth in such a case is Kuukkanen (2021).

[54] See Kuhn (1973); "Scientific Knowledge as Historical Product," fall 1981, TSKP 5.42. See *The Last Writings*, 100–101, for Mladenovic's reconstruction of Kuhn's positions concerning theory choice in his projected Chap. 8, which was never written.

[55] *The Last Writings*, 259.

for a long while after *Structure*, until the years in which he wrote the postscript to that book and related papers. Despite that, the broader perspective always remained latent in his mind. For example, when he delayed, time and again, the writing of *Structure*, it was that broader perspective, only discerned in part at that moment, that hindered the book's progress. And when he tried to make sense of paradigms, that broader perspective, developed in parallel through his model of paradigm-based learning, helped to make sense of them. He continued trying to look for it for the rest of his life. In a sense, Kuhn fits the mold of the Conantian strategist and of the canon of junior fellows of the Harvard Society of Fellows. His views indeed looked beyond the point he had found them, in this case in the philosophy of science, and he attempted to transform the basics of his discipline. At the end of his lifetime, he was still fighting with that project.

An interesting aspect of his point of view was the way he developed his own brand of an evolutionary perspective for the philosophy of science. Conversations with Hempel, such as the one seen above in this Epilogue, or with Michael Ghiselin, helped him to observe the open issues of his mixture of a rupturist perspective with an evolutionary one—not unusual in biology after the Punctuated Equilibrium view of Stephen Jay Gould and Niles Eldredge.[56] Kuhn expected to make them compatible. However, as he explained by letter in 1972, the Darwinian analogy to scientific evolution cannot be considered flawless. He wrote:

> I do mean my appeal to evolution to be a good deal more than a metaphor in the sense that I believe scientific knowledge does evolve, that its present nature can be understood only through a knowledge of its past history, and that competition for survival has played throughout an essential role. On the other hand, I do not suppose for a moment that detailed biological theory for the evolution of organisms can simply be moved over into the study of science wholesale.[57]

The idea that scientific evolution does not have a teleological nature is one of the parallelisms that work well, but the tree of disciplines throughout history is peculiar, and the episodes of "reunification of the sciences," out of which new specialties emerge every so often, are not strange—Kuhn mentions these.[58] In any case, for him the tree of disciplines is similar to "a layman's diagram for a biological evolutionary tree"—and note the cautionary reference to the layperson's perspective on the latter.[59] That helps us to see that the Darwinian analogy fits Kuhn's perspective very well without being perfect. Science, as a cultural activity, evolves according to the characteristics of its own kind.[60]

---

[56] See Eldredge and Gould (1972). See also his exchange with Ghiselin (December 1986–February 1987) in TSKP 23.29. Ghiselin mentioned the Eldredge–Gould paper in his last letter to Kuhn (Ghiselin to Kuhn, 20 February 1987, TSKP 23.29).

[57] Kuhn to George Ballester, 9 June 1972, TSKP 4.6.

[58] See Kuhn (1991a, 97).

[59] Kuhn (1991a, 98).

[60] On Kuhn's evolutionary views, see Wray (2011, Part II), Kuukkanen (2012; 2021), and Haufe (2022).

K. Brad Wray's qualification of Kuhn's theory as *evolutionary* and *social* at the same time is particularly appropriate.[61] In a way, it looks redundant as, in the absence of a goal that accounts for the non-Kuhnian perspective of "evolution-toward-what-we-wish-to-know," the explanation of the evolutionary tree for science requires a basic social dynamics such as the one Kuhn provides. But in another sense, it is not, because it need not be prima facie clear which of those qualifications, social and evolutionary, follows from the other, and in Kuhn's case the evolutionary dynamics follows from a social explanation of revolutions and its meaning for the progress of science. The evolutionary dimension of Kuhn's views evolved itself over time. He made it clearer in the mature stage of his thinking that the resulting disciplines that emerge from a revolutionary episode may involve a cultural equivalent of "speciation," not of "mutation," as he expounded in "The Road since *Structure*."[62] This vision is somewhat implicit in views that we observe in the last section of *Structure*, when he states that the only way in which the kind of mature sciences that he studies there cover a growingly higher number of problems is by the branching of disciplines.[63] What's more, in *Structure* he already notes that it involves a fading "communication with other groups, both scientific and lay."[64] Incommensurability thus plays a positive role, which he emphasizes in 1990.[65] This is an aspect of the evolutionary tree of the sciences that the social mechanism of scientific communities explains very well.

His mature lexical model, which accounts for the nature of the link that holds members of scientific groups together—with all the complexity that he added to such groups, especially as regards their relationship to scientific revolutionary change, in the 1960s—gives a detailed account of this social mechanism and thus for this evolutionary view of his theory of scientific change. That mature model makes it easier to understand how a revolutionary process, when understood in detail, produces that kind of evolutionary tree. Moreover, it justifies some turns in the way of constructing a philosophy of science, such as the one that leads Kuhn to defend his own developmental perspective for that field.[66] That mature lexical model turned out to be one based on the notion of a structured kind set (non-incompatible with the former, of course) in *Plurality*, and thus the book was intended to offer the foundations for such a theory. As I have repeated often in this book, that project, though liable to reconstruction on the basis of his many documents and publications, remained a work in progress. This book is intended to provide the reader with more facts in order to make that reconstruction easier. Now, we must bring this biographical reconstruction to its end.

---

[61] See Wray (2011).

[62] Kuhn (1991a, 98). The expression "evolution-toward-what-we-wish-to-know" is from *Structure*, 170

[63] *Structure*, 169.

[64] See Footnote 63.

[65] Kuhn (1991a, 98–99). On this group of themes, see the references in fn. 60, and esp. Haufe (2022).

[66] See Kuhn (1991a, 95–96).

The last half-decade of Kuhn's life was a mixture of happiness, promising and fruitful work in philosophy, and, unfortunately, some fatal health problems. In May 1991, Kuhn, about to be made emeritus at MIT the following August, had made some progress on his ideas since 1987. He was already working on the book. In April 1993 he was still working on it at a moderate pace. Unfortunately, a malignant tumor was diagnosed in August or September 1993, and he underwent radiotherapy treatment during the remaining months. In the meantime, he saw from a distance how Feyerabend's health weakened and how he sadly passed away in February 1994. At that time, Kuhn was recovering from his illness and the prospects of returning to his work on the book seemed to be good, but he had to stop again after cataract surgery. In September, however, he had written the first two chapters in its present form (both were finished on September 19, 1994). Around a year later, he had already written three more chapters: Chaps. 3–4 on February 28, 1995, and Chap. 5 on September 24, 1995.[67] Then, he began Chap. 6, although he had to stop in October because of some breathing problems. He visited Greece in October, where he gave a long and now famous interview (an invaluable source of data and insight into his biography for this book),[68] and then Italy. Yet, in February 1996, he had breathing problems again, which led to the discovery that the cancer had come back, this time with fatal consequences. He did not work on the book again, and passed away on June 17, 1996.[69]

The Plurality of Worlds remained unpublished for 26 years. Few people knew its contents. Now that its pages are publicly available, and that archival resources are often consulted and its contents described, commented on, and sometimes published, a new, deeper idea of the lifelong project that Kuhn constructed step by step is now possible. The unfinished book has shown many interesting ideas about his views, but also two complementary visions. The first is that there is still work to do to see how far Kuhn's developmental perspective can reach, and if it is powerful enough to turn into a much broader tool for the philosophical interpretation of science, even more than it already is. The second vision refers to the importance of an inquiry into the big project that he developed during his entire career from its initial steps. Aspects of that book, such as in Chap. 4, show the intimate connection with ideas that can be observed 45 years before in texts such as the Lowell Lectures, for instance. It does not mean that Kuhn always worked on one and the same theory. His global theory evolved and the details that changed in time were important, as we have already seen in this book. However, how he worked on different aspects of his vision—from the more historically-laden vision of revolutionary change in science to the social perspective of scientific research in *Structure*, and then to the underlying explanatory model of the training and working of the scientific mind and language—shows the

[67] See *The Last Writings*, 131, 166, 190, 223, 259.

[68] See Interview 1995, 255.

[69] This paragraph is based on the following letters: Kuhn to Hoyningen-Huene, 13 May 1991, 14 April 1993, 11 September 1993, 15 December 1993, 26 May 1994, 25 June 1995, 8 August 1995, 28 November 1995, and 10 April 1996, and Hoyningen-Huene to Kuhn, 3 December 1993, 27 April 1994, and 14 June 1995. All of them in PHA.

very different areas in which his global theory still can and must be developed. His biographical itinerary not only shows an inquiring, critical mind at work and a global view that was studied in detail in different phases, but also the virtues of naturalism and pluralism as a good approach to method in the philosophy of science.

Credits

A phrase on T. S. Kuhn by P. W. Anderson (Interview with Alexei Kojevnikov, 30 March 1999) is here reproduced in Chap. 1 with permission of the Niels Bohr Library & Archives, American Institute of Physics (College Park, MD, USA).

An extract of a letter by W. V. Quine to W. Dennes is here reproduced in Chap. 4 with permission of Douglas B. Quine, PhD.

A line from a letter from K. R. Popper to T. S. Kuhn is here reproduced in Chap. 6 with kind permission of the Karl Popper Archives, University of Klagenfurt, Austria (copyright © 2008 University of Klagenfurt/Karl Popper Archives. All rights reserved).

Figure 8.4 is here reproduced from Juan V. Mayoral (2024b): "Kuhn's 'Does Knowledge "Grow"?': Second Thoughts on the Nature of Epistemic Progress," in *Rethinking Thomas Kuhn's Legacy*, Yafeng Shan (ed.). Boston Studies in the Philosophy and History of Science, vol. 345. Cham: Springer, 301

J. V. Mayoral, *Thomas Kuhn*, Springer Biographies,
https://doi.org/10.1007/978-3-031-86512-1

References

Kuhn's Published Writings[1]

1945. *A Précis of General Education in a Free Society, with Commentaries by William Allen Neilson, Claude M. Fuess, George D. Stoddard, William H. Cowley, Augustus N. Hand, and Thomas S. Kuhn.* Cambridge, Mass.: Harvard Student Council, n.d.

Kuhn, T.S., and J.H. Van Vleck. 1950. A Simplified Method of Computing the Cohesive Energies of Monovalent Metals. *Physical Review* 79: 382–388.

1950. An Application of the W.K.B. method to the Cohesive Energy of Monovalent Metals. *Physical Review* 79: 515–519.

1951a. A Convenient General Solution of the Confluent Hypergeometric Equation, Analytic and Numerical Development. *Quarterly of Applied Mathematics* IX: 1–16.

1951b. Newton's '31st Query' and the Degradation of Gold. *Isis* 42: 296–298.

1952a. Robert Boyle and the Structural Chemistry in Seventeenth Century. *Isis* 43: 12–36.

1952b. Reply to Boas. *Isis* 43: 123–124.

1952c. The Independence of Density and Pore-size in Newton's Theory of Matter. *Isis* 43: 364–365.

1953a. Review of A. R. Hall, Ballistics in the Seventeenth Century: A Study in the Relations of Science and War with Reference Principally to England. *Isis* 44: 284–285.

1953b. Review of Joseph F. Scott, The Scientific Work of René Descartes (1596–1650) and Albert G. A. Balz, Descartes and the Modern Mind. *Isis* 44: 285–287.

1953c. Review of Herbert Dingle, The Scientific Adventure: Essays in the History and Philosophy of Science, *Speculum*. 28: 879–880.

1954a. Review of Franklin L. Baumer, *Main Currents of Western Thought: Readings in Western European Intellectual History from the Middle Ages to the Present. Isis* 45: 100.

1954b. Review of Galileo Galilei, *Dialogue on the Great World Systems*, English Translation by Thomas Salusbury, Revision and notes of Giorgio de Santillana, and of Galileo Galilei, *Dialogue Concerning the Two Chief World Systems—Ptolemaic & Copernican*, English Translation by Stillman Drake, Foreword by Albert Einstein. *Science* 119: 546–547.

1955a. Carnot's Version of "Carnot's Cycle." *American Journal of Physics* 23: 91–95.

1955b. La Mer's Version of "Carnot's Cycle." *American Journal of Physics* 23: 387–389.

1958a. Newton's Optical Papers. In *Isaac Newton's Papers and Letters on Natural Philosophy*, ed. I. B. Cohen and R. E. Schofield. Cambridge, Mass.: Harvard University Press.

1958b. The Caloric Theory of Adiabatic Compression. *Isis* 49: 132–140.

[1] In this list, I only include the references that are mentioned in this book.

J. V. Mayoral, *Thomas Kuhn*, Springer Biographies,
https://doi.org/10.1007/978-3-031-86512-1

1959a. The Essential Tension: Tradition and Innovation in Scientific Research. In *The Third (1959) University of Utah Conference on the Identification of Scientific Talent*, ed. C. W. Taylor. Salt Lake City: University of Utah Press. Reprinted in Essential Tension as Essay 9 (q. v.).

1959b. Energy Conservation as an Example of Simultaneous Discovery. In *Critical Problems in the History of Science*, ed. M. Clagett. Madison, Wisconsin: University of Wisconsin Press. Reprinted in Essential Tension as Essay 4 (q. v.).

1959c. Review of Lynn Thorndike, *A History of Magic and Experimental Science*, vols. VII-VIII. *Manuscripta* 3: 53–57.

1960. Engineering Precedent for the Work of Sadi Carnot. *Actes du IXe Congrès International d'Historie des Sciences*, vol. 1, Barcelona: Hermann & Cie. Also published in Archives internationales d'histoire des sciences. 39 (1960): 251–255.

1961a. Sadi Carnot and the Cagnard Engine. *Isis* 52: 567–574.

1961b. The Function of Measurement in Modern Physical Science. *Isis* 52: 161–193. Reprinted in Essential Tension as Essay 8 (q. v.).

1962. The Historical Structure of Scientific Discovery. *Science* 136: 760–764. Reprinted in *Essential Tension* as Essay 7 (q. v.).

1963a. The Function of Dogma in Scientific Research. In *Scientific Change*, ed. A.C. Crombie. New York: Basic Books.

1963b. Discussion. In *Scientific Change*, ed. A.C. Crombie. New York: Basic Books.

1964. A Function for Thought Experiments. in *L'aventure de la science: Mélanges Alexandre Koyré*, vol. 2. Paris Hermann. Reprinted in *Essential Tension* as Essay 10 (q. v.).

1967. The Turn to Recent Science. *Isis* 58: 409–419.

Kuhn, John L. Heilbron, Paul Forman and Lini Allen. 1967. *Sources for the History of Quantum Physics: An Inventory and Report*. Philadelphia: The American Philosophical Society.

1968. The History of Science. In *International Encyclopedia of the Social Sciences*, vol. 14, ed. D. L. Sills. New York: Macmillan. Reprinted in *Essential Tension* as Essay 5 (q. v.).

Heilbron, John L., and Kuhn. 1969. The Genesis of the Bohr Atom. *Historical Studies in the Physical Sciences* 1: 211–290.

1970a. Logic of Discovery or Psychology of Research?, In *Criticism and the Growth of Knowledge*, ed. I. Lakatos and A. Musgrave. Cambridge: Cambridge University Press (q. v.). Reprinted in Essential Tension as Essay 11.

1970b. Reflections on My Critics. In *Criticism and the Growth of Knowledge*, ed. I. Lakatos and A. Musgrave. Cambridge: Cambridge University Press. Reprinted in The Road as Chapter 6 (q. v.).

1971a. Notes on Lakatos. In *PSA 1970: In Memory of Rudolf Carnap. Proceedings of the 1970 Biennial Meeting of the Philosophy of Science Association*, ed. R. C. Buck and R. S. Cohen. Boston Studies in the Philosophy of Science, vol. 8. Dordrecht: D. Reidel Publishing Company.

1971b. The Relations between History and the History of Science. *Daedalus* 100: 271–304. Reprinted in Essential Tension, Essay 6 (q. v.).

1973. Objectivity, Value Judgment, and Theory Choice. In *Essential Tension*, Essay 13.

1974. Second Thoughts on Paradigms. In *The Structure of Scientific Theories*, ed. F. Suppe. Urbana: University of Illinois Press. Reprinted in *Essential Tension*, Essay 12 (q. v.).

1975a. Tradition mathématique et tradition expérimentale dans le développement de la physique. *Annales* 30: 975–998.

1975b. The Quantum Theory of Specific Heats: A Problem in Professional Recognition. In *Proceedings of the XIV International Congress for the History of Science 1974*, vol. 1. Tokyo: Science Council of Japan.

1976a. Mathematical versus Experimental Traditions in the Development of Physical Science. *Journal of Interdisciplinary History* 7: 1–31. Reprinted in *Essential Tension*, Essay 3 (q. v.).

1976b. Theory Change as Structure Change: Comments on the Sneed Formalism. *Erkenntnis* 10: 179–199. Reprinted in *The Road*, Chapter 7 (q. v.).

1976c. The Relations between History and Philosophy of Science. In *Essential Tension*, Essay 1.

1978. Newton's Optical Papers. In *Isaac Newton's Papers and Letters on Natural Philosophy*, 2nd ed, ed. I. B. Cohen and R. E. Schofield. Cambridge, Mass.: Harvard University Press.
1979a. Alexandre Koyré & the History of Science: On an Intellectual Revolution. *Encounter* 34: 67–69.
1979b. Metaphor in Science. In *Metaphor and Thought*, ed. A. Ortony. Cambridge: Cambridge University Press. Reprinted in *The Road*, Chapter 9 (q. v.).
1979c. Foreword. In L. Fleck, *Genesis and Development of a Scientific Fact*. R. K. Merton and T. J. Trenn (eds.). English translation by Trenn and F. Bradley. Chicago: The University of Chicago Press.
1980a. Einstein's Critique of Planck. In *Some Strangeness in the Proportion: A Centennial Symposium to Celebrate the Achievements of Albert Einstein*, ed. H. Woolf. Reading, Mass.: Addison-Wesley.
1980b. The Halt and the Blind: Philosophy and History of Science. *The British Journal for the Philosophy of Science* 31: 181–192.
1983a. Commensurability, Comparability, Communicability. In *PSA 1982: Proceedings of the 1982 Biennial Meeting of the Philosophy of Science Association*, vol. 2, ed. P. D. Asquith and T Nickles. East Lansing, Michigan: Philosophy of Science Association. Reprinted (with a postscript) in *The Road*, Chapter 2 (q. v.).
1983b. Reflections on Receiving the John Desmond Bernal Award. *4S Review* 1: 26–30.
1984. Revisiting Planck. *Historical Studies in the Physical Sciences* 14: 231–242. Reprinted as "Afterword: Revisiting Planck," in *Black-Body Theory and the Quantum Discontinuity, 1894–1912*, 2nd ed. Chicago: The University of Chicago Press, 1987 (q. v.).
1986. The Histories of Science: Diverse Worlds for Diverse Audiences. *Academe* 72: 29–33.
1987. What Are Scientific Revolutions? In *The Probabilistic Revolution, Vol. I: Ideas in History*, ed. L. Krüger, L. J. Daston and M. Heidelberger. Cambridge, Mass.: The MIT Press. Reprinted in *The Road*, as Chapter 1 (q. v.).
1989. Possible Worlds in the History of Science. In *Possible Worlds in Humanities, Arts and Sciences*, ed. S. Allén. Berlin: Walter de Gruyter. Reprinted in *The Road* as Chapter 3 (q. v.).
1990. Dubbing and Redubbing: The Vulnerability of Rigid Designation. In *Scientific Theories*, ed. C. W. Savage. Minnesota Studies in the Philosophy of Science, vol. XIV. Minneapolis: University of Minnesota Press.
1991a. The Road since *Structure*. In *PSA 1990: Proceedings of the Biennial Meeting of the Philosophy of Science Association*, ed. A. Fine, M. Forbes, and L. Wessels. East Lansing, Michigan: Philosophy of Science Association. Reprinted in *The Road*, Chapter 4 (q. v.)
1991b. The Natural and the Human Sciences. In *The Interpretive Turn: Philosophy, Science, Culture*, ed. D.R Hiley, J.F. Bohman, and R. Shusterman. Ithaca: Cornell University Press. Reprinted in *The Road*, Chaper 10 (q. v.).
1992. *The Trouble with the Historical Philosophy of Science*, Robert and Maurine Rothschild Distinguished Lecture. Cambridge, Mass.: Department of History of Science, Harvard University. Reprinted in *The Road* as Chapter 5 (q. v.).
1993. Afterwords. In *World Changes: Thomas Kuhn and the Nature of Science*, ed. P. Horwich. Cambridge, Mass.: The MIT Press. Reprinted in *The Road* as Chapter 11 (q. v.).
1999. Remarks on Incommensurability and Translation. In *Incommensurability and Translation: Kuhnian Perspectives on Scientific Communication and Theory Change*, ed. R. Rossini Favretti, G. Sandri and R. Scazzieri. Cheltenham: Edward Elgar.
2014. Rhetoric and Liberation. *Poroi* 10: 1–10.
2024. Appendix: Transcription of Thomas S. Kuhn's 'Does Knowledge "Grow"? Edited by J.V. Mayoral. In *Rethinking Thomas Kuhn's Legacy*, ed. Y. Shan. Boston Studies in the Philosophy and History of Science, vol. 345. Cham: Springer.

Other References

Abbott, Andrew. 2016. *Structure* as Cited, *Structure* as Read. In *Kuhn's Structure of Scientific Revolutions at Fifty*, ed. Robert J. Richards and Lorraine Daston. Chicago: The University of Chicago Press.

Andersen, Hanne. 2001. *On Kuhn*. Belmont, Cal.: Wadsworth.

Andersen, Hanne, Peter Barker, and Xiang Chen. 2006. *The Cognitive Structure of Scientific Revolutions*. Cambridge: Cambridge University Press.

Anderson, Philip. 1999. Interview of Philip W. Anderson by Alexei Kojevnikov. 30 March 1999. College Park, MD USA: Niels Bohr Library & Archives, American Institute of Physics. Available at: www.aip.org/history-programs/niels-bohr-library/oral-histories/23362-1.

Andresen, J. 1999. Crisis and Kuhn. *Isis* 90: S43–S67.

[Anonymous]. 1948. The Society of Fellows: The Story of Fifteen Years. *Harvard Alumni Bulletin* 50: 593–599.

Austin, John L. 1946. Other Minds. *Proceedings of the Aristotelian Society* Supplementary Volume, 20: 148–187. Reprinted in his *Philosophical Papers*, 3rd ed., ed. J.O. Urmson and G.J. Warnock. Oxford: Clarendon, 1979 (q. v.).

Austin, John L. 1956–1957. A Plea for Excuses. *Proceedings of the Aristotelian Society* NS 57: 1–30. Reprinted in his *Philosophical Papers*, 3rd ed., ed. J.O. Urmson and G.J. Warnock. Oxford: Clarendon, 1979 (q. v.).

Axtell, James. 2006. *The Making of Princeton University: From Woodrow Wilson to the Present*. Princeton: Princeton University Press.

Ayer, Alfred J. 1946. *Language, Truth and Logic*, 2nd ed. London: Victor Gollancz Ltd., 1960 (orig. ed. 1936).

Barber, Bernard. 1970. Introduction. In *L. J. Henderson on the Social System: Selected Writings*, ed. B. Barber. Chicago: The University of Chicago Press.

Barker, Peter. 2021. The Copernican Revolution since Kuhn. In *Interpreting Kuhn: Critical Essays*, ed. K. Brad Wray. Cambridge: Cambridge University Press.

Barnes, S.B., and R.G.A. Dolby. 1970. The Scientific Ethos: A Deviant Viewpoint. *Archives Européennes De Sociologie* 11: 3–25.

Bartlett, Paul D. 1983. James Bryant Conant, March 26, 1893–February 11, 1978. National Academy of Sciences. *Biographical Memoirs* 54: 90–125.

Beller, Mara. 1999. *Quantum Dialogue: The Making of a Revolution*. Chicago: The University of Chicago Press.

Beltrán, Antonio. 1995. *Revolución científica, Renacimiento e historia de la ciencia*. Madrid: Siglo XXI.

Bethell, J.T. 1998. *Harvard Observed: An Illustrated History of the University in the Twentieth Century*. Cambridge, Mass.: Harvard University Press.

Bird, Alexander. 2000. *Thomas Kuhn*. Chesham: Acumen.

Bird, Alexander. 2002. Kuhn's Wrong Turning. *Studies in History and Philosophy of Science* 33: 443–463.

Black, Max. 1946. *Critical Thinking: An Introduction to Logic and Scientific Method*. New York: Prentice-Hall.

Boas Hall, Marie. 1952. Newton and the Theory of Chemical Solution. *Isis* 43: 123.

Boas Hall, Marie. 1999. Recollections of a History of Science Guinea Pig. *Isis* 90: S68–S83.

Bokulich, Alisa, and Federica Bocchi. 2024. Kuhn's "Fifth Law of Thermodynamics": Measurement, Data, and Anomalies. In *Kuhn's The Structure of Scientific Revolutions at 60*, ed. K.B. Wray. Cambridge: Cambridge University Press.

Bonner, T.N. 2002. *Iconoclast: Abraham Flexner and A Life in Learning*. Baltimore: The Johns Hopkins University Press.

Boring, Edwin G. 1963. Eponym as Placebo. In *History, Psychology, and Science: Selected Papers*, ed. R.I. Watson and D.T. Campbell. New York: John Wiley and Sons.

Born, Max. 1944. *Experiment and Theory in Physics*. Cambridge: Cambridge University Press.

Borradori, Giovanna. 1994. Paradigms of Scientific Revolution: Thomas S. Kuhn (an interview with Kuhn by G. Borradori). In *The American Philosopher*, English Translation by R. Crocitto. Chicago: The University of Chicago Press.

Bortolini, Matteo. 2011. The "Bellah Affair" at Princeton: Scholarly Excellence and Academic Freedom in America in the 1970s. *The American Sociologist* 42: 3–33.

Boyer, Carl B. 1959. Commentary on the Papers of Thomas S. Kuhn and I. Bernard Cohen. In *Critical Problems in the History of Science*, ed. M. Clagett. Madison, Wisconsin: The University of Wisconsin Press.

Braithwaite, Richard B. 1953. *Scientific Explanation.* Cambridge: Cambridge University Press.

Bridgman, Percy W. 1922. *Dimensional Analysis.* New Haven (Conn.): Yale University Press.

Bridgman, Percy W. 1927. *The Logic of Modern Physics.* Nueva York: Macmillan.

Bridgman, Percy W. 1938. Operational Analysis. *Philosophy of Science* 5: 114–131.

Bridgman, Percy W. 1945. The Prospect for Intelligence. *The Yale Review* 54: 444–461.

Bridgman, Percy W. 1947. Science and Freedom: Reflections of a Physicist. *Isis* 37: 128–131.

Brinton, C. Crane. 1952. *The Anatomy of Revolution*, 1st ed. New York: W. W. Norton & Co., 1938. Rev. ed., New York: Prentice-Hall, 1952.

Brock, William H. 2000. *The Chemical Tree: A History of Chemistry.* New York: W. W. Norton & Co.

Brown, Louis. 1999. *A Radar History of World War II: Technical and Military Imperatives.* Bristol: Institute of Physics Publishing.

Browning, Robert. 1864. Rabbi Ben Ezra. In *Dramatis Personae*, London: Chapman and Hall. Reprinted in *Poetical Works, 1833–1864*, I. Jack (ed.), Oxford: Oxford University Press, 1970 (q. v.).

Bruner, Jerome S., and Leo Postman. 1949. On the Perception of Incongruity: A Paradigm. *Journal of Personality* 18: 206–223.

Brush, Stephen G. 2000. Thomas Kuhn as a Historian of Science. *Science and Education* 9: 39–58.

Buchwald, Jed Z. 1992. Kinds and the Wave Theory of Light. *Studies in History and Philosophy of Science* 23: 39–74.

Buchwald, Jed Z. 2010. A Reminiscence of Thomas Kuhn. *Perspectives on Science* 18: 279–283.

Buderi, Robert. 1996. *The Invention that Changed the World: How a Small Group of Radar Pioneers Won the Second World War and Launched a Technological Revolution.* New York: Simon & Schuster.

Butterfield, Herbert. 1949. *The Origins of Modern Science, 1300–1800.* London: G. Bell and Sons.

Butterfield, Herbert. 1965. *The Whig Interpretation of History.* London: W. W. Norton.

Campbell, K.M. 1989. The Hessian Hills School, 1930–1941: Social Reconstructionism in Practice. *Journal of the Midwest History of Education Society* 17: 78–86.

Carey, Susan. 2022. Becoming a Cognitive Scientist. *Annual Review of Developmental Psychology* 4: 1–19.

Carnap, Rudolf. 1928a. *Der logische Aufbau der Welt.* Berlin: Weltkreis.

Carnap, Rudolf. 1928b. *Scheinprobleme in der philosophie.* Berlin: Weltkreis.

Carnap, Rudolf. 1934. *Logische Syntax der Sprache.* Vienna: Springer.

Carnap, Rudolf. 1950. Empiricism, Semantics, and Ontology. *Revue Internationale De Philosophie* 4: 20–40.

Carnap, Rudolf. 1963. Intellectual Autobiography. In *The Philosophy of Rudolf Carnap*, ed. P.A. Schilpp. The Library of Living Philosophers, vol. XI. La Salle, Illinois: Open Court.

Carnap, Rudolf. 2003. *The Logical Structure of the World and Pseudoproblems in Philosophy.* English Translation by R.A. George. Chicago: Open Court.

Carus, André W. 2007. *Carnap and Twentieth-Century Thought: Explication as Enlightenment.* Cambridge: Cambridge University Press.

Cassidy, David C. 2011. *A Short History of Physics in the American Century.* Cambridge, Mass.: Harvard University Press.

Cavell, Stanley. 2010. *Little Did I Know: Excerpts from Memory.* Stanford: Stanford University Press.

Cedarbaum, Daniel G. 1983. Paradigms. *Studies in History and Philosophy of Science* 14: 173–213.

Cohen, I. Bernard. 1974. History and the Philosopher of Science. In *The Structure of Scientific Theories*, ed. F. Suppe. Urbana: University of Illinois Press.

Cohen, I. Bernard. 1984. A Harvard Education. *Isis* 75: 13–20.

Cohen, I. Bernard. 1999. *Howard Aiken: Portrait of a Computer Pioneer*. Cambridge, Mass.: The MIT Press.

Cohen, I. Bernard., and Robert E. Schofield, eds. 1978. *Isaac Newton's Papers and Letters on Natural Philosophy*. Cambridge, Mass: Harvard University Press.

Cohen, Morris R., and Ernest Nagel. 1934. *An Introduction to Logic and Scientific Method*. New York: Harcourt Brace and Co.

Cohen-Cole, Jamie. 2014. *The Open Mind: Cold War Politics and the Sciences of Human Nature*. Chicago: The University of Chicago Press.

Collodel, Matteo. 2014. A Note on Kuhn and Feyerabend at Berkeley (1956–1964). Unpublished draft (22 February 2014). Available at: https://www.academia.edu/3683558/A_Note_on_Kuhn_and_Feyerabend_at_Berkeley_1956_1964_6th_version_February_2014_.

Collodel, Matteo, and Eric Oberheim. 2024. Introduction. In *Feyerabend's Formative Years. Volume 2. Feyerabend on Logical Empiricism, Bohm & Kuhn: Correspondence and Unpublished Paper*, ed. M. Collodel and E. Oberheim. Vienna Circle Institute Library, 10. Cham: Springer.

Collodel, Matteo, and Eric Oberheim, eds. 2024. *Feyerabend's Formative Years. Volume 2. Feyerabend on Logical Empiricism, Bohm & Kuhn: Correspondence and Unpublished Paper*. Vienna Circle Institute Library, 10. Cham: Springer.

Committee on the Objectives of a General Education in a Free Society. 1945. *General Education in a Free Society: Report of the Harvard Committee*, Introduction J: B. Conant, Cambridge, Mass.: Harvard University Press.

Conant, James B. 1944. Science and the National Welfare. *The Journal of Higher Education* XV: 399–406, 454.

Conant, James B. 1947a. *On Understanding Science: An Historical Approach*. New Haven, Conn: Yale University Press.

Conant, James B. 1947b. The University and the State. *The Journal of Higher Education* XVIII: 281–286.

Conant, James B. 1948a. Introduction. In *Harvard Case Histories in Experimental Science*, vol. 1, ed. James B. Conant and Leonard K. Nash. Cambridge, Mass.: Harvard University Press.

Conant, James B. 1948b. *Robert Boyle's Experiments in Pneumatics*. Cambridge, Mass.: Harvard University Press. Reprinted in J.B. Conant and L.K. Nash (eds.), *Harvard Case Histories in Experimental Science*, 2 vols. Cambridge, Mass.: Harvard University Press, 1948, 1957. Vol. 1, Case 1.

Conant, James B. 1948c. *The Overthrow of the Phlogiston Theory: The Chemical Revolution of 1775–1789*. Cambridge, Mass.: Harvard University Press. Reprinted in J.B. Conant and L.K. Nash (eds.), *Harvard Case Histories in Experimental Science*, 2 vols. Cambridge, Mass.: Harvard University Press, 1948, 1957. Vol. 1, Case 2 (q. v.).

Conant, James B. 1952. Foreword. In *General Education in Science*, ed. I.B. Cohen and F.G. Watson. Cambridge, Mass.: Harvard University Press.

Conant, James B. 1957. George Sarton and Harvard University. *Isis* 48: 301–305.

Conant, James B. 1970. *My Several Lives: Memoirs of a Social Inventor*. New York: Harper and Row.

Conant, James B. 1977. Theodore William Richards, January 31, 1868–April 2, 1928. National Academy of Sciences. *Biographical Memoirs* 44: 251–286.

Conant, James B., and Newton H. Black. 1920. *Practical Chemistry*. New York: Macmillan (rev. ed. 1929).

Conant, James B., and Newton H. Black. 1937. *New Practical Chemistry*. New York: Macmillan (rev. ed. 1946).

Conant, Jennet. 2017. *Man of the Hour: James B. Conant, Warrior Scientist*. New York: Simon and Schuster.

Condon, Edward U. 1938. Mathematical Models in Modern Physics. *Journal of the Franklin Institute* 225: 255–261.
Cooley, John C. 1942. *A Primer of Formal Logic*. New York: Macmillan.
Cosslett, Tess, ed. 1984. *Science and Religion in the Nineteenth Century*. Cambridge: Cambridge University Press.
Crane, Tim. 1992. The Non-conceptual Content of Experience. In *Aspects of Psychologism*. Cambridge, Mass.: Harvard University Press, 2014.
Creath, Richard (ed.). 1990. *Dear Carnap, Dear Van: The Quine-Carnap Correspondence and Related Work*. Berkeley, Cal.: University of California Press.
Cremin, L.A. 1961. *The Transformation of the School: Progressivism in American Education, 1876–1957*. Nueva York: Alfred A. Knopf.
Croce, Paul J. 1995. *Science and Religion in the Era of William James*. Chapel Hill: The University of North Carolina Press.
Croly, H. 1909. *The Promise of American Life*. prólogo de Michael McGerr, Boston: Northeastern University Press, 1989.
Crombie, Alistair C., ed. 1963. *Scientific Change*. New York: Basic Books.
Cross, Stephen J., and William R. Albury. 1987. Walter B. Cannon, L. J. Henderson, and the Organic Analogy. *Osiris* 2nd Series 3: 165–192.
de Laplace, Pierre-Simon. 1814. *Essai philosophique sur les probabilités*. Paris: Mme. Ve. Courcier.
Demopoulos, William. 2013. Three Views of Theoretical Knowledge. In *Logicism and Its Philosophical Legacy*. Cambridge: Cambridge University Press.
Devlin, William J. 2021. Kuhn and the Varieties of Incommensurability. In *Interpreting Kuhn: Critical Essays*, ed. K. B. Wray. Cambridge: Cambridge University Press.
Dewey, John. 1899. *The School and Society*. Chicago: The University of Chicago Press. Reprinted in *John Dewey: The Middle Works, 1899–1924*, vol. 1, ed. J.A. Boydston, Introduction by J. R. Burnett. Carbondale: Southern Illinois University Press, 1976.
Dewey, John. 1916. *Democracy and Education*. New York: Macmillan. Reprinted in *John Dewey: The Middle Works, 1899–1924*, vol. 9, ed. J. A. Boydston. Introduction by Sidney Hook, Carbondale: Southern Illinois University Press, 1980 (q. v.).
Dewey, John. 1922. *Human Nature and Conduct*. New York: Henry Holt and Co., 1922. Reprinted in *John Dewey: The Middle Works, 1899–1924*, vol. 14, ed. J.A. Boydston, Introduction by M. G. Murphey, Carbondale: Southern Illinois University Press, 2008 (q. v.).
Dewey, John. 1930. From Absolutism to Experimentalism, In *Contemporary American Philosophy: Personal Statements*, New York: Macmillan. Reprinted in *John Dewey: The Later Works, 1925–1953*, vol. 5, G.P. Adams, W.P. Montague and J.A. Boydston, Introduction by P. Kurtz, Carbondale: Southern Illinois University Press, 1984 (q. v.).
Dewey, John. 1931. *Philosophy and Civilization*. Nueva York: Minton, Balch, and Co.
Dewey, John. 1932. Review of *Universities*, by A. Flexner. *International Journal of Ethics* 42: 331–332.
Dewey, John. 1938. *Logic: The Theory of Inquiry*. New York: Henry Holt and Co. Reprinted in *John Dewey: The Later Works, 1925–1953*, vol. 12, ed. J.A. Boydston. Introduction by Ernest Nagel, Carbondale: Southern Illinois University Press, 2008.
Dewey, John, and Arthur Bentley. 1949. *Knowing and the Known*. Boston: Beacon Press, 1949. Reprinted in *John Dewey: The Later Works, 1949–1952*, vol. 16, ed. J.A. Boydston. Introduction by T. Z. Lavine, Carbondale: Southern Illinois University Press, 1989.
Diggins, John P. 1998. Max Eastman. In *A Companion to American Thought*, ed. R.W. Fox and J.T. Kloppenberg. Oxford: Blackwell (orig. 1995).
Driver-Linn, Erin. 2003. Where Is Psychology Going? Structural Fault Lines Revealed by Psychologists' Use of Kuhn. *American Psychologist* 58: 269–278.
Earman, John. 1993. Carnap, Kuhn, and the Philosophy of Scientific Methodology. In *World Changes: Thomas Kuhn and the Nature of Science*, ed. P. Horwich. Cambridge, Mass.: The MIT Press.
Eaton, Ralph M. 1931. *General Logic*. New York: Charles Scribner's Sons.

Eldredge, Niles, and Stephen Jay Gould. 1972. Punctuated Equilibria: An Alternative to Phyletic Gradualism. In *Models in Paleobiology*, ed. T.J.M. Schopf. San Francisco: Freeman, Cooper.
Finkelstein, Louis. 1947. Sol M. Stroock. *Publications of the American Jewish Historical Society* 37: 479–480.
Firth, Roderick. 1949. Sense-data and the Percept Theory, Part I. *Mind* 58: 434–465.
Firth, Roderick. 1950. Sense-data and the Percept Theory" Part II. *Mind* 59: 35–56.
Fleck, Ludwik. 1935. *Entstehung und Entwicklung einer wissenschaftlichen Tatsache: Einführung in die Lehre vom Denkstill und Denkkollektiv*. Basel, Switzerland: Benno Schwabe & Co.
Fleck, Ludwik. 1979. *Genesis and Development of a Scientific Fact*, ed. R.K. Merton and T.J. Trenn, Trans. T. J. Trenn and F. Bradley, Foreword by T. S. Kuhn, Chicago: The University of Chicago Press.
Flexner, A. 1916. A Modern School. *American Review of Reviews* 53: 465–474.
Flexner, A. 1930. *Universities: American, English, German*. Oxford: Oxford University Press.
Flexner, A. 1960. *An Autobiography*. Introduction by A. Nevins, Nueva York: Simon & Schuster.
Forman, Paul. 1971. Weimar Culture, Causality, and Quantum Theory, 1918–1927: Adaptation by German Physicists and Mathematicians to a Hostile Intellectual Environment. *Historical Studies in the Physical Sciences* 3: 1–115.
Forrester, J. 2007. On Kuhn's Case: Psychoanalysis and the Paradigm. *Critical Inquiry* 33: 782–819.
Friedman, Michael. 1999. *Reconsidering Logical Positivism*. Cambridge: Cambridge University Press.
Friedman, Michael. 2003. Kuhn and Logical Empiricism. In *Thomas Kuhn*, ed. T. Nickles. Cambridge: Cambridge University Press.
Frisch, Mathias. 2014. *Causal Reasoning in Physics*. Cambridge: Cambridge University Press.
Fuller, Steve W. 2000. *Thomas Kuhn: A Philosophical History for Our Times*. Chicago: The University of Chicago Press.
Fuller, Steve W. 2002. *Social Epistemology*, 2nd ed. Bloomington, Indiana: Indiana University Press, (orig. ed., 1988).
Fuller, Steve W. 2006. The Philosophical Buck Stops Here. *Philosophy of the Social Sciences* 36: 355–366.
Galison, Peter. 1981. Kuhn and the Quantum Controversy. *The British Journal for the Philosophy of Science* 32: 71–85.
Galison, Peter. 1995. Context and Constraints. In *Scientific Practice: Theories and Stories of Doing Physics*, ed. J. Z. Buchwald. Chicago: The University of Chicago Press.
Galison, Peter. 1997. *Image and Logic: A Material Culture of Microphysics*. Chicago: The University of Chicago Press.
Galison, Peter. 2016. Practice All the Way Down. In *Kuhn's Structure of Scientific Revolutions at Fifty*, ed. R.J. Richards and L. Daston. Chicago: The University of Chicago Press.
Gattei, Stefano. 2008. *Thomas Kuhn's "Linguistic Turn" and the Legacy of Logical Empiricism: Incommensurability, Rationality and the Search for Truth*. Aldershot: Ashgate.
Gibson, Roger F. 1996. Quine, Wittgenstein and Holism. In *Wittgenstein and Quine*, ed. R.L. Arrington and H.J. Glock. London: Routledge.
Gillispie, Charles C. 1962. The Nature of Science. *Science* 138: 1251–1253.
Gillispie, Charles C. 1999. Apologia Pro Vita Sua. *Isis* 90: S84–S94.
Godfrey-Smith, Peter. 2003. *Theory and Reality: An Introduction to the Philosophy of Science*. Chicago: The University of Chicago Press.
Goodman, Nelson. 1954. *Fact, Fiction, and Forecast*. London: The Athlone Press.
Gordon, H. Scott. 1991. *The History and Philosophy of Social Science*. London: Routledge.
Gross, Neil. 2008. *Richard Rorty: The Making of an American Philosopher*. Chicago: The University of Chicago Press.
Guerlac, Henry. 1961. *Lavoisier—The Crucial Year: The Background and Origin of His First Experiments on Combustion in 1772*. New York: Gordon and Breach, 1990.
Guerlac, Henry. 1987. *Radar in World War II*, vol. 1 (Secs. A–C), The History of Modern Physics, 1800–1950, vol. 8 (2 vols.), American Institute of Physics, New York: Tomash Publishers.

Hacking, Ian. 1993. Working in a New World: The Taxonomic Solution. In *World Changes: Thomas Kuhn and the Nature of Science*, ed. P. Horwich. Cambridge, Mass.: The MIT Press.

Hacking, Ian. 2012. Introductory Essay. In T.S. Kuhn, *The Structure of Scientific Revolutions*, 50th Anniversary Edition. Chicago: The University of Chicago Press.

Hagstrom, Warren O. 1965. *The Scientific Community*. Carbondale: Southern Illinois University Press.

Hamlin, Christopher. 2016. The Pedagogical Roots of the History of Science: Revisiting the Vision of James Bryant Conant. *Isis* 107: 282–308.

Hanson, Norwood R. 1958. *Patterns of Discovery: An Inquiry into the Conceptual Foundations of Science*. Cambridge: Cambridge University Press.

Harman, Peter M. 1982. *Energy, Force, and Matter: The Conceptual Development of Nineteenth-Century Physics*. Cambridge: Cambridge University Press.

Harvey, Joy. 1999. History of Science, History and Science, and Natural Sciences: Understanding Teaching of the History of Science at Harvard, 1938–1970. *Isis* 90: S270–S294.

Haufe, Chris. 2022. *How Knowledge Grows: The Evolutionary Development of Scientific Practice*. Cambridge, Mass.: The MIT Press.

Heilbron, John L. 1968. Quantum Historiography and the Archive for History of Quantum Physics. *History of Science* 7: 90–111.

Heilbron, John L. 1998. Thomas Samuel Kuhn, 18 July 1922–17 June 1996. *Isis* 89: 505–515.

Hellman, C. Doris. 1957. Review of *The Copernican Revolution: Planetary Astronomy in the Development of Western Thought*, by Thomas S. Kuhn. *Renaissance News* 10: 217–220.

Hempel, Carl G. 1942. The Function of General Laws in History. *The Journal of Philosophy* 39: 35–48. Reprinted in *Aspects of Scientific Explanation*. New York: Free Press, 1965 (q. v.).

Hempel, Carl G. 1952. *Fundamentals of Concept Formation in Empirical Science*, in *International Encyclopedia of Unified Science*, vol. II, 7. Chicago: The University of Chicago Press.

Hempel, Carl G. 1958. The Theoretician's Dilemma: A Study in the Logic of Theory Construction. In *Aspects of Scientific Explanation*. New York: Free Press, 1965.

Hempel, Carl G. 1965b. *Aspects of Scientific Explanation*. New York: Free Press.

Hendel, Charles W., ed. 1927. *Hume: Selections*. New York: Charles Scribner's Sons.

Henderson, Lawrence J. 1970. Sociology 23 Lectures. In *L. J. Henderson on the Social System: Selected Writings*, ed. Barber, B. Chicago: The University of Chicago Press.

Henle, Paul, Horace M. Kallen, and Susanne K. Langer, eds. 1951. *Structure, Method and Meaning: Essays in Honor of H. M. Sheffer*. New York: The Liberal Arts Press.

Heyl, Barbara S. 1968. The Harvard "Pareto Circle." *Journal of the History of the Behavioral Sciences* 4: 316–334.

Hershberg, James G. 1993. *James B. Conant: Harvard to Hiroshima and the Making of Nuclear Age*. Nueva York: Alfred A. Knopf.

Hoddeson, Lillian, and Vicki Daitch. 2002. *True Genius: The Life and Science of John Bardeen*. Washington, D. C.: Joseph Henry Press.

Hollinger, David A. 1998. Afterwords. In *History at Berkeley: A Dialog in Three Parts*, by G. A. Brucker, H. F. May and D. A. Hollinger, Chapters in the History of the University of California, No. 7, Berkeley: Center for Studies in Higher Education and Institute of Governmental Studies, University of California, Berkeley.

Hollis, Martin. 1970. The Limits of Irrationality. In *Rationality*, ed. B. R Wilson. Oxford: Blackwell (orig. 1967).

Holton, Gerald. 1999. Some Lessons from Living in the History of Science. *Isis* 90: S95–S116.

Homans, George C. 1986. *Coming to My Senses: The Autobiography of a Sociologist*. New Brunswick: Transaction Books (orig. 1984).

Homans, George C., and Orville T. Bailey. 1959. The Society of Fellows, Harvard University, 1933–1947. In *The Society of Fellows*, ed. C. Brinton. Cambridge, Mass.: Harvard University Press.

Horwich, Paul, ed. 1993. *World Changes: Thomas Kuhn and the Nature of Science*. Cambridge, Mass.: The MIT Press.

Howson, Colin, ed. 1976. *Method and Appraisal in the Physical Sciences: The Critical Background to Modern Science, 1800–1905*. Cambridge: Cambridge University Press.

Hoyningen-Huene, Paul. 1993. *Reconstructing Scientific Revolutions: Thomas S. Kuhn's Philosophy of Science*, Foreword by T. S. Kuhn, English Translation by A. T. Levine, Chicago: The University of Chicago Press.

Hoyningen-Huene, Paul. 1995. Two Letters of Paul Feyerabend to Thomas S. Kuhn on a Draft of The Structure of Scientific Revolutions. *Studies in History and Philosophy of Science* 26: 353–387.

Hoyningen-Huene, Paul. 2006a. More Letters by Paul Feyerabend to Thomas S. Kuhn on Proto-Structure. *Studies in History and Philosophy of Science* 37: 610–632.

Hoyningen-Huene, Paul. 2006b. Context of Discovery versus Context of Justification and Thomas Kuhn. In *Revisiting Discovery and Justification: Historical and Philosophical Perspectives on the Context Distinction*, ed. J. Schickore and F. Steinle. Berlin: Springer.

Hoyningen-Huene, Paul. 2012. Philosophical Elements in Thomas Kuhn's Historiography of Science. *Theoria* 27: 281–292.

Hoyningen-Huene, Paul. 2015. Kuhn's Development Before and After *Structure*. In *Kuhn's Structure of Scientific Revolutions—50 Years On*, ed. W. J. Devlin and A. Bokulich, Boston Studies in the Philosophy and History of Science, vol. 311. Dordrecht: Springer.

Hoyningen-Huene, Paul. 2021. The Genealogy of Thomas Kuhn's Metaphysics. In *Interpreting Kuhn: Critical Essays*, ed. K.B. Wray. Cambridge: Cambridge University Press.

Hufbauer, Karl F. 1997. Kuhn's Discovery of History (1940–1958). Draft of a Paper presented at the Symposium *The Legacy of Thomas S. Kuhn*, Dibner Institute for the History of Science and Technology, MIT, 20–22 November 1997.

Hufbauer, Karl F. 2012. From Student of Physics to Historian of Science: T. S. Kuhn's Education and Early Career, 1940–1958. *Physics in Perspective* 14: 421–470.

Hull, David L. 2001. Studying the Study of Science Scientifically. In *Science and Selection: Essays on Biological Evolution and the Philosophy of Science*. Cambridge: Cambridge University Press.

Hull, David L., Peter D. Tessner and Arthur M. Diamond. 1978. Planck's Principle. *Science* 202: 717–723.

Hume, David. 1902. *Enquiries Concerning the Human Understanding and Concerning the Principles of Morals*, 2nd ed., ed. L.A. Selby-Bigge. Oxford: Oxford University Press.

Irzik, Gürol. 2012. Kuhn and Logical Positivism: Gaps, Silences, and Tactics of SSR. In *Kuhn's The Structure of Scientific Revolutions Revisited*, ed. V. Kindi and T. Arabatzis. London: Routledge.

Irzik, Gürol., and Teo Grünberg. 1995. Carnap and Kuhn: Arch Enemies or Close Allies? *British Journal for the Philosophy of Science* 46: 285–307.

Irzik, Gürol., and Teo Grünberg. 1998. Whorfian Variations on Kantian Themes: Kuhn's Linguistic Turn. *Studies in History and Philosophy of Science* 29: 207–221.

Isaac, Joel. 2012a. *Working Knowledge: Making the Human Sciences from Parsons to Kuhn*. Cambridge, Mass: Harvard University Press.

Isaac, Joel. 2012b. Kuhn's Education: Wittgenstein, Pedagogy, and the Road to *Structure*. *Modern Intellectual History* 9: 89–107.

Jacobs, Struan. 2006–2007. Michael Polanyi and Thomas Kuhn: Priority and Credit. *Tradition and Discovery* 33: 25–36.

Jacobs, Struan. 2009. Thomas Kuhn's Memory. *Intellectual History Review* 19: 83–101.

Jacobs, Struan. 2010. J. B. Conant's Other Assistant: Science as Depicted by Leonard K. Nash, Including Reference to Thomas Kuhn. *Perspectives on Science* 18: 328–351.

Kaiser, David. 2016. Thomas Kuhn and the Psychology of Scientific Revolutions. In *Kuhn's Structure of Scientific Revolutions at Fifty*, ed. R.J. Richards and L. Daston. Chicago: The University of Chicago Press.

Kemble, Edwin C. 1938. Operational Reasoning, Reality, and Quantum Mechanics. *Journal of the Franklin Institute* 225: 263–275.

Kemble, Edwin C. 1952. The Role of Philosophy in a General Education Course in Physical Science. In *General Education in Science*, ed. I.B. Cohen and F.G. Watson. Foreword by James Bryant Conant, Cambridge, Mass.: Harvard University Press.

Kemble, Edwin C. 1962. Interview of Edwin C. Kemble by Thomas S. Kuhn. 11 May 1962. Niels Bohr Library & Archives, American Institute of Physics, College Park, MD USA. Available at: https://www.aip.org/history-programs/niels-bohr-library/oral-histories/4702-1

Kemble, Edwin C., Francis Birch, and Gerald Holton. 1970. Percy Williams Bridgman. In *Dictionary of Scientific Biography*, vol. 1, ed. C.C. Gillispie. New York: Charles Scribner's Sons, 1981.

Kevles, Daniel. 1995. *The Physicists: The History of a Scientific Community in Modern America*, 2nd ed. Cambridge, Mass.: Harvard University Press.

Kindi, Vasso P. 1995. Kuhn's *The Structure of Scientific Revolutions* Revisited. *Journal for General Philosophy of Science* 26: 75–92.

Kindi, Vasso P. 2005. Should Science Teaching Involve the History of Science? An Assessment of Kuhn's View. *Science and Education* 14: 721–731.

Kindi, Vasso P. 2021. Kuhn, the Duck, and the Rabbit—Perception, Theory-Ladenness, and Creativity in Science. In *Interpreting Kuhn: Critical Essays*, ed. K. Brad Wray. Cambridge: Cambridge University Press.

Kitcher, Philip. 1998. Tom Kuhn—An Appreciation. *Biology and Philosophy* 1–4.

Klein, Martin J., Abner Shimony, and Trevor J. Pinch. 1979. Paradigm Lost? A Review Symposium. *Isis* 70: 429–440.

Koffka, Kurt. 1935. *Principles of Gestalt Psychology*. New York: Harcourt, Brace & World.

Koyré, Alexandre. 1940. *Études galiléennes*. Paris: Hermann.

Koyré, Alexandre. 1973a. *Études d'histoire de la pensée scientifique*. Paris: Gallimard.

Koyré, Alexandre. 1973b. *The Astronomical Revolution*. English Translation by R. E. W. Maddison. London: Methuen (orig. ed. 1961).

Koyré, Alexandre. 1978. *Galileo Studies*. English Translation by John Mepham. Hassocks, Sussex: The Harvester Press.

Kridel, C. 2013. Social Reconstructionism or Child-centered Progressivism? Difficulties Defining Progressive Education from the PEA's 1939 Documentary Film, School. *American Educational History Journal* 40: 279–295.

Kripke, Saul. 1972. Naming and Necessity. In *Semantics of Natural Language*, ed. D. Davidson and G. Harman. Dordrecht: D. Reidel Publishing Company. Revised and enlarged edition: *Naming and Necessity*. Oxford: Blackwell, 1980.

Kuhn, Jehane. 2011. A Consistent Man. *Constructivist Foundations* 6: 138.

Kuukkanen, Jouni-Matti. 2010. Kuhn on Essentialism and the Causal Theory of Reference. *Philosophy of Science* 77: 544–564.

Kuukkanen, Jouni-Matti. 2012. Revolution as Evolution: The Concept of Evolution in Kuhn's Philosophy. In *Kuhn's Structure of Scientific Revolutions Revisited*, ed. V. Kindi and T. Arabatzis. London: Routledge.

Kuukkanen, Jouni-Matti. 2021. Truth, Incoherence, and the Evolution of Science. In *Interpreting Kuhn: Critical Essays*, ed. K.B. Wray. Cambridge: Cambridge University Press.

La Mer, Victor K. 1949. Some Current Misconceptions of N. L. Sadi Carnot's Memoir and Cycle. *Science* 109: 598.

La Mer, Victor K. 1954. Some Current Misinterpretations of N. L. Sadi Carnot's Memoir and Cycle. *American Journal of Physics* 22: 20–27.

La Mer, Victor K. 1955. Some Current Misinterpretations of N. L. Sadi Carnot's Memoir and Cycle. II. *American Journal of Physics* 23: 95–102.

Lakatos, Imre. 1970. Falsification and the Methodology of Scientific Research Programmes. In *Criticism and the Growth of Knowledge*, ed. I. Lakatos and A. Musgrave. Cambridge: Cambridge University Press.

Lakatos, Imre. 1971. History of Science and Its Rational Reconstructions. In *PSA 1970: In Memory of Rudolf Carnap. Proceedings of the 1970 Biennial Meeting of the Philosophy of Science*

Association, ed. R. C. Buck and R. S. Cohen. Boston Studies in the Philosophy of Science, vol. 8. Dordrecht: D. Reidel Publishing Company.
Lakatos, Imre, and Alan Musgrave, eds. 1970. *Criticism and the Growth of Knowledge*. Cambridge: Cambridge University Press.
Lakatos, Imre, and Paul K. Feyerabend. 1999. *For and Against Method: Including Lakatos's Lectures on Scientific Method and the Lakatos–Feyerabend Correspondence*, ed. M. Motterlini. Chicago: The University of Chicago Press.
Langer, Susanne K. 1937. *Introduction to Symbolic Logic*. Boston: Houghton Mifflin.
Larvor, Brendan. 2003. Why Did Kuhn's *Structure of Scientific Revolutions* Cause a Fuss? *Studies in History and Philosophy of Science* 34: 369–390.
Lewis, Clarence I. 1918. *A Survey of Symbolic Logic*. Berkeley, Cal.: University of California Press.
Lewis, Clarence I. 1929. *Mind and the World-Order: Outline of a Theory of Knowledge*. New York: Charles Scribner's Sons.
Lewis, Clarence I. 1946. *An Analysis of Knowledge and Valuation*. La Salle, Ill.: Open Court.
Lewis, Clarence I. 1951. Notes on the Logic of Intension. In *Structure, Method and Meaning: Essays in Honor of H. M. Sheffer*, ed. P. Henle, H.M. Kallen and S.K. Langer. New York: The Liberal Arts Press.
Lewis, Clarence I., and Cooper H. Langford. 1932. *Symbolic Logic*, 2nd ed., 1959. New York: Dover.
Lucas, George R., Jr. 1989. *The Rehabilitation of Whitehead, Albany*. New York: SUNY Press.
Macbeath, A.M. 1962. Meaning of Measurement in the Sciences [Review of *Quantification: A History of the Meaning of Measurement in the Natural and Social Sciences*, ed. Harry Woolf. Indianapolis: The Bobbs-Merrill Company, Inc., 1961]. *Nature* 196: 506.
Mach, Ernst. 1882. The Economical Nature of Physical Inquiry. In *Popular Scientific Lectures*, 3rd ed., English translation by T. J. McCormack, Chicago: Open Court, 1898.
Maier, Anneliese. 1949. *Die Vorläufer Galileis im 14. Jahrhundert*. Roma: Edizioni di "Storia e Letteratura."
Marcum, James A. 2005. *Thomas Kuhn's Revolution: An Historical Philosophy of Science*. London: Continuum.
Marcum, James A. 2015a. *Thomas Kuhn's Revolutions: A Historical and an Evolutionary Philosophy of Science?* London: Bloomsbury.
Marcum, James A. 2015b. The Evolving Notion and Role of Kuhn's Incommensurability Thesis. In *Kuhn's Structure of Scientific Revolutions—50 Years On*, ed. W.J. Devlin and A. Bokulich. Boston Studies in the Philosophy and History of Science, vol. 311. Cham: Springer.
Masterman, Margaret. 1970. The Nature of a Paradigm. In *Criticism and the Growth of Knowledge*, ed. I. Lakatos and A. Musgrave. Cambridge: Cambridge University Press.
Matthiessen, F.O. 1941. *American Renaissance: Art and Expression in the Age of Emerson and Whitman*. Oxford: Oxford University Press.
Mayoral, Juan V. 2005. Thomas S. Kuhn y la función de los instrumentos científicos en el desarrollo de las ciencias físicas. *Éndoxa* 19: 373–424.
Mayoral, Juan V. 2008a. Kuhn and Whitehead. In *Handbook of Whiteheadian Process Thought*, vol. 1, ed. M. Weber and W. Desmond. Frankfurt: Ontos.
Mayoral, Juan V. 2008b. On Understanding the Scientific Mind: A Study of Conant's Influence on Kuhn. *Paradigmas: Revista de filosofía y de humanidades* 7: 40–58.
Mayoral, Juan V. 2009. Intensions, Belief and Science: Kuhn's Early Philosophical Outlook (1940–1945). *Studies in History and Philosophy of Science* 40: 175–184.
Mayoral, Juan V. 2011. Hacia una reinterpretación de la ciencia normal: Kuhn y la física de su tiempo (1940–1951). *Asclepio* 63: 221–248.
Mayoral, Juan V. 2013. Las Conferencias Lowell de Kuhn: Un estudio crítico. *Theoria* 78: 459–476.
Mayoral, Juan V. 2015. Significado, conocimiento y creencia en Kuhn: La influencia de Wittgenstein y Austin. In *Wittgenstein: La superación del escepticismo*, ed. D. Pérez Chico and J.V. Mayoral. Madrid: Plaza y Valdés.

Mayoral, Juan V. 2017a. Mundos fenoménicos y léxicos científicos: El relativismo lingüístico de Thomas Kuhn. *Revista de Filosofía* 42: 117–134.

Mayoral, Juan V. 2017b. The Given, The Pragmatic *A Priori* and Scientific Change. In *Pragmatism in Transition*, ed. P. Olen and C. Sachs. London: Palgrave.

Mayoral, Juan V. 2017c. *Thomas S. Kuhn: La búsqueda de la estructura.* Zaragoza: Prensas de la Universidad de Zaragoza.

Mayoral, Juan V. 2020. Kuhn's *Structure*: Revolutionary Thinking in Turbulent Times. *Metascience* 29: 249–255.

Mayoral, Juan V. 2022. The Signposts of Inquiry: Dewey's *Logic* and Kuhn's Philosophy of Science. *Análisis: Revista de Investigación Filosófica* 9: 351–379.

Mayoral, Juan V. 2023a. The Normal and the Revolutionary: Kuhn's Conversations with Rorty. *International Studies in the Philosophy of Science* 36: 105–120.

Mayoral, Juan V. 2023b. Kuhn's Reconstruction of *Structure*: The Theoretical Background. In *Perspectives on Kuhn: Contemporary Approaches to the Philosophy of Thomas Kuhn*, ed. L. Giri, P. Melogno and H. Miguel. Western Ontario Series in the Philosophy of Science, vol. 84, Cham: Springer.

Mayoral, Juan V. 2024a. Chris Haufe's *How Knowledge Grows. BJPS Review of Books.* Available at: https://www.thebsps.org/reviewofbooks/mayoral-on-haufe/

Mayoral, Juan V. 2024b. Kuhn's "Does Knowledge 'Grow'?": Second Thoughts on the Nature of Epistemic Progress. In *Rethinking Thomas Kuhn's Legacy*, ed. Yafeng Shan. Boston Studies in the Philosophy and History of Science, vol. 345. Cham: Springer.

Mayoral, Juan V. 2024c. Review of The Last Writings of Thomas S. Kuhn, Edited by Bojana Mladenovic. *Journal for General Philosophy of Science* 55: 171–175.

McGerr, M. 1998. Progressivism. In *A Companion to American Thought*, ed. R.W. Fox and J.T. Kloppenberg, 548–550. Oxford: Blackwell (orig. 1995).

Melogno, Pablo. 2019. The Discovery–Justification Distinction and the New Historiography of Science: In Thomas Kuhn's Thalheimer Lectures. *HOPOS: The Journal of the International Society for the History of Philosophy of Science* 9: 152–178.

Melogno, Pablo. 2023. Kuhn's "The Natures of Conceptual Change": The Search for a Theory of Meaning and the Birth of Taxonomies (1980–1994). *International Studies in the Philosophy of Science* 36: 87–103.

Melogno, Pablo, and Leandro Giri. 2023. Towards a Genealogy of Thomas Kuhn's Semantics. *Perspectives on Science* 31: 385–404.

Merton, Robert K. 1938. Science, Technology and Society in Seventeenth Century England. *Osiris* 4: 360–632.

Merton, Robert K. 1942. The Normative Structure of Science. In *The Sociology of Science: Theoretical and Empirical Investigations*, ed. N.W. Storer. Chicago: The University of Chicago Press, 1973.

Merton, Robert K. 1977. *The Sociology of Science: An Episodic Memoir.* Carbondale, IL: Southern Illinois University Press.

Mink, Louis. 1966. The Autonomy of Historical Understanding. *History and Theory* 5: 24–47.

Mladenovic, Bojana. 2017. *Kuhn's Legacy: Epistemology, Metaphilosophy, and Pragmatism.* New York: Columbia University Press.

Mladenovic, Bojana. 2022. Editor's Introduction. In *The Last Writings of Thomas S. Kuhn: Incommensurability in Science*, ed. B. Mladenovic. Chicago: The University of Chicago Press.

Molenski, Martin X., SJ. 2006–2007. Polanyi vs. Kuhn: Worldviews Apart. *Tradition and Discovery* 33: 8–24.

Moore, George E. 1925. A Defence of Common Sense. In *Contemporary British Philosophy: Personal Statements*, 2nd Series, ed. J.H. Muirhead. London: Allen and Unwin. Reprinted in *Philosophical Papers*. London: Allen and Unwin, 1959 (q. v.).

Moore, George E. 1939. Proof of an External World. *Proceedings of the British Academy* 25: 273—300. Reprinted in *Philosophical Papers*. London: Allen and Unwin, 1959 (q. v.).

Moyer, Albert E. 1991. P. W. Bridgman's Operational Perspective on Physics, Part. I: Origins and Development. *Studies in History and Philosophy of Science* 22: 237–258.

Murdoch, John E. 1987. Alexandre Koyré and the History of Science in America: Some Doctrinal and Personal Reflections. *History and Technology* 4: 71–79.

Murphey, Murray G. 2005. *C. I. Lewis: The Last Great Pragmatist*. Albany, New York: SUNY Press.

Nelson, J.S., A. Megill, and D.N. McCloskey, eds. 1987. *The Rhetoric of the Human Sciences: Language and Argument in Scholarship and Public Affairs*. Madison, Wisconsin: The University of Wisconsin Press.

Neuber, Matthias. 2011. Feigl's 'Scientific Realism.' *Philosophy of Science* 78: 165–183.

Newton, Isaac. 1730. *Opticks: or, a Treatise of the Reflections, Refractions, Inflections and Colours of Light*, 4th ed. London: William Innys.

Newton-Smith, William H. 1981. *The Rationality of Science*. London: Routledge.

Nye, Mary Jo. 1999. *Before Big Science: The Pursuit of Modern Chemistry and Physics, 1800–1940*. Cambridge, Mass.: Harvard University Press.

Nye, Mary Jo. 2011. *Michael Polanyi and His Generation: Origins of the Social Construction of Science*. Chicago: The University of Chicago Press.

Oberheim, Eric. 2005. On the Historical Origins of the Contemporary Notion of Incommensurability: Paul Feyerabend's Assault on Conceptual Conservatism. *Studies in History and Philosophy of Science* 36: 363–390.

Oberheim, Eric, and Paul Hoyningen-Huene. 2013. The Incommensurability of Scientific Theories. In *The Stanford Encyclopedia of Philosophy* (Spring 2013 Edition), ed. E.N. Zalta. Available at: http://plato.stanford.edu/archives/spr2013/entries/incommensurability/.

Ornstein, Martha. 1928. *The Rôle of Scientific Societies in the Seventeenth Century*. Foreword by J. H. Robinson. Chicago: The University of Chicago Press.

Pearson, Karl. 1900. *The Grammar of Science*, 2nd ed. London: Adam and Charles Black.

Perrillo, J. 2016. Between the School and the Academy: The Struggle to Promote Teacher Research at Columbia University's Lincoln School, 1917–1935. *History of Education Quarterly* 56: 90–114.

Piaget, Jean. 1927. *Le causalité physique chez l'enfant*. Paris: Alcan.

Piaget, Jean. 1928. *Judgment and Reasoning in the Child*. English Translation by M. Warden, London: Kegan Paul.

Piaget, Jean. 1930. *The Child's Conception of Physical Causality*. English Translation by M. Gabain, London: Kegan Paul.

Pinto de Oliveira, José Carlos. 2017. Thomas Kuhn, The Image of Science and the Image of Art: The First Manuscript of *Structure*. *Perspectives on Science* 25: 746–765.

Polanyi, Michael. 1958. *Personal Knowledge: Towards a Post-Critical Philosophy*, Rev. ed. London: Routledge, 1962.

Polanyi, Michael. 1968. Logic and Psychology. *American Psychologist* 23: 27–43.

Popper, Karl R. 1956. Three Views Concerning Human Knowledge. In *Contemporary British Philosophy: Personal Statements*, vol. III, ed. H.D. Lewis. London: Allen & Unwin. Reprinted in *Conjectures and Refutations: The Growth of Scientific Knowledge*. New York: Basic Books, 1962.

Popper, Karl R. 1959. *The Logic of Scientific Discovery*. London: Routledge, 2002 (First German ed., 1935).

Popper, Karl R. 1983. *Realism and the Aim of Science*, ed. W.W. Bartley, III. London: Routledge.

Popper, Karl R. 1994. Models, Instruments, and Truth. In *The Myth of the Framework*, ed. M.A. Notturno. London: Routledge.

Porter, Theodore M. 2004. *Karl Pearson: The Scientific Life in a Statistical Age*. Princeton: Princeton University Press.

Psillos, Stathis. 1999. *Scientific Realism: How Science Tracks Truth*. London: Routledge.

Psillos, Stathis. 2011a. Choosing the Realist Framework. *Synthese* 180: 301–316.

Psillos, Stathis. 2011b. On Reichenbach's Argument for Scientific Realism. *Synthese* 181: 23–40.

Putnam, Hilary. 1975. The Meaning of "Meaning." In *Language, Mind and Knowledge*, ed. K. Gunderson. Minnesota Studies in the Philosophy of Science, vol. VII. Minneapolis: University of Minnesota Press. Reprinted in *Mind, Language and Reality: Philosophical Papers, vol. 2*. Cambridge: Cambridge University Press.

Putnam, Hilary. 2001. Hans Reichenbach: Realist and Verificationist. In *Future Pasts*, ed. J. Floyd and S. Shieh. Oxford, Oxford University Press.

Quine, Willard V.O. 1951. Two Dogmas of Empiricism. *Philosophical Review* 60: 20–43. Reprinted in Quine, *From a Logical Point of View: 9 Logico-Philosophical Essays*, 2nd ed. Cambridge, Mass.: Harvard University Press, 1980.

Quine, Willard V. O. 1960. *Word and Object*. Cambridge, Mass: The MIT Press.

Quine, Willard V.O. 1980. *From a Logical Point of View: 9 Logico-Philosophical Essays*, 2nd ed. Cambridge, Mass.: Harvard University Press (orig. 1953).

Quine, Willard V. O. 1985. *The Time of My Life: An Autobiography*. Cambridge, Mass: The MIT Press.

Quine, Willard V.O. 2008. Two Dogmas in Retrospect. In *Confessions of a Confirmed Extensionalist and Other Essays*, ed. D. Follesdal and D.B. Quine. Cambridge, Mass.: Harvard University Press.

Ravetz, Jerome R. 1971. *Scientific Knowledge and Its Social Problems*. Oxford: Clarendon.

Read, Rupert. 2012. *Wittgenstein among the Sciences*, ed. S. Summers. Farnham: Ashgate.

Reese, William J. 2001. The Origins of Progressive Education. *History of Education Quarterly* 41: 1–24.

Regis, Ed. 1987. *Who Got Einstein's Office? Eccentricity and Genius at the Institute for Advanced Study*. Cambridge, Mass.: Perseus Publishing.

Reichenbach, Hans. 1938. *Experience and Prediction: An Analysis of the Foundations and the Structure of Knowledge*. Introduction by A. Richardson. Notre Dame, Notre Dame University Press, 2006.

Reichenbach, Hans. 1951. *The Rise of Scientific Philosophy*. Berkeley, Cal.: University of California Press.

Reisch, George A. 1991. Did Kuhn Kill Logical Empiricism? *Philosophy of Science* 58: 264–277.

Reisch, George A. 1995. *A History of the International Encyclopedia of Unified Science*. PhD Dissertation, Chicago: The University of Chicago.

Reisch, George A. 2005. *How the Cold War Transformed Philosophy of Science: To the Icy Slopes of Logic*. Cambridge: Cambridge University Press.

Reisch, George A. 2012. The Paranoid Style in American History of Science. *Theoria* 75: 323–342.

Reisch, George A. 2016. Aristotle in the Cold War: On the Origins of Thomas Kuhn's *The Structure of Scientific Revolutions*. In *Kuhn's Structure of Scientific Revolutions at Fifty*, ed. R.J. Richards and L. Daston. Chicago: The University of Chicago Press.

Reisch, George A. 2019. *The Politics of Paradigms: Thomas S. Kuhn, James B. Conant, and the Cold War "Struggle for Men's Minds."* Albany, NY: SUNY Press.

Reisch, George A., and Adam Tamas Tuboly. 2025. A Forgotten Chapter in the History of Sociology of Science: Philipp Frank's 'Good Nose.' In *The Legacy of the Vienna Circle*, ed. G. Schiemer. Vienna Circle Institute Yearbook, vol. 30. Cham: Springer.

Reuben, Julie A. 1996. *The Making of the Modern University: Intellectual Transformation and the Marginalization of Morality*. Chicago: The University of Chicago Press.

Richards, Theodore W. 1898. Progress in Physical Chemistry. *Science* 8: 721–727.

Richardson, Alan. 2007. "That Sort of Everyday Image of Logical Positivism": Thomas Kuhn and the Decline of Logical Empiricist Philosophy of Science. In *The Cambridge Companion to Logical Empiricism*, ed. A. Richardson and T. Uebel. Cambridge: Cambridge University Press.

Rockefeller, Steven C. 1991. *John Dewey: Religious Faith and Democratic Humanism*. Nueva York: Columbia University Press.

Rorty, Richard M. 1979. *Philosophy and the Mirror of Nature*. Princeton: Princeton University Press.

Rorty, Richard M. 1985. Solidarity or Objectivity? In *Post-Analytic Philosophy*, ed. J. Rajchman and C. West. New York: Columbia University Press. Reprinted in *Objectivity, Relativism, and Truth: Philosophical Papers, Vol. 1*. Cambridge: Cambridge University Press, 1991.

Rorty, Richard M. 1987. Science and Solidarity. In *The Rhetoric of the Human Sciences: Language and Argument in Scholarship and Public Affairs*, ed. J.S. Nelson, J.S., A. Megill and D.N. McCloskey. Madison, Wisconsin: The University of Wisconsin Press. Reprinted in *Objectivity, Relativism, and Truth: Philosophical Papers, Vol. 1*. Cambridge: Cambridge University Press, 1991.

Rorty, Richard M. 1989. *Contingency, Irony, and Solidarity*. Cambridge: Cambridge University Press.

Rosenthal, Sandra B. 2007. *C. I. Lewis in Focus: The Pulse of Pragmatism*. Bloomington: Indiana University Press.

Ross, Dorothy. 1991. *The Origins of American Social Science*. Cambridge: Cambridge University Press.

Rossiter, Margaret W. (ed.). 1999. *Catching Up with the Vision: Essays on the Occasion of the 75th Anniversary of the Founding of the History of Science Society*. A Supplement to *Isis*, 90. Chicago: The University of Chicago Press.

Rudolph, John L. 2002. *Scientists in the Classroom: The Cold War Reconstruction of American Science Education*. New York: Palgrave.

Rury, John L. 1985. American School Enrolment in the Progressive Era: An Interpretive Inquiry. *History of Education* 14: 49–67.

Russell, Bertrand. 1912–1913. On the Notion of Cause. *Proceedings of the Aristotelian Society* 13:1–26. Reprinted in *Mysticism and Logic and other Essays*, London: George Allen & Unwin, 1917 (q. v.).

Russell, Bertrand. 1924. *An Introduction to Mathematical Philosophy*, 2nd ed. London: George Allen and Unwin.

Russell, Bertrand. 1929. *Our Knowledge of the External World*. New York: W. W. Norton and Co.

Russell, Bertrand. 1938. *Principles of Mathematics*, 2nd ed. New York: W. W. Norton and Co.

Russell, Bertrand, and Alfred N. Whitehead. 1925. *Principia Mathematica*, 2nd ed., vol. 1. Cambridge: Cambridge University Press.

Ryan, Alan. 1997. *John Dewey and the High Tide of American Liberalism*. Nueva York: W. W. Norton & Co. (orig. 1995).

Salmon, Wesley C. 1990. The Appraisal of Theories: Kuhn Meets Bayes. In *PSA: Proceedings of the Biennial Meeting of the Philosophy of Science Association, 1990*, vol. 2., ed. A. Fine, M. Forbes and L. Wessels. East Lansing: Philosophy of Science Association.

Salmon, Wesley C. 2005. *Reality and Rationality*. ed. P. Dowe and M.H. Salmon. Oxford, Oxford University Press.

Sankey, Howard. 1993. Kuhn's Changing Concept of Incommensurability. *British Journal for the Philosophy of Science* 44: 775–791. Reprinted in his *Rationalism, Relativism and Incommensurability*, Aldershot: Ashgate.

Sankey, Howard. 1994. *The Incommensurability Thesis*. Aldershot: Ashgate.

Santayana, George. 1900. The Poetry of Barbarism. In *Interpretations of Poetry and Religion*. New York: Charles Scribner's Sons.

Sarton, George. 1927–1948. *Introduction to the History of Science*, 3 vols. Baltimore: Williams and Wilkins Co.

Sarton, George. 1936. *The Study of the History of Science*. Cambridge, Mass.: Harvard University Press.

Sarton, George. 1950. The History of Science in the Carnegie Institution. *Osiris* 9: 624–638.

Sarton, George. 1954. Reminiscences of a Pioneer. *Osiris* 11: 108–118.

Saveth, Edward N. 1988. Education of an Elite. *History of Education Quarterly* 28: 367–386.

Schiappa, Edward. 2014. Thomas Kuhn and POROI, 1984. *Poroi* 10: 1–3.

Schweber, S.S. 1986. The Empiricist Temper Regnant: Theoretical Physics in the United States, 1920–1950. *Historical Studies in the Physical Sciences* 17: 55–98.

Senior, James K. 1958. The Vernacular of the Laboratory. *Philosophy of Science* 25: 163–168.
Servos, John W. 1986. Mathematics and the Physical Sciences in America, 1880–1930. *Isis* 77: 611–629.
Servos, John W. 1990. *Physical Chemistry from Ostwald to Pauling: The Making of a Science in America*. Princeton: Princeton University Press.
Seth, Suman. 2013. Quantum Physics. In *The Oxford Handbook of the History of Physics*, ed. J.Z. Buchwald and R. Fox. Oxford: Oxford University Press.
Shils, Edward A., and Henry A. Finch (eds.). 1949. *Max Weber on the Methodology of the Social Sciences*. Glencoe, Ill.: The Free Press.
Slater, John C. 1963. Interview of John C. Slater by Thomas S. Kuhn and John H. Van Vleck. 3 October 1963. Niels Bohr Library & Archives, American Institute of Physics, College Park, MD USA. Available at: https://www.aip.org/history-programs/niels-bohr-library/oral-histories/4892-1.
Smith, Crosbie. 2003. Force, Energy, and Thermodynamics. In *The Cambridge History of Science, vol. 5: The Modern Physical and Mathematical Sciences*, ed. M.J. Nye. Cambridge: Cambridge University Press.
Sneed, Joseph D. 1971. *The Logical Structure of Mathematical Physics*. Dordrecht: D. Reidel Publishing Company.
Solís, Carlos. 1994. Alexandre Koyré y la historia de la ciencia. Introduction to A. Koyré, *Pensar la ciencia*, Spanish Translation by A. Beltrán. Barcelona: Paidós.
Solís, Carlos. 1997. La revolución kantiana de Kuhn. *Éndoxa* 9: 5–30.
Solís, Carlos, and Manuel Sellés. 2005. *Historia de la ciencia*. Madrid: Espasa.
Sopka, Katherine. 1988. *Quantum Physics in America: The Years Through 1995*. The History of Modern Physics, vol. 10. Foreword by A. Pais, Introduction by G. Holton. New York: Tomash Publishers.
Stadler, Friedrich. 2007. The Vienna Circle: Context, Profile, and Development. In *The Cambridge Companion to Logical Empiricism*, ed. A. Richardson and T. Uebel. Cambridge: Cambridge University Press.
Stegmüller, Wolfgang. 1976. *The Structure and Dynamics of Theories*. English Translation by William Wohlhueter. New York: Springer.
Swerdlow, Noel M. 2004. An Essay on Thomas Kuhn's First Scientific Revolution, The Copernican Revolution. *Proceedings of the American Philosophical Society* 148: 64–120.
Synnott, Marcia G. 1979. *The Half-Opened Door: Discriminations and Admissions at Harvard, Yale, and Princeton, 1900-1970*, 2nd ed., New Brunswick: Transaction Publishers, 2010.
Te Heesen, Anke. 2020. Thomas S. Kuhn, Earwitness: Interviewing and the Making of a New History of Science. *Isis* 111: 86–97.
Temple, Frederick. 1884. *The Relations between Religion and Science*. London: Macmillan.
Thackray, Arnold, and Robert K. Merton. 1972. On Discipline Building: The Paradoxes of George Sarton. *Isis* 63: 472–495.
Thackray, Arnold, and Robert K. Merton. 1976. George Alfred Léon Sarton. In *Dictionary of Scientific Biography*, vol. 12, ed. C.C. Gillispie. New York: Charles Scribner's Sons.
Timmins, Adam. 2013. Why Was Kuhn's *Structure* More Successful than Michael Polanyi's Personal Knowledge. *HOPOS: The Journal of the International Society for the History of Philosophy of Science* 3: 306–317.
Treviño, A. Javier. 2001. Introduction: The Theory and Legacy of Talcott Parsons. In *Talcott Parsons Today: His Theory and Legacy in Contemporary Sociology*, ed. A.J. Treviño. Foreword by N. J. Smelser. Lanham: Rowman and Littlefield.
Van Dongen, Jeroen. 2020. In Europe. *Physics in Perspective* 22: 3–25.
Van Evra, James W. 1966. *A History of Some Aspects of the Theory of Logic, 1850-Present*. PhD Dissertation, Michigan State University.
Van Fraassen, Bas C. 2002. *The Empirical Stance*. New Haven, Conn.: Yale University Press.
Vleck, Van, and H. John. 1971. Reminiscences of the First Decade of Quantum Mechanics. *International Journal of Quantum Chemistry* 5: 3–20.

Von Wright, Georg H. 1972. Wittgenstein on Certainty. In *Problems in the Theory of Knowledge*, ed. G.H. von Wright. The Hague: Nijhoff.

Walter, Maila L. 1990. *Science and Cultural Crisis: An Intellectual Biography of Percy Williams Bridgman (1882–1961)*. Stanford, Cal: Stanford University Press.

Warner, Sam B., Jr. 1984. *Province of Reason*. Cambridge, Mass: Harvard University Press.

Weber, Max. 1930. *The Protestant Ethic and the Spirit of Capitalism*, English Translation by T. Parsons, Introduction by A. Giddens. London: Routledge.

Weber, Max. 1978. *Economy and Society*, ed. G. Roth and C. Wittich. California: University of California Press.

Weber, Max. 1988. Soziologische Grundbegriffe. In *Gesammelte Aufsätze zur Wissenschaftslehre*, ed. J. Winckelmann. Tübingen: Mohr (orig. 1921).

Werner, Heinz. 1948. *Comparative Psychology of Mental Development*, Rev. ed. Foreword by G. W. Allport, New York: International Universities Press.

Westbrook, Robert B. 1991. *John Dewey and American Democracy*. Ithaca: Cornell University Press.

Westfall, Richard S. 2014. Five Questions on the Scientific Revolution. *Galilæana* 11: 29–35.

Westman, Robert S. 1994. Two Cultures or One? A Second Look at Kuhn's *The Copernican Revolution*. *Isis* 85: 79–115.

White, Morton. 1999. *A Philosopher's Story*. University Park, Pennsylvania: The Pennsylvania State University Press.

Whitehead, Alfred N. 1913. *Axioms of Projective Geometry*. Cambridge: Cambridge University Press.

Whitehead, Alfred N. 1917. *The Organisation of Thought*. London: Williams and Norgate.

Whitehead, Alfred N. 1920. *The Concept of Nature*. Cambridge: Cambridge University Press, 1955.

Whitehead, Alfred N. 1925. *Science and the Modern World*. New York: Macmillan.

Whitehead, Alfred N. 1927. *Symbolism: Its Meaning and Effect*. New York: Macmillan.

Whitehead, Alfred N. 1929. *Process and Reality: An Essay in Cosmology*, Revised edition, ed. D.R. Griffin and D.W. Sherbourne. New York: Free Press, 1978.

Whorf, Benjamin L. 1950. *Four Articles on Metalinguistics*. Washington: Foreign Service Institute, Department of State.

Whorf, Benjamin L. 1956. *Language, Thought, and Reality: Selected Writings of Benjamin Lee Whorf*, ed. J.B. Carroll. Cambridge, Mass.: The MIT Press.

Wiggins, David. 1980. *Sameness and Substance*. Oxford: Blackwell.

Wigner, Eugene P., and Frederick Seitz. 1933. On the Constitution of Metallic Sodium. *Physical Review* 43: 804–810.

Wigner, Eugene P., and Frederick Seitz. 1934. On the Constitution of Metallic Sodium. II. *Physical Review* 46: 509–524.

Willey, Basil. 1934. *The Seventeenth Century Background: Studies in the Thought of the Age in Relation to Poetry and Religion*. London: Chatto & Windus.

Williams, Donald C. 1947. *The Ground of Induction*. Cambridge, Mass: Harvard University Press.

Williams, Michael. 2004. Wittgenstein, Truth and Certainty. In *Wittgenstein's Lasting Significance*, ed. M. Kölbel and B. Weiss. London: Routledge.

Willison, George F., ed. 1940. *Let's Make a Play: Twelve Plays by Children*. New York: Harper and Brothers.

Winters, E.A. 1967. Man and His Changing Society: The Textbooks of Harold Rugg. *History of Education Quarterly* 7: 493–514.

Wittgenstein, Ludwig. 1953. *Philosophical Investigations*, 4th ed, ed. P.M.S. Hacker and J. Schulte. English Translation by G.E.M. Anscombe, P.M.S. Hacker and J. Schulte. Oxford: Blackwell, 2009.

Wittgenstein, Ludwig. 1958. *The Blue and Brown Books: Preliminary Studies for the "Philosophical Investigations."* Oxford: Blackwell.

Wittgenstein, Ludwig. 1969. *On Certainty*, ed. G.E.M. Anscombe and G.H. von Wright. English Translation by D. Paul and G.E.M. Anscombe. Oxford: Blackwell.

Wood, Oscar P. 1953. Review of Henle, Kallen and Langer (eds.) *Structure, Method and Meaning: Essays in Honor of H. M. Sheffer*, *Mind* 62: 396–405.

Woodger, Joseph H. 1939. *The Technique of Theory Construction*, in *International Encyclopedia of Unified Science*, vol. II, 5. Chicago: The University of Chicago Press.

Woolf, Harry. 1961a. The Conference on the History of Quantification in the Sciences. *Isis* 52: 133–134.

Woolf, Harry (ed.). 1961b. *Quantification: A History of the Meaning of Measurement in the Natural and Social Sciences*. Indianapolis: Bobbs–Merrill.

Wray, K. Brad. 2003. Is Science Really a Young Man's Game? *Social Studies of Science* 33: 137–149.

Wray, K. Brad. 2011. *Kuhn's Evolutionary Social Epistemology*. Cambridge: Cambridge University Press.

Wray, K. Brad. 2016. The Influence of James B. Conant on Kuhn's *Structure of Scientific Revolutions*. *HOPOS: The Journal of the International Society for the History of Philosophy of Science* 6: 1–23.

Wray, K. Brad. 2021. *Kuhn's Intellectual Path: Charting The Structure of Scientific Revolutions*. Cambridge: Cambridge University Press.

Wray, K. Brad. 2024. Co-authorship in Chemistry at the Turn of the Twentieth Century: The Case of Theodore W. Richards. *Foundations of Chemistry* 26: 75–88.

Zachary, G. Pascal. 1997. *Endless Frontier: Vannevar Bush, Engineer of the American Century*. Cambridge, Mass.: The MIT Press, 1999.

Zammito, John H. 2004. *A Nice Derangement of Epistemes: Post-Positivism in the Study of Science from Quine to Latour*. Chicago: The University of Chicago Press.

Zangwill, Andrew. 2021. *A Mind Over Matter: Philip Anderson and the Physics of the Very Many*. Oxford: Oxford University Press.

Zuckerman, Harriet, and Robert K. Merton. 1972. Age, Aging, and Age Structure in Science. In *The Sociology of Science: Theoretical and Empirical Investigations*, ed. N.W. Storer. Chicago: The University of Chicago Press, 1973.

Author Index

J. V. Mayoral, *Thomas Kuhn*, Springer Biographies,
https://doi.org/10.1007/978-3-031-86512-1

Subject Index

J. V. Mayoral, *Thomas Kuhn*, Springer Biographies,
https://doi.org/10.1007/978-3-031-86512-1

V

W

The manufacturer's authorised representative in the EU is Springer Nature Customer Service Centre GmbH, Europaplatz 3, 69115 Heidelberg, Germany. If you have any concerns regarding our products, please contact ProductSafety@springernature.com

Printed and bound by CPI Group (UK) Ltd, Croydon, CR0 4YY
01/10/2026
02225968-0018